C0-AUD-861

THE BOOK OF ENOCH

or

I ENOCH

STUDIA IN VETERIS TESTAMENTI PSEUDEPIGRAPHA

EDIDERUNT

A. M. DENIS et M. DE JONGE

VOLUMEN SEPTIMUM

MATTHEW BLACK

THE BOOK OF ENOCH
or
I ENOCH

LEIDEN
E. J. BRILL
1985

THE BOOK OF ENOCH

or

I ENOCH

A New English Edition

WITH COMMENTARY AND TEXTUAL NOTES

BY

MATTHEW BLACK

IN CONSULTATION WITH JAMES C. VANDERKAM

WITH AN APPENDIX ON THE 'ASTRONOMICAL' CHAPTERS (72-82)
BY OTTO NEUGEBAUER

LEIDEN
E. J. BRILL
1985

ISBN 90 04 07100 8

CONTENTS

INDEXES

PREFACE

In addition to his pioneering textual work on the Ethiopic Enoch (Charles [1906]), R. H. Charles produced two editions of I Enoch, the second, the 1912 Clarendon Enoch, a revision of the first (Charles [1893]). Although frequently criticised, especially for some of its more adventurous conjectures, the 1912 Clarendon *Enoch* has, nevertheless, proved to be by far the most popular of modern language editions. It has, in fact, virtually come to be regarded, not only in the Anglo-Saxon world, as the standard edition of I Enoch. Now, however, after seventy years and the accumulation of fresh knowledge and numerous Enoch studies (including Milik's *editio princeps* of the Aramaic fragments) it has become sadly out of date.

One of the reasons for the wide-spread use of the 1912 Charles, apart from its sound basis in textual matters and its dignified English, has undoubtedly been its bold and imaginative efforts to recover the ideas of the original, even to the restoration of its putative Aramaic or Hebrew basis. Since the textual and philological instruments available for such a task are mainly the often confused and extensively corrupted Ethiopic texts, such an objective might, at first sight, seem an unrewarding, if not hopeless, enterprise. The continued use, however, of the 1912 Charles is a measure of the success of Charles's efforts: there were some successful and some unsuccessful guesses, but there were many more textually and philologically sound observations and insights. The original Enoch had by no means been lost beyond recovery in the jungle of Ethiopic variations and corruptions.

The present edition has had the same objective as Charles: students of Jewish apocalyptic are interested only in the recovery of the original Enoch, so far as that can be reasonably deduced from the Geʿez recensions and in the light of the extant Greek and Aramaic fragments. Moreover, it must not be forgotten that by far the greater part of the text of I Enoch, even as represented in the Ethiopic canonical version, does not require any critical surgery to yield its plain and obvious meaning; the main problems have to do with a minority of *cruces interpretum*, even though these are more numerous in Enoch than in the more protected texts of canonical scripture.

In view of this common objective it seemed best, instead of attempting to produce an entirely new version, to try to build on Charles's revised version of 1912; and this I have endeavoured to do. So far from seeming to be taking liberties with the work of a renowned scholar, I feel I am paying tribute to the enduring character of his work. And for kindly allowing me to do so, I am grateful to the Oxford University Press.

In the commentary and textual notes, the latter a new feature (Charles combines his textual notes with his commentary), my indebtedness to my learned predecessors is obvious on virtually every page. For contemporary advice and help I am immensely grateful to Professor J. C. VanderKam of the Department of Philosophy and Religion in the North Carolina State University. Dr. VanderKam's own forthcoming monograph, *Enoch and the Growth of an Apocalyptic Tradition* (to be published in the CBQ monograph series), I have had the privilege of reading in manuscript. A pupil of Professor Thomas Lambdin of Harvard, and at present engaged in preparing a critical edition of the Ethiopic *Book of Jubilees*, Dr. VanderKam has been especially helpful in checking my Ethiopic, in particular the transcriptions in the commentary and textual notes. Professor Ephraim Isaac of Bard College, New York, has also been most obliging by making available off-prints and typescripts of his notes on Enoch, together with his translation of the important new Ethiopic Tana 9 manuscript. The 'astronomical book' or 'section' (Chh. 72-82) raised problems which were beyond my competence, and Professor Otto Neugebauer, of the Institute of Advanced Studies in Princeton, kindly offered to undertake responsibility for both translation and commentary of this section. This study entitled 'The "Astronomical" Chapters of the Ethiopic Book of Enoch (72-82)' (with additional notes from myself on the Aramaic astronomical fragments) appeared originally in the transactions of the Royal Danish Academy of Sciences and Letters (Det Kongelige Danske Videnskabernes Selskab, Matematisk-fysiske Meddelelser 40:10). I am not only grateful to Professor Neugebauer for undertaking this portion of the book, but also to the Danish Academy, for kindly allowing me to republish it (with slight adjustments) as Appendix A.

For substantial grants towards the costs of publication I have to thank the British Academy, the University of St. Andrews Publications and Muniments Committee and the Carnegie Trust for the Universities of Scotland. I have also to thank the Leverhulme Trust for the award of an Emeritus Fellowship in 1978 to enable me to complete this work, and the Institute for Advanced Studies in Princeton for enrolling me as a member for the academic year 1977-78, when I was able to undertake full-time research on the Enoch problem.

One of my former students, Mrs. Eve Macfarlane, has been responsible for the typing of the manuscript and has increased my indebtedness to her by also proof-reading it for consistency of style and by preparing a bibliography etc. and an index. My former secretary Miss M. C. Blackwood also prepared several preliminary typed drafts of different sections, and I am most grateful to her too for her careful deciphering and typing of my hand-written manuscript.

January 14th, 1983 Matthew Black

ABBREVIATIONS, SYMBOLS AND BIBLIOGRAPHY

Bibliographical Abbreviations

Barr, 'Aramaic-Greek Notes'	J. Barr, 'Aramaic-Greek Notes on the Book of Enoch' Part I JSS 23 (1978), 184-198, Part II JSS 24 (1979), 179-192.
BDB	Brown, Driver and Briggs, *A Hebrew and English Lexicon of the Old Testament*, Oxford 1962.
Black, *An Aramaic Approach*	M. Black, *An Aramaic Approach to the Gospels and Acts*, 3rd. ed., Oxford 1967.
Black, *Scrolls and Christian Origins*:	M. Black, *The Scrolls and Christian Origins*, Edinburgh 1962
Bousset-Gressmann[4]:	*Die Religion des Judentums im späthellenistischen Zeitalter*, verfasst von D. Wilhelm Bousset in dritter verbesserter Auflage herausgegeben von H. Gressmann, 4. photomechanisch gedrückte Auflage mit einem Vorwort von Eduard Lohse, Handbuch zum N.T. 21, Tübingen 1966.
C-B	*The Last Chapters of Enoch in Greek* by Campbell Bonner *Studies and Documents VIII*, London 1937.
Charles [1983]	*The Book of Enoch translated from Professor Dillmann's Ethiopic Text*, edited ... by R. H. Charles, Oxford 1893.
Charles [1906]	Anecdota Oxoniensia *The Ethiopic Version of the Book of Enoch*, edited ... by R. H. Charles, Oxford 1906.
Charles	R. H. Charles, *The Book of Enoch*, Oxford 1912
Coppens, *Le Fils d'Homme*	J. Coppens, *Le Fils d'Homme Vétéro- et Intertestamentaire*, Leuven, 1983.
Cowley, *Aramaic Papyri*:	*Aramaic Papyri of the Fifth Century B.C.*, edited ... by A. Cowley, Oxford 1923.
DJD	Discoveries in the Judaean Desert, Vols. I-V.
Dalman, *Words of Jesus*:	Gustaf Dalman, *The Words of Jesus*, Edinburgh 1902.
Dexinger	Ferdinand Dexinger, *Henochs Zehnwochenapokalypse und Offene Probleme der Apokalyptikforschung*, Brill, Leiden 1977.
Dillmann	A. Dillmann, *Das Buch Henoch*, Leipzig 1853.
Dillmann *Aeth.*	A. Dillmann, *Liber Henoch Aethiopice*, Lipsiae 1851.
Dillmann, *Gramm.*	*Ethiopic Grammar* by August Dillmann, second edition enlarged and improved by C. Bezold, translated by James A. Crichton, London 1907.
Dillmann, *Lex.*	Chr. Fr. Augusti Dillmann *Lexicon Linguae Aethiopicae*, Lipsiae MDCCCLXV.
Dillmann SAB 1892:	A. Dillmann, 'Über den neugefundenen griechischen Text des Henoch-Buches, Sitzungsberichte der Akademie zu Berlin, 1892, 1039-54, 1079-92.

Driver, *Aramaic Documents*: *Aramaic Documents of the Fifth Century* ... ed. by G. R. Driver, Oxford 1954.

Driver, *Judaean Scrolls*: G. R. Driver, *The Judaean Scrolls*, Oxford 1965.

Fitzmyer, *Genesis Apocryphon*: J. A. Fitzmyer, *The Genesis Apocryphon of Qumran Cave I*, Rome 1971.

Fitzmyer, *A Wandering Aramean*: *A Wandering Aramean: Collected Aramaic Essays*, Scholars Press 1979.

Flemming *Das Buch Henoch*, *Äthiopischer Text*, herausgegeben von Joh. Flemming, TU, NF Bd. 7, Leipzig 1902.

Flemming-Radermacher: *Das Buch Henoch* herausgegeben von J. Flemming und L. Radermacher, GCS, Leipzig 1901.

G-K *Gesenius-Kautzsch Gesenius' Hebrew Grammar* ... edited by E. Kautzsch ... revised by A. E. Cowley, Oxford 1910.

Gesenius, *Thes.* G. Gesenii *Thesaurus* ... *Linguae Hebraeae et Chaldaeae Veteris Testamenti*, Lipsiae 1829.

P. Grelot, 'La Légende d'Hénoch': 'La Légende d'Hénoch dans les apocryphes et dans la Bible: son origine et signification' in RSR xlvi (1958) 5-26, 181-210.

H-R *A Concordance to the Septuagint* ... by ... Edwin Hatch ... and Henry A. Redpath, Oxford MDCCCXCVII.

Hengel, *Judaism and Hellenism*: M. Hengel, *Judaism and Hellenism: Studies in their Encounter in Palestine during the Early Hellenistic Period*, SMC Press, 1974.

Hoftijzer Charles-F. Jean-Jacob Hoftijzer, *Dictionnaire des Inscriptions Sémitiques de l'Ouest*, Leiden 1965.

Kautzsch, *Die Apokryphen und Pseudepigraphen*: *Die Apokryphen und Pseudepigraphen des Alten Testaments*, ... herausgegeben von E. Kautzsch, Tübingen 1900.

Knibb M. A. Knibb, *The Ethiopic Book of Enoch*, 2 vols., Oxford 1978.

Kraeling, *Brooklyn Papyri*: *The Brooklyn Museum Aramaic Papyri*, ... ed. ... by Emil G. Kraeling, New Haven 1953.

Kuhn, *Konkordanz*: K. G. Kuhn, *Konkordanz zu den Qumrantexten*, Göttingen 1960.

Levy CW J. Levy, *Chaldäisches Wörterbuch über die Targumim*, Leipzig 1881.

Levy NHCW J. Levy, *Neuhebräisches und Chaldäisches Wörterbuch über die Talmudim und Midraschim*, Leipzig 1876.

Liddell and Scott: Liddell and Scott, *Greek-English Lexicon*, Oxford 1968.

Lods A. Lods, *Le Livre d'Hénoch, Fragments Grecs* ... traduit et annotés par Adolphe Lods, Paris 1892.

Löw, *Flora* I. Löw, *Die Flora der Juden*, 4 vols., Wien und Leipzig 1924-34.

Martin *Le Livre d'Hénoch* traduit sur le Texte Éthiopien par François Martin, Paris 1906.

Milik J. T. Milik, *The Books of Enoch*, Oxford 1976.

Milik, 'Hénoch au Pays des Aromates': in RB, Tom. 65 (1958), 70-77.

Nestle-Aland[26]: Nestle-Aland *Novum Testamentum Graece*, 26, neu bearbeitete Auflage, Stuttgart 1979.

Nickelsburg, 'Enoch 97-104':	George W. E. Nickelsburg, 'Enoch 97-104: A Study of the Greek and Ethiopic Texts' in *Armenian and Biblical Studies* ed. by M. E. Stone, Jerusalem, St. James Press, 1976, 90-156.
Nöldeke	T. Nöldeke, *Syrische Grammatik*, Leipzig 1898 (reprint Darmstadt 1966).
Noth, *Die israel. Personennamen*	M. Noth, *Die israelitische Personennamen im Rahmen der gemeinsemitischen Namengebung*, Stuttgart 1928.
Nötscher, *Zur theologischen Terminologie der Qumran-Texte*:	F. Nötscher, *Zur theologischen Terminologie der Qumran-Texte*, Bonner Biblische Beiträge 10, Bonn 1956.
P. Sm.	Payne Smith, *Thesaurus Syriacus*, Oxford MDCCCLXXIX.
Schmidt, 'Original Language'	Nathaniel Schmidt, 'The Original Language of the Parables of Enoch' in *Old Testament and Semitic Studies in Memory of William Rainey Harper*, ed. by R. F. Harper et al. (Chicago 1908) Vol. ii, 329-49.
Schodde	G. H. Schodde, *The Book of Enoch*, Andover 1882.
Schürer, *History*	Emil Schürer, *The History of the Jewish People in the Age of Jesus Christ* (175 B.C.-A.D. 135): A New English Version, revised and edited by G. Vermes and F. Millar, (Edinburgh, Vol. 1 1973, Vol. 2 1979).
Torrey, *Notes*	C. C. Torrey, 'Notes on the Greek Texts of Enoch', JAOS 62 (1942) 52-60.
Ullendorff, '*An Aramaic "Vorlage"*':	Edward Ullendorff, 'An Aramaic "Vorlage" of the Ethiopic Text of Enoch', Atti del Convegno Internazionale di Studi Etiopici (Academia Nazionale dei Lincei. Problemi attuali di scienza e di cultura 48), Roma 1960, 259-67.
Wendt, K., *Lib. Nat.*:	*Das Mashafa Milad* (*Liber Nativitatis*) CSCO Scriptores Aethiopici Tom. 41-44 (1962).
Yadin, *Scroll of the War*:	Y. Yadin, *The Scroll of the War of the Sons of Light against the Sons of Darkness*, Oxford 1962.

Sigla and other Symbols

⌜ ⌝	a supralinear bracket indicates a probable addition or expansion of the text by the Greek or Ethiopic translator.
[ ]	a square bracket indicates an omission in the text or a putative Greek variant.
« »	a double bracket indicates that the word(s) enclosed should probably be omitted.
ditt.	dittograph
En$^{a\ b\ c\ etc.}$	refer to the Aramaic fragments as in Milik's editio princeps.
Eth.	Ethiopic
Eth I	the earliest recension of the Ethiopic text in the manuscripts Tana 9 (9a), g, g^1, m, q, t, u
Eth$^{g\ m\ t\ etc.}$	refer to manuscripts of Eth. I. Sigla as in Charles xxi-xxiv except for $_1$g now g^1, $_1$a now a^1, $_1$b now b^1

Eth II	the mediaeval recension of the Ethiopic text in all manuscripts with the exception of Eth I (Tana 9 (9a), g, g^1, m, q, t, u)
Eth^M	Ethiopic majority text (with both Eth I and Eth II support)
G	Gizeh Papyrus text
G^2	Gizeh Papyrus duplicate of Chh. 19.3-21.9
G^{vat}	Vatican Greek tachygraph
G^b	Chester Beatty Papyrus ed. C. Bonner.
G^{2069}	Oxyrhynchus Papyrus XVII
h m t	homoioteleuton
italics	in the Translation indicate an emended or problematical text
Sync $Sync^1$ $Sync^2$	Georgius Syncellus's Greek text (ed. Dindorf)

Additional Select Bibliography

Andel, C. P. van, *De Structuur van de Henoch-Traditie in het Nieuwe Testament*, Utrecht 1955.
Appel, H., *Die Komposition des äthiopischen Henochbuches*, Gütersloh 1906.
Baldensperger, W., *Die messianisch-apokalyptischen Hoffnungen des Judenthums*, Strassburg 1903.
Bietenhard, H., *Die Himmlische Welt im Urchristenthum und Spätjudenthum*, Tübingen 1951.
Blass-Debrunner, *A Greek Grammar of the New Testament and other Early Christian Literature*, F. Blass and A. Debrunner, translated by R. W. Funk, Cambridge 1961.
Bouriant, U., 'Fragments Grecs du Livre d'Énoch': in *Mémoires Publiés par les Membres de la Mission archéologique Française au Caire*, Paris 1892, 93f.; 228-232 (A. Lods).
Burkitt, F. C., *Jewish and Christian Apocalypses*, London 1914.
Charles, R. H., *The Apocalypse of Baruch*, London 1896.
Cooke, S. A., *A Text-Book of North-Semitic Inscriptions*, Oxford 1903.
Creation, Christ and Culture, Studies in Honour of T. F. Torrance, ed. by R. W. A. McKinney, Edinburgh 1976.
Dieterich, A., *Nekyia*, Leipzig 1893.
Donner, H. - Röllig, W., *Kanaanäische und Aramäische Inschriften*, Wiesbaden 1962.
Donum Gentilicium: New Testament Studies in honour of David Daube, ed. by C.K. Barrett, E. Bammel and W. D. Davies, Oxford 1978.
Dupont-Sommer, A., *The Essene Writings from Qumran*, Oxford 1961.
Glasson, T. F., *The Second Advent: the Origins of the New Testament Doctrine*, Oxford 1965.
Grelot, P. *L'Espérance juive à l'heure de Jésus* (Paris, 1978)
Hartman, Lars, *Asking for a Meaning: A Study of I Enoch 1-5*, Lund 1979.
Hartman, Lars, *Prophecy Interpreted, Conjectanea Biblica*, New Testament Series I, Lund 1966.
Hofmann, J. C. K., 'Über die Entstehungszeit des Buches Henoch' ZDMG 6 (1852) 87-91.
Hooker, M. D. and Wilson, S. G., *Paul and Paulinism*; *Essays in honour of C. K. Barrett* (London, SPCK, 1982)
Hooker, M. D., *The Son of Man in Mark*, London, SPCK, 1967.
Hunt., A. S., *The Oxyrhynchus Papyri* Part XVII, London 1927.
Jansen, H. L., *Die Henochgestalt: Eine vergleichende religionsgeschichtliche Untersuchung*, Oslo 1939.
Jeremias, J., *Theophanie: die Geschichte einer alttestamentlichen Gattung* (Wissenschaftliche Monographien zum Alten und Neuen Testament), Bd. 10, Neukirchen-Vluyn 1965.
Köstlin, K. R., *Über die Entstehung des Buches Henoch*, Tüb. Theol. Jahrbuch, 15 (1856) 240-279, 370-386.
Lee, J. A. L., *A Lexical Study of the Septuagint Version of the Pentateuch*, Scholars Press, 1983.
Lietzmann, H., *Der Menschensohn*, Leipzig, 1896.
Manson, T. W., *Some Reflections on Apocalyptic*, in *Aux Sources de la Tradition Chrétienne Mélanges* ... à M. Goguel, Neuchâtel 1950, 139-145.
Manson, T. W., *Studies in the Gospels and Epistles*, Manchester 1962.
Milik, J. T., *Ten Years of Discovery in the Wilderness of Judaea*, London, SCM, 1959.

Montgomery, J. A., *Aramaic Incantation Texts from Nippur*, Philadelphia 1913.

Neugebauer, O., *Ethiopic Astronomy and Computus*, Österr. Akad. d. Wiss., Philos.-hist. Kl. SB 347 (1979).

Nickelsburg, G. W. E., *Resurection, Immortality and Eternal Life in Intertestamental Judaism* Harvard Theological Studies XXVI, Cambridge, Mass., 1972.

Rossell, W. H., *A Handbook of Aramaic Magical Texts*, Shelton Semitic Series No. 2, New Jersey 1953.

Rowley, H. H., *The Relevance of Apocalyptic*, London 1963.

Schulthess, F., *Grammatik des christlich-palästinischen Aramäisch*, Tübingen 1924.

Sjöberg, E., *Der Menschensohn im äthiopischen Henochbuch*, Lund; Gleerup 1946.

Suter, D. W., *Tradition and Composition in the Parables of Enoch*, Scholars Press, 1979.

Volz, P., *Die Eschatologie der jüdischen Gemeinde im neutestamentlichen Zeitalter*, Tübingen 1934 (reprinted 1966).

Wacker, M. T., *Weltordnung und Gericht: Studien zu I Henoch 22*, Würzburg, 1982.

Weise, M., *Kultzeiten und Kultischer Bundesschluss in der 'Ordensregel' vom Toten Meer*, Leiden 1961.

Zuntz, G., 'Notes on the Greek Enoch', JBL 61 (1942), 193f., 'The Greek Text of Enoch', JBL 63 (1944), 53f., 'Enoch on the Last Judgement', JThS 45 (1944), 161f.

Abbreviated References

Periodicals

Bib.	Biblica
BWANT	Beiträge zur Wissenschaft vom alten und neuen Testament
CBQ	Catholic Biblical Quarterly
CSCO	*Corpus Scriptorum Christianorum Orientalium*
CSEL	*Corpus Scriptorum Ecclesiasticorum Latinorum*
ET	Expository Times
HThR	Harvard Theological Review
JA	Journal Asiatique
JAOS	Journal of the American Oriental Society
JBL	Journal of Biblical Literature
JNES	Journal of Near Eastern Studies
JSS	Journal of Semitic Studies
JThSt	Journal of Theological Studies
NTS	New Testament Studies
Nov. Test.	Novum Testamentum
PEQ	Palestine Exploration Quarterly
RB	Revue Biblique
RSR	Recherches de Science Religieuse
Semitica	Semitica
Str.B.	H. L. Strack - P. Billerbeck, *Kommentar zum Neuen Testament aus Talmud und Midrash*
ThBl	Theologische Blätter
ThWNT	Theologisches Wörterbuch zum Neuen Testament
TS	Theological Studies
VT	Vetus Testamentum
ZAW	Zeitschrift für die alttestamentliche Wissenschaft
ZDMG	Zeitschrift der deutschen morgenländischen Gesellschaft
ZNTW	Zeitschrift für die neutestamentliche Wissenschaft
ZThK	Zeitschrift für Theologie und Kirche

Biblical Books

Old Testament

Gen.	Genesis
Exod.	Exodus
Lev.	Leviticus
Num.	Numbers
Dt.	Deuteronomy
Jos.	Joshua
Jg.	Judges
Ru.	Ruth
1 Sam./1 Kg.	1 Samuel
2 Sam./2 Kg.	2 Samuel
1 Kg./3 Kg.	1 Kings
2 Kg./4 Kg.	2 Kings
1 Chr.	1 Chronicles
2 Chr.	2 Chronicles
Ezr.	Ezra
Neh.	Nehemiah
Est.	Esther
Job	Job
Ps.	Psalms
Prov.	Proverbs
Ec.	Ecclesiastes
Ca.	(Canticles) Song of Solomon
Isa.	Isaiah
Jer.	Jeremiah
Lam.	Lamentations
Ezek.	Ezekiel
Dan.	Daniel
Hos.	Hosea
Jl.	Joel
Am.	Amos
Ob.	Obadiah
Jon.	Jonah
Mic.	Micah
Nah.	Nahum
Hab.	Habakkuk
Zeph.	Zephaniah
Hag.	Haggai
Zech.	Zechariah
Mal.	Malachi

New Testament

Mt.	Matthew
Mk.	Mark
Lk.	Luke
Jn.	John
Ac.	Acts of the Apostles
Rom.	Romans
1 C.	1 Corinthians
2 C.	2 Corinthians
Gal.	Galatians
Eph.	Ephesians
Phil.	Philippians
Col.	Colossians
1 Th.	1 Thessalonians
2 Th.	2 Thessalonians
1 Tim.	1 Timothy
2 Tim.	2 Timothy
Tit.	Titus
Phm.	Philemon
Heb.	Hebrews
Jas.	James
1 Pet.	1 Peter
2 Pet.	2 Peter
1 Jn.	1 John
2 Jn.	2 John
3 Jn.	3 John
Jude	Jude
Rev.	Revelation.

Apocrypha

1 Esd.	1 Esdras
2 Esd.	2 Esdras
Tob.	Tobit
Jdt.	Judith
Wis.	The Wisdom of Solomon
Sir.	Ecclesiasticus
Bar.	Baruch
Letter of Jeremiah	
Man.	The Prayer of Manasseh
1 Mac.	1 Maccabees
2 Mac.	2 Maccabees
3 Mac.	3 Maccabees

Pseudepigrapha

Jub.	The Book of Jubilees
1 En.	1 Enoch
Test.	The Testaments of the XII Patriarchs (separately noted)
Sib. Or.	The Sibylline Oracles
2 Bar.	2 Baruch or the Syriac Apocalypse of Baruch.
3 Bar.	3 Baruch
Ass. Mos.	The Assumption of Moses
2 En.	2 Enoch, or The Book of the Secrets of Enoch
Pss. Sol.	The Psalms of Solomon

INTRODUCTION

I. Problems of Text and Interpretation

In his magisterial volumes on *Judaism and Hellenism*[1] Martin Hengel links the earliest parts of I Enoch with the Book of Daniel as together marking the zenith (der Höhepunkt) of Jewish apocalyptic of the intertestamental period. Certainly, if Daniel can be called the classic Jewish apocalypse, 1 Enoch does not come far behind, and may even have equal claim to classical status in intertestamental literature; and if the so-called Parables of Enoch did, in fact, precede the Gospels, they constitute a remarkable *praeparatio evangelica*. Ernst Käsemann maintained that 'apocalyptic ... was the mother of all Christian theology';[2] he meant primitive Christian apocalyptic, but to this 1 Enoch was a chief contributor. It is not surprising to find the Book of Enoch cited as scripture by the author of the Epistle of Jude (vs. 14-15). As a source of background evidence for the study of the emerging Christian church and its literature, the Book of Enoch is a star witness.

For the scholar who is prepared to tackle the problems of this influential book, the main difficulty—and the most formidable deterrent—is the fact that it has only survived, for the greater part of the work, in an Ethiopic version and in a rébarbative form of text, at times in not a little confusion and disorder, and where frequent corruptions tend to obscure genuine variant readings. It is true, substantial portions have also survived in a Greek translation,[3] and, more recently, fragments of the original Aramaic have been found at Qumran;[4] and these have contributed greatly to our knowledge and understanding of the book. But formidable difficulties remain: the surviving Greek portions do not cover more than a fraction—roughly about a third—of the whole work, and have also reached us in manuscripts which have suffered equally severely in transmission;[5] and the Aramaic fragments amount disappointingly to no more than 5% of the Ethiopic book,[6] at times representing little more than an identifiable word or letter.

---

[1] *Judentum und Hellenismus*, Tübingen, 1973, 330 (S.C.M., London, 1974, 180).

[2] ZThK LIX (1962), 284, cf. ZThK LVII (1960), 162f. The thesis stirred up a lively debate in systematic as well as Biblical theological circles; see the discussion by Ebeling and Fuchs in ZThK LVIII, 1961, Bultmann in *Apophoreta*, Festschrift for E. Haenchen, Berlin, 1964 and J. Moltmann, *Theologie der Hoffnung*, Munich, 1977, 120f.

[3] See *Apocalypsis Henochi Graece*, ed. M. Black, Leiden, 1970.

[4] Editio princeps by J. T. Milik, *The Books of Enoch, Aramaic Fragments of Qumrân Cave 4*, Oxford, 1976.

[5] See especially *Apocalypsis Henochi Graece*, 8.

[6] Review by E. Ullendorff-M. Knibb in Bulletin of the School of Oriental and African Studies, 40 (1977), 601.

Considerable progress has been made in the editing and elucidation of the Eth. version since the editio princeps of Laurence in 1821[7] and the pioneering work of Dillmann.[8] The two main critical editions, those of Charles[9] and Flemming[10] sorted out the mss into two main groupings, an older recension made on the basis of a Greek version (Charles's α text, Flemming's Group I), and a later recension of the Ethiopic (Charles's β text, Flemming's Group II), the latter evidently designed to be an ecclesiastical Vulgate or 'canonical' text of a book which had its own place in the Eth. canon of the Old Testament. The editions of Charles[1912] and Flemming (and the translation of Flemming-Radermacher[11] and the 1912 Charles version) were based on the Group I recension.

The most recent edition, that of Michael Knibb,[12] consists of a photographic reproduction of a single ms (Rylands Eth. MS 23), a typical representative of the Ethiopic Vulgate, accompanied by a literal English translation and an apparatus consisting mainly of readings from the oldest group of Eth. mss identified by Flemming and Charles (the sigla of these editors have been changed and there has been a fresh collation of all the Eth I mss known to Knibb (2, 36)): this is accompanied by a collation of the Greek texts and of the Aramaic fragments, the latter deciphered by Knibb from photographs at his disposal; two mss of the ecclesiastical text were freshly collated together with the recently identified Group I ms Tana 9 (Tana 9[a]), the existence of which was first reported in the catalogue of E. Hammerschmidt, *Äthiopische Handschriften vom Ṭānāsee ... Verzeichnis der orientalischen Handschriften in Deutschland*, xx, 1 (Wiesbaden, 1973).[13]

---

[7] *Maṣḥafa Henok Nabiy, The Book of Enoch the prophet. An Apocryphal production, supposed to have been lost for ages; but discovered at the close of the last century in Abyssinia; and now first translated from an Ethiopic MS. in the Bodleian Library*, by Richard Laurence ... Oxford ... 1821.

[8] *Liber Henoch Aethiopice* (Lipsiae, 1851), *Das Buch Henoch* (Leipzig, 1853).

[9] *The Ethiopic Version of the Book of Enoch edited from twenty-three MSS together with the Fragmentary Greek and Latin Versions* (Oxford, 1906).

[10] *Das Buch Henoch: Äthiopischer Text herausgegeben von Dr. Joh. Flemming* (TU, NF Bd. 7, Leipzig, 1902).

[11] *Das Buch Henoch* herausgegeben von J. Flemming and L. Radermacher, GCS, Leipzig, 1901.

[12] *The Ethiopic Book of Enoch*: *A New Edition in the Light of the Aramaic Dead Sea Fragments*, Vol. 1 *Text and Apparatus*; Vol. 2 *Introduction, Translation and Commentary*, Oxford, 1978.

[13] A translation based on this manuscript is being prepared for the Duke-Doubleday series of Pseudepigrapha by Dr. Ephraim Isaac of Bard College, New York, and a facsimile edition of the text has been promised (see J. H. Charlesworth, 'The SNTS Pseudepigrapha Seminars at Tübingen and Paris on the Books of Enoch', in NTS, 25(1979), 319). Dr. Isaac has kindly made his translation available to me.

Work on the variants of this ms has shown its unique value (see Appendix C, Analysis of Preferred Readings, 422 f.). In the opinion of Knibb (2, 36 n. 34) 'the discovery and collation of any further Ethiopic manuscripts of Enoch is unlikely to add in any very significant way to our knowledge of the Ethiopic text'.

No discoveries or studies since the appearance of the Flemming and Charles editions have substantially altered their basic division of the mss into the two groupings or recensions (here, following Flemming and Knibb designated Eth I and Eth II). It is true, that division is not always clear cut; as Flemming noted, Eth I mss are often found on the side of Eth II, although the opposite, Eth II mss supporting Eth I, occurs less frequently. (Flemming, IX, cf. further below, 6). Knibb has qualified the high estimate placed by Charles and Flemming on Eth I over against Eth II mss: 'it is important that the value of the Eth I manuscripts should not be over-emphasised' (2, 32); Eth I is also extensively corrupted (ibid.). '*In many cases*, as Charles and Flemming recognized, the original Ethiopic text has survived not in Eth I, but in Eth II manuscripts' 2.35 (italics mine). (Charles, in fact, wrote (Charles[1906], xxii) '... it is noteworthy that *in a limited number of cases* β preserves the original text where α is secondary' (italics mine). Flemming enumerated 6 cases where Eth II went correctly with the Greek version (Flemming, X n. 1, but cf. Charles[1906], xxii, n. 1).) A more far-reaching claim for the value of Eth II mss has been made recently by G. W. E. Nickelsburg; and this is discussed in detail in Appendix C, *Analysis of Preferred Readings*, and the results of this reassessment of the comparative value of the two recensions given later in this Introduction (6f.).

It is impossible to date these two recensions (the earliest Eth. mss belong to the 15th and 16th centuries), and, in any case, the ecclesiastical Vulgate may be the end result of a progressive revision.[14] The dating of the Greek portions, however, (Chh. 1-32, 6th c. A.D., 97-107, 4th c.)[15] can provide us with a rough *terminus a quo* for the Eth. translation. It was not improbably Syrian monophysite immigrants, condemned at the Council of Chalcedon in 451 A.D., who started up or stimulated this literary activity in Ethiopia.[16]

The discovery of the Aramaic fragments has settled, once and for all, the vexed question of the semitic original—at any rate for the bulk of the book, with the notable exception of the 'Parables of Enoch'.[17] The fact that there was an original Aramaic version—in possibly more than one recension[18]—does not rule out the possibility of the existence of a Hebrew Enoch, and evidence for such a book (or parts of one) is not lacking, although insufficient in quantity to make an estimate of its coverage possible.[19] The language of the original *Grundschrift* of the 'Parables' is still

---

[14] Cf. Charles[1906], xxii, Knibb 2, 28.

[15] Cf. *Apocalypsis Henochi Graece*, 7f.

[16] Cf. B. M. Metzger, *The Early Versions of the New Testament, Their Origin, Transmission and Limitations* Oxford, 1977, 221.

[17] For the 'coverage' of the book by the Aram. frgs. see especially Milik, 5.

[18] See below, 13, 21, 111, 262f., 289.

[19] See *Apocalypsis Henochi Graece*, 6, with special reference to the 'Noah' apocalypses.

being debated, though no one seems to doubt that it was a Greek *Vorlage* which the Ethiopic translated, with or without the help of a copy of the semitic (in my view Hebrew) original.[20] In view of the Jewish-Syrian origin of the literary revival in Ethiopia in the 5th century,[21] the availability of Hebrew or Aramaic originals might even be taken for granted and their total absence surprising.

For all parts of the book there is general agreement that the Ethiopic is a tertiary version, a translation of a Greek *Vorlage*, itself rendering an Aramaic and/or Hebrew *Grundschrift*.[21a] There is no reason to doubt that this Greek version, like other such Greek translations of intertestamental writings, was made by Christian scribes for Christians,[22] in some cases probably for Jewish-Christian congregations.

The discovery of portions of a Greek version of Chh. 1-32 and the existence of quotations in the Fathers, particularly the extensive quotations in the writings of George Syncellus,[23] led to the first important advance in the critical study of the book. The close connection of the Ethiopic version with the Gizeh text (G), even to the sharing of the same corruptions, was early recognised.[24] Comparison made it clear that ʿE was translated from a MS which was also the parent or ancestor of G^{g}'[25] (or, one ought perhaps to add, 'the descendant of an ancestor of G'). The underlying Greek text of the Ethiopic version was, in fact, at times identical, though generally—in additions, omissions or variant readings—exhibiting evidence of stemming from a different recension of the same ancestor. The Syncellus citations, on the other hand, had drawn on an independent Greek version, in some respects more faithful to its semitic base.[26]

In a comparison and collation with the Ethiopic texts variants were noted in most editions; Charles (but also following Dillmann and his contemporary Flemming) again and again supplied a putative Greek equivalent for Ethiopic variants from the Greek version. Knibb has supplied a separate *apparatus criticus* in Vol. 1 to accommodate the Greek variants (the Aramaic fragments are included in his notes to the translation). When the newly discovered Chester Beatty papyrus of Chh. 97-107 was edited, Campbell Bonner,

---

[20] See below, 185f.

[21] Cf. above, 3 n. 16.

[21a] That an Aram. *Enoch* as well as a Greek version was available to the Eth. translators is argued by E. Ullendorff ('An Aramaic "Vorlage"' and M. Knibb (2.37f.) (see below, 185). A detailed critique of this theory by J. C. VanderKam is appearing in the forthcoming Festschrift for Thomas Lambdin, edited by D. M. Golomb, B. Halpern and S. T. Hollis.

[22] Cf. *Apocalypsis Henochi Graece*, 8.

[23] See *Apocalypsis Henochi Graece*, 7f.

[24] Dillmann, SAB 1893, 1040, Charles[1906], xiiif., Charles xviii, Flemming-Radermacher, 2.

[25] Charles, loc. cit.

[26] Cf. Flemming-Radermacher, loc. cit., and Charles, xvii.

though himself unacquainted with Ethiopic, consulted an eminent American Ethiopic scholar (W. H. Worell, C-B vii,22) in his use of the Ethiopic manuscripts, again occasionally suggesting a possible Greek equivalent, though confining himself, for the most part, to equivalents in English translation. Hitherto, however, no systematic attempt has been made to try to recover genuine Greek textual variants from the forbidding mass of Ethiopic variations. So many of the latter, however, can be ignored, since they consist of 'inner Ethiopic' variations, differences in orthography, or frequent haplographies and dittographies (a feature also of the Greek version), of word order, grammatical 'improvements' and plain corruptions. In spite of the bewildering variety of such Ethiopic variations, a close study of the Eth. apparatus, in particular in the Charles[1906] and the Flemming editions (now supplemented by Knibb) uncovers a rich vein of significant *variae lectiones*, significant for the translator, even if it is only the restoration of a καί or of a correct tense, but, in particular, meaningful for the exegete; the sifting in this way of the wheat, as it were, from the chaff proves to be a most rewarding exercise, opening up a fresh avenue of approach to the Enoch text, and proving itself to be of no less value than the newly discovered Aramaic fragments, if only because so much more of the text of Enoch is actually covered by the Greek version.

Genuine *variae lectiones* of this kind are generally identifiable without too much difficulty. There are times, however, when an 'inner Ethiopic' rendering, in particular a free translation or paraphrase, can be mistaken for an original variant reading, and, vice versa, a genuine variant taken to be an 'inner-Ethiopic' rendering. Obviously allowance must be made for such ambivalent cases, especially where the sense of the passage is affected.

An attempt was first made by me, following the precedent of Campbell Bonner, to compile an apparatus of such variants in English, but this was eventually abandoned in favour of a reconstruction of all putative Greek variants into their equivalent in Biblical (or Septuagint) Greek, as had been sporadically attempted by earlier editors. This proved, in fact, to be a much more reliable, and, indeed, a more accurate method of reporting these variants. Where the Greek version was available in the earlier and later chapters, the recognition of such 'Greek' *variae lectiones* presented few difficulties, always provided ambivalent cases were recognised. But even in the 'Parables' where no Greek version exists, the variants in the Ethiopic mss, which had a claim to a Greek origin, were not difficult to identify. In both these parts of the Book care has been taken only to give putative Greek equivalents of Ethiopic variants where the Greek vocabulary, grammar and syntax are attested in Biblical Greek. To prevent any possible confusion with the extant Greek texts, and their variants (e.g. in Syncellus quotations), all such 'Greek' variants are included *in square brackets* in the commentary, and, where it seemed necessary or desirable, accompanied in the textual notes by the Ethiopic variant which is there being translated. The apparatus also includes the Aramaic fragments, to which I had direct access.[27] Attempts to go beyond the Greek to the original Aramaic are confined to passages where the Aramaic is extant or where there are *cruces interpretum*; and these latter cases, in particular, are dealt with mainly in the commentary.

These *Textual Notes* represent a new departure in the treatment of 1 Enoch and its problems. Hitherto the *apparatus criticus* has either been attached to editions of the Ethiopic text and compiled in Ethiopic, or, as in the Charles edition of 1912 or the Campbell Bonner edition of the Greek papyrus, included in the commentary. The only other editions which adopted a similar method were those of Martin[28] (with an apparatus in French) and Flemming-Radermacher (with an apparatus in German).[29]

It was no surprise to find, as a result of this collation, complete confirmation of the superiority of the Eth I text-type, in all parts of the book. Analysis of the readings adopted for Chh. 1-36, however, yielded one fresh result: it was no longer Eth^{g}—as Charles and Flemming had concluded—but the recently discovered Eth$^{tana\ (tana^{a})}$ which was *facile princeps* as the best single representative of Eth I in these chapters.[30] Since the editions of Flemming and Charles, a Greek version of Chh. 97.6-107.3 has been discovered and edited: a collation of Ethiopic variants with this text has led to a similar result, the superiority of Eth I, with Eth^{tana} again as its best representative. Although no Greek texts have survived for the 'Parables', a survey and analysis of the preferred readings here show Eth^{tana} again the dominant partner of the Eth I text. This older Eth I recension is the only secure foundation for any edition of 1 Enoch which can make any claim to represent, even at third remove, the words of the original semitic author(s).

Eth I manuscripts, however, do not enjoy a complete monopoly of correct readings. A limited number of these are to be found in all parts of the book in Eth II manuscripts, but these are seldom of much consequence—the Ethiopic Vulgate is an extremely poor text. Charles's judgement in this connection, which was based on a first-hand acquaintance with the texts, is entirely justi-

---

[27] By the kind permission of the late Père Roland de Vaux and with the encouragement of the late Professor Sir Godfrey Driver, and a grant from the Pilgrim Trust, I was able to spend a semester in 1957 in the former Palestine Institute and the British School of Archaeology in Jerusalem, working, with J. T. Milik, to whom the Aramaic fragments had been assigned, on the copying and decipherment of the fragments.

[28] *Le Livre d'Hénoch traduit sur le Texte Éthiopien* par Francois Martin (Paris, 1906).

[29] Above, 2, n. 11.

[30] See Appendix C, *Analysis of Preferred Readings*, 422f.

fied: 'The first (form of the text, α) is represented by gg[1]mqtu (and in some degree by *n*), ... the second (β) ... owes its origins to the labours of native scholars of the sixteenth and seventeenth centuries ... The result of these labours has been *on the whole disastrous*; for these scholars had neither the knowledge of the subject-matter nor yet critical materials to guide them as to the form of text. Hence *in nearly every instance where they have departed from the original unrevised text they have done so to the detriment of the book*' (italics mine).[31]

Since the textual foundation of any fresh translation and commentary can only be the Eth I text, the best editions of which are still those of Charles and Flemming (supplemented by Knibb), and since Charles himself had already produced in his 1912 edition a revised version of his original 1893 translation, a revision based on this text, it seemed preferable to seek to update, by a third 'recension', the Charles 1912 translation rather than to attempt to produce an entirely new English version. Besides, the Charles[1912] translation has established itself, in spite of criticism of many points of detail, as the standard English version of the Book of Enoch, and one which has widely commended itself for the quality of its English, no less than for its sound textual basis. The revision, nevertheless, in this new English version has been thorough, with many changes introduced in the light of the new evidence. I have also 'modernised' some of Charles's expressions, revised his headings, but retained, for the most part, the poetic format, where parallelism of lines and clauses pointed to such a structure in the original. Much of the latter is necessarily conjectural, but it seemed justified as a method of indicating the nature and style of the original semitic composition.

In the commentary I am deeply indebted to my predecessors, in particular to the pioneering work of August Dillmann, on whom Charles, Schodde, Martin, Knibb also freely draw. Many Biblical parallels and allusions, references to other apocalyptic writings or patristic sources come from Dillmann and Charles; and these have been 'updated' wherever new editions seem to be required. A special feature of the commentary, however, is the extension of these Biblical references and allusions, including phraseology which is clearly Biblical in inspiration; in fact, as in the Qumran scrolls, the original writings, Aramaic and Hebrew, are steeped in Biblical references and allusions; their language is that of the Hebrew Bible, and it is the Hebrew scriptures which supplied the chief source of their inspiration. So far as the theological ideas of 1 Enoch are concerned, many also explicitly developed, as in the 'Parables', I have sought, as far as possible, without undue repetition of previous work, to bring these also up-to-date in the light of the most recent theological studies.

---

[31] Charles[1906], xxi.

II. The Structure of I Enoch

Gunkel described I Enoch as 'ein wüstes Durcheinander verschiedener Traditionen',[32] and Beer went further and spoke of a 'Labyrinth ... chaotisch zusammengewirbelten Stoffe.'[33] Certainly, what the Book presents to the reader is a bizarre variety of often disparate and overlapping traditions, containing units of narrative and discourse which could be as early as the second century B.C. (the date assigned to the oldest portions of the Aramaic),[34] side by side with redactional supplements etc. which are centuries later—each and all of them traceable to 'authors' of different periods and persuasions. Not surprisingly, in view of this, I Enoch has now attracted the attention of 'tradition-historical', 'form critical' and 'redaction-historical' approaches to its problems; and these could have an increasingly important contribution to make to their elucidation.[35]

Although the person of Enoch himself and the elaboration of his Biblical 'biography' (Gen. 5.21-24) does invest the Book of Enoch with a formal unity, there is one corpus of traditions within it which relate directly to Noah, not to Enoch. These used to be described as the 'Noah interpolations' (most notably the sections at Chapters 6-11, 54.7-55.2, 60, 65-69.25 and 106-108), all deriving from an earlier 'Book of Noah' (Charles, xlvi). They are more accurately to be described as 'Noah apocalypses', but it is a mistake to regard them as a 'foreign body' within the Enoch saga. If there is one dominant theme which, next to Enoch himself, 'unifies' the Book of Enoch, it is that of the Last Judgement, 'the Great Judgement', and of this the Deluge was not only the precursor but, in fact, the First Great Judgement (cf. 2 Pet. 3.5-7). Cf. Dillmann, 89. The Deluge and the Last Judgement are twin events, and belong together in the Enoch story. Moreover, Enoch does at times have a part to play in the Noah apocalypses (cf. especially, 106.7f.); and, in one case (60.1f., below, 225) what began its existence as a Noah vision has evidently been clumsily adapted to become a vision of Enoch. At the same time, we now possess Hebrew fragments of an original Noah apocalypse to confirm the evidence of Jub 10.13 (cf. 21.10) for the existence of a Book of Noah (see below 319), so that earlier editors were not so far wrong in maintaining that the 'author' or 'author-redactor' of I Enoch drew freely on such sources in his

---

[32] *Schöpfung und Chaos* (1895), 289.

[33] *Das Buch Henoch*, 224.

[34] Milik, 5f.

[35] See, e.g. Dexinger, 102f. and for a recent review of tradition-historical and form critical studies of Enoch, with special reference to Chapters 1-36, see M.-T. Wacker, *Weltordnung und Gericht: Studien zu 1 Henoch 22* (Würzburg, 1982), 16f. Dr Wacker also discusses the 'history of religions' approach to Enoch.

composition of the Book; and we owe it to him that these Noah apocalyptic traditions have not perished.

I Enoch has reached us in the Ethiopic version in the traditionally Hebrew pentateuchal arrangement of five 'books', like the five books of the Psalter, the five Megilloth, the five divisions of the Pirke Aboth etc. According to Milik (4. f.) the pentateuchal structure was inherited by the Ethiopic from the Qumran Enoch, but with one notable difference: the Aramaic Enoch had, as its second 'Book', not the Book of the Parables of the Ethiopic version, but a new 'Book of the Giants', a sequel to the Watcher legend, fragments of some ten or twelve manuscripts of which have been found among the 4Q material from Qumran (Milik, ibid.), evidence of its wide popularity in Essene circles. Milik maintains that this 'Book of the Giants' was the second book in the original Aramaic Enoch Pentateuch. The work achieved a similar popularity as at Qumran in certain heretical Christian circles; it was canonised by the Manichaeans, so that, not surprisingly, it disappeared altogether from mainstream Christianity (and no doubt, also, like so much else from Qumran, in 'orthodox' Judaism as well). It was removed, Milik concludes, from the Enoch Pentateuch and replaced by the Book of the Parables, a Christian work, composed originally in Greek, and of a date not earlier than the third century A.D. (Milik, 95). See below, 183, for Milik's reconstruction.

The question of the character and date of the Parables is one of central importance in Enoch studies, in view, in particular, of its 'Son of Man' figure. The problem is discussed fully later in the Introduction to the Parables in the commentary (181 f.). Meantime, we must ask: Did the so-called 'Book of the Giants' in fact constitute a second Book of Enoch, or was it simply, originally, a part of the Watcher legend itself, i.e. of the Book of the Watchers?

It must be said at once that Milik's identification of these fragments on the Giants with the surviving Sogdian fragments of the Manichaean book is a brilliant piece of detective work. There seems little doubt too that these fragments belong to the original Aramaic Watcher legend and consist of a further elaboration of the story of the 'bastard' offspring of the Watchers, the Giants, already told in the Book of the Watchers. Enoch is specifically mentioned in the fragments (Milik, 305), and the first copy 4QGiants[a] (the only complete copy published) was evidently written by the same scribe as was responsible for the third copy of the Aramaic Enoch (4QEn[c]): '... it is ... quite certain that 4QEnGiants[a] formed part of the same scroll as that of En[c]' (Milik, 310), a scroll which supplied fragments of Chapters 1-36, containing the Watcher legend (as well as 83-90, the Dream-Visions and 91-108, the 'Epistle of Enoch'). Milik (ibid.) concludes: ... 'our copy (i.e. 4QGiants[a]) ... would have come after the first part', i.e. after Chapters 1-36. The case for regarding the 'Book of the Giants" as a sequel to the 'Book of the Watchers' seems conclusive. But are we obliged to assume that, since it became a canonical book for the Manichaeans, these fragments come from what constituted a separate 'second Book of Enoch'? As we shall see shortly, the so-called 'Book of the Watchers', i.e. the story of the fall of the Watchers and of their bastard children, the Giants, ends at 16.4, and not, as Milik supposes, at Chapter 36. Chapters 17-36, describing Enoch's journeys and visions of heaven and hell, form a quite new and separate apocalypse. If the elaboration of the Giants' story belongs to the Watcher legend, then it must be a part of the first 'Book of Enoch', not an entire second book.

One further consideration tells conclusively against the theory of an original pentateuchal Enoch. The extant 'astronomical' and calendrical fragments from Qumran point to a vastly larger amount of original Aramaic texts of this nature than has survived and been reproduced in the Ethiopic 'astronomical' Book, Chapters 72-82: 'The complete text of this book in Aramaic was ... very long, so that copies of it *filled voluminous scrolls*, and this explains why it was never included with the other Enochic writings on the same strip of parchment' (Milik 8, italics mine). Some Greek fragments from the 'astronomical Book', overlapping with the Ethiopic, were identified by Milik, and of these he writes (19): 'It can be seen clearly now that the Egyptian Jews responsible for the translation from Aramaic were at pains to shorten the voluminous, prolix, and terribly monotonous original.' The latter was, in fact, 'freely adapted' as well as 'abridged.' Comparison of the Ethiopic 'astronomical' section with the Greek and Aramaic fragments points to a similar conclusion for the Ethiopic: what we have in Chapters 72-82 are representative 'astronomical' and calendrical excerpts, translated from abridged and adapted Greek excerpts, replacing the tediously long Aramaic calendrical calculations and 'astronomical' speculations, no doubt selected and 'edited' to represent the main aspects of this Enochian tradition, the movements of the sun, the phases of the moon, the 'rose' of the winds, the variations of the weather etc. Moreover, since the invention and elaboration of the calendar was traditionally attributed to Enoch, no Enoch corpus could afford to exclude these 'astronomica' etc. Visions of the heavenly bodies, of the movements of the stars and the winds, are already recorded in Chapters 33-36, a section firmly attested for the Aramaic Enoch (Milik, 203f., 234f.). The 'astronomical' Book, Chapters 72-82 is a further development of this same Enochian theme, perhaps even elaborating on the details of Chapters 33-36 (Comm. 180). The Parables of Enoch also include their own astronomica (e.g., Chapters 43-44, 59).

If this is a correct explanation of the genesis of the 'Astronomical Book of Enoch', then it is manifestly an artificial, originally Greek, versional creation, translated and extracted from a mass of original Aramaic material. There never existed in Aramaic a third astronomical 'Book of Enoch'; all that we in fact have is an abridged selection of calendrical and 'astronomical' translation pieces put together from the vast Aramaic corpus of such texts by some later editor. By the same token, we must go on to ask: if the astronomical section was 'tailor-made' (though by unskilled 'cutters') to constitute a third 'Book of Enoch', and we can only be certain of three other sections or 'Books', the Watcher legend and Enoch's Journeys, the Dream-Visions and the 'Epistle of Enoch', may not the Book of the Parables have also been 'cobbled together'—to mix metaphors—at the same time as the 'astronomical' Book, and from similar Greek translation sources, to form an Enochic Pentateuch, probably by some Jewish Christian translator-redactor? The origin and date of the Parables may be problematical, but they almost certainly owe their presence in I Enoch to Christian interest. Was the Ethiopic Enoch Pentateuch a translation of a primary Greek Enoch Pentateuch, put together by the creative efforts of the first Christian redactor of Enoch? And since Jude 14 seems to be drawing on an existing version of the Greek Enoch (see below 109), that redaction may well have been completed as early as the beginning of the second century A.D.

The last part of the Book consists of Chapters 91-108, at the beginning of which there are two clearly definable literary units, a short Nature Poem (93.11-14) and the famous Apocalypse of Weeks (93.3-10, 91.11-17, surviving in Ethiopic in a dislocated text, but with the correct verse order in the Aramaic). That both these pieces of literary composition had an anterior existence (the first from a sapiential source), and have been adapted and incorporated in the larger Book as it has now reached us, is nowadays a generally accepted view (but see below, 286, 288).

Chapters 92, 93.1-2, where the Aramaic fragment contains a reference to a 'writing', almost certainly a 'letter' or 'epistle' i.e. a literary *epistole*, of Enoch, is usually taken to refer to the whole of Chapters 91-108; and since the Greek version of 97-107 (the Chester Beatty text) concludes with the subscription Επιστολη Ενωχ, this entire section of the Enoch corpus has come to be designated 'The Epistle of Enoch'. But did Enoch's *epistole* at 92, 93.1-2 originally have such a comprehensive and vast content, an entire Book of Enoch; or is this too the creation of the Greek redactor-editor? A study of the Aramaic text of these verses uncovers an interesting fact. While the Aramaic fragments at 92, 93.1-2 present difficulties, the reference at 93.1-2 to Enoch 'giving over', 'handing over' his 'letter' or *epistole* to his son Methuselah makes it more than probable that the letter in question was a literary *epistole* of the type found in the apocrypha, e.g. the 'Letter of Jeremiah' or the shorter *epistole* at 2 Macc. 1. 1, and not a designation of the whole of the extended assortment of traditions from Chapter 91-108. How much of the paraenesis which accompanies or follows the *epistole* could have been in the original it is impossible to say; it could have been a long or a short *epistole*, perhaps consisting of the Apocalypse of Weeks and nothing else. But if this hypothesis is sound, we would then require to explain the Greek title Επιστολη Ενωχ for Chapters 91 (97) to 107 as an invention of a Greek redactor. The Greek version of 100.6 does refer to 'the words of this *epistole*', but the original כתבא or סיפרא could mean simply 'writing' or 'book' and refer to the Book of Enoch as a whole, as the ambiguous Ethiopic maṣḥaf 'writing, book, epistle' does at 108.1. Was it perhaps from the Greek version of 100.6 that the Greek redactor took his subscription?

III. The Component Elements of I Enoch

The Book of Enoch is like an intricately devised jig-saw puzzle, or rather a collection of such puzzles, in which, after the main component pieces have been put together to make a whole picture (or a series of them), there still remain elements unaccounted for which baffle the most ingenious attempts to fit them into a coherent whole. No account of the component elements of this complicated and at times, in the versions which have transmitted it, even confused material (the description 'this stupid book'[36] can really only apply to the work of transmission), can ever hope to be more than approximate and subject to necessary revision with the discovery of new evidence, the discarding in the light of it of outworn theories, and the fresh insights which renewed study produces. Thus, with the discovery of the Aramaic fragments, the lengthy controversies about the original language, at least of the greater part of the Enoch corpus, have been laid to rest, though the possibility of an ancient Hebrew Enoch is by no means finally settled. (Comm, 187). The following critical survey of the main elements which have gone to the composition of the Enoch corpus is designed mainly as a guide to the reader to assist him in his consultation of the Translation and Commentary. But there is no Ariadne's thread to lead him through the Enochian labyrinth.

Chapter 1.1-9 *Prooimium and Central Theme: the Great Judgement*

Chapter 1.1-9 is the first identifiable literary unit, fragmentarily preserved in Aramaic, and consisting mostly of a description, in a discourse (παραβολή) attributed to Enoch himself, of an impending theophany and universal judgement, referred to elsewhere as 'the Great Judgement'. It forms an extremely fitting Prooimium (so Dillmann, 90) to the whole Enoch corpus, since the idea of a final Judgement is one of its dominant themes, if not the central theme of the whole book. The passage is modelled chiefly on

---

[36] J. Y. Campbell, in 'The Origin and Meaning of the Term Son of Man' in JThS XLVIII (1947), 148 (the remark is attributed to H. L. Goudge).

Dt. 33.1-2, where a verse announcing the blessing of Moses is followed by the description of a Sinai theophany. But the Enoch apocalyptic prophecy does not look back to a Sinai event in the past, but to a theophany from Sinai in the future, culminating in a fearful Judgement (v. 9). This climactic verse is quoted by Jude (14) in what looks like an abridged form of the Greek version which has survived. Was Jude (and perhaps 2nd Pet.) quoting from or referring to an ancestor of our Greek text of En. 1-36? (Comm. on 10.12, 14, 18.11).

Chapters 2-5: *A Nature Homily*

Chapter 2 introduces a second theme, the contrast between the harmony and regularity of nature, the movements of the heavenly bodies, the orderly succession of the seasons, with Israel's defiance of the divine ordinances, a defiance which can only lead to her 'eternal execration' (5.4-5). Later, at 18.15 and 21.3, we learn that there are also seven 'wandering stars' (cf. Jude 13) which, like the angel-Watchers, disobey the divine ordinances and meet the same fate. Later still, at Chapter 80, the theme is developed of unnatural disturbances in the world of nature. Such ideas are a commonplace of different cultures; they receive noble expression in Hooker's *Ecclesiastical Polity*, i, 16.8: 'Of Law there can be no less acknowledged than that ... her voice is the harmony of the world: all things in heaven and earth do her homage, the very least as feeling her care, the greatest as not exempted from her power.' The underlying idea that a breach of divine law leads to a condign punishment becomes another leitmotiv in the Enoch apocalypses. Indeed, together with the idea of a final Judgement, Dr. M.-T. Wacker regards it as constituting the central theme of I Enoch, as the title of her recent study implies, 'Weltordnung und Gericht' (above, 8).

Chapters 1-5 together thus provide a most appropriate introduction to the Book of Enoch; and these chapters are, in fact, headed Einleitung by Dillmann (90). Apart, however, from the theme of judgement, the two parts (1,2-5) have very little in common, and there is new evidence that a poem on the seasons taken from the 'astronomical' and calendrical scrolls has been utilised by the author of this nature homily (Comm., 111, 419). Nevertheless, the composition of chapters 1-5 as a single literary unit in the original Aramaic is firmly attested by En^{c}1i, where 1.9 is immediately followed by 2.1 (Milik, 184). But the new evidence does show that the Aramaic author of the nature homily was also himself a redactor of earlier material, and the calendrical scrolls are probably among the oldest (Milik, 8).

Chapter 6-16.4 *The Legend of the Watchers*

This long section introduces the most famous and the most frequently cited part of the Enoch corpus,[37] one of special interest in view of its connection with Gen. 6.1-4. Watchers or celestial vigilantes, the highest ranking echelon of angelic beings, descended to corrupt the 'daughters of men' and produce a race of 'bastards', the Giants, known as the Nephilim, from whom were to proceed demonic spirits to plague mankind till the Last Day. An embassy of Enoch on their behalf to the Most High proved unavailing: their bastard offspring perished in internecine conflict, which their fathers witnessed, while they themselves were confined beneath the earth until the Last Judgement.

That this legend probably circulated in written form independently of its present context in I Enoch seems highly probable; the evidence of the Aramaic fragments, dated to the second century B.C., linking 6.1 with 5.9 (the end of the nature homily) is not as certain as the connection of 2.1 with 1.9 (Milik, 165). But we need not doubt that the Watcher legend formed an integral part, indeed a central piece, in the earliest extant Aramaic text of the opening section of I Enoch. Traditionally the Enoch tale is regarded as an elaboration of Gen. 6.1-4: Milik makes an attractive case for reversing this judgement, and, since literary interdependence is certain, borrowing of ideas and phraseology is by Genesis from Enoch and not the other way round. Gen. 6.1-4 certainly does look like a hebraised excerpt from the Aramaic narrative, possibly to provide a mythological beginning (the celestial origin of evil) to the Noah saga (Comm., 124).

Within the Aramaic text and the versional tradition there is evidence of the elaboration and growth of the legend. In the names of the dekadarchs, the leaders of the two hundred Watchers, an earlier source is probably being used (6.7f), but if so, it has been well integrated within the structure of the Aramaic text (cf. Milik, 150). On the other hand, the names given by Syncellus (7.1-2) to the sons of the Giants, viz. Nephilim, and to the sons of the Nephilim, Elioud, are imaginative elaborations either by Syncellus himself or from his Greek source (the Manichaeans take the further step of providing additional personal names for the Giants in their canonical book). The scrolls refer to 'the Giants, the Nephilim' which Milik renders wrongly as 'the Giants and the Nephilim', thus perpetuating the Syncellus mistake of identifying the Nephilim with the sons of the Giants; the LXX version of Num. 13.33 (and cf. Gen. 6.4) rightly renders Hebrew נפילים by γίγαντας. (Comm. on 7.2, 15.11, 16.1, contrast Milik, 299, 240, L1. 20-21, 305-306). In this respect Milik is repeating the view of previous commentators, assuming that the names given by Syncellus to the grandsons and great grandsons of the Watchers are an original feature of the myth. The *vox obscura* Nephilim is best explained as a

[37] See Charles 14 for a list, to which we can now add the knowledge of the myth in Gnostic circles; cf. Yvonne Jannsens 'La Thème de la Fornication des Anges' in *The Origin of Gnosticism: Colloquium of Messina*, ed. by Ugo Bianchi (Leiden, 1967), 488-495. Cf. Charles, xcii.

nomen proprium, whose etymology has been forgotten, for the mythical race of the Giants. Syncellus's name Elioud for the great grandsons of the Watchers is probably to be explained as having arisen by a misunderstanding of Aramaic ילוד 'offspring (of the Watchers)', no doubt read from a corrupt Aramaic text.

Chapters 17-36 *Enoch's Journeys*

This longer section of the Enoch saga consists of a series of journeys of the patriarch, accompanied by his *angelus interpres*, which are a remarkable blend of geographical and botanical fact, based possibly on actual travels in the east (Milik suggests that the author may have been a traveller involved in the trade in aromatics), combined with mythological and apocalyptic fantasy. The result is the second of the most famous of the Enoch traditions—the forerunner if not the prototype of Dante's Paradiso and Inferno—the seer's revelations, in the course of his terrestrial and extra-terrestrial journeys, of the secrets of the heavenly realms and of the nether world, of Hades or Sheol and of Paradise.

It begins with a journey to the West in a chapter (17) which is manifestly drawing on Greek mythology (cf. Charles, 38), but including 'astronomical' and meteorological phenomena (17.3-18.5), probably from an original 'astronomical' source. The climax of this first part of Enoch's sojourn in the West is his vision of the seven mountains and the mountain-throne of God (18.6f.), leading up to the sight of a fiery abyss in which the seven 'wandering stars' and the Watchers are incarcerated until the Judgement. (Chapter 20 is a curious digression giving a list of the names of the seven archangels and their functions.) Chapter 21 resumes with the same spectacular visions, the fate of the seven stars and the Watchers, though the latter are now confined to a different fiery abyss from that of the stars.

Chapter 22 describes, in another quarter of the nether regions, the abodes, or promptuaria, of the departed, where they too are confined until Judgement Day. Chapter 23-25 resumes the theme of the seven mountains and the mountain-throne of God, which now includes a vision of the Tree of Life among the 'fragrant trees' surrounding the mountain-throne. Chapter 26 goes on to describe a new Jerusalem 'in the middle of the earth', and beside it the Vale of Gehenna, reserved for blasphemers and their judgement (27.2-3).

From earth's centre the journey moves to the East (28-32), in which its famous perfumes and spices are accurately described, the whole leading up to a description of Paradise, the 'Paradise of righteousness', and within it the 'Tree of Knowledge.' The closing chapters 33-36 introduce familiar 'astronomical' and meteorological phenomena in two concluding journeys to the North and South.

Charles, 46 (following Dillmann, but contrast Martin) thinks of two journeys of Enoch in the West, but what, in fact, we seem to have in the section 17-24 is duplicate accounts (or redactions) of the same visions, separated by chapter 20 on the seven archangels. Version A comprises chapters 17-19.3, version B 21.1-25.7. Thus 24.1-2 is a duplicate version of 18.6-9, 21.1-6 of 18.12-16, 21.7-10 of 18.11. While in version A Enoch is shown a vision of the seven fiery mountains encircling the mountain-throne of God the emphasis appears to fall on the fate of the 'wandering stars' and the Watchers in the fiery abyss. Version B begins with the latter, but now with a separate fiery abyss for each, and then goes on to the classic vision of the abodes of the departed, where, like the disobedient stars and the Watchers, the souls of the dead, whether righteous or not, are also to be kept till the Last Judgement. It closes with the vision of the seven mountains, the mountain-throne of God and the Tree of Life.

The extant Aramaic fragments have no light to shed on these duplicate accounts. Only a word or two have been identified and correctly assigned by Milik to 18.8-12, 15 (Milik, 200, 228). More substantial fragments, however, are extant for chapters 22-32. The emphasis of version A in chapters 17-18 on the prison of the stars and the Watchers suggests that it could be older than the fuller version B; version A is more germane to the Watcher legend, while version B looks like a later expansion to include the section on the promptuaria of the departed, but modelled on the account of the emprisonment of the stars and the Watchers. The vision of the mountain-throne of God in version A is similarly expanded to include the Tree of Life.

In the interpretation of chapter 22 on the abodes of the dead, there has been much confusion on the number of the promptuaria and concerning the identity and fate of some of the souls of the departed. The text speaks first of four compartments or 'hollows' (perhaps 'dungeons' or 'strong-holds') (v. 2), then of three (v. 9), and interpreters proceed to find three classes of the departed in vs. 9-13, and then try to account for the missing fourth. Dillmann suggested that the fourth class comprised the souls of men at vv. 5-7 of which Abel was a prototype. But the Aramaic text has now confirmed the conjecture of Charles that one spirit or soul only is mentioned in these verses, namely that of Abel; the plural in the Greek text is a translator's alteration. We must, similarly, set aside the view of Milik (231, L1 7-8, following Dillmann) that vs. 5-7 refer to 'the special compartment in the abode of the dead reserved for men borne away by violent death.' The vision of Sheol for all who suffer death by violence, whether justly or unjustly, is described at v. 12. It is a separate, if related, episode which is here singled out for prominent mention, the first murder of an innocent man (so Lods, 176). Charles removed the discrepancy by emending the text at v. 2 from 'four' to 'three' and from 'three' to 'two' and the emendation is still defended.[38] A study of the text, however, in the light of the Aramaic fragments, shows that the assumption that there are three categories described at vv. 9-13 is false: there are, in fact, four, (1) the righteous, (2) the sinners, here, as elsewhere not just the unrighteous but the enemies of the righteous, their oppressors who have escaped condemnation in this life, but are ripe for judgement; (3) those who have suffered death by violent means, whether justly or unjustly, their cases still to be heard; and finally (4) a class who were not completely debased, but whose sin was that they consorted with the sinners, here a probable reference to the 'fellow travellers', the 'quislings' in Israel under the Seleucids or Roman oppressors. And the condign punishment to be meted out to this last class is that they are not to be 'awakened' from the sleep of death to confront their Judge; they are to remain for ever in their dark dungeon in Sheol, and forego for ever the hope of resurrection. Next, and the main clue to the removal of the apparent discrepancy: at v. 2 a distinction is made between the three dungeons that are dark, and the one that is illuminated (although the latter may also have been dark, like all of Sheol, only furnished with a spring of pellucid water). This fourth promptuarium is unmistakably the one which is to receive the righteous; the three are for all the other souls of men. When therefore, we read at v. 9 'those three' the reference can only be to the three groups listed at 10-13; 'yonder one' (deictic οὗτος) at 9b refers to the fourth abode of the souls of the righteous.

At 24.4f., 25.4 mention is made of the Tree of Life, to be distinguished from the Tree of Knowledge situated in the Paradise of righteousness in the East (32.3). The implication seems to be that this tree flourished in a paradise on the slopes of the mountain-throne of God in the West (or the North-West). Certainly at 77.3 we read of a Paradise of righteousness situated in the North, and the reference at 24.4, 25.4 could be to the same Paradise. We seem then to have two different traditions about the location of the Paradise of righteousness, for according to 33.2, where it is so named, it lies in the East, whereas at 77.3, where it is also named 'Paradise of righteousness', it is located in the North. We have to do with two independent traditions, a Jewish western Illysium, but with Sheol-like features, and an eastern Eden-evidence of further hellenistic influence on the author of this part of Enoch.

---

[38] Cf. M.-T. Wacker, *Weltordnung und Gericht*, 107f., F. Nötscher, *Altorientalischer und alttestamentlicher Auferstehungsglauben* (1926), 276. G. W. E. Nickelsburg, *Resurrection, Immortality and Eternal Life in Intertestamental Judaism*, Harvard Theological Studies XXVI (Cambridge, Mass. 1972), 137. The passage has been dealt with extensively (see Nickelsburg and Wacker, loc. cit.)

Chapters 37-71: *The Parables of Enoch*

En. 37-71, preserved only in an Ethiopic version in manuscripts chiefly of the sixteenth and seventeenth centuries, consists of three 'parables' or 'discourses' attributed to Enoch, 37-44, 45-57, 58-69. (The word rendered 'parable' (Aram. 4Q En. 1.2, מתלה = Heb. משל) means, in this context, 'oration' or 'discourse'; and since Enoch's discourses are about matters celestial, 'apocalypses' or 'revelations' is a free but not inaccurate rendering.) The 'Son of Man' is introduced at 46.1-6 and 48. The three 'parables' include a Noah apocalypse (60), and a short discourse attributed to the archangel Michael (68.2 (67.12)-69.29). The concluding section, a kind of epilogue, takes up again the main theme of the patriarch's celestial journeyings and brings him finally to the Palace of God in the 'heaven of heavens' (70-71.17), where the climax of the apocalypse is an angelic revelation to Enoch (71.14):

'You are the Son of Man who is born for righteousness,
And righteousness abides upon you,
And the righteousness of the Chief of Days forsakes you not.'[39]

The plain meaning of these words is that Enoch himself is divinely designated or called to an even higher celestial role than he already enjoys as immortalised Patriarch, 'scribe of righteousness', namely, the role of the heavenly Man of his earlier visions (46,48). Obviously it is important to decide if this apocalyptic concept is Jewish or Christian (or Jewish-Christian), as Milik maintains, and—no less vital an issue—if it can be shown that it came from a Jewish source which was pre-Christian in inspiration, origin and date. A decision, one way or the other, is laid upon the conscience of every interpreter of the Parables. It is one, moreover, which depends on an understanding of the character and composition, also on the view taken of the date, of the Book of the Parables. These problems are fully discussed below in the Introduction to the Parables (181f.)

Chapters 72-82 *The 'Astronomical' Chapters*

The extant Aramaic source material on which the redactor of this section is drawing and the nature of its composition have been discussed above (10f.). A full treatment of the section, with translation and commentary and additional notes on the Aramaic fragments, are given below in Professor Otto Neugebauer's Appendix A.

Within the section there are three 'intrusions of non-astronomical material' (relevant only in so far as they deal with an apocalyptic vision of the perversion of nature and the heavenly bodies), viz. chapters 80.2-8, 81.1-10,

---

[39] For R. H. Charles's surgical operation on the text to fit in with his 'messianic' theory, see M. Black, 'The "Son of Man" in the Old Biblical Literature', in ET LX (1948-49), 12, and 'The Eschatology of the Similitudes of Enoch', in JThST, N.S., Vol. III, Pt. 1 (April 1952), 4.

and 82.1-3, the first developing the theme of the perversion of nature, the second that of the 'heavenly tablets', into the secrets of whose contents Enoch alone has been initiated, a theme resumed at 93.2, 103.2, 106.19 (cf. also 47.3). Chapter 81.5-6 serve as an introduction to the 'Epistle' of Enoch: Enoch is miraculously transported back to earth to spend a year imparting this secret lore to Methuselah, which is to be written down and passed on to later generations of mankind (cf. 82.1-3; 85.2, 91.1-2, 108.1). For the implications of this descent to earth from Paradise and his subsequent 'elevation' again for the 'biography' of Enoch, see below, on 12.4 (Comm., 142).

Chapters 83-90 *The Dream-Visions of Enoch*

The section on the Dream-Visions of Enoch represents the longest and the most self-consistent part of the Book. It has also been preserved in not unsubstantial fragments in the original Aramaic, consisting of four manuscripts, the oldest probably dating c. 175 B.C. (Milik, 41). It comprises a 'zoomorphic' history of the world, in effect for the author, of Israel, in a series of visions in which the principal protagonists are symbolised by animals, the whole set within the framework of the instruction of Enoch of his son Methuselah (cf. 83.1, 85.1-2).

The First Dream-Vision is of the Deluge, concluding with a prayer of thanksgiving (83-84). The Second Dream-Vision begins the zoomorphic history from the creation of Adam, 'a white bull' and his descendants by Eve, 'a heifer', all similarly represented (85). Chapters 86-88 present a zoomorphic version of the Watcher legend. Thereafter the visions follow in historical sequence in chapters 89-90:

Chapter 89: 1- 9 The Deluge and the deliverance of Noah.
10-27 From the death of Noah to the Exodus.
28-40 Israel in the desert, the giving of the Law and the entrance into Palestine.
41-50 From the period of the Judges to the building of the Temple.
51-67 The Kingdoms of Israel and Judah to the destruction of Jerusalem.
68-71 From the destruction of Jerusalem to the Return from the Exile: the first period of Gentile hegemony (the rule of the seventy shepherds).
72-77 The second period, from the time of Cyrus to Alexander the Great.

Chapter 90: 1- 5 Third period: from Alexander the Great to the Seleucids.
6-19 Fourth period: the Maccabaean Revolt.

20-42 The Last Judgement: the new Jerusalem, the new Eden, and the Second Adam.

'There is no difficulty about the critical structure of this Section. It is the most complete and self-consistent of all the Sections ...' (Charles, 179). While this long section does present fewer difficulties with regard to structure and composition than other parts of Enoch, it does seem surprising that it should contain two versions of the Deluge, the first occupying the prominent position as the First Dream-Vision, the second told a few chapters later in its proper chronological sequence. There can be no question here of editorial duplication as at chapters 18-24; both accounts of the Deluge are quite independent, and, indeed, may be held to supplement each other. Probably no significance can be attached to the absence of any fragment of the First Dream-Vision in the Aramaic manuscripts, especially as its text has every mark of semitic origin. As we have already noted, the Noah saga, in particular the catastrophe of the Deluge as the first Great Judgement, is a prominent theme in I Enoch; and this may perhaps account for the Deluge as the First Dream-Vision in this section of the book. There must, however, have been several extra-Biblical stories of the Deluge in circulation, so that it is possible that the First Dream-Vision was one of these, quite independent of the later vision, and one to which the author (or a redactor) gave this place of prominence in this section of Enoch.

Chapters 86-88, the zoomorphic vision of the fall and fate of the Watchers (they are falling 'stars' metamorphosed into 'bulls') has not been 'interpolated' into the history (Milik, 43). It follows in chronological sequence on chapter 85, since the events recorded at Gen. 6.1-4 fall within the patriarchal period. The author of this allegorical version of the Watcher legend may well have been acquainted with the Aramaic 'Book of the Watchers', as some shared phraseology as well as common ideas seem to suggest (Charles, 179).

Like its visionary counterpart, the Apocalypse of Weeks, which is also a historical schema of the history of Israel, the Dream-Visions are of cardinal importance for the dating of these classic and extensive portions of the Enoch tradition: 90.28 like 93.7 envisions a new Temple (and a new Jerusalem), and this suggests a date of composition prior to the recapture of Jerusalem and the rededication of the Temple in 165 B.C. In both visions history for the authors ends with their account of the struggles of the Maccabees; thereafter apocalyptic or transcendental eschatology takes over.

Milik has argued, against earlier interpreters, that the 'white bull' at the climax of the Second Dream-Vision is a symbol, not of the Messiah, but of the 'second Adam', corresponding to the first Adam, represented by a white bull. This is, in fact, so obvious that it is surprising that it has not previously been emphasised; it is clearly of some theological importance. But need this view of the symbolism totally rule out 'messianic' implications? The new

Adam may have been the apocalyptic writer's idea of the coming Messiah (as it became that of Paul), a Messiah the Gentiles will fear and revere (90.37), as they were expected to fear the traditional Davidic King-Messiah. The identity of the 'black buffalo' (v. 38) (Dillmann's brilliant conjecture that ῥῆμα renders ראם, Aram. ראמא) is obscure (cf. Charles, 216f.) Is it too fanciful to see in the symbol of the buffalo with the huge black horns a reference to the descendants of Ham, brought, like the Gentiles, within the new human family? Milik suggests that it represents the first of the eschatological patriarchs, but this could have been a member of the negroid race from the South, from Egypt and the Nile valley, Nubia or Ethiopia.[40]

Chapters 91-108: *The 'Epistle' of Enoch*

As we have seen (above, 11), chapters 91-108 are generally subsumed under the title 'The Epistle of Enoch' (Milik, 47f.), a title derived from the subscription at the end of the Chester-Beatty Greek papyrus. Whatever its title (if it ever had one), its most precious legacy is the section generally known as the Apocalypse of Weeks (93.3-10, 91.11-17).

The whole of this last component of I Enoch (91-108) is again a conglomerate of different traditions, but for the most part consisting of extensive paraenesis, but including another Nature Poem (93.11-14), and closing with a commission of Enoch to Methuselah and his family (105), a description of the miraculous birth of Noah (106-107), and a later short apocalyptic piece (108).

Chapters 91.1-10, 18-19 are characteristic Enoch paraenesis, as Enoch addresses Methuselah and his family, some parts of which are unmistakably elaborations of themes from the Apocalypse of Weeks, the second part of which is found at 91.11-17. Chapters 92.1-10 and possibly 93.1-2, are connected, in some way, with an 'epistle' of Enoch (and may include parts of it), the whole probably a much shorter composition than the traditional contents of Enoch's Epistle (chapters 91-108). See above, 11 and Comm., 285.

The Apocalypse of Weeks is discussed fully in the introduction to it in the Commentary (287f.). That it is a separate piece of earlier vintage than the context of paraenesis in which it has been placed, is generally agreed, an indication of the hand of the redactor at the Aramaic stage of the growth of the traditions. Fortunately, it has been substantially preserved in Aramaic, to appear as a composition in the finest traditions of ancient semitic poetic craftmanship. Although there are some gaps for the earlier verses, the extant Aramaic text preserves the verses in their original correct order. (The poem has been divided in two and the order of the pieces reversed in the Ethiopic

---

[40] For both these points, 'the white bull' and 'black buffalo', see my essay in *Crèation, Christ and Culture*, 19.f. and Milik, 45.

version.). As we have already seen (above, 20) this classic apocalyptic poem is specially important for the question of date, at least for its own composition. Above all its eschatology, its vision of a universal Judgement and of a new heaven and a new earth, now enshrined in an ancient Aramaic text of the second century B.C., has provided a classic documentary foundation in pre-Christian Judaism for the eschatology of the New Testament, even down to the phraseology of Rev. 21.1: 'And on it (the Tenth Week) the first heaven shall pass away and a new heaven will appear ...'.

The remainder of the 'Epistle of Enoch' consists of repetitious paraenesis, including monotonously repeated elegiacs or dirges against 'the sinners', dwelling, for the most part, but with variations of expression, on the leitmotif of the ineluctable fate of 'the sinners' and the blessedness or the rewards and destiny of the righteous, who receive the assurance, not of rewards in this life, but of a portion in the life to come: '... the spirits of you righteous who have died will live ... And their spirit shall not perish ...' (103.1-4, cf. also 108.12-14).

Chapter		
Chapter	94:	Admonitions to the righteous and woes to the sinners (cf. Isa. 5.8f.).
	95	Enoch's grief: fresh woes against the sinners.
	96	Grounds for hope for the righteous: woes for the wicked.
	97	Evils in store for sinners and possessors of unrighteous wealth.
	98	The self-indulgence of the sinners: sin originated by man: all sin recorded in heaven: woes for the sinners.
	99	Woes for the godless, the law-breakers: plight of sinners in the last days: further woes.
	100	The sinners destroy each other: the coming Judgement: further woes for the sinners.
	101	Exhortation to the fear of God: all nature fears him but not the sinners.
	102	Terrors of the Day of Judgement: the misfortunes of the righteous.
	103	Assurances for the righteous.
	104	Further assurances for the righteous: admonitions of the sinners.
	105	The Commission of Enoch to Methuselah and his family.
	106-107	The Birth of Noah.
	108	Another writing of Enoch.

All that has survived in Aramaic of this long Enoch paraenesis are a few fragments from 94.1 and 104.13-106.2 (Milik, 269f., 206f.). But Enoch's commission to Methuselah in chapter 105, which has been lost in the Greek version (it has survived in the Ethiopic so that it once did exist in Greek) has been preserved in Aramaic, unfortunately in no more than a sufficient number of words and letters to make the identification certain (Milik, 208, Ll. 20-21, cf. Knibb, 2, 243). The whole chapter has up till now been regarded as a Christian interpolation, since the Ethiopic text (hitherto our only authority) has been construed since Dillmann as a 'Word of the Lord', which concludes with 'For I and My Son will be united with them for ever in the paths of uprightness in their lives ...' (Charles). These words, however, can also be construed as part of the *oratio recta* of Enoch (Comm., 318f.) where the reference is to Enoch himself and his son: 'For I and my son (Methuselah) will be united (or associated) with them for ever in the paths of uprightness and peace.' The reference could then be to the legacy of Enoch transmitted through Methuselah to mankind in his writings. (Comm., 319).

Chapters 106-107 on the miraculous birth of Noah and of which fragments of an Aramaic text have been preserved, almost certainly do come from a Noah apocalypse. The last chapter 108 is clearly a later addition to the Enoch corpus by a writer anxious to develop ideas about Gehenna from the older Book at chapters 18, 21, and adding a paraenesis which seems directed especially to an ascetic group (vs. 7-9). But it concludes with a noble poetic piece, possibly from an earlier original source, recounting, in language and imagery inspired by Dan. 12.3, the splendours of immortality awaiting the righteous, but not overlooking the condign fate of the sinners.

TRANSLATION

THE BOOK OF ENOCH [1]

THE FIRST VISION OF ENOCH (CHH. 1-36)

CHAPTER 1

Prooimium and Central Theme: The Great Judgement

(1) The words of blessing, according to which Enoch blessed the righteous elect who, on the day of tribulation, are to destroy all the godless. (2) And he took up his discourse and said: '[Oracle of Enoch], a righteous man whose eyes were opened by God, and who saw a vision of the Holy One in heaven, which the angels showed me, and from the words of the [watchers and] holy ones I heard all; and I understood what I saw; not for this generation, but for a generation remote do I speak. (3) And concerning the elect I now speak, and about them I take up my discourse:

'The great Holy One shall come forth from his dwelling,
(4) And the eternal God shall tread upon the earth upon Mount Sinai,
And he shall appear from his camp,
And reveal himself in the power of his might from the highest heaven.
(5) And all shall be afraid, and the watchers shall quake,
And *they shall seek to hide themselves* in all the corners of the earth.
And all the ends of the earth shall be shaken.
And great fear and trembling shall seize them to the ends of the earth,
(6) And the lofty mountains shall be shaken;
they shall fall and be disintegrated;
And the high places shall be laid low so that the hills are dissolved;
They shall melt like wax before the fiery flame.

---

[1] Maṣḥafa Hēnoch Eth^{q} adds 'the vision which he saw'. Eth$^{m\ h\ n}$ (Charles,[1906] 2, n. 1 cf. Flemming 1, n. 2) add '(the Book) of the holy Saviour of the world; (the Book) of the prophet Enoch'. The words have probably been added by a scribe seeking to attribute the book to Christ as well as to Enoch ('Saviour of the world' is attested as a designation for Christ; Dillmann, *Lex.*, 1113). It seems very unlikely that any scribe would ever have applied this title to Enoch himself (see, however, my article 'The Eschatology of the Similitudes of Enoch' in JThSt, N.S., III, Pt. 1, 4 n. 2). Eth^{ryl} has a colophon: 'In the Name of the Lord God, merciful and gracious long-suffering and of great compassion and righteous. It was I copied the Book of the prophet Enoch—may his blessing and the gift of his assistance be with his servant, the son of George (Walda Giyorgis) for ever and ever, Amen'.

(7) And the earth shall be rent in sunder, and all that is upon the earth shall perish,
And there shall be a universal judgement.
(8) But with the righteous he will make peace, and he will protect the elect, and mercy shall be upon them,
And they shall all belong to God.
And he shall show them favour and bless them all, and he will assist all and help us;
And light shall appear upon them, and he will make peace with them.
(9) Behold! he comes with ten thousand holy ones to execute judgement upon all,
And he will destroy all the ungodly and convict all flesh of all the works of their ungodliness
Which they have ungodly committed, and of all the arrogant and hard words which sinners have uttered against him.'

A Nature Homily (Chh. 2-5)

CHAPTER 2

(1) Consider all [his works] and observe the works (of creation) in heaven,
How the heavenly luminaries do not change their paths *in the conjunction of their orbits*,
How each of them rises and sets in order, at its appointed time,
And at their fixed seasons they appear, and do not violate their proper order.
(2) Observe the earth and consider his works which have been wrought in it,
That from the first to the last no work of God is changed, and all is made manifest to you.
(3) Observe the signs of summer and winter, how the whole earth is filled with water, and clouds and dew and rain fall down upon it.

CHAPTER 3

(1) Observe that all trees become withered in appearance, and all their leaves are shed, except fourteen trees, which do not shed their leaves, but remain *without* (*their leaves*) *being renewed* up to two or three years.

CHAPTER 4

(1) And again, observe the signs of summer, how the heat (of the sun) in them becomes a burning and scorching heat; and you seek shelter and shade

before it because of the heat, and the earth burns with the scorching heat, and you cannot tread on the ground or on the rocks because of its heat.

CHAPTER 5

(1) Consider all trees; in all of them green leafage appears and covers them; and all their fruit appears in glorious splendour. Examine and consider all these works (of creation) and reflect that the God who lives for ever and ever has created all these works. (2) And all his works which he has made for ever attend on him year by year; and all his works serve him and do not change, but all perform his commands. (3) See how the seas and rivers together perform and do not change their tasks by abandoning his commands. (4) But you have changed your works, and have not been steadfast nor done according to his commandments, but you have transgressed against him, and spoken proud and hard words with your impure mouths against his majesty: (you) hard in your heart, you shall have no peace. (5) Then you will curse your days, and the years of your life will perish, and the years of your perdition will be multiplied under an everlasting curse; you shall not have mercy or peace. (6) Then you shall leave your names, as an everlasting curse, for all the righteous; and by you (i.e. by your name) shall all who curse curse, and all sinners and ungodly shall curse by you; and all who are without sin shall rejoice, and they shall have remission of sins, and all mercy and peace and tranquility; salvation they shall have, a fair light, ⌜and they shall inherit the earth⌝. But for all you sinners there shall be no salvation, but upon all of you dissolution (and) execration. (7) And for the elect there will be light and joy and peace, and they will inherit the earth: but for you, the godless, there will be execration. (8) Then shall wisdom be given to the elect, and all of them shall live and shall sin no more, either through sinning unwittingly or from pride: but those who have wisdom will be humble. In an intelligent man it (wisdom) is illumination, and to a prudent man it is understanding; and they shall not err. (9) And they shall not be condemned all the days of their lives, nor shall they die by the fury of (his) anger, but they shall complete the number of the days of their lives, and their lives shall be increased in peace; and the times *of their festivals* will be filled with joy and lasting peace during all the days of their lives.

The Legend of the Watchers (Chh. 6-16)

CHAPTER 6

(1) And it came to pass, when the children of men had multiplied, in those days there were born to them beautiful and comely daughters. (2) And

watchers, children of heaven, saw them and desired them, and lusted after them; and they said one to another: 'Come, let us choose for ourselves wives from the daughters of earth, and let us beget us children. (3) And Semhazah, who was their leader, said to them: 'I fear you will not want to do this deed, and I alone shall pay the penalty for a great sin. (4) And they all answered him and said: 'Let us all swear an oath, and bind one another with imprecations that we shall not depart, any of us, from this plan until we carry it out and do this deed'. (5) Then they all swore together and bound one another with imprecations. (6) And they were two hundred who descended in the days of Jared on the summit of Mount Hermon; and they called the mount Hermon, because they swore and bound one another with imprecations upon it. (7) And these are the names of their leaders: ŠEMḤAZAH who was their chief; ʾARʿTEQIF, second to him; RAMTʾEL, third to him; KOKABʾEL, fourth to him; [ʾURʾEL], fifth to him; RAʿMʾEL, sixth to him; DANʾEL, seventh to him; ZIQʾEL, eighth to him; BARAQʾEL, ninth to him; ʿASAʾEL, tenth to him; ḤERMONI, eleventh to him; MATRʾEL, twelfth to him; ʿANANʾEL, thirteenth to him; SITHWAʾEL, fourteenth to him; ŠIMŠʾEL, fifteenth to him; SAHRʾEL, sixteenth to him; TAMMʾEL, seventeenth to him; TURʾEL eighteenth to him; YAMMʾEL, nineteenth to him; ZEHORʾEL, twentieth to him. (8) These are the leaders and their dekadarchs.

CHAPTER 7

(1) These (leaders) and all the rest (of the two hundred watchers) took for themselves wives from all whom they chose; and they began to cohabit with them and to defile themselves with them, and they taught them sorcery and spells and showed them the cutting of roots and herbs. (2) And they became pregnant by them and bore great giants of three thousand cubits; and there were [not] born upon earth off-spring [which grew to their strength]. (3) These devoured the entire fruits of men's labour, and men were unable to sustain them. (4) Then the giants treated them violently and began to slay mankind. (5) They began to do violence to and to attack all the birds and the beasts of the earth and reptiles [that crawl upon the earth], and the fish of the sea; and they began to devour their flesh, and they were drinking the blood. (6) Thereupon the earth made accusation against the lawless ones.

CHAPTER 8

(1) Asael taught men to make swords of iron and breast-plates of bronze and every weapon for war; and he showed them the metals of the earth, how

to work gold, to fashion [adornments] and about silver, to make bracelets for women; and he instructed them about antimony, and eye-shadow, and all manner of precious stones and about dyes and *varieties of adornments*; and the children of men fashioned them for themselves and for their daughters and transgressed; (2) and there arose much impiety on the earth and they committed fornication and went astray and corrupted their ways. (3) Semhazah taught spell-binding and the cutting of roots; Hermoni taught the loosing of spells, magic, sorcery and sophistry. Baraqel taught the auguries of the lightning; Kokabiel taught the auguries of the stars; Zikiel taught the auguries of fire-balls; Arteqif taught the auguries of earth; Simsel taught the auguries of the sun; Sahrel taught the auguries of the moon. And they all began to reveal secrets to their wives. (4) Then the giants began to devour the flesh of men, and mankind began to become few upon the earth; and as men perished from the earth, their voice went up to heaven: 'Bring our cause before the Most High, and our destruction before the glory of the Great One'.

CHAPTER 9

(1) Then Michael, Sariel, Raphael and Gabriel looked down from the sanctuary in heaven, and they saw much blood shed on the earth; and the whole earth was full of godlessness and violence which men were committing against it. (2) [And they went in] and said to [the angels]: 'The voice and cry of the children of earth are ascending to the gates of heaven. (3) Now to you, the holy ones of heaven, the souls of men are making their suit, complaining with groans and saying: 'Bring our case before the Most High, and our destruction before the glory of the Great One'. (4) Then [Raphael] and Michael [and Sariel and Gabriel] went in and said to the Lord of the ages: 'Thou art our great Lord, Lord of the ages, Lord of lords and God of gods and King of the ages; and thy glorious throne is for all generations of eternity, and thy Name is holy and great and praised unto all ages. (5) For thou art he who has created all things and hast power over all; and all things are revealed and unconcealed before you; and thou seest all things and there is nothing can be hid from you. (6) Thou seest what Asael has done, what he has introduced and taught, wrong-doing, and sins upon the earth, and all manner of guile in the land; that he revealed the eternal mysteries prepared in heaven and made them known to men, and *his abominations the initiates among the children of men make for themselves.* (7) Semhazah instructed men in spell-binding, (he) whom thou hast appointed ruler of all spell-binders. (8) And they cohabited with the daughters of ⌜the men of the⌝ earth, and had intercourse with them, and they were defiled by the females, and revealed to them all manner of sins, and taught them to make hate-

charms. (9) And now behold! the daughters of men brought forth from them sons, giants, bastards; and *much blood was spilled upon the earth*, and the whole earth was filled with wickedness. (10) And now, behold! the souls of mortal men are crying and making their suit to the gates of heaven; and their groaning has ascended, and they cannot escape the wrongs that are being done on the earth. (11) But thou knowest all things before they come to pass, and thou seest them and hast let them alone; and thou dost not say to us what we should do with regard to them on account of these things'.

CHAPTER 10

(1) Then the Most High said and the great Holy One spoke up and sent Sariel to the son of Lamech, saying: (2) 'Go to Noah and say to him in my Name: 'Hide yourself', and show him the End that is approaching; that the earth will be completely destroyed; and ⌜tell him⌝ that a Deluge is about to come on the whole earth, to destroy all things from the face of the earth. (3) And now instruct the righteous one what to do, and the son of Lamech, that he may save his life and escape for all time; and from him a plant shall be planted and established for all generations for ever'. (4) And to Raphael he said, 'Go, Raphael, and bind Asael; fetter him hand and foot and cast him into darkness; make an opening in the desert which is in ⌜the desert of⌝ Dudael, and there go and cast him in. (5) And place upon him jagged and rough rocks, and cover him with darkness and let him abide there for all time, and cover his face that he may not see the light. (6) And on the day of the great judgement he will be led off to the blazing fire. (7) Heal the earth which the watchers have ruined, and announce the healing of the earth, that I shall heal its wounds and that the children of men shall not altogether perish on account of the mysteries which the watchers have disclosed and taught the children of men. (8) The whole earth has been devastated by the works of the teaching of Asael; record against him all sins.' (9) And to Gabriel the Lord said: 'Go, Gabriel, to the giants, (their) bastard off-spring, the children of fornication, and destroy (those) sons of the watchers from among the sons of men. Muster them (for battle), and send them, one against the other, in a battle of destruction. (10) Length of days shall not be theirs: they shall all request (it) of you, but no petition shall be granted to their fathers on their behalf, that they should not expect to live an eternal life, but that each one of them should live five hundred years'. (11) And the Lord said to Michael: 'Go, Michael, make it known to Semhazah and the others who, with him, were united with the daughters of men, to defile themselves with them in their uncleanness. (12) And when their sons shall be slain, and they see the destruction of their beloved ones, bind them for seventy generations in valleys of the earth, until the great day of their judgement and

the time of the end, until the judgement which is for ever and ever becomes absolute. (13) Then they shall be dragged off to the fiery abyss in torment, and in a place of incarceration they shall be imprisoned for all time. (14) And everyone who is consumed by lust and is corrupted, from now on will be bound together with them and at the (fixed) time [of the judgement which] I shall judge, they shall perish for all generations for ever. (15) I shall destroy all the spirits of the bastard offspring of the watchers, because they wrong mankind. (16) I shall destroy all iniquity from upon the face of the earth, and every evil work shall come to an end; and there shall appear the plant of righteousness; and it shall be a blessing, and deeds of righteousness shall be planted with joy for ever. (17) And now all the righteous shall escape, and shall live till they beget thousands; and all the days of your youth and of your old age you shall fulfil in peace. (18) Then shall the whole earth be tilled in righteousness, and it shall all be planted with trees, and filled with blessing. (19) And all luxuriant trees will be planted in it; and they will plant vines in it, and the vine which they plant will produce a thousand measures of wine, and of all seed which is sown upon it, each seah will produce a thousand seah; and every seah of olives will produce up to ten baths of oil. (20) And as for you, cleanse the earth from all uncleanness, and from all injustice and from all sinfulness and godlessness; and all the unclean things that have been wrought ⌜on the earth⌝ remove from the earth. (21) And all the children of men are to become righteous and all nations shall serve and bless me, and all shall worship me. (22) And the whole earth shall be freed from all defilement and from all uncleanness, and wrath and castigation: and I shall not again send a Deluge upon it unto generations of generations and for ever.

CHAPTER 11

(1) And at that time I shall open the treasures of blessing that are in heaven, to send them down upon the earth, upon the work and labour of the children of men. (2) Then peace and righteousness shall be united for all the days of eternity, and for all generations of eternity.'

CHAPTER 12

(1) And before these things Enoch was taken up, and none of the children of men knew where he had been taken up, or where he was or what had happened to him. (2) But his dealings were with the watchers, with the holy ones, in his days. (3) And I, Enoch, was standing blessing the Lord of majesty, and the King of the ages, and behold! watchers of the great Holy One were calling me and saying to me: (4) 'Enoch, scribe of righteousness, go, declare to the watchers of heaven who have left the high heaven and the

holy, eternal Sanctuary and have defiled themselves with women; and they themselves do as the children of earth do, and have taken to themselves wives: (say) 'You have wrought great destruction on the earth; (5) and you shall have no peace or forgiveness'. (6) And inasmuch as they delight in their children, the slaughter of their beloved ones they shall see, and over the destruction of their children they shall lament and make supplication without end: but they shall have neither mercy nor peace.

CHAPTER 13

(1) And Enoch went and said to Asael: 'You shall have no peace: a severe sentence has been issued against you that you should be bound'. (2) Nor shall forbearance, petition nor mercy be yours, on account of the wrongs you have taught, and all the deeds of godlessness, wrong-doing and sin which you showed to the children of men.' (3) Then I went and spoke to them all together, and they were afraid, and fear and trembling seized them. (4) And they besought me to draw up for them a memorial and petition that they might obtain forgiveness, and that I should read their memorial and petition before the Lord of heaven. (5) For they themselves were unable any longer to speak (to him) nor to lift up their eyes to heaven for shame for the sins for which they were condemned. (6) Then I wrote out their memorial and petition and their requests, with reference to their spirits and the deeds of each one of them, and with regard to their requests that they might obtain forgiveness and *restoration*. (7) And I went off and sat down by the waters of Dan in the land of Dan, which is south-west of Hermon; and I was reading the memorial of their requests until I fell asleep. (8) And behold! dreams came to me, and visions fell upon me, until I lifted up my eyes to the gates of heaven, and I saw visions of wrath and reproof, and a voice came, saying: 'Speak to the sons of heaven to rebuke them'. (9) And I woke up and went to them; and they were all seated, assembled together and mourning, in Abel-maim, which is between Lebanon and Senir, with their faces covered. (10) And I recounted before them all the visions I had seen in dreams; and I began speaking words of truth and visions and reprimanding the watchers of heaven.

CHAPTER 14

(1) The record of the words of truth of the reprimand of the eternal watchers, in accordance with the command of the great Holy One in the dream which I saw. (2) I saw in my dream what I am now telling with a tongue of flesh, with the breath of my mouth, which the Great One has given to the children of men to converse therewith, and to understand with their mind. (3) As he has endowed, fashioned and created the children of men to

understand words of insight, so me he has endowed, fashioned and created to reprimand the watchers, the children of heaven. (4) I wrote down your petition, (you) watchers, but in a vision it was revealed to me (that), forasmuch as your petition will not be granted to you all the days of eternity, sentence will be made final, by decree, upon you; (5) that from now on you shall no longer ascend to heaven throughout all ages; and it has been ordered to bind you in bonds in the earth for all the days of eternity; (6) and that before these things you shall see the destruction of your beloved ones, *and of all their sons and their flocks*; *and that you will have no heirs to them*; and they will fall before you by the sword *in total destruction*; (7) since your petition on their behalf will not be granted, nor on your own behalf; for all your petitioning and pleading, not a single word *will be implemented* from the document which I have written.

(8) And it was shown to me thus in a vision: Behold! clouds were calling me in my vision, and dark clouds were crying out to me; fire-balls and lightnings were hastening me on and driving me, and winds, in my vision, were bearing me aloft, and they raised me upwards and carried and brought me into the heavens. (9) And I went in till I drew near to a wall, built of hailstones, with tongues of fire surrounding it on all sides; and it began to terrify me. (10) And I entered into the tongues of fire and drew near to a large house built of hailstones; and the walls of the house were like tesselated paving stones, all of snow, and its floor was of snow. (11) Its upper storeys were, as it were, fireballs and lightnings, and in the midst of them (were) fiery Cherubim, *celestial watchers.* (12) And a flaming fire was around *all* its walls, and its doors were ablaze with fire. (13) And I entered into that house, and it was hot as fire and cold as snow; and there were no delights in it; horror overwhelmed me, and trembling took hold of me.(14) And shaking and trembling, I fell on my face. And I saw in a vision, (15) and behold! another house greater than that one and its door was completely opened opposite me; and it (the second house) was all constructed of tongues of fire. (16) And in every respect it excelled in glory and honour and grandeur that I am unable to describe to you its glory and grandeur. (17) And its floor was of fire, and its upper chambers were lightnings and fire-balls, and its roof was of blazing fire. (18) And I beheld and saw therein a lofty throne; and its appearance was like the crystals of ice and the wheels thereof were like the shining sun, and (I saw) *watchers*, *Cherubim.* (19) And from underneath the throne came forth streams of blazing fire, and I was unable to look on it. (20) And the glory of the Great One sat thereon, and his raiment was brighter than the sun, and whiter than any snow. (21) And no angel was able to enter this house, or to look on his face, by reason of its splendour and glory; and no flesh was able to look on him. (22) A blazing fire encircled him, and a great fire stood in front of him, so that none who surrounded him

could drawn near to him; ten thousand times ten thousand stood before him. He had no need of counsel; in his every word was a deed. (23) And *the watchers and holy ones* who draw near to him turn not away from him, by night or by day, nor do they depart from him. (24) As for me, till then I had been prostrate on my face, trembling, and the Lord called me with his own mouth and said to me: 'Come hither, Enoch, and hear my word'. (25) And there came to me one of the holy angels and he raised me up and brought me to the door, and I bowed my face low.

CHAPTER 15

(1) And he spoke up and said to me: 'Fear not, Enoch, righteous man and scribe of righteousness. Come hither and hearken to my voice. (2) Go and say to the watchers of heaven who have sent you to intercede on their behalf: 'It is you who should be petitioning on behalf of men, and not men on your behalf. (3) Why have you left the high heaven and the eternal Holy One, and lain with women, and defiled yourselves with the daughters of men and taken to yourselves wives, and acted like the children of earth and begotten giants for sons. (4) But you were holy, spirits that live forever, yet you defiled yourselves with the blood of women, and have begotten (children) by the blood of flesh; and you lusted after the daughters of men and have produced flesh and blood, just as they do who die and perish. (5) It was for this reason I gave them females that they might impregnate them and thus produce children by them, that *pregnancy* should never fail them upon the earth. (6) But as for you, you formerly were spirits that live for ever and do not die for all generations for ever. (7) And for this reason I did not provide wives for you, because for celestial spirits heaven is their dwelling-place.

(8) And now the giants, who have been produced from spirits and flesh, shall be called mighty spirits upon the earth, and on the earth shall be their dwelling-place. (9) Evil spirits shall come forth from their bodies, for from men they have come, and from the holy watchers is the beginning of their creation and ⸢the beginning⸣ of their origins. Evil spirits they shall be called upon the earth. (10) As for celestial spirits, heaven shall be their dwelling-place: but for terrestrial spirits, born upon the earth, on the earth shall be their dwelling-place. (11) But the *vicious* spirits (issuing) from the giants, the Nephilim — they inflict harm, they destroy, they attack, they wrestle and dash to the ground, *causing injuries*; they eat nothing, but fast and thirst and produce hallucinations, and they collapse. (12) And these spirits will rise against the sons of men and women, from whom they came forth.

CHAPTER 16

(1) From the day of the time of the slaughter, destruction and death of the giants, the Nephilim, the spirits which came forth from their bodies will go on destroying, uncondemned. In such ways they will destroy until the day of the end, until the great judgement, in which the great aeon will be completed. (2) And now [say] to the watchers, who were once in heaven, who sent you to petition on their behalf: (3) 'You were in heaven, and there was no secret that was not revealed to you; and unspeakable secrets you know, and these you made known to women in your hardness of heart; and by these secrets females and mankind multiplied evils upon the earth'. (4) Say to them therefore, 'You shall have no peace'.

Enoch's Journeys (Chh. 17-36)

CHAPTER 17

(1) And I was taken and brought to a certain place, where those who were there were like blazing fire, and when they wished, they took on the appearance of men. (2) And they brought me to a place of dark storm-clouds and to a mountain whose summit reached to heaven. (3) And I saw the places of the luminaries and the chambers of the stars and of thunder-peals, to the uttermost reaches, where (I saw) a bow of fire, arrows and their quivers, and a fiery sword and all the lightning-flashes. (4) And they brought me to *subterranean waters* and to the fire of the West which takes hold of all the settings of the sun. (5) And we came to a river of fire whose fire was flowing like water and was pouring into the great Ocean towards the West. (6) And I saw the great rivers, and I reached ⌜the great river and⌝ the great darkness, and I went and came whither no flesh goes. (7) And I saw *wintry regions of storm-clouds*, and the outpouring from the abyss of all the waters. (8) And I saw the mouth of all the rivers of the earth and the mouth of the abyss.

CHAPTER 18

(1) I saw the storehouses of all the winds, and I saw that by them he set in order all created things and the foundations of the earth. (2) And I saw the corner-stone of the earth; and I saw the four winds that support the earth and the firmament of heaven. (3) And I saw that the winds spread out the heights of heaven, and are stationed between heaven and earth; they are the pillars of heaven. (4) And I saw the winds which cause the sky to turn and the orb of the sun to revolve, and all the stars. (5) I saw the winds on the

earth carrying the clouds; I saw the paths of the angels; I saw at the ends of the earth the firmament of the heavens above. (6) I proceeded on and saw a place which burns night and day where there were seven mountains of precious stones, three towards the east and three lying towards the south. (7) And of those which were to the east, one was of stones of varied hues, and another of pearls and another of stones of the colour of antimony; those towards the south were of carnelians. (8) And the one in the middle reached up to the heavens, like the throne of God, and it was of *emeralds*, and the top of the throne was of sapphires. (9) And I saw a blazing fire beyond those mountains. (10) And I saw there a place *beyond the ends of the earth*, and there the heavens came to an end. (11) And I saw a deep abyss of the earth with pillars of heavenly fire, and I saw in it pillars of fire descending which were immeasurable in either height or depth. (12) And beyond this abyss I saw a place which had no firmament of heaven above, nor foundation of earth beneath it; there was no water in it and no birds, but it was a waste and horrible place. (13) And there I saw seven stars like great burning mountains, concerning which, when I enquired, (14) the angel said to me: 'This place is the end of the heavens and the earth; this has become a prison for the stars and the hosts of heaven. (15) And the stars which rotate in the fire, these are they which transgressed the commandment of the Lord at the beginning of their rising, because they did not come forth at their proper times. (16) And he was wroth with them, and he incarcerated them until the time of the completion of the punishment for their sins, in ten thousand years.'

CHAPTER 19

(1) And Uriel said to me: 'Here the angels who had intercourse with women will abide, and their spirits, taking on many forms, will harm men and lead them astray, to sacrifice to demons as to gods, until the great judgement, in which they will be finally judged. (2) And the wives of the angels who transgressed shall become *sirens*.' (3) And I, Enoch, alone saw the visions, all things that exist: and no one of men shall see as I saw.

CHAPTER 20

(1) And these are the names of the holy angels who keep watch. (2) Uriel, one of the holy angels, namely, the one in charge of thunder and *earthquake*. (3) Raphael, one of the holy angels, who is over the spirits of men. (4) Raguel, one of the holy angels, who tends the hosts of the luminaries. (5) Michael, one of the holy angels, who has been put in charge of the blessings to come to the people (of Israel), ⌈and over the people.⌉ (6) Sariel, one of the holy angels, who is in charge of the spirits which *lead*

men astray in the spirit. (7) Gabriel, one of the holy angels, who is in charge of Paradise, the serpents (Seraphim?) and the Cherubim. (8) Remiel, one of the holy angels, whom God set over those who rise (from the dead).

CHAPTER 21

(1) And I went on to a formless void, (2) and there I saw a terrible thing—neither heaven above nor a firmly founded earth, but a place empty and terrible. (3) And there I saw seven stars of heaven bound together, prostrate on it, like great mountains, and burning in fire. (4) Then I said: 'For what iniquity have they been bound, and why have they been cast down here?' (5) Then Uriel, one of the holy angels who was with me and a leader among them, spoke to me, saying: 'About whom, Enoch, do you ask and concerning whom do you seek out the truth? (6) These are those among the stars of heaven who transgressed the commandment of the Lord, and were bound here till the completion of ten thousand years, the period of their punishment'. (7) And from thence I went to another place, more terrible than the former; and I saw dreadful things—a great fire there burning and blazing, and that place had an opening (reaching down) right to the abyss, filled with columns of great fires descending; neither their extent nor their size was I able to see nor to discern. (8) Then I said: 'How fearful is this place and terrible to look on'. (9) Then Uriel, one of the holy angels who was with me, spoke up and said to me: 'Enoch, why are you so afraid and terrified?' And I replied: '[I am terrified] on account of this fearful place and before this terrible spectacle'. (10) And he said to me: 'This place is the prison of the angels: here they will be imprisoned for an eternity'.

CHAPTER 22

(1) And from thence I was transported to another place; and he (Uriel) showed me towards the west, a large and lofty mountain of flint-hard rock, (2) and four hollow places in it, deep and wide and very bare; three of them were dark, and one bright, with a spring of water in its midst. And I said: 'How bare are those hollows, and deep and dark to view'. (3) Then Raphael, one of the holy angels who was with me, spoke up and said to me: 'These hollows are (there) in order that the spirits of the souls of the dead should be gathered together (in them); for this very purpose they were fashioned that here all the souls of men should be assembled. (4) And behold, these (hollows) are pits, fashioned in this way for their incarceration, until the day they will be judged, until the time of the last day, of the great judgement which will be exacted from them.' (5) There I saw the spirit of a dead man making complaint; and his lamenting reached up to heaven as he cried and complained. (6) Then I asked Raphael, the watcher and holy one

who was with me, and I said to him: 'This spirit—whose is it whose voice is thus going forth and complaining to heaven?'. (7) And he answered me, saying: 'This is the spirit that came forth from Abel whom Cain, his brother, slew: and he will bring accusations against him until his seed perishes from the face of the earth, and from the offspring of men his seed is destroyed'. (8) Then I asked about all the hollows, why they are separated, one from the other. (9) And he answered me, saying: 'These three have been made that the spirits of the dead might be separated. And yonder one was separated off for the spirits of the righteous, one in which there is a spring of *pellucid* water. (10). And that (hollow) there was fashioned for the spirits of the sinners, when they die and are buried in the earth, but judgement has not been carried out upon them in their lives. (11) Here their spirits shall be set apart, for this great torment, till the great day of judgement, (the day) of scourgings and torment of the accursed everlastingly, *to the end that there be* retribution on their spirits: there they shall be bound for ever. (12) And that (third) hollow has been separated off for the spirits of those who have accusations to bring and information to lay with regard to their destruction, when they were murdered in the days of the sinners. (13) And that (fourth hollow) has been fashioned for the spirits of men who were not wholly lawless, but with the lawless they collaborated. Because those here enduring suffering receive a lesser punishment than they (the lawless), their spirits will not be punished on the day of the *righteous* judgement, but neither will they be awakened, from here?' (14) Then I blessed the Lord of glory and said: 'Blessed be the Judge of righteousness, the Lord of glory and righteousness, the everlasting Lord'.

CHAPTER 23

(1) And from thence I was transported to another place, to the west, right to the ends of the earth. (2) And I saw a blazing fire which ran without resting and did not flag in its course, *holding to it equally* by day and by night. (3) And I asked, saying: 'What is this (fire) that has no rest?'. (4) Then Raguel, one of the holy angels who was with me, answered me and said to me: 'This stream of fire is the fire in the west which tends all the luminaries of heaven'.

CHAPTER 24

(1) And from thence I was transported to another place on the earth, and he showed me mountains of fire which burned day and night. (2) And I went beyond them and saw seven magnificent mountains, each differing from the other, whose stones were priceless for their beauty, and all valuable and their appearance glorious and fair in form; three of them (facing) towards the

east, one planted on the other, three to the south, one on the other, with deep, crooked ravines, no one of which approached any other. (3) And there was a seventh mountain between these (mountains) and it surpassed (them) in height, (being) like the seat of a throne; and fragrant trees surrounded it. (4) And there was amongst them a tree with a fragrance such as I have never at any time smelt, and no tree among them, nor any other, flourished like it; it had a fragrance sweeter than all spices, and its leaves and flowers and wood never wither; its fruit is beautiful and resembles the clusters of the date-palms. (5) Then I said: 'How beautiful is this tree and fragrant, and (how) fair are its leaves, and (how) very lovely to the eye are its blossoms'. (6) Then Michael, one of the holy angels who was with me—and he was a leader of them—answered me, (25.1) and said to me:

CHAPTER 25

(1) (24.6 Then Michael) ... said to me: 'Enoch, why do you ask and why do you marvel at the fragrance of this tree, and why do you desire to learn the truth?' (2) Then I, Enoch, answered him, saying: 'About everything I desire to learn, but most of all about this tree'. (3) And he answered me, saying: 'This high mountain which you see, whose summit is like a throne of the Lord, is the throne on which the great Holy One sits, the Lord of glory, the everlasting King, when he descends to visit the earth in blessing. (4) And as for this fragrant tree—no flesh shall be allowed to touch it until the great judgement, when there will be a recompense for all and a consummation for ever. (5) Then, ⌈to the righteous and pious⌉ its fruit shall be given to the elect *for food*; and it shall be transplanted to a sacred place beside the temple of the Lord, the everlasting King. (6) Then will they be glad exceedingly and rejoice and enter the holy place; and they shall bring (as an offering) into it its sweet-smelling odours in their very bones; and they shall live a longer life upon earth just as your fathers did; and in those days tortures, pains, labours and blows will not touch them'. (7) Then I blessed the Lord of glory, the King of the ages, because he has prepared such things for the righteous, and fashioned such things, and promised to give them to them.

CHAPTER 26

(1) And from thence I was transported to the middle of the earth, and I saw a blessed place, in which were trees, with saplings surviving and burgeoning from a felled tree. (2) And there I saw a sacred mount and coming forth from beneath the mount, from the east side, a stream; and its descent was towards the south. (3) And I saw towards the east another mount higher than the first (lit. this one); and in the midst of them a valley deep and narrow, and through it a stream ran alongside this (higher)

mount. (4) And to the west thereof was another mount, lower than it and of no great height, and a valley at its foot between them, deep and dry, and other valleys deep and dry at the farthest parts of the three mounts. (5) And all the valleys were deep and narrow, of flint-hard rock, and no tree was planted in them. (6) And I marvelled at the rocky ground and I marvelled at the valley; indeed, I marvelled exceedingly.

CHAPTER 27

(1) Thereupon I said: 'Why is this land blessed, and all of it full of trees, whereas that valley is accursed?'. (2) Then Uriel, one of the holy angels who was with me, answered me and said to me: 'This accursed valley is for those who are accursed for ever: here shall all the accursed be gathered together who utter with their lips unseemly speech against the Lord, and speak hard words against his glory. Here they shall be assembled, and here shall be their place of judgement. (3) In the last days there shall take place in them the spectacle of righteous judgement in the presence of the righteous, (a spectacle) for all time; (and) here will the pious bless the Lord of glory, the everlasting King. (4) In the days of their judgement they will bless him according as he has imparted mercy to them.' (5) Then I blessed the Lord of glory, and his glory I made known and I praised (him) in a manner worthy of his majesty.

CHAPTER 28

(1) And from thence I went on into the midst of a mountainous part of the desert; and I saw *the Arabah*, and [a place] by itself, (2) full of trees and plants, with water gushing forth upon it from above, (3) (then) being carried by a copious aqueduct approximately towards the north-west; and from all sides it was taking up water and dew.

CHAPTER 29

(1) Again, from thence I went into another place in the desert and withdrew to the east of this mountainous region. (2) And there I saw *juniper trees*, redolent of the fragrance of incense and myrrh; and their trees were like almond trees.

CHAPTER 30

(1) And beyond them, I went far to the east; and I saw another vast place, valleys with perennial streams; (2) and I saw sweet cane of a fragrance like camel-hay. (3) And by the banks of those valleys I saw the fragrant cinnamon; and beyond the valleys I was transported on towards the east.

CHAPTER 31

(1) And I was shown other mountains, and also in them I saw trees, from which came forth the resin which is called styrax and galbanum. (2) And beyond those mountains I was shown another mountain, to the east of the ends of the earth, and on it were lign-aloe trees; and all the trees in it were full of aloes and it resembles the shell of the almond nut. (3) When *an incision is made* in those trees, a pleasant odour comes from them; when that bark is ground, it is the sweetest of all fragrances.

CHAPTER 32

(1) And beyond these [mountains], approximately northwards, east of them, I was shown other mountains, full of choice nard, aromatic shells, cardamon and pepper. (2) And from thence I was transported eastwards of all these mountains, far from them to the east of the earth; and I crossed the Erythraean sea and withdrew very far from it; and I passed over the (great) darkness, far from it. (3) And I passed on beside the Paradise of righteousness, and I saw from afar, superior to those (former) trees, trees, more numerous and large, growing there, sweet-smelling, tall, exceedingly beautiful and glorious; and (I saw) the tree of knowledge, of the fruit of which those who partake understand great wisdom. (4) That tree is in height like the cypress, and its leaves are like (the leaves of) the carob tree, and its fruit is like the clusters of the vine, shining brightly; and its fragrance penetrates far away from the tree. (5) Then I said: 'How beautiful is this tree, and how pleasing is its appearance!'. (6) Then the holy angel who was with me, Raphael, answered me and said: 'This is the tree of knowledge of which your father of old and your mother of old before you ate; and they learned knowledge, and their eyes were opened, and they knew that they were naked, and they were driven out of the garden.'

CHAPTER 33

(1) And from thence I went to the ends of the earth and saw there great beasts, and each differed from the other; and birds also (which) differed in their appearance and beauty and call, each differing from the other. (2) And to the east of these beasts I saw the ends of the earth whereon the heavens rest, and the gates of the heavens opened. (3) And I saw how the stars of heaven come forth, and I counted the gates out of which they proceed, and I wrote down all their places of exit, each one individually, according to their number and their names, according to *their conjunctions* and positions, their seasons and their months, as Uriel, one of the watchers who was with me, showed me. (4) He showed all things to me and wrote it

down, and also their names he wrote down for me and their ordinances and their companies.

CHAPTER 34

(1) And from thence I went northwards to the ends of the earth, and there I saw great and glorious works (of creation) at the ends of the whole earth. (2) And there I saw three gates of heaven open in heaven: through each of them north winds go forth; when they blow, there is cold, hail, frost, snow, dew and rain. (3) And from one gate they blow for blessing: but when they blow through the other two gates, it is with violence and affliction upon the earth, and violently they blow.

CHAPTER 35

(1) And from thence I went towards the west to the ends of the earth, and I saw there gates opened, just as I had seen in the east, as many gates as there are outlets.

CHAPTER 36

(1) And from thence I was transported to the south, to the ends of the earth, and there I saw three gates of the heavens opened; and there came forth from thence the south winds and dew and rain and *a* (*hot*) *wind*. (2) And from thence I was transported towards the east, to the ends of the earth, and I was shown there three gates of the heavens opened towards the east, and above them smaller gates. (3) Through each of these smaller gates there pass the stars of heaven and proceed to the west in the paths which have been shown them. (4) And when I saw, I blessed continually, and I will bless the Lord of glory, who has wrought great and glorious wonders, that he might show the grandeur of his work to his angels, to spirits, and to mankind, that they might glorify his work, and that every one of his creatures might see the work of his might, and glorify the great work of his hands, and bless him for ever.

THE PARABLES OR SECOND VISION

The First Parable (Chh. 37-44)

CHAPTER 37

(1) The Second Vision which he saw, a vision of wisdom, which Enoch saw, (Enoch) son of Jared, son of Malalel, son of Cainan, son of Enos, son of

Seth, son of Adam. (2) The following is the beginning (the sum) of the words of wisdom which I spoke up and uttered, saying to those who dwell on the earth: 'Hear, you men of old, and consider, you men of latter days, the words of the Holy One which I shall utter in the presence of the Lord of spirits. (3) The former first it is right to mention, but from those in latter days we shall not withhold the beginning (sum) of wisdom. (4) Until now there has not been imparted (to anyone) from the Lord of spirits such wisdom as I have received, according to my powers of understanding (and) the good pleasure of the Lord of spirits, from whom has been given to me the lot of eternal life. (5) Now three 'parables' came to me, and I spoke up and said to those who dwell on earth:

CHAPTER 38

(1) The first Parable:
When the congregation of the righteous shall appear,
And sinners shall be judged for their sins
And driven from the face of the earth.
(2) And when the Righteous One shall appear before the righteous and elect whose works depend upon the Lord of spirits,
And light shall appear to the righteous and elect who dwell on the earth;
Where will then be a dwelling-place for sinners
And where a resting-place for those who have denied the Name of the Lord of spirits?
It were better for them never to have been born.
(3) When their secrets are revealed to the Righteous One, he will judge sinners,
And the godless will be driven from the presence of the righteous and elect.
(4) From then on those who possess the earth will be neither powerful nor exalted;
And they shall not be able to behold the face of the holy,
For the light of the Lord of spirits shall shine on the faces of the holy and righteous and elect.
(5) Then shall mighty kings be destroyed,
And they shall be given over into the hands of the righteous and the holy.
(6) And thenceforward no one shall plead for mercy before the Lord of spirits,
For life for them is finished.

CHAPTER 39

(1) And it shall come to pass in those days that *exalted ones from heaven*

shall descend to resemble the children of the elect and holy; and their seed shall become mingled with that of the children of men. (2) In those days Enoch received writings of anger and wrath and writings (to bring) panic and tumult. There shall be no mercy for them, says the Lord of spirits. (3) And at that time clouds and a storm-wind snatched me up from the face of the earth and set me down at the ends of heaven.

(4) And there I saw another vision of the dwellings of the righteous and the resting-places of the holy.

(5) There my eyes saw their dwellings with the angels
And their resting-places with the holy ones;
And they were petitioning and supplicating and interceding on behalf of the children of men.
Righteousness flowed like water before them, and mercy like dew upon the earth;
Thus it will be among them for ever and ever.

(6) And in that place my eyes saw the Elect One of righteousness and faithfulness,
And there will be righteousness in his days;
The righteous and elect will be innumerable before him for ever and ever.

(7) And I saw their abode beneath the wings of the Lord of spirits,
And all the righteous and elect were radiant like the brightness of fire before him;
And their mouths were full of blessing,
And their lips praised the Name of the Lord of spirits;
And righteousness shall not fail before him.

(8) There I desired to dwell and my soul longed for that abode;
There had my lot been assigned before,
For thus it had been resolved about me before the Lord of spirits.

(9) And in those days I praised and exalted the Name of the Lord of spirits with blessings and praises;
For he has established me in blessedness and splendour, according to the good pleasure of the Lord of spirits.

(10) And my eyes looked long at that place, and I blessed him and praised him, saying:
Blessed is he and blessed may he be from the beginning and for ever.

(11) And in his sight there is no end. He knew before ever the world was created what was to happen for ever,
And for generation after generation what was to come.

(12) Those who sleep not bless thee: they stand before thy glory and bless, praise and extol, saying: 'Holy, holy, holy, is the Lord of spirits: he fills the earth with spirits'. (13) And here my eyes saw all those who sleep not: they stand before him and bless and say: 'Blessed be thou, and blessed be the

Name of the Lord⸢of spirits⸣ for ever and ever'. (14) And my countenance was changed until I was unable to look.

CHAPTER 40

(1) And after that I saw thousands upon thousands and myriads upon myriads, beyond number and reckoning, who stood before the glory of the Lord of spirits. (2) And on the four sides of the Lord of spirits I saw four presences, different from those that sleep not, and I learned their names, for the angel that went with me and showed me all secret things made their names known to me. (3) And I heard the voices of those four presences as they uttered praises before the Lord of glory. (4) The first voice blessed the Lord of spirits for ever and ever. (5) And the second voice I heard blessing the Elect One and the elect ones who depend upon the Lord of spirits. (6) And the third voice I heard as it was petitioning and supplicating on behalf of the dwellers upon earth, and was pleading in the Name of the Lord of spirits. (7) And I heard the fourth voice fending off the satans and forbidding them to come before the Lord of spirits to accuse those who dwell upon the earth. (8) After that I asked the angel of peace who went with me, who showed me all secrets: 'Who are these four presences which I have seen, and whose words I have heard and written down?' (9) And he said to me: 'This first is Michael, the merciful and long-suffering, and the second, who is set over all the diseases and all the wounds of the children of men, is Raphael: and the third, who is set over all the powers, is Gabriel: and the fourth, who is set over the repentance (leading) to hope of those who will inherit eternal life, is named Phanuel'. (10) And these are the four angels of the Lord of spirits and the four voices I heard in those days.

CHAPTER 41

(1) And after that I saw all the secrets of the heavens, and how a kingdom is divided and the actions of men are weighed in the balance. (2) There I saw the dwelling-places of the elect and the congregations of the godly, and my eyes saw there all the sinners being driven from thence which deny the Name of the Lord of spirits, and being dragged off; and they could not abide because of the punishment which proceeds from the Lord of spirits. (3) And there my eyes saw the secrets of the lightning and of the thunder, and the secrets of the winds, how they are divided to blow over the earth, and the secrets of the clouds and dew; and there I saw whence they come forth in that place, and from there the dust of the earth is saturated. (4) And there I saw closed storehouses, and from them the winds are distributed, the storehouse of hail and winds, and the storehouse of mist and the storehouse of

the clouds; and clouds from it (i.e. the storehouse of the clouds) abide over the earth from the beginning of the world. (5) And I saw the storehouses of the sun and moon, whence they (the sun and moon) go forth and whither they return, and their glorious return, and how one is more glorious than the other; and from their revolutions (I saw) the festivals; and how they do not leave their courses and neither lengthen nor reduce their courses, but they keep faith with one another in the covenant by which they abide. (6) And first the sun goes forth and traverses his path according to the commandment of the Lord of spirits, and his Name will endure for ever. (7) And after that I saw the hidden and visible (lit. open) path of the moon, and she accomplishes the course of her path in that place by day and by night—the one holding a position opposite to the other before the Lord of spirits.

And they give thanks and praise and rest not;
For unto them is their thanksgiving rest.
(8) For the sun has many an orbit for blessing or for curse,
And the course of the path of the moon is light to the righteous,
And darkness to sinners in the Name of the Lord of spirits
Who made a separation between light and darkness,
And strengthened the spirits of the righteous,
In the name of his righteousness.

(9) For no angel hinders and no power is able to hinder; for he will appoint a Judge for them all and he will judge them all before him.

CHAPTER 42

(1) Wisdom found no place where she might dwell,
And her dwelling-place came to be in heaven.
(2) Wisdom went forth to make her dwelling among the children of men,
And found no dwelling-place:
Wisdom returned to her place,
And became established among the angels.
(3) And Iniquity went forth from her chambers;
Those she did not seek she found,
And dwelt among them,
(Welcome) as rain in a desert and dew on a thirsty land.

CHAPTER 43

(1) And I saw lightnings besides and the stars of heaven, and I saw how he called them all by their names and they hearkened unto him. (2) And I saw how they are weighed in a righteous balance according to their amount of light, according to the width of their spaces and the day of their appearing,

and how their movement produces lightning, and their motions according to the number of the angels, and (how) they keep faith with each other. (3) And I asked the angel who went with me who showed me secret things: 'What are these?'. (4) And he said to me: 'The Lord of spirits has shown you a parable pertaining to them (lit. their parable): these are the names of the holy who dwell on the earth and believe on the Name of the Lord of spirits for ever and ever'.

CHAPTER 44

(1) Also another phenomenon I saw with regard to the lightnings: how some of the stars arise and become lightnings; and they are unable to remain with them (i.e. the stars).

The Second Parable (Chh. 45-57)

CHAPTER 45

(1) And this is the Second Parable concerning those who deny the Name of the Lord of spirits and the *testimony* of the holy ones.

(2) And into the heaven they shall not ascend,
And on the earth they shall not come:
Such shall be the lot of sinners
Who have denied the Name of the Lord of spirits,
Who are thus kept for the day of suffering and tribulation.
(3) On that day my Elect One shall sit on the throne of glory
And he shall bring their works to the test,
There shall be no place of rest for them,
And their souls shall become heavy within them when they see my elect ones,
And those who have made supplication to my glorious Name.
(4) On that day I will cause my Elect One to dwell among them,
And I will transform the heaven and make it an eternal blessing and light;
(5) And I will transform the earth and make it a blessing:
And I will cause my elect ones to dwell upon it;
But sinners and evil-doers shall not set foot thereon.
(6) For I have provided for and satisfied my righteous ones with peace,
And have caused them to dwell before me:
But for sinners, with me there is judgement impending,
So that I shall destroy them from the face of the earth.

CHAPTER 46

(1) And there I saw One who had a head of days,
And his head was white like wool,
And with him was another whose countenance had the appearance of a man,
And his face was full of graciousness, like one of the angels.

(2) And I asked the angel who went with me and showed me all the secret things, concerning yonder Son of Man, who he was, and whence he was, (and) why he went with the Chief of Days? (3) And he answered and said unto me:

This is the Son of Man to whom belongs righteousness,
And righteousness dwells with him;
And all the treasures of that which is hidden he reveals
Because the Lord of spirits has chosen him,
And whose cause before the Lord of spirits triumphs by uprightness for ever.

(4) And the Son of Man whom you have seen
Shall rouse up the kings and the mighty from their couches and the strong from their thrones,
And he shall loosen the loins of the powerful and break the teeth of sinners.

(5) And he shall cast down the kings from their thrones and kingdoms
Because they do not extol and praise him,
Nor with humble gratitude acknowledge whence the sovereignty was bestowed on them.

(6) And the faces of the powerful *shall change colour* and be covered with shame,
And darkness shall be their dwelling and worms shall be their beds,
And they shall have no hope of rising from their beds
Because they do not extol the Name of the Lord of spirits.

(7) And these are they who rule the stars of heaven,
And raise their hands against the Most High,
And tread upon the earth and occupy it.
And all their deeds manifest unrighteousness,
And their power rests upon their riches,
And their faith is in the gods which they have made with their own hands,
And they deny the Name of the Lord of spirits.

(8) And they persecute the houses of his congregation,
And the faithful who depend upon the Name of the Lord of spirits.

CHAPTER 47

(1) And at that time the prayers of the righteous shall ascend,
And the blood of the righteous one from the earth before the Lord of spirits.
(2) In those days the holy ones who dwell above in the heavens
Shall join together with one voice and supplicate and pray and praise
And give thanks and bless the Name of the Lord of spirits,
On account of the blood of the righteous which has been shed,
And that the prayer of the righteous may not be in vain before the Lord of spirits,
That judgement may be executed on their behalf, and that their long-suffering may not be for ever.
(3) In those days I saw the Chief of Days when he seated himself upon the throne of his glory,
And the books of the living were opened before him:
And all his host which is in heaven above and his council stood before him,
(4) And the hearts of the holy were filled with joy;
Because the number of the righteous had been reached,
And the prayer of the righteous had been heard,
And the blood of the righteous one had been avenged before the Lord of spirits.

CHAPTER 48

(1) And in that place I saw the fountain of righteousness which was inexhaustible:
And around it were many fountains of wisdom;
And all the thirsty drank of them and were filled with wisdom,
And their dwellings were with the righteous and holy and elect.
(2) And at that time the Son of Man was named in the presence of the Lord of Spirits
And his name before the Chief of Days;
(3) And before the sun and the 'signs' were created
Before the stars of the heavens were made,
His name was named before the Lord of spirits.
(4) He shall be a staff to the righteous whereon to stay themselves and not fall,
And he shall be the light of the Gentiles,
And the hope of those who are troubled in their hearts.
(5) All who dwell on earth shall fall down and worship before him,

And will glorify and bless and celebrate with song the Name of the Lord of spirits.
(6) And for this reason he has been chosen and hidden from everlasting before him and for ever;
(7) And the wisdom of the Lord of spirits will reveal him to the holy and righteous;
For he will preserve the portion of the righteous,
Because they loath and despise this world of unrighteousness,
And loath all its works and ways in the Name of the Lord of spirits:
For in his Name they will be saved and he will become the avenger of their lives.
(8) At that time downcast in countenance shall the kings of the earth become,
And the strong who occupy the land, on account of the deeds of their hands;
For on the day of their anguish and affliction they shall not save themselves,
(9) And I will give them over into the hands of my elect ones:
As stubble in the fire so shall they burn before the face of the holy,
As lead in the water shall they sink before the face of the righteous,
And no trace of them shall any more be found.
(10) And on the day of their affliction there shall be rest on the earth,
And before them they shall fall and not rise again:
And there shall be no one to take them by his hands and raise them up:
For they have denied the Lord of spirits and his Anointed One.
And blessed be the Name of the Lord of spirits.

CHAPTER 49

(1) For wisdom is poured out like water,
And glory fails not before him for evermore;
(2) For he is mighty in all the secrets of righteousness,
And unrighteousness shall disappear as a shadow and have no continuance;
Because the Elect One standeth before the Lord of spirits,
And his glory is for ever and ever,
And his might to generations of generations.
(3) And in him dwells the spirit of wisdom,
And the spirit which gives insight,
And the spirit of understanding and of might,
And the spirit of those who sleep in righteousness.
(4) And he shall judge the secret things,
And none shall be able to utter an idle word before him;

For he is the Elect One before the Lord of spirits according to his good pleasure.

CHAPTER 50

(1) And in those days a change shall take place for the holy and elect,
And the light of days shall remain upon them,
And glory and honour shall return to the holy,
(2) On the day of affliction in which evil shall have been heaped up against the sinners.
But the righteous shall be victorious in the Name of the Lord of spirits:
And he will cause the Gentiles to witness it, that they may repent and forego the works of their hands.
(3) They shall have no honour before the Lord of spirits,
Yet through his Name shall they be saved,
And the Lord of spirits will have compassion on them, for his compassion is great.
(4) And he is righteous in his judgement.
And in the presence of his glory unrighteousness shall assuredly not maintain itself:
At his judgement whoever does not repent shall perish before him.
(5) And from henceforth I shall have no mercy on them, saith the Lord of spirits.

CHAPTER 51

(1) And in those days shall the earth give back that which has been entrusted to it,
And Sheol shall give back that which has been committed to it,
And Abaddon shall repay that which it owes.
(2) And he shall choose the righteous and holy from among them,
For the day has drawn nigh that they should be saved.
(3) And the Elect One shall in those days sit on his throne,
And all the secrets of wisdom shall come forth from the meditation of his mouth,
For the Lord of spirits hath appointed him and hath glorified him.
(4) And in those days shall the mountains leap like rams,
And the hills also shall skip like lambs satisfied with milk,
And all will become angels in heaven.
(5) Their faces will shine with joy, for in those days the Elect One shall arise and the earth shall rejoice,
And the righteous shall dwell upon it and the elect shall go and walk thereon.

CHAPTER 52

(1) And after those days in that place where I had seen all the visions of the things that are in secret—for I had been carried off in a whirlwind and they had borne me towards the west— (2) there my eyes saw all the secrets of heaven which are destined to happen upon earth—(I saw) a mountain of iron, and a mountain of copper, and a mountain of silver, and a mountain of gold, and a mountain of soft metal and a mountain of lead. (3) And I asked the angel who went with me, saying, 'What are those things which I have seen in secret?'. (4) And he said to me: 'All these things which you have seen shall serve the dominion of his Anointed One, that he may be powerful and mighty on the earth'. (5) And the angel of peace answered, saying to me: 'Wait a little and all the secrets shall be revealed to you which the Lord of spirits *has prepared.* (6) And these mountains which your eyes have seen, the mountain of iron, the mountain of copper, and the mountain of silver, and the mountain of gold, and the mountain of soft metal, and the mountain of lead, all these shall be, in the presence of the Elect One, as wax before the fire, and like water which streams down from above. These mountains shall become powerless under his feet.

(7) And it shall come to pass in those days that they shall not be saved.
Either by gold or by silver, and none be able to escape.
(8) And there shall be no iron for war,
Nor covering for a breastplate.
Bronze shall be of no service, and tin shall be useless and of no account,
And lead shall not be sought.
(9) And all these things shall be rejected and destroyed from the surface of the earth,
When the Elect One shall appear before the Lord of spirits.

CHAPTER 53

(1) There my eyes saw a deep valley and its entry was open, and all who dwell on land and sea and the islands shall bring to him gifts and presents and tokens of homage, but that deep valley shall not become full.

(2) Their hands *fashion idols*, but all sinners who work iniquity *shall be swallowed up*,
And from before the Lord of spirits sinners shall perish from off the face of his earth;
Who shall not abide, but shall be exterminated for ever and ever.

(3) For I saw all the angels of punishment continually preparing all the (iron) fetters of Satan. (4) And I asked the angel of peace who went with me: 'For whom are they preparing these (iron) fetters?'

(5) And he spoke up and said to me: 'They prepare these for the kings and mighty of this earth, that they may thereby be destroyed.

(6) And after this the Righteous and Elect One shall appear, and his congregations from now on shall not be subject to interdict in the Name of the Lord of spirits.

(7) And these mountains shall become *level land* before his face
And the hills shall become like a fountain of water,
And the righteous shall have rest from the oppression of sinners.

CHAPTER 54

(1) And I turned and looked to another part of the earth, and saw there a
deep valley with burning fire. (2) And they brought the kings and the
potentates, and cast them into this deep valley. (3) And there my eyes saw
that their fetters were being fashioned, iron chains of incalculable weight.
(4) And I asked the angel of peace who went with me, saying: 'For whom are
these fetters being prepared?'. (5) And he said to me: 'Those are being
prepared for the host of Azazel, so that they may take them and cast them
into the depths of hell, and they shall cover over them with rough stones,
as the Lord of spirits commanded. (6) And Michael, and Gabriel, and
Raphael, and Phanuel — they shall take hold of them on that great day, and
cast them on that day into the burning furnace, that the Lord of spirits may
exact retribution from them for their unrighteousness in becoming subject to
Satan and leading astray those who dwell on the earth. (7) And at that time
the punishment of the Lord of spirits shall go forth, and all the chambers of
waters which are above the heavens, and all the fountains of waters which are
beneath the earth will be opened. (8) And all the waters shall be united with
waters: that which is above the heavens is masculine, and the water which is
beneath the earth is feminine. (9) And all who dwell on earth and those
who dwell under the ends of heaven shall be wiped out. (10) And inasmuch
as they acknowledged their unrighteousness which they have wrought on the
earth, for this reason they shall perish.

CHAPTER 55

(1) And after that the Chief of Days repented and said: 'In vain have I
destroyed all who dwell on the earth'. (2) And he swore by his great Name:
'Henceforth I will not do so to all who dwell on the earth, and I will set a sign
in the heaven: and it shall be a pledge of good faith between me and them for
ever, so long as the heavens are above the earth. (3) And this is my
command [with regard to the host of Azazel]: when I am pleased to seize
them by the hand of the angels on the day of tribulation and pain in the face
of that anger and wrath of mine, I shall cause my anger and wrath to abide

on them, says the Lord, the Lord of spirits. (4) You mighty kings who occupy the earth, you are destined to behold my Elect One, as he sits on the throne of glory and judges Azazel, and all his associates and all his hosts in the Name of the Lord of spirits.

CHAPTER 56

(1) And I saw there hosts of angels of punishment, as they went, holding scourges and fetters of iron and bronze. (2) And I asked the angel of peace who went with me, saying: 'To whom are these who hold the scourges going?'. (3) And he said to me: 'To their elect and beloved ones, that they may be cast in the chasm of the abyss of the valley.

(4) And then that valley shall be filled with their elect and beloved,
And the days of their lives shall be at an end,
And the days of their leading astray shall not thenceforward be reckoned.
(5) And in those days shall the angels assemble,
And they shall send forth their chiefs eastwards to the Parthians and the Medes,
To stir up the kings (there) so that a spirit of unrest shall come upon them,
And they shall rouse them from their thrones,
And they will break forth as lions from their lairs,
And as hungry wolves *from their dens*.
(6) And they shall come up and tread underfoot the land of my elect ones,
And the land of my elect ones shall be before them a trampling-place and a highway.
(7) But the city of my righteous ones shall prove an obstacle to their horses,
And they shall incite to warring with one another,
And fail to keep faith with one another;
So that a man shall not trust his brother, nor a son his father or his mother.
Till there will be corpses without number *of their men*;
Their punishment will not be in vain.
(8) In those days Sheol shall open her mouth and they shall be engulfed therein,
Their destruction Sheol shall not remit,
But sinners shall be swallowed up from the presence of the elect.

CHAPTER 57

(1) And it came to pass after this that I saw a host besides of chariots, with men riding thereon, and coming on the winds from the east and from the west, until the middle of the day. (2) And the sound of the noise of their

chariots was heard, and when this commotion took place, the holy ones from heaven observed it, and the pillars of the earth were moved from their place, and the sound thereof was heard from the ends of heaven to the ends of earth, in a single day. (3) And they shall all fall down and worship the Lord of spirits. And this is the end of the Second Parable.

THE THIRD PARABLE (CHH. 58-71)

CHAPTER 58

(1) And I began to speak the third parable concerning the righteous and the elect.
(2) Blessed are you, you righteous and elect,
For glorious shall be your lot.
(3) And the righteous shall be in the light of the sun,
And the elect in the light of eternal life:
The days of their life shall be unending,
And the days of the holy without number.
(4) And they shall seek the light and find righteousness with the Lord of spirits;
There shall be peace to the righteous in the Name of the eternal Lord.
(5) And thereafter it shall be said to the holy
That they should seek out *in his Name* the secrets of righteousness (and) the portion of faithfulness,
For it (righteousness) shall shine like the sun upon the earth and darkness shall pass away.
(6) And there shall be light inexhaustible,
And to a limit of days they shall not come.
For the former darkness shall have been destroyed,
And the light established before the Lord of spirits,
And the light of uprightness established for ever before the Lord of spirits.

CHAPTER 59

(1) In those days my eyes saw the secrets of the lightnings, and the luminaries and their ordinances; and they flash for blessing or a curse as the Lord of Spirits wills. (2) And there I saw the secrets of the thunder; and when it crashes in heaven above, the sound thereof is heard among the habitations of earth. There were shown to me the thunder-peals, for peace and blessing or for curse, according to the command of the Lord of spirits. (3) And after that all the secrets of the lights and lightnings were shown to me, and they lighten for blessing and for plenty.

CHAPTER 60

(1) In the year five hundred, in the seventh month, on the fourteenth (day) of the month in the life of Enoch. In that Parable I saw how a mighty quaking made the heaven of heavens to quake, and the host of the Most High, and the angels, thousands upon thousands and myriads upon myriads, were disquieted with a great disquiet. (2) And the Chief of Days sat on the throne of his glory, and the angels and the righteous ones stood around him.

(3) And a great trembling seized me, and fear took hold of me,
And my loins gave way, and dissolved were my reins,
And I fell upon my face.

(4) And Michael sent another angel from among the holy ones and he raised me up, and when he had raised me up my spirit returned; for I had not been able to endure the look of this host, and the commotion and the quaking of the heaven. (5) And Michael said to me: 'What have you seen that you are so shaken? Until today has lasted the day of his mercy; and he has been merciful and long-suffering towards those who dwell on the earth: (6) but when the day, and the power, and the punishment, and the judgement come, which the Lord of spirits has prepared for those who worship not the righteous Judge, and for those who deny the righteous Judge, and for those who take his Name in vain—that day is prepared, for the elect a covenant, but for sinners a visitation.' (7) And on that day two monsters will be separated from one another, a female monster called Leviathan, to dwell in the abyss of the ocean over the fountains of the waters. (8) But the male is named Behemoth, who covers with his belly an empty wilderness named Duidain(?), on the east of the Garden where the elect and righteous dwell, where my great-grandfather was taken up, the seventh from Adam, the first man whom the Lord of spirits created. (9) And I besought the other angel that he should show me the might of these monsters, how they were separated in one day and cast, the one into the abysses of the sea, and the other into the dry land of the wilderness. (10) And he said to me: 'Son of man, herein you do seek to know what is hidden.' (11) And there spoke to me the other angel who went with me and revealed to me things in secret: things first and last, in heaven and in the heights, and beneath the earth in the depths, at the ends of heaven and in the foundations of the heavens, and in the storehouses of the winds; (12) and how the wind-spirits are divided, and how they are weighed and how *the gates* of the winds are numbered, each according to the wind-spirit's powers; and the intensity of the phases of the moon, according to its right strength; and the division of the stars according to their names, and every division which is made. (13) And thunders according to where they roll, and all the divisions that are made among the lightnings that it may lighten, and their hosts that they may at once obey. (14) For the thunder

has pauses in the interval given to its peal; and the thunder and lightning are inseparable, but by the operation of one and the same spirit, the two of them proceed inseparably; (15) for when the lightning flashes the thunder utters its voice, and the spirit makes (the peal) cease at its appointed time, dividing equally between them (i.e. between thunder peals and lightning flashes); for the storehouses of their occurrences are as the sand; each of them at its proper time is held in with a bridle and turned back by the power of the spirit, and likewise pushed forward in accordance with the many regions of the earth. (16) And the spirit of the sea is masculine and strong, and according to the might of its strength draws it back with a rain, and in like manner it is driven forward and disperses upon all the shores of the land. (17) And the spirit of the hoar-frost is an angel *of ill omen*, but the spirit of the hail is a good angel. (18) And the spirit of the snow, because of its power, never fails; and it has a spirit all to itself, and what rises from it is like smoke, and its name is frost. (19) And the spirit of the storm-cloud is not associated with them in their storehouses, but it has a special storehouse; for its course has the Glory in it, both in light and in darkness, in winter and in summer; and in its storehouse is an angel. (20) And the spirit of the dew has its dwelling at the ends of the heaven, and is connected with the storehouses of the rain, and its course is in winter and summer: and clouds and the storm-clouds are associated, and the one gives to the other. (21) And when the spirit of the rain moves out from its storehouse, the angels come and open the storehouse and lead it out, and when it is diffused over the whole earth, it unites with all the waters on the earth.[1] (22) For the waters are for those who dwell on the earth, for they are nourishment for the ground from the Most High who is in heaven: therefore, on this account, there is a measure for the rain, and the angels *measure it* (23) And all these things I saw towards the Garden of the righteous. (24) And the angel of peace who was with me said to me: 'These two monsters are to be ready for the great day of the Lord to be feasted, so that the punishment of the Lord of spirits may fall upon them and not come forth in vain. Children shall be slain with their mothers, and sons with their fathers. (25) When the punishment of the Lord of spirits falls (ceases) upon all, thereafter there shall be judgement with mercy and forbearance.

CHAPTER 61

(1) And I saw in those days how long cords were given to two angels, and they took to themselves wings and flew, and they went towards the

[1] Or read '... it unites with all the waters on the earth. And if it always united with the waters on the earth, (22) *water would be plentiful for the inhabitants of the earth*.'

north. (2) And I asked the angel who was with me, saying to him: 'Why have they taken those cords and gone off?' And he said to me, 'They have gone to measure.' (3) And the angel who went with me said to me:

'These are they who will bring the lots of the righteous,
And the lines of the righteous to the righteous,
That they may stay themselves on the Name of the Lord of spirits for ever and ever.

(4) The elect shall begin to dwell with the elect,
And these are the measures which shall be given to faithfulness,
And to those *who hold fast to righteousness.*

(5) And those cord-measures shall reveal all that is hidden in the depths of the earth,
Those who have been destroyed by the desert,
And those who have been devoured by the fish of the sea and by wild beasts,
That they will return and stay themselves on the day of the Elect One;
For none shall be destroyed before the Lord of spirits,
And none can be destroyed.

(6) And all in heaven above received a command and power
And one voice and one illumination like to fire.

(7) And the First of voices they blessed,
And they extolled and lauded with wisdom,
And skilled they were in utterance, and with *the spirit of the living creatures.*

(8) And the Lord of spirits placed the Elect One on the throne of glory.
And he shall judge all the works of the holy ones in heaven above,
And in the balance shall their deeds be weighed.

(9) And when he shall lift up his countenance
To judge their hidden ways in the Name he utters of the Lord of spirits,
And their footsteps according to the righteous judgement of the Lord Most High,
Then shall they all with one voice celebrate and bless,
And glorify and extol and sanctify the Name of the Lord of spirits.

(10) And all the hosts of the heavens shall call out, and all the holy ones above, and the host of the Lord, the Cherubin, Seraphin, and Ophannin, and all the angels of power and all the angels of dominions and the Elect One, and the other powers on the earth and over the water (11) on that day, and they shall raise one voice, and bless and glorify and sanctify and exalt in the spirit of faithfulness, and in the spirit of wisdom, and in the spirit of patience, and in the spirit of mercy, and in the spirit of judgement and of peace, and in the spirit of goodness, and shall all say with one voice: 'Blessed is he, and may the Name of the Lord of spirits be blessed for ever and ever'.

(12) All who sleep not in heaven above shall bless him:
All the holy ones who are in heaven shall bless him,
and all the elect who dwell in the Garden of Life:
And every spirit of light who is able to bless, and glorify, and extol and hallow thy blessed Name,
And all flesh which shall glorify exceedingly and bless thy Name for ever and ever.
(13) For great is the mercy of the Lord of spirits, and he is long-suffering,
And all his works and all his mighty deeds, as many as he has accomplished,
He has revealed to the righteous and elect
In the Name of the Lord of spirits.

CHAPTER 62

(1) And then the Lord of spirits commanded the kings and the mighty and the exalted, and those who possess the earth, and said: 'Open your eyes and lift up your horns, if you can, to acknowledge the Elect One.
(2) And the Elect One sat on the throne of his glory,
And the spirit of righteousness was poured out upon him,
And the word of his mouth slays all the sinners,
And all the unrighteous are destroyed from before his face.
(3) And there shall stand up on that day all the kings and the mighty,
And the exalted and those who possess the earth,
And they shall see him and recognise him, because he sits on the throne of his glory,
And righteousness is judged before him, and no lying word is spoken before him.
(4) Then shall pain come upon them as upon a woman in travail,
⌜And bringing forth is painful for her⌝,
When her child enters the mouth of the womb,
And she has pain in bringing forth.
(5) And one portion of them shall look on the other,
And they shall be terrified,
And they shall be downcast of countenance,
And pain shall seize them,
When they see that child of woman sitting on the throne of his glory.
(6) And the mighty kings and all who possess the earth shall glorify and bless and extol him who reigns over all, the One who hides himself.
(7) For from the beginning the Son of Man was hidden,
And the Most High preserved him in the presence of his (heavenly) host,
And revealed him to the elect.

(8) And the congregation of the elect and holy shall be sown,
And all the elect shall stand before him on that day.
(9) And all the mighty kings and the exalted and those who rule the earth
Shall fall down before him on their faces,
And worship and set their hope on that Son of Man,
And petition him and supplicate for mercy from him.
(10) But the Lord of spirits shall then so affright them
That they shall hastily go forth from his presence,
And their faces shall be filled with shame,
And the darkness shall grow deeper on their faces.
(11) And he will deliver them to the angels of punishment,
To execute vengeance on them because they have oppressed his children and his elect.
(12) And they shall be a spectacle for the righteous and for his elect who will exult over them,
Because the wrath of the Lord of spirits rests upon them,
And his sword will be drunk from them.
(13) And the righteous and elect shall be saved on that day,
And they shall never henceforward see the face of the sinners and unrighteous.
(14) And the Lord of spirits shall abide by them,
And with that Son of Man they shall eat,
And lie down and rise up for ever and ever.
(15) And the righteous and elect shall be raised up from the earth,
And they shall cease to be of downcast countenance;
They shall be clothed with garments of glory,
(16) And this shall be your garment, a garment of life from the Lord of spirits,
And your garments shall not grow old, nor your glory pass away before the Lord of spirits.

CHAPTER 63

(1) In those days the mighty kings who possess the earth shall implore his angels of punishment to whom they were delivered up to grant them a brief respite that they might fall down and worship before the Lord of spirits, and confess their sins before him. (2) And they shall bless and glorify the Lord of spirits, and say:

'Blessed is the Lord of spirits and the Lord of kings,
And the Lord of the mighty and the Lord of the exalted,
And the Lord of glory and the Lord of wisdom.

(3) Every secret thing will be brought to light;
Thy power is from generation to generation,
And thy glory for ever and ever.
Unfathomable are all thy secrets and innumerable,
And thy righteousness is beyond reckoning.
(4) We have now learned that we should glorify
And bless the Lord of kings and him who rules over all kings.'
(5) And they shall say:
'Would that we had a respite to glorify, praise and give thanks before thy glory!
(6) And now we long for a brief respite and find it not:
We pursue it but obtain it not:
And light has vanished from before us,
And darkness shall be our dwelling-place for ever and ever.
(7) For before him we have not given thanks,
Nor glorified the Name of the Lord of spirits, nor glorified our Lord in all his works,
But our trust has been in the sceptre of our kingdom, and in our own glory.
(8) But in the day of our suffering and tribulation he saves us not,
And we find no respite to give thanks,
Because our Lord is true in all his works, and in his judgements and his justice,
And his judgements show no respect of persons.
(9) And we pass away from before his face on account of our works,
And all our sins are justly reckoned up.'
(10) Then will they say to them: 'Our souls are sated with ill-gotten gains,
But they will not keep us from going down to the grave, from the flames of the *pit* of Sheol.'
(11) And after that their faces shall be filled with darkness
And shame before that Son of Man,
And they shall be driven from his presence,
And the sword shall remain before his face in their midst.

(12) Thus spake the Lord of spirits: 'This is the ordinance and judgement, with respect to the mighty and the kings and the exalted and those who possess the earth, before the Lord of spirits'.

CHAPTER 64

(1) And other presences I saw hidden in that place. (2) I heard the voice of the angel saying: 'These are the angels who descended from heaven to the

earth, and revealed what was hidden to the children of men, and led the children of men astray into committing sin'.

CHAPTER 65

(1) And in those days, I, Noah, saw that the earth had been brought low and its destruction was nigh. (2) And I set out from thence and went to the ends of the earth, and cried aloud to my great-grandfather Enoch: and I said three times with an embittered voice, 'Hear me, hear me, hear me'. (3) And I said to him: 'Tell me what it is that is being done on the earth that the earth is so afflicted and shaken, lest perchance I shall perish with it.' (4) And thereupon at once there was a great commotion on the earth, and a voice was heard from heaven, and I fell on my face. (5) And Enoch my great-grandfather came and stood by me, and said to me: 'Why have you cried to me with a bitter cry and weeping? (6) A command has gone forth from the presence of the Lord of spirits concerning those who dwell on the earth that this must be their end, because they have learned all the secrets of the angels, and all the wrong-doing of the satans, and all their secret powers, and all the powers of those who practise sorcery, and the power of spells, and the power of those who make molten images of every created thing; (7) and further how silver comes forth out of the dust of the earth, and how soft metal originates in the earth. (8) For lead and tin are not produced from the earth like the former (i.e. silver): it is a mine that produces them, and an angel stands therein and the angel *purifies* (*them*).' (9) And after that my great-grandfather Enoch took hold of me by my hand and raised me up, and said to me: 'Go, for I have asked the Lord of spirits about this commotion on the earth. (10) And he (the Lord of spirits) said to me: 'Because of their iniquities their judgement shall be fully carried out and shall not *be withheld* before me; because of the *sorceries* which they have invented and learned, the earth and those who dwell upon it shall be destroyed. (11) And as for those (watchers) that they will have no return (to heaven) for ever, because they have shown them (mankind) what was hidden, and they have been condemned: but not so for you, my son,—the Lord of spirits knows that you are innocent and guiltless of this reproach concerning the secrets.

(12) And he has established your name to be among the holy ones,
And will preserve you from those who dwell on the earth,
And he has established your seed in righteousness both for sovereignty and for great honours,
And from your seed shall proceed a fountain of the righteous and holy without number for ever'.

CHAPTER 66

(1) And after that he showed me the angels of punishment who were prepared to come and let loose all the powers of the waters which are beneath the earth in order to bring judgement and destruction on all who abide and dwell on the earth. (2) And the Lord of spirits gave commandments to the angels who were going forth, that they should not show their power but keep watch (i.e. over the waters), for those angels were over the powers of the waters. (3) And I went away from the presence of Enoch.

CHAPTER 67

(1) And in those days the word of the Lord came to me, and he said to me: 'Noah, behold your lot has come up before me, a lot without blame, a lot of love and uprightness. (2) And now the angels are building a wooden (vessel), and when they have completed this work I will place my hand upon it and preserve it, and there shall come forth from it the seed of life, and a transformation shall take place so that the earth will not be void of inhabitants. (3) And I will establish your offspring before me for ever and ever, and I will spread abroad those who dwell with you upon the face of the earth: it (your offspring) *shall not be barren* on the face of the earth but it shall be blessed and multiply upon the earth in the Name of the Lord'.

(4) And he will imprison those angels who have revealed iniquity (to mankind) in that burning valley which my great-grandfather Enoch had formerly shown to me in the west among the mountains of gold and silver and iron and soft metal and tin. (5) And I saw that valley in which there was a great convulsion and the waters were in turmoil. (6) And when all this took place, from that fiery molten (lit. soft) metal and from the convulsion which shook them (the waters) in that place, there was produced a smell of sulphur, and it was combined with those waters, and that valley of the angels who had led astray (mankind) burned beneath that land. (7) And through its ravines proceed rivers of fire, where these angels are punished who had led astray the inhabitants of the earth.

(8) But those waters in those days shall serve the kings and the mighty and the exalted, and those who occupy the earth, for the healing of the body, but for the punishment of the spirit. Their spirits are full of lust, so that their bodies are punished, for they have denied the Lord of spirits; and they see their punishment daily, and yet they believe not in his Name. (9) And great as was the burning of their bodies (with lust), so will be the price they will pay in their spirits for ever and ever; for before the Lord of spirits none shall utter a lying word. (10) For the judgement shall come upon them, because they believe in the lust of their body and deny the spirit of the Lord (11) And these waters themselves shall undergo a change in those days; for when those

angels are punished in those waters, these water-springs shall change their temperature, and when they ascend, this water of the springs shall change and become cold. (12) And I heard Michael speaking up and saying: 'This punishment wherewith the angels are punished is a warning for the kings and the mighty who possess the earth. (13) Because these waters of punishment minister to the healing of the body of the kings and the lust of their body; therefore they will not see and will not believe that those waters will change and become a fire which burns for ever.'

CHAPTER 68

(1) And thereafter he (Michael) gave me instruction in all the things that are secret in the book of my great-grandfather Enoch, (the book of) the parables which had been given to him; and he (Enoch) had composed them for me in the words of the Book of the Parables. (2) And on that day Michael answered Raphael and said: 'The power of the spirit seizes me and makes me tremble because of the severity of the sentence regarding the secrets, the punishment of the angels. Who can endure the severity of the sentence which is to be executed and not be filled with fear before it?' (3) And Michael answered again, and said to Raphael: 'Who is there whose heart is not faint for fear at it (the punishment), and whose reins do not quiver with fear at this word of punishment that has gone forth against them from those whom they *in like manner afflicted*'. (4) 'And it shall come to pass when they shall stand before the Lord of spirits',[1] Michael said thus to Raphael, 'they will receive no mercy in the sight of the Lord, for the Lord of spirits has been angry with them because they *have fashioned images of the Lord*. (5) Accordingly, there shall come upon them the sentence regarding the secrets for ever and ever; for neither idol nor man shall be accorded his (God's) portion. But they by themselves will receive their condemnation for ever and ever.'

CHAPTER 69

(1) And thereafter the sentence (upon them) will terrify and make them to tremble, inasmuch as this has been shown to those who occupy the land. (2) And behold the names of those angels ⌈and these are their names⌉: the first of them is Semyaza, the second Arestiqifa, and the third Armen, and the fourth Kokabiel, the fifth Turiel, the sixth Rumyal, the seventh Danyal, the eighth Neqael, the ninth Baraqyal, the tenth Azazel, the

---

[1] Or 'But it came to pass, when he stood before the Lord of spirits, the holy Michael spoke as follows to Raphael ...' (Knibb).

eleventh Armaros, the twelfth Bataryal, the thirteenth Basasyal, the fourteenth Hananel, the fifteenth Turiel, and the sixteenth Simpsiel, the seventeenth Yetriel, the eighteenth Tumiel, the nineteenth Turiel, the twentieth Rumyal, the twenty-first Azazel. (3) And the following are their archangels with their names, their centurions, their quinquegenarii and their dekadarchs. (4) The name of the first Yeqon: that is the one who led astray (all) the sons of the angels, and brought them down to the earth, and led them astray through the daughters of men. (5) And the second was named 'Asbe'el: he counselled the sons of the holy angels in an evil plan, and led them astray so that they defiled their bodies with the daughters of men. (6) And the third was named Gādre'el: he it is who showed the children of men all lethal blows, and he led Eve astray, and showed the weapons of death to the sons of men, the shield and the coat of mail, and the sword for the battle ⌜and all the weapons of death to the children of men⌝. (7) And by him they have gone forth upon those who dwell on earth from that day and for evermore. (8) And the fourth was named Pēnēmue: he taught the children of men that bitter was sweet and sweet bitter, and he showed them all the secrets of their sophistry. (9) And he instructed mankind in writing with ink and paper, and thereby there are many who have gone astray from eternity to eternity and until this day. (10) For men were not born in order, in this way, to give confirmation to their good faith with pen and ink. (11) For men were not otherwise created than like the angels, that they should continue righteous and pure, and death, which destroys everything, would not have touched them, but through this knowledge of theirs they are perishing, and through this power it (death) is consuming us. (12) And the fifth was named Kasdeya': this is he who showed the children of men all the evil afflictions from spirits and demons, and the diseases of the embryo in the womb, so that it miscarries, and the afflictions of the soul, the bites of the serpent, and the afflictions which come through the noontide heat, (and) the son of the serpent named Tabā'et. (13) And it was he who reckoned up the gematria (lit. the number) of the Chief of Days for Kasbeel, who revealed the sum of the oath to the angels when he dwelt above in glory, and its name is BIQA. (14) This (satan) told Michael to show him the hidden Name, that they might pronounce it in the oath, so that those who revealed all that was secret to the children of men might tremble before that Name and oath. (15) And this is the power of this oath, for it is powerful and strong, and he (God) had placed this oath 'AKA' in the hand of Michael. (16) And these are the secrets of this oath: 'Through his oath the firmament and the heavens were suspended before the world was created and for ever. (17) And through it the earth was founded upon the waters,

And from the secret recesses of the mountains comes *sweet water*,
From the creation of the world and unto eternity.

(18) And through that oath the sea was created and its foundations;
For the time of its wrath he placed for it the sand as a barrier,
And it does not pass beyond its boundary from the creation of the world and to eternity.
(19) And through that oath are the depths made fast,
And abide and stir nor from their place from eternity to eternity.
(20) And through that oath the sun and moon complete their course,
And deviate not from their ordinance from eternity to eternity.
(21) And through that oath the stars complete their course,
And he calls them by their names,
And they answer him from eternity to eternity.
(22) And likewise, with regard to the waters, to their winds, and to all spirits and their courses from all regions of spirits.'

(23) And there are kept the storehouses of the thunder-peals and the flashes of the lightning: and there are kept the storehouses of the hail and of the hoar-frost, and the storehouses of the storm-cloud, and the storehouses of the rain and of the dew. (24) And all those give thanks and praise before the Lord of spirits, and glorify (him) with all their power; and their sustenance is in all (their) thanksgiving; they will praise and glorify and extol the Name of the Lord of spirits for ever and ever.

(25) And this oath is binding upon them,
And by it they shall be kept, and they shall keep to their paths,
And their course shall not be spoiled.

Close of the Third Parable (Chh. 69.26-71)

(26) And great joy was theirs,
And they blessed and glorified and extolled,
Because the name of the Son of Man had been revealed to them.
(27) And he sat on the throne of his glory,
And the sum of judgement was given to the Son of Man,
And he will cause the sinners to pass away and be destroyed from off the face of the earth,
And those who have led the world astray, (28) with chains shall they be bound,
And in their assembling-place of destruction shall they be imprisoned,
And all their works vanish from the face of the earth.
(29) And from henceforth there shall be nothing corruptible,
For that Son of Man has appeared,
And has seated himself on the throne of his glory,
And all evil shall pass away and depart from before his face,

And the word of the Son of Man shall be strong before the Lord of spirits,
This is the third Parable of Enoch.

CHAPTER 70

(1) And it came to pass thereafter that the name *of a son of man* (i.e. Enoch) *was raised up to the Lord of spirits* from those who dwell on the earth. (2) And he was raised aloft on a chariot of the spirit, and his name was bruited abroad among them. (3) And from that day I was no longer numbered among them; and he placed me between two regions, between the north and the west, where the angels took the cords to measure for me the place for the elect and righteous. (4) And there I saw the first fathers, and the righteous who from everlasting dwell in that place.

CHAPTER 71

(1) And it came to pass after this that my spirit was translated,
And it ascended into the heavens:
And I saw the sons of the holy angels.
They were treading on flames of fire:
Their garments were white ⌜and their robes⌝
And the light of their countenances (shone) like snow.
(2) And I saw two streams of fire,
And the light of that fire shone like hyacinth,
And I fell on my face before the Lord of spirits.
(3) And the angel Michael, one of the archangels, seized me by my right hand,
And lifted me up and led me forth into all the secret things,
And he showed me all the secrets of mercy and ⌜he showed me⌝ all the secrets of righteousness.
(4) And he showed me all the secrets of the ends of the heavens,
And all the storehouses of the stars, and all the luminaries,
Whence they proceed in the presence of the holy ones.
(5) And he translated my spirit, and I, Enoch, was in the heaven of heavens,
And I saw there, in the midst of those luminaries, *a house as it were* built of hailstones,
And among those hailstones tongues of fire of the *living creatures*.
(6) And my spirit saw the girdle which was encircling the house with fire;

On its four sides were streams filled with the fire of the *living creatures*,
And they girt that house.
(7) And around it were Seraphin, Cherubin, and Ophannin;
And these are they who sleep not and guard the throne of his glory.
(8) And I saw angels who could not be counted,
Thousands upon thousands, and myriads upon myriads, encircling that house,
And Michael, and Raphael, and Gabriel, and Phanuel,
And the holy angels who are above the heavens, go in and out of that house.
(9) And Michael, and Gabriel, Raphael and Phanuel,
And many holy angels without number,
Came forth from that house.
(10) And with them the Chief of Days,
His head was white and pure as wool,
And his raiment indescribable.
(11) And I fell on my face,
And my whole body became weak from fear,
And my spirit was transformed;
And I cried with a loud voice,
With the spirit of power,
And blessed and glorified and extolled.
(12) And these blessings which went forth out of my mouth were well
pleasing before the Chief of Days. (13) And the Chief of Days came with
Michael and Gabriel, Raphael and Phanuel, and thousands and myriads of
angels without number. (14) And that angel (Michael) came to me and
greeted me with his voice and said to me:
'You are the Son of Man who is born for righteousness,
And righteousness abides upon you,
And the righteousness of the Chief of Days forsakes you not'.
(15) And he said to me:
'He proclaims to you peace in the name of the world to come;
For from hence has proceeded peace since the creation of the world,
And so shall it be to you for ever and for ever and ever.
(16) And all shall walk in your ways, since righteousness never forsakes you:
With you will be their dwelling-places, and with you their inheritance,
And they shall not be separated from you for ever and ever.
(17) And so there shall be length of days with the Son of Man,
And the righteous shall have peace and an upright way
In the Name of the Lord of spirits for ever and ever.'

CHAPTERS 72-80.1

For these chapters, see Appendix A 'The "Astronomical" Chapters of the Ethiopic Book of Enoch', 386-419

CHAPTER 80.2-8

THE PERVERSION OF THE HEAVENLY BODIES

(2) And in the days of the sinners years shall become shorter,
And their seeds shall be late in their lands and fields,
And all work on the earth shall be changed,
And shall not appear in its time:
And the rain shall be held back
And the heavens shall be shut up.
(5b) And drought shall come in the remotest regions of the Great Waggon to the west.
(3) And in those times the fruits of the earth shall be late,
And shall not grow in their time
And the fruits of the trees shall be withheld in their time.
(4) And the moon shall alter her order
And not appear at her (proper) time.
(5) And in those days she shall appear in the heavens,
And shall shine more brightly than accords with the order of (her) light.
(6) And many leaders of the stars shall stray from the commandments (of God),
And those shall change their orbits and tasks,
And not appear at the seasons commanded them.
(7) And the whole law of the stars shall be closed to the sinners,
And the thoughts of those who dwell on earth shall err concerning them,
And they shall turn away from all their ways,
And they shall err and think them to be gods.
(8) And evil shall be multiplied upon them,
And punishment shall come upon them so as to destroy all.

THE HEAVENLY TABLETS AND THE MISSION OF ENOCH (CHH. 81-82.3)

CHAPTER 81

(1) And he (Uriel) said to me:
'Look, Enoch, at the book of the heavenly tablets,
And read what is written on them, and acquaint yourself with every single thing'.

(2) And I looked at everything in the heavenly tablets, and I read everything that was written, and I understood everything, and I read the book, and everything that was written in it, of all the deeds of mankind and of all the children of flesh on the earth, unto the generations of eternity. (3) And forthwith I blessed the great Lord, the King of glory for ever, in that he has made all the works of the world,

And I extolled the Lord because of his patience,
And I wept ⌜on earth⌝ because of the sons of Adam.

(4) And thereafter I said: 'Blessed is the man who dies, righteous and good,
About whom no record of unrighteousness is written down,
And against whom no sin has been found on the day of judgement.'

(5) And these seven holy ones brought me and placed me on earth before the door of my house, and said to me: 'Make everything known to your son Methuselah, and show all your children that no flesh is righteous in the sight of the Lord, for he created them. (6) For one year we shall leave you with your children until you again shall have given them your last charges, so that you may teach your children and record (it) for them, and testify to all your children, and in the second year you shall be taken from their midst.

(7) Let your heart be strong,
For the good shall proclaim righteousness to the good;
The righteous with the righteous shall rejoice,
And shall wish one another well.

(8) But sinner shall die with sinner,
And the apostate shall sink along with the apostate.

(9) And those who practise righteousness shall *not* die because of the deeds of men,
Nor be destroyed because of the deeds of the godless.'

(10) And in those days they ceased to speak with me; and I came to my people, blessing the Lord of eternity.

CHAPTER 82.1-3

(1) And now, my son Methuselah, all these things I am recounting to you and writing down for you; and I have revealed to you everything, and given you writings of all these things. Keep, my son Methuselah, the writings of your father's hand, that you may deliver them to the generations of eternity. (2) Wisdom I have given to you and to your children, and to those who will be your children, that they may transmit it to their children, and to generations of generations for ever, to whoever is endowed with wisdom; and they shall celebrate all the wise. Wisdom shall slumber, (3) (but) in their mind those who have understanding shall not slumber, but they shall hearken with

their ears that they may learn this wisdom, and it shall be better for those that partake of it than rich food.

CHAPTERS 82.4-20

For these verses, see Appendix A 'The "Astronomical" Chapters of the Ethiopic Book of Enoch', 386-419

THE DREAM-VISIONS (CHH. 83-90)

First Dream-Vision: the Deluge

CHAPTER 83

(1) And now I will show you, my son Methuselah, all my visions which I have seen, recounting them before you. (2) Two visions I saw before I took a wife, and the one was quite unlike the other: the first when I was learning to write: the second, before I took your mother to wife—I saw a terrible vision; and I made supplication about them to the Lord. (3) I had laid me down in the house of my grandfather Malalel, and I saw in a vision (how) the heavens tottered and shook and fell on the earth. (4) And when it fell to the earth I saw how the earth was swallowed up in a deep abyss, and mountains *crashed down on* mountains, and hills sank down on hills, and tall trees were torn up from their roots, and hurled down and sank into the abyss. (5) And thereupon a word came into my mouth, and I lifted up (my voice) to cry aloud and said: 'The earth is destroyed'. (6) And my grandfather Malalel roused me as I lay by him, and said to me: 'What ails you that you cry so, my son, and why do you make such moaning?' (7) And I recounted to him the whole vision which I had seen, and he said to me: 'A terrible thing you have seen, my son, and horrible indeed is your dream-vision as to the secrets of the sins of the whole earth, and (that) it must sink into the abyss and be destroyed with a great destruction. (8) And now, my son, arise and make petition to the Lord of glory, since you are true, that a remnant may remain on the earth, and that he may not wipe out the whole earth. (9) My son, from heaven all this will come upon the earth and upon the earth there will be great destruction.' (10) After that I arose and prayed and implored and besought, and wrote down my prayer for the generations of the world, and I will show everything to you, my son Methuselah. (11) And when I had gone forth below and seen the heaven, and the sun rising in the east, and the moon setting in the west, and *stars on the wane* and the whole earth, and everything as he had known (it) in the beginning, then I blessed the Lord of judgement and extolled him because he had made the sun to go forth from the windows of the east, so that it ascends and rises on the face of the heavens, and sets out and goes in the path which has been shown to it.

CHAPTER 84

(1) And I lifted up my hands in righteousness and blessed the great Holy One, and spoke with the breath of my mouth, and with the tongue of flesh, which God has made for the children of mankind, that they should speak therewith, ⌜and he gave them breath and a tongue and a mouth that they should speak therewith⌝.

(2) Blessed be thou, O Lord, King,
Great and mighty in thy greatness, Lord of the whole creation of the heavens,
King of kings and God of the whole world.
And thy dominion and sovereignty and greatness abide for ever and ever,
And throughout all generations thy dominion:
And all the heavens are thy throne for ever,
And the whole earth thy footstool for ever and ever.
(3) For thou hast made and thou rulest all things,
And nothing is too hard for thee;
Wisdom leaves thee not nor is she moved
From the seat of thy throne and from thy presence.
And thou knowest and seest and hearest everything,
And there is nothing hidden from thee, ⌜for thou seest everything⌝.
(4) And how the angels of thy heaven are doing wrong,
And upon mankind abideth thy wrath until the great day of judgement.
(5) And now, O God and Lord and great King,
I implore and pray that thou mayest fulfil for me my prayer,
That thou wouldest leave me a posterity upon earth;
And not wipe out all mankind
And make the earth without inhabitant,
So that there should be an eternal destruction.
(6) And now, my Lord, wipe out from upon the earth the flesh which aroused thy wrath,
But flesh of righteousness and uprightness establish as a seed-bearing plant for ever,
And hide not thy face from the prayer of thy servant, O Lord.

SECOND DREAM VISION: THE ZOOMORPHIC HISTORY OF THE WORLD (CHH. 85-90)

CHAPTER 85

(1) And after this I saw in another dream, and I will show the whole dream to you, my son. (2) And Enoch lifted up (his voice) and spoke to his son Methuselah: 'To you, my son, will I speak: hear my words—incline your ear

to the dream-vision of your father. (3) Before I took to wife your mother Edna, I saw in a vision on my bed and behold a bull came forth on the earth, and that bull was white; and after it came forth a heifer, and along with it came forth two bull-calves, one of them black and the other red. (4) And the black bull-calf gored the red one and chased him upon the earth, and from then on I could not see that red bull-calf. (5) But the black bull-calf grew up and had intercourse with a heifer; and I saw that many oxen issued from him, resembling him; and they were following after him. (6) And that first cow went from the presence of that first bull in order to seek the red bull-calf, but found him not, and lamented with a great lamentation over him and sought him. (7) And I looked until that first bull had intercourse with her and quieted her, and from that time onward she cried out no more. (8) And after that she bore another white bull, and after him she bore many black bulls and cows. (9) And I saw in my sleep that white bull, and it likewise grew up and became a great white bull, and from him issued many white bulls, and they resembled him. (10) And they began to beget many white bulls which resembled them, one following the other.

CHAPTER 86

(1) And again I saw with my eyes as I slept, and I saw the heaven above, and behold a single star fell from heaven, and it *became transformed* (into a bull), and it fed and pastured among those oxen. (2) And after this I saw the large and black oxen, and behold, they all *destroyed* their stalls and their pastures and their calves, and they began *to butt* one another. (3) And again I saw in the vision, and looked towards the heavens, and behold I saw many stars descend and cast themselves down from heaven beside that first star, and like it they became bulls amongst those cattle, and ⌜with them⌝ pastured among them. (4) And I looked at them and saw, and behold they all let out their members, like horses, and began to mount the cows of the oxen; and they all became pregnant and bore elephants, camels and asses. (5) And all the oxen feared them and were affrighted at them, and began to bite with their teeth and to devour and to gore with their horns. (6) And they began to devour these oxen; and behold all the children of the earth began to tremble and quake before them and to flee.

CHAPTER 87

(1) And again I saw them, and they began to gore each other and to devour each other, and the earth began to cry aloud. (2) And I raised my eyes again to heaven, and I saw in the vision, and behold there came forth from heaven beings who were like white men: and four went forth from that place, and three with them. (3) And those three that had last come forth

grasped me by my hand and took me up, away from the children of the earth, and raised me up to a lofty place, and showed me a tower high above the earth, and all the high places were smaller. (4) And they said to me: 'Remain here till you see everything that befalls those elephants, camels, and asses, and the stars and all the oxen.'

Punishment of the Fallen Angels by the Archangels

CHAPTER 88

(1) And I saw one of those who had come forth first, and he seized that first star that had fallen from the heavens, and bound it hand and foot and cast it into an abyss: now that abyss was narrow and deep, and closed in and dark. (2) And one of them drew a sword, and gave it to those elephants and camels and asses: then they began to smite one another, and the whole earth quaked because of them. (3) And as I was looking in the vision, behold, then, one of those four who had come forth stoned them from heaven, and he gathered and took all the many stars whose privy members were like those of horses, and bound them all hand and foot, and cast them in an abyss of the earth.

The Deluge and the Deliverance of Noah

CHAPTER 89.1-9

(1) And one of those four went to one of the white oxen, and he instructed him in a mystery, trembling as he was: he was born a bull and became a man, and built for himself a great vessel and dwelt thereon: and three bulls dwelt with him on that vessel and it covered over them. (2) And again I raised my eyes to heaven and saw a lofty roof, with seven sluices therein, and these sluices flowed with much water into an enclosure. (3) And I looked again, and behold chambers were opened up on the ground in that great enclosure, and the water began to pour forth and rise upon the ground; and I saw that enclosure till all its ground was covered with water. (4) And water, darkness and thick cloud increased upon it; and I looked at the height of the water, and the water had risen above the height of the enclosure, and was pouring over the enclosure, and it rose over the earth. (5) And all the cattle of that enclosure were gathered together until I saw how they sank and were swallowed up and perished in the water. (6) But the vessel floated on the water, while all the oxen and elephants and camels and asses sank to the bottom with all the animals, so that I could no longer see them, and they were not able to escape, (but) perished and sank into the depths. (7) Again I watched in the vision until those sluices ceased from that lofty roof, and the springs of the (underground) chambers were

stopped, and other deeps were at the same time opened up. (8) Then the water flowed down into these, till the earth was uncovered; and the vessel settled down on the earth, and the darkness retired and the light came. (9) But the white bull which had become a man came out of the vessel, and the three bulls with him, and one of those three bulls was white like that bull, and one of them was red as blood, and one black. And the white bull departed from them.

From the Death of Noah to the Exodus

CHAPTER 89.10-27

(10) And they began to beget beasts of the field and birds, so that there arose from them every variety of species: lions, leopards, wolves, dogs, hyenas, wild-boars, foxes, rock-badgers, swine, *wild ostriches*, vultures, kites, eagles and ravens; and among them was born a white bull. (11) And they began to bite one another; and that white bull which was born amongst them begat a wild ass and a white bull with it, and the wild asses multiplied. (12) But the bull-calf which was sired by him begat a black wild boar and a white ram of the flock; and the wild boar begat many boars, and the ram begat twelve sheep. (13) And when the twelve sheep had grown up, they gave up one of them to the asses, and those asses in turn gave up that sheep to the wolves, and that sheep grew up among the wolves. (14) And the ram led all the eleven sheep to dwell with it and to pasture with it among the wolves: and they multiplied and became many flocks of sheep. (15) And the wolves began to terrify them, and they oppressed them until they had destroyed their little ones, and they cast their young into a river of deep water: but those sheep began to cry aloud on account of their little ones, and to complain to their Lord. (16) And a sheep which had been saved from the wolves fled and came to the wild asses; and I saw the sheep as they lamented and cried, and they besought their Lord with all their might, till the Lord of the sheep descended at the call of the sheep from a lofty abode, and came to them and pastured them (17) And he called the sheep which had escaped from the wolves, and spoke with it concerning the wolves, that it should warn them not to harm the sheep. (18) And the sheep went to the wolves according to the word of the Lord, and another sheep met it and went with it, and the two went and entered together into the assembly of those wolves, and spoke with them and warned them not to harm the sheep from henceforth. (19) And thereafter I saw the wolves, and how they oppressed the sheep exceedingly with all their power; and the sheep cried aloud. (20) And their Lord came to the sheep and began to smite those wolves: and the wolves began to make lamentation; but the sheep became quiet and forthwith ceased to cry out. (21) And I saw the sheep until they departed from the wolves; but the eyes of

the wolves were blinded, and those wolves went forth in pursuit of the sheep with all their forces. (22) But the Lord of the sheep went with them, leading them, and all his sheep followed him: and his face was glorious and his appearance fearful and splendid. (23) But the wolves began to pursue those sheep till they encountered them at a sea of reeds. (24) And the sea of reeds was divided, and the water stood on this side and on that before their faces, and their Lord led them and placed himself between them and the wolves. (25) And while those wolves had not yet seen the sheep, they proceeded into the midst of the sea of reeds, and the wolves followed the sheep and the wolves ran after them into that sea of reeds. (26) And when they saw the Lord of the sheep, they turned to flee before his face, but the sea of reeds gathered itself together, and suddenly it became as it had been created, and the water swelled and rose till it covered those wolves. (27) And I saw till all the wolves who pursued those sheep perished and were drowned.

Israel in the Desert, the Giving of the Law, and the Entrance into Palestine

CHAPTER 89.28-40

(28) But the sheep passed on from that water and went forth into a wilderness, where there was no water and no grass; and they began to open their eyes and to see; and I saw the Lord of the sheep pasturing them and giving them water and grass, and the ram was going and leading them. (29) And the ram ascended to the summit of a lofty crag, and the Lord of the sheep sent it to them. (30) And after that I saw the Lord of the sheep who stood before them, and his appearance was powerful, majestic and fearful, and all those sheep saw him and were afraid before him. (31) And they were all fearful of him and trembling. And they cried out to ⌜the sheep which led them and to⌝ the other sheep which was with them, saying: 'We are not able to stand up before our Lord nor to look on him'. (32) And the sheep which led them ascended a second time to the summit of the crag, but the eyes of the sheep began to be blinded and they strayed from the way which he had shown them, without that sheep knowing about them. (33) And the Lord of the sheep was exceedingly wrathful against them, and that sheep got to know of it, and went down from the summit of the crag, and came to the sheep and found the greatest part of them blinded in their eyes and fallen away. (34) And when they saw it they feared and trembled at its presence, and desired to return to their folds. (35) And that sheep took other sheep with it, and fell upon those sheep which had fallen away and then began to slay them; and the sheep feared its presence, and thus that sheep brought back those sheep that had fallen away, and they returned to their folds. (36) And I looked in the vision till that sheep was tranformed and became a man and

built a tabernacle for the Lord of the sheep, and he made all the sheep to stand at that tabernacle. (37) And I looked until the sheep which had gone to meet that sheep which led the sheep fell asleep: and I saw till all the great sheep had perished and little ones arose in their place, and they came to a pasture, and approached a stream of water. (38) Then that sheep, their leader which had become a man, withdrew from them and fell asleep, and all the sheep sought it and cried on account of it with a great crying. (39) And I looked till they left off their crying for that sheep and crossed that stream of water; and there arose other sheep as leaders in the place of those who had fallen asleep, and they led them. (40) And I looked till the sheep came to a goodly place, and a pleasant and glorious land, and I looked till those sheep were satisfied; and the tabernacle was in the midst of them in the pleasant land.

From the Period of the Judges to the Building of the Temple

CHAPTER 89.41-50

(41) And sometimes their eyes were opened, and sometimes blinded, till another sheep arose and led them and brought them all back, and their eyes were opened. (42) And the dogs and the foxes and the wild boars began to devour those sheep till the Lord of the sheep raised up another sheep, a ram from them, to lead them. (43) And that ram began to butt with his horns on this side and on that those dogs, foxes and wild boars till he had destroyed them all. (44) And the sheep whose eyes were opened saw that that ram (Saul), who was in the midst of the sheep, *had abandoned his lead*, and had begun to butt those sheep and trampled upon them, and began to walk in a path that was not straight. (45) And the Lord of the sheep sent the sheep (Samuel) to another sheep (David) and raised it to become a ram and to lead the sheep instead of that ram which *had abandoned its lead*. (46) And it went to it and spoke to it alone, and raised up that ram and made it the prince and leader of the sheep; but during all these things those dogs oppressed the sheep.(47) And the former ram pursued the latter ram, and that latter ram arose and fled before it; and I looked until those dogs brought down the former ram. (48a) And the second in command (David) to that ram (Saul) arose and led the little sheep. (49) And those sheep grew and multiplied; but all the dogs, and foxes, and wild boars feared, and fled from it, and that ram butted and killed all the wild beasts, and those wild beasts had no longer any power among the sheep and robbed them no more of anything. (48b) And that ram begat many sheep and fell asleep; and a little sheep became a ram in its stead, and became prince and leader of those sheep. (50) And a house great and broad was built for those sheep; (and) a tower

lofty and great was built on the house for the Lord of the sheep, and that house was low, but the tower was elevated and was lofty; and the Lord of the sheep stood on that tower and they offered a full table before him.

The Kingdoms of Israel and Judah to the Destruction of Jerusalem

Chapter 89.51-67

(51) And still I saw those sheep that again they strayed and went many ways, and forsook their house, and the Lord of the sheep called some from amongst the sheep and sent them to the sheep, but the sheep began to slay them. (52) And one of them was saved and not slain, and it fled and cried aloud over the sheep; and they sought to slay it, but the Lord of the sheep saved it from the hands of the sheep, and brought it up and made it to dwell with me. (53) And many other sheep he sent to those sheep to testify to them and lament over them (54) And after that I saw, when they forsook the house of the Lord of the sheep and his tower, they fell away entirely, and their eyes were blinded; and I saw the Lord of the sheep how he wrought much slaughter upon them in their pastures until those sheep invited that slaughter and betrayed his place. (55) And he gave them over into the hands of the lions and leopards, and wolves and hyenas, and into the hand of the foxes, and to all the wild beasts, and those wild beasts began to tear in pieces those sheep. (56) And I saw that he forsook their house and their tower and gave them all into the hands of the lions, to tear and devour them, into the hands of all the wild beasts. (57) And I began to cry aloud with all my power, and to call to the Lord of the sheep, and to reveal to him, concerning the sheep that they were being devoured by all the wild animals. (58) But he remained unmoved, though he saw it, and rejoiced that they were devoured and swallowed up and carried off, and gave them up into the hands of all the wild beasts for food. (59) And he called seventy shepherds, and drove off those sheep to them that they might pasture them, and he spoke to the shepherds and their servants: 'Let each of you henceforward pasture the sheep one by one, and everything that I command you, do. (60) And I will deliver them up to you exactly numbered, and tell you which of them are to be destroyed—and these you are to destroy. And he handed over to them those sheep. (61) And he called [a watcher, one of the seven white ones], and said to him: 'Observe and mark everything that the shepherds will do to those sheep; for they will destroy from among them more than I have commanded them. (62) And all the excess and the destruction which shall be wrought by the shepherds, record, (namely) how many they destroy according to my command,and how many they destroy of their own volition: and all the destruction of every individual shepherd record with regard to

them (63) And read it out before me by exact number how many they destroy, and how many were delivered over to them for destruction, that I may have this as a testimony against them, that I may know all the deeds of the shepherds, to reckon them up and see what they do, whether or not they abide by my command which I have commanded them. (64) But they must not know it, and you shall not declare it to them, nor reprove them, but only record against each individual all the destruction of the shepherds, of each one by one in his time, and bring it all up before me.' (65) And I looked until when those shepherds pastured in their season, and they began to slay and to destroy more than they were bidden, and they abandoned those sheep into the hands of the lions. (66) And the lions and the leopards ate and devoured the greater part of those sheep, and the wild boars ate along with them; and they burnt that tower and demolished that house. (67) And I became exceedingly grieved on account of the tower, and because that house of the sheep had been demolished, and thereafter I was unable to see whether those sheep entered that house.

First Period of the Shepherds—from the Destruction of Jerusalem to the Return from the Exile(?)

CHAPTER 89.68-71

(68) And the shepherds and their servants delivered over those sheep to all the wild beasts, to devour them, and each one of them received in his time a definite number: and as for each of them, [that white watcher] shall write down in a book how many of them he destroys. (69) And each one slew and destroyed many more than was decreed for them: and I began to weep and lament very much on account of those sheep. (70) And thus in the vision I saw that one who wrote, how he wrote down every one that was destroyed by those shepherds, day by day, and carried up and laid down and showed that whole book to the Lord of the sheep—(even) everything that they had done, and all that each one of them had made away with, and all that they had given over to destruction. (71) And the book was read before the Lord of the sheep, and he took the book from his hand ⌜and read it⌝ and sealed it and laid it (safely) away.

Second Period: from the Time of Cyrus to Alexander the Great

CHAPTER 89.72-77

(72) And thereafter I saw while the shepherds pastured for twelve periods; and behold, three of those sheep returned and came and entered and began to

build up all that had fallen down of that house: but the wild boars hindered them, so that they could not. (73) And they began again to build as before, and they raised up that tower, and it was named the high tower; and they began again to place a table before the tower, but all the bread on it was polluted and not pure. (74) And above all the eyes of those sheep were blinded so that they saw not, and (the eyes of) their shepherds likewise; and they were delivered up in excessive numbers to their shepherds for destruction; and they trampled the sheep with their feet and devoured them. (75) And the Lord of the sheep remained unmoved till they became scattered among all the wild sheep; and they had intercourse with them; and they (the shepherds) did not save them out of the hands of the beasts. (76) And the one who wrote the book carried it up, and showed it and read out (their) presumptuous deeds to the Lord of the sheep, and implored him on their account, and besought him as he showed him all the doings of their shepherds, and gave testimony before him against all the shepherds. (77) And he took the book and laid it down beside him and departed.

Third Period: from Alexander the Great to the Seleucids

Chapter 90.1-5

(1) And I saw till the time that in this manner thirty-five shepherds undertook the pasturing (of the sheep), and they all completed, each individually, their period as did the first; and others received them into their hands to pasture for their period, each shepherd in his own period. (2) And after that I saw in my vision all the birds of heaven coming, the eagles, the vultures, the kites, the ravens; but the eagles led all the birds; and they began to devour those sheep, and to peck out their eyes and to devour their flesh. (3) And the sheep cried out because their flesh was being devoured by the birds, and as for me I cried and lamented in my dream over that shepherd who pastured the sheep. (4) And I looked until those sheep were devoured by the ⌜dogs and⌝ eagles and kites, and they left neither flesh nor skin nor sinew at all on them till only their bones stood there: and their bones too fell to the earth and the sheep became few. (5) And I looked until the time that twenty three shepherds had undertaken the pasturing and had completed in their several periods fifty-eight periods.

Fourth Period: the Maccabean Revolt

Chapter 90.6-19

(6) And behold, a *few* rams were born to those white sheep and they began to open their eyes to see, and to cry to the sheep; (7) and they importuned

them, but they did not hearken to what they said to them, but were exceedingly deaf, and their eyes were exceedingly and terribly blinded. (8) And I saw in the vision how the ravens swooped down on the rams and seized *the leader* of the rams, and dashed the sheep in pieces and devoured them. (9) And I looked until horns grew upon those rams, but the ravens broke their horns; and I saw till there sprouted a great horn of one of the sheep, and their eyes were opened. (10) And it had regard to them, and their eyes were opened, and it cried to the sheep, and the rams saw it and all ran to it. (11) And what is more, those eagles and vultures and ravens and kites still kept tearing the sheep and swooping down upon them to devour them; and the sheep *suffered*, and the rams lamented and cried out.

(12) And those ravens strove and fought with it and sought to lay low its horn, but they did not prevail over it. (13) And I watched them till the shepherds and eagles and those vultures and kites came; and they cried to the ravens that they should break the horn of that ram, and they strove and fought with it, and it strove with them and cried that its help might come. (14) And I looked till that man who wrote down the names of the shepherds and carried (them) up into the presence of the Lord of the sheep, came, and he helped it and saved it, and *he cared for it* in every way: he had come down for the help of that ram. (15) And I saw till the Lord of the sheep came to them in wrath, and all who saw him fled, and they all fell down *in darkness* before his face. (16) All the eagles and vultures and ravens and kites were gathered together and brought with them all the wild sheep; they all came together and helped each other to break that horn of the ram. (17) And I saw that man who wrote the book according to the command of the Lord, till he opened that book concerning the destruction which those twelve last shepherds had wrought, and he showed before the Lord of the sheep that they had destroyed much more than their predecessors. (18) And I looked till the Lord of the sheep came to them, and took in his hand the staff of his wrath; and he smote the earth, and the earth burst asunder; and all the beasts and all the birds of heaven fell away from those sheep and were submerged in the (cleft) earth, and it (the earth) covered them. (19) And I saw till a great sword was given to the sheep, and the sheep proceeded against all the beasts of the field to slay them, and all the beasts and the birds of heaven fled before their face.

The Judgement, the New Jerusalem and the New Eden

CHAPTER 90.20-42

(20) And I saw till a throne was erected in the pleasant land, and the Lord of the sheep sat thereon, and *the white one* took the sealed books and opened those books before the Lord of the sheep. (21) And the Lord called those

men, the seven chief white ones, and commanded that they should bring before him, ⌜beginning with the first star that led the way⌝, *all* the stars whose genitals were like the genitals of horses, and the first star which fell first, and they brought them all before him. (22) And he said to that man who wrote before him, being one of those seven white ones, ⌜and said to him⌝: 'Seize those seventy shepherds to whom I delivered the sheep, and who took them (and) by themselves slew more than I commanded them'. (23) And behold they were all bound, I saw, and they all stood before him. (24) And the judgement was first held over the stars, and they were judged and found guilty, and went to the place of condemnation, and they were cast into an abyss, full of flaming fire, and full of pillars of fire. (25) And those seventy shepherds were judged and found guilty, and they were cast into that fiery abyss. (26) And I saw at that time how a like abyss was opened in the midst of the earth, full of fire, and they brought those blinded sheep, and they were all judged and found guilty and cast into this fiery abyss, and they burned; now this abyss was on the south side of that house. (27) And I saw those sheep burning, and their bones burning. (28) And I stood up to see till the old house was removed; and all the columns were brought out, and all the pillars and ornaments of the house were at the same time wrapped up along with it, and it was taken out and put in a place in the south of the land. (29) And I looked till the Lord of the sheep brought a new house greater and loftier than that first and raised it up in place of the first which had been removed: all its columns were new, and its ornaments were new and larger than those of the first, the old one which he had taken away; and the Lord of the sheep was in the midst of it. (30) And I saw all the sheep which had been left, and all the birds of heaven, falling down and doing homage to those sheep and making petition to them and obeying them in everything. (31) And thereafter those three who were clothed in white and had seized me by my hand, who had taken me up before, and the hand of that ram also seizing hold of me, led me in and set me down in the midst of those sheep *who were without condemnation* (32) And those sheep were all white, and their wool was abundant and clean. (33) And all that had been destroyed and dispersed, and all the beasts of the field, and all the birds of heaven, assembled in that house, and the Lord of the sheep rejoiced with great joy because they were all good and had returned to his house. (34) And I looked till they laid down that sword, which had been given to the sheep, and they brought it back into the house, and it was sealed before the presence of the Lord, and all the sheep were invited into that house, but it held them not. (35) And the eyes of them all were opened, and they saw well, and there was not one among them that did not see. (36) And I saw that that house was large and broad and very full. (37) And I saw that a white bull was born, with large horns, and all the beasts of the field and all the birds of the air feared him and made petition to him all the time. (38) And I saw till all their species were tranformed, and they all became white bulls; and the first among them became a buffalo, and that buffalo became a great animal with great black horns on its head; and the Lord of the sheep rejoiced over them and over all the oxen. (39) And I slept in their midst: and I awoke and saw everything. (40) This is the vision which I saw while I slept, and I awoke and blessed the Lord of righteousness and gave him glory. (41) Then I wept with a great weeping, and my tears stayed not till I could no longer endure it: when I saw, they flowed on account of what I had seen; for everything shall come and be fulfilled, and all the deeds of men, each according to his destiny, were shown to me. (42) On that night I remembered the first dream, and because of it I wept and was troubled—because I had seen that vision'.

CHAPTER 91.1-10, 18-19

ENOCH ADDRESSES METHUSELAH AND HIS FAMILY

(1) And now, my son Methuselah, call to me all your brothers
And gather together to me all the children of your mother,
For a voice calls me and a spirit is poured out upon me,
That I may show you everything that shall befall you for ever.

(2) And thereupon Methuselah went and summoned to him all his brothers and assembled his relatives. (3) And he (Enoch) spoke to all the children of righteousness and said:

Hear, sons of Enoch, all the words of your father,
And hearken aright to the voice of my mouth;
For I admonish you and say unto you, beloved ones,
Love rectitude and walk therein.

(4) And draw not nigh to rectitude with a double heart,
And associate not with those of a double heart,
But walk in righteousness, my children,
And it shall guide you in good paths,
And righteousness shall be your companion.

(5) For I know that wrong-doing is bound to become strong on the earth,
And a great chastisement is to be accomplished on the earth,
And all oppression is to come to an end;
It shall be up-rooted from its foundations,
And its whole structure shall disappear.

(6) Oppression shall again reach its peak upon the earth,
And all work(er)s of oppression and wrong-doing and wickedness *shall possess it all*, in double measure.

(7) And when oppression and sin and blasphemy and wrong-doing in all manner of deeds increase,
And apostasy and wickedness and uncleanness increase,
A great chastisement shall come from heaven upon all these,
And the holy Lord will come forth with wrath and chastisement,
To execute judgement on the earth.
(8) In those days wrong-doing shall be up-rooted from its foundations,
And the foundations of oppression together with deceit,
And they shall be destroyed from under heaven.
(9) And everything shall be given over, the idols of the heathen and (their) towers, to the blazing fire;
And they shall remove them from the whole earth;
They (the heathen or their idols) shall be cast into the judgement of fire,
And shall perish in wrath and in a grievous judgement that is for ever.
(10) And the righteous shall arise from their sleep,
And wisdom shall arise and be given unto them.

For 91.11-17 see below, 87

(18) And now I tell you, my children, and I show you the paths of righteousness and the paths of wrong-doing;
I will show them to you again,
That you may know what will come to pass.
(19) And now, hearken unto me, my children,
And walk in the paths of righteousness,
And walk not in the paths of wrong-doing;
For all who walk in the paths of oppression shall perish everlastingly.

ENOCH'S EPISTLE (CHH. 92,93.1-2)

CHAPTER 92

(1) [The Epistle of Enoch which] he wrote and gave to his son Methuselah. Enoch, skilled scribe and wisest of men, and *the chosen of the sons of men and judge of all the earth*, to all my children and to later generations, to all dwellers on earth who observe uprightness and peace.
(2) Be not grieved in your spirit on account of the times,
For the great Holy One has appointed times for all things.
(3) And the righteous shall awake from sleep,
He shall arise and proceed in the ways of righteousness,
And all his paths and conversation shall be in eternal goodness and grace.
(4) And he (the great Holy One) will be gracious to the righteous and give him eternal uprightness,

And he will give him power so that he shall execute judgement with goodness and righteousness,
And he shall walk in eternal light.
(5) And sin shall perish in darkness for ever,
And shall no more then appear from that day for evermore.

CHAPTER 93.1-2

(1) And after he had given over his Epistle (to Methuselah),
Enoch took up his discourse, saying:
(2) Concerning the children of righteousness and the eternal elect
Sprung from the plant of righteousness and uprightness,
These things will I recount and make known to you, my children:
I, Enoch, was shown in a heavenly vision,
And from the words of the watchers and holy ones I came to know everything,
And from *the tablets* of heaven I read and understood everything.

CHAPTER 93.11-14

A NATURE POEM

(11) For who is there of all the sons of men who is able to know what are [the ordinances of heaven]; or to who is there of all the sons of men who is able to hear the voice of the Holy One without being troubled; and who can think his thoughts, or what man is there who can behold all the works of heaven, (12) [or show forth] the glory which [they are declaring]; or who can see his breath or spirit and be able to return and tell about it; or who can ascend and discern all their ends (i.e. of the heavens) or devise or create things like them? (13) Or who is there of all the sons of men who can know what is the length and breadth of the whole earth; or who is there to whom has been shown all its extent and its shape? (14) And is there any man who can discern the length of the heavens and what is their height and how they are supported, and how great is the number of the stars and where all the luminaries come to rest?

CHAPTERS 93.3-10, 91.11-17

THE APOCALYPSE OF WEEKS

(93.3) And then Enoch took up his discourse and said:
I was born the seventh in the First Week;
And till my time justice was delayed.
(4) And thereafter there shall arise the Second Week

In which falsehood and violence shall spring up;
And in it will be the former End; and in it a man shall be saved.
And after it is ended, oppression shall increase,
And he (Noah) shall make a law for sinners.
(5) And thereafter, in the Third Week, at its close,
A man shall be chosen as a plant of righteous judgement;
And *his posterity* shall come forth as a plant of eternal righteousness;
(6) And thereafter, in the Fourth Week, at its close,
A vision of holy ones and of righteousness shall be revealed,
And a law for generations upon generations, and the court (of the tabernacle) be made for them.
(7) And thereafter, in the Fifth Week, at its close,
The House of glory and dominion shall be built for ever.
(8) And thereafter, in the Sixth Week, all in it will become blind,
And the hearts of all of them shall godlessly forsake wisdom.
And in it a man shall ascend; and at its close,
The House of dominion shall be burnt with fire,
And in it the whole people (and) the captains of the host shall be dispersed.
(9) And thereafter, in the Seventh Week, a perverse generation shall arise,
And many shall be its misdeeds and all its doings shall be apostate.
(10) And at its close the elect shall be chosen,
As witnesses to righteousness, from the eternal plant of righteousness,
To whom shall be given seven-fold wisdom and knowledge ⌜concerning all his creation⌝.
(91.11) And they will uproot the foundations of oppression,
And the structure of falsehood therein to destroy it *utterly*.
(12) And thereafter there shall arise the Eighth Week of righteousness,
In which a sword will be given to all the righteous,
To execute a righteous judgement on all the wicked,
And they will be given over into their hands.
(13) And at its close they shall acquire possessions righteously,
And there shall be built the royal House of the Great One, in splendour,
For all generations for ever.
(14) And thereafter, the Ninth Week will arise
In which a righteous judgement will be revealed
For all the children of the whole earth;
And all work(er)s of iniquity shall vanish from all of the whole earth,
And they will be cast into the eternal pit,
And all men shall look to the true, eternal path.

(15) And thereafter will arise the Tenth Week,
In the seventh part of which an everlasting judgement and the (decreed) time of the great judgement
Will be exacted from all the watchers of heaven.
(16) And in it the first heaven shall pass away,
And a new heaven shall appear,
And all the powers of heaven will shine and rise for ever and ever, with seven-fold light.
(17) And thereafter there shall be many Weeks; to all their number there shall be no end for ever,
In which they shall practise goodness and righteousness;
And sin shall be no more *seen* for ever.
(17b(?) And the righteous shall awake from their sleep,
And they shall arise and walk in the paths of righteousness;
And unrighteousness shall altogether cease,
And the earth will be at rest from oppression, for all generations for ever.

ΕΠΙΣΤΟΛΗ ΕΝΩΧ (CHH. 94-108)

CHAPTER 94

Admonitions to the Righteous and Woes to the Sinners (cf. Isa. 5.8f.)

(1) And now I say unto you, my sons, love righteousness and walk therein;
For the paths of righteousness are worthy of acceptance,
But the paths of unrighteousness shall suddenly be destroyed and fail.
(2) And to illustrious men of a generation to come shall the paths of wrong-doing and death be revealed,
And they shall hold themselves afar from them,
And shall not follow them.
(3) And now I say unto you, the righteous:
Walk not in the paths of wickedness, nor in the paths of death,
And draw not nigh to them lest you be destroyed.
(4) But seek and choose for yourselves righteousness, and a life of goodness,
And walk in the paths of peace,
That you may live and prosper.
(5) And hold fast to my precepts in the meditation of your heart,
And let not my words be effaced from your hearts;
For I know that sinners will tempt men to entreat wisdom evilly,
And no place will be found for her,
And trials will in no way decrease.

(6) Woe to those who build up unrighteousness and wrong-doing,
And lay deceit as a foundation;
For they shall be suddenly overthrown,
And they shall have no peace.
(7) Woe to those who build their houses with sin;
For from all their foundations shall they be overthrown,
And by the sword shall they fall.
And those who acquire gold and silver, in judgement they shall suddenly perish.
(8) Woe to you, you rich, for you have trusted in your riches,
And from your riches you will be parted,
Because you have not remembered the Most High in the days of your riches.
(9) You have committed blasphemy and unrighteousness,
And have become ready for the day of slaughter,
And for the day of darkness and for the day of the great judgement.
(10) Thus I speak and declare unto you:
He who hath created you will overthrow you,
And for your fall there will be no compassion,
And your Creator will rejoice in your destruction.
(11) And your righteous ones in those days shall be
An object of contempt for the sinners and the godless.

CHAPTER 95

Enoch's Grief: Fresh Woes against the Sinners

(1) Oh that mine eyes were *a fountain of waters*
That I might weep over you,
And pour out my tears ⌜as a *fountain of waters*,⌝
That so I might rest from the trouble in my heart.
(2) May hatred and wickedness be yours,
That judgement may come upon you, sinners.
(3) Fear not the sinners, you righteous;
For again the Lord will deliver them into you hands,
That you may execute judgement upon them according to your will.
(4) Woe to you who pronounce anathemas,
That they may be loosed:
Remedies will be lacking to you because of your sin.
(5) Woe to you who requite your neighbour with evil;
For you shall be requited according to your works.
(6) Woe to you, lying witnesses,

And to those who weigh out injustice,
For suddenly you shall perish.
(7) Woe to you, sinners, for you persecute the righteous;
For you (the righteous) will be delivered up ⸢and persecuted⸣ unjustly,
And their yoke will lie heavily upon you.

CHAPTER 96

Grounds for Hope for the Righteous: Woes for the Wicked

(1) Be hopeful, you righteous; for the sinners shall suddenly perish before you.
And you shall have power over them as you will.
(2) And in the day of tribulation of the sinners,
Your young ones shall mount and rise as eagles,
And higher than that of vultures shall be your nest,
And you shall ascend and enter the crevices of the earth,
And the clefts of the rock, all times, like rock-badgers, from before the lawless;
And they shall groan because of you and weep like desert-owls.
(3) But as for you, fear not, you that have suffered;
For healing shall be yours,
And a bright light shall shine on you,
And a voice of rest you shall hear from heaven.
(4) Woe unto you, you sinners, for your riches make you appear righteous,
But your hearts convict you of being sinners,
And this word shall be a testimony against you for a memorial of (your) evil deeds.
(5) Woe to you who eat the finest of the wheat,
And drink *new wine, the choicest of the wine*,
And tread underfoot the poor in your might.
(6) Woe to you who *drink waters that at all times fail*,
For suddenly retribution will be required of you,
And you will be exhausted and wither away,
Because you have forsaken the fountain of life.
(7) Woe to you who work unrighteousness and deceit and blasphemy,
It shall be a memorial against you for evil.
(8) Woe to you, you mighty, who with might oppress the righteous;
For the day of your destruction is coming.
In those days many and good days shall come to the righteous—at the day of your judgement.

CHAPTER 97

Evils in Store for Sinners and Possessors of Unrighteous Wealth

(1) Believe, you righteous, that the sinners will become an object of reproach
And perish in the day of violence.
(2) Be it known unto you (sinners) that the Most High is mindful of your destruction,
And that the angels of heaven will rejoice over your destruction.
(3) What will you do, you sinners,
And whither will you flee on that day of judgement,
When you hear the voice of the prayer of the righteous.
(4) And as for you, you shall fare like unto them,
(You) against whom this word will be a testimony:
'You have been associated with the sinners'.
(5) And in those days the prayers of the righteous shall reach unto the Lord,
And for you the days of your judgement shall come.
(6) And the complete account of your iniquities will be read out before the great Holy One,
And your faces will be covered with shame.
And he will do away with all the deeds *which partook of iniquity*.
(7) Woe to you, sinners, in the midst of the sea or on land—
Their memory of you will be ill for you.
(8) Woe to you who acquire gold and silver (but) not by righteousness, and say:
'We have become exceedingly rich and acquired possessions;
And we have obtained everything we have wished.
(9) And now let us do whatever we desire,
For we have gathered silver, and filled our storehouses,
And many are the goods in our houses'.
(10) But like water they will be poured out;
You are deceived, for your riches shall not abide for you,
But they will quickly *be taken away from you*;
For you have acquired it all unjustly,
And you will be delivered up to a great curse.

CHAPTER 98

The Self-Indulgence of the Sinners. Sin Originated by Man: All Sin Recorded in Heaven. Woes for the Sinners

(1) And now I swear unto you, to the wise and to the foolish,

That you will see many iniquities on the earth.
(2) For you men will put on finery more than women,
And bright colours more than maidens,
In royalty and in grandeur and in power.
But silver and gold and purple and priceless things *are destined to pass away*,
And your household goods will be poured away like water.
(3) Because they (you) have neither knowledge nor wisdom,
In this wise you shall perish together, with all your possessions and all your glory and splendour,
And in shame and desolation and great slaughter
Your spirits shall be cast into the furnace of fire.
(4) I swear to you, sinners, that *slavery* was not sent on earth,
But men of themselves created it,
And those who practise it shall become subject to a great curse;
(5b) For it was not ordained that there be slave, male or female,
From above it (slavery) did not proceed (lit., was not given), but arose from oppression.
(5c) So also injustice was not sent (lit., given) from above, but (springs) from transgression,
(5d) Likewise, neither was a woman created sterile,
(5a) And childlessness has not been imparted (by God) to a woman, but (comes) because of the deeds of her hands;
(5e) By her own iniquities she is condemned to childlessness, and childless she will die.
(6) I swear to you, sinners, by the great Holy One,
That all your wicked deeds are revealed in heaven,
And none of your unrighteous acts is covered up and hidden.

(7) And do not imagine to yourselves or think that no sins are known or seen in heaven, and are not written down in the presence of the Most High. (8) From now on you must know that all your iniquities will be written down day by day until the day of your judgement.

(9) Woe to you, fools, for you will perish in your folly; and to the wise you do not look and good will not come to you, but evil will encompass you.

(10) And now know that you have been prepared for the day of destruction: and do not hope to be saved, you sinners; rather you shall depart and die; for you are unaware, although you are prepared for the day of the great judgement, for the day of tribulation and great distress for your spirits.

(11) Woe to you, obstinate of heart who work wickedness and eat blood. Whence is it that you have good things to eat and drink and are satisfied? Surely it is from all the good things which the Lord the Most High has placed in abundance on the earth—you shall have no peace. (12) Woe to you who

love the deeds of unrighteousness; wherefore do you hope for prosperity for yourselves? Know that you shall be delivered into the hands of the righteous and they shall behead and slay you, and will not spare you.

(13) Woe to you who rejoice in the tribulation of the righteous; for no grave shall be dug for you. (14) Woe to you who seek to set at nought the words of the righteous; for you shall have no hope of (eternal) life. (15) Woe to you who write lying words and words of heresy; they write their lies so that men apostatize and cause others to fall away; (16) you yourselves fall away and you shall not have peace, but, of a sudden, you shall die.

CHAPTER 99

Woes for the Godless, the Law-Breakers: Plight of Sinners in the Last Days: Further Woes

(1) Woe to you who cause apostasies,
And by your deceitful works receive honour and glory.
You shall surely perish; there will be no good life for you.
(2) Woe to you who alter the words of truth,
And pervert the everlasting law,
And count themselves to be without sin;
They shall be swallowed up in the earth.
(3) At that time you should prepare, you righteous, to present your petitions as a memorial,
And to place them as a testimony before the angels,
That they may bring the sins of the wicked before the Most High to remind him (of them).
(4) At that time the peoples will be thrown into confusion,
And the kindreds of the peoples will rise in revolt on the day of destruction of wickedness.
(5) At that time women bearing children will miscarry,
And they will expose and abandon their infant babes:
And pregnant women will abort and those who are giving suck will cast away their children.
They will not return to their children nor to their sucklings,
Nor will they spare their loved ones.
(6) Again I swear to you, sinners,
That (your) sin is destined for the day of unceasing bloodshed.
(7) And for those who worship stones and carve images of silver—and gold and wood, stone and clay, and serve phantoms, demons, abominations, evil spirits and all (kinds of) idols, not according to knowledge—no help will be obtained from them.
(8) They will fall into apostasy by reason of the folly of their hearts,

And their eyes will be blinded because of the fear in their hearts;
And by the visions of their dreams (9) —by them they will go astray and become fearful.
For all their works (idols?) in vain they have fashioned (served?) and have worshipped stones,
In a trice they will perish.
(10) And at that time blessed are all who receive the words of the wise and learn them,
To observe the commandments of the Most High and walk in the paths of his righteousness,
and do not apostatise with apostates; for they will be saved.
(11) Woe to you who plot evil against your neighbour;
For you shall be slain in Sheol.
(12) Woe to you who use false and deceitful weights
And who *prosper* upon the earth;
For because of this an end will be made of them (you).
(13) Woe to you who build your houses, (but) not by your own labour,
And who construct every building with stones and bricks of iniquity;
Woe to you, you shall have no peace.
(14) Woe to you who reject the foundations and the eternal heritage of their fathers,
And whose spirits will go in pursuit of apostasy—
You shall have no rest.
(15 Woe to you who work unrighteousness and assist wrong-doing
And slay their neighbours, until the day of the great judgement.
(16) For then he shall overthrow your glory, and put evil into your hearts;
And he will arouse his fierce anger to destroy you all with the sword.
And all the righteous will remember your wrong-doing.

CHAPTER 100

The Sinners Destroy each other: The Coming Judgement: Further Woes for Sinners

(1) And at that time, in one and the same place, fathers shall attack their children,
And brother with one another shall fall in (mutual) destruction,
Until their blood flows like a river.
(2) For a man shall not stay his hand from his son(s) nor from his son's sons,
Nor from his beloved one to slay him;
Nor the sinner from his dear one nor from his brother;

From dawn until sunset they shall slay one another together.
(3) And the horse shall walk up to its breast in the blood of sinners,
And the chariot shall be submerged to its axles.
(4) And angels shall descend into (their) hiding-places on that day,
And will gather together into the one place all who were aiding and abetting wrong-doing;
And the Most High will arise on that day,
To exact a great judgement from all the sinners.
(5) And he will set a guard from the holy angels over all the righteous and holy:
They will guard them as the apple of an eye,
Until there is an end to all wickedness and to all sin.
And thereafter the pious will sleep a pleasant sleep,
And there will no longer be one to terrify them.
(6) Then the wise among men shall perceive,
And the children of earth shall understand all the words of this book,
And they shall recognise that their riches will not avail to save them,
In the downfall of wrong-doing.
(7) Woe to you, sinners, when you afflict the righteous on the day of severe tribulation,
And burn them up with fire,
For you will be requited according to your deeds.
(8) Woe to you, you obstinate of heart,
Who watch in order to devise wickedness:
Fear shall lay hold on you,
And there shall be none to help you.
(9) Woe to all you, sinners, on account of the words of your mouth,
And on account of the works of your hands;
For the service (works) *of the Holy One* you have forsaken:
You shall burn in blazing fire.

(10) And now know that from the angels he will enquire into your deeds, from heaven and from the sun and from the moon and from the stars, with regard to your sins, because, upon the earth, you executed judgement towards the righteous. (11) And he will summon to testify against you all the clouds and mist and dew and rain; for they shall all be withheld from you so as not to fall upon you, and they shall be watchful over your sins. (12) Now, therefore, give gifts to the rain that it be not withheld from falling on you, and to the dew and cloud and mist, pay gold and silver that they may come down (upon you). (13) Whenever snow and hoar-frost and its cold, and the winds and their frost, and all their afflictions drive down upon you, you will be unable to stand up to their afflictions.

CHAPTER 101

Exhortation to the Fear of God: All Nature Fears Him, but not the Sinners

(1) Contemplate, therefore, children of men, the works of the Most High, and fear to do evil before him. (2) If he closes the windows of heaven, and withholds the rain and the dew from descending on the earth because of you, what then will you do? (3) And if he sends his anger upon you because of all your deeds, will you not be the ones who plead with him? (But) because you uttered with your months proud and harsh words against his majesty, you will have no peace. (4) Behold the sea-captains who sail upon the sea, how their ships are tossed about by the waves and are shaken by the storms. (5) They are in distress, and on this account they are afraid because all their goods and property they have shipped abroad by sea with them, and they are anxious in mind that the sea will swallow them up and they will perish in it. (6) Are not the entire sea and all its waters and all its turbulence the work of the Most High, and he has determined their bounds and confined it and fenced it round with sand. (7) And at his rebuke it becomes afraid and dries up, and the fish die and everything in it. But you, sinners on the earth, do not fear him. (8) Did he not create heaven and earth and everything that is in them? And who gave skill and wisdom to all who travel upon the seas? (9) Do not the sea-captains fear the seas? Yet sinners fear not the Most High.

CHAPTER 102

Terrors of the Day of Judgement: The Misfortunes of the Righteous

(1) And at that time, if he should hurl against you a wave (tempest?) of blazing fire, where will you flee and be saved? And when he raises his voice, with a mighty sound, against you, (2) will you not be shaken and afraid?

(3a) And the angels shall execute what has been commanded them,
(3b) And heaven and all the (heavenly) luminaries shall shake and tremble in great alarm,
(3c) And shall seek to hide themselves before the presence of the glory of the Great One;
(3d) And the whole earth shall shake and tremble and be disturbed,
(3e) And the children of earth shall tremble and quail.
As for you, sinners, accursed shall you be for ever, and you shall have no peace.
(4) Fear not, souls of the righteous who have died,

(5) And grieve not because your souls have gone down to Sheol in tribulation,
And that, in your lives, your body of flesh did not obtain (a reward) according to your piety;
For the days of your lives were the days of sinners and of the accursed on the earth;
But *wait patiently* for the day on which is the judgement of sinners, and for the day of cursing and punishment.
(6) And when you die, then sinners will say about you:
'The righteous die like us, and what did they gain by their (good) works?
(7) Behold, like us, they die in grief and darkness,
And what advantage do they have over us?
(8) Hereafter shall they be saved and rise from the dead ⌜and see (the light) for ever⌝?
Truly, they too, behold! have died, and thereafter shall not see the light for ever.
(9) Therefore, it is well for us to eat and drink and rob and do wrong, to plunder and to acquire wealth and see good days.
(10) Consider then those who are truly righteous,
In what manner their end falls out;
For no unrighteousness has been found in them till they die.
(11) But they perished and became as though they had not been;
And their souls descended into Sheol in tribulation.'

CHAPTER 103

ASSURANCES FOR THE RIGHTEOUS

(1) And now I swear to you, the righteous, by the glory of the Great One and by the honour of his sovereignty, and by his magnificence I swear to you—

(2) That I understand the following mystery,
For I have read the tablets of heaven;
And I have seen the book(s) of the holy ones,
And I have found inscribed and written therein concerning you (i.e. the righteous):
(3) That all good things and joy and honour are prepared
And written down for the spirits of the righteous dead,
And manifold good shall be given to you in recompense for your labours,
And your lot will be superior to the lot of the living.
(4) And the spirits of you righteous who have died will live and rejoice and be glad,

And their spirit shall not perish, nor their memorial from before the face of the Great One
Unto all the generations of the ages: wherefore no longer fear their reproaches.
(5) Woe to you, deceased sinners,
When you die with your ill-gotten wealth, then your associates will say about you:
'Happy are the sinners, all their days they enjoyed prosperity,
(6) And now they have died in prosperity and wealth,
And they have not seen tribulation or murder in their lives;
And they have died in splendour, and judgement has not been executed on them during their lives'.
(7) Know that their souls will be made to descend into Sheol,
(8) And they shall be afflicted in great tribulation, and in darkness and in the toils of death and in a blazing fire.
And to the great judgement their souls will come,
And the great judgement shall be for all generations for ever.
Woe to you, for you shall have no peace.
(9) Were you not saying, you righteous who were pious in your life-time?
'In the days of our tribulation we have toiled laboriously and experienced every trouble,
And have met with much evil and been consumed,
And have become few and lost heart.
(10) And we perish and there is no one to help us by word or deed;
We are slain, and no protector of any kind have we found,
We are ground down ⌈and perish⌉, and have no expectation of survival from one day to the next.
(11) We hoped to become the head and have become the tail:
We have toiled laboriously, but were not masters of the fruits of our toil,
And we have become the food for sinners,
And lawless men have laid their yoke heavily upon us.
(12) They have had dominion over us that hated us,
And to those that goaded us and beheaded us, we have bowed our necks,
But they have shown us no mercy.
(13) We sought to depart from them that we might escape and be at rest,
But we found no place where we might flee and be safe from them.
(14) And we made complaint to the rulers in our distress,
And we cried out against those who slandered and oppressed us,
But they did not receive our petitions
And would not hearken to our voice.

(15) And they (the rulers) did not help us, finding no fault with those who oppressed and devoured us:
But they supported them against us, and assisted those who oppressed and devoured us and made us few;
And they concealed their wrong-doing,
And did not remove from us the yoke of those who devoured us:
And they scattered us and murdered us and made us few;
And they were not informed about our murdered ones (murderers?) and concealed our murder;
Nor were they reminded of the sins of sinners,
That they had lifted up their hands against us'.

CHAPTER 104

Further Assurances for the Righteous: Admonitions of the Sinners

(1) I swear unto you, that in heaven the angels remember you for good before the glory of the Great One, and your names are written before the glory of the Great One. (2) Be of good courage, for aforetime you were worn down by evils and afflictions, but now you shall shine and appear as the lights of heaven, and the portals of heaven shall be opened unto you. (3) And your cry will be heard, and your judgement for which you cry will also appear; for vengeance shall be required from the rulers for all your affliction, and from all who have helped those who plundered you. (4) Be of good courage, and do not abandon your hope; for you shall have great joy as the angels of heaven. (5) Are you about to commit iniquity, are you about to hide yourselves on the day of the great judgement? Are you to be found out to be like sinners, and is the everlasting judgement to be required of you for all generations for ever? (6) But now fear not, you righteous, when you see the sinners growing strong and prospering: be not companions with them, but keep afar from all their evil-doings; for you shall become companions of the angels of heaven. (7) And if you sinners say: 'None of our sins will be investigated or written down'—every one of your sins will be written down day by day. (8) And now I declare unto you that light and darkness, day and night, behold all your sins. (9) Be not godless in your hearts, and lie not and pervert not the words of truth, nor charge with lying the words of the great Holy One, nor count on your idols; for all your false gods (lit. lies) and all your godlessness issue not in righteousness but in great sin. (10) And now I know this secret, that sinners will alter and pervert and much distort the words of truth, and will speak wicked words, and lie, and fashion great graven images, and write books according to their words (i.e. of the 'graven

images') and in their names. (11) Would that they would write truthfully all my words in their languages, and neither alter nor omit aught from my words but write them all down truthfully—all that I first testified against them. (12) And again another secret I know, that my books shall be given to the righteous and the pious and the wise to become a cause of joy and uprightness and much wisdom. (13) And to them shall the books be given, and they shall believe in them and rejoice over them, and all the righteous shall be recompensed, who have learned from them all the paths of righteousness.

CHAPTER 105

THE COMMISSION OF ENOCH TO METHUSELAH AND HIS FAMILY

(1) Then—⌜it is a word of the Lord⌝—they (the righteous) shall summon and testify to the children of⌞earth by their wisdom. Explain to them, because you are their teachers and *leaders* over all the earth; (2) for I and my son (Methuselah) will be united with them for ever in the paths of uprightness in their lives; and you shall have peace. Rejoice you children of uprightness. Amen.

CHAPTER 106

THE BIRTH OF NOAH (CHH. 106-107)

(1) After a time I took a wife for Methuselah my son, and she bore a son and called his name Lamech (saying) 'Brought low has righteousness been to this day'. And when he came to maturity, (Methuselah) took for him a wife and she bore him a child. (2) And when the child was born, his body was whiter than snow and redder than the flower of the rose; the tresses of the hair of his head were all white and like white wool, and thick and glorious; and when he opened his eyes the house shone like the sun. (3) And when he was taken up from the hands of the midwife, he opened his mouth and blessed the Lord of righteousness. (4) And his father Lamech was afraid of him and fled, and came to his father Methuselah and said to him: (5) 'I have begotten a strange son; he is not like mankind, but resembles the children of the angels of heaven; and his type is different and he is not like us, and his eyes are like the rays of the sun, and his countenance is glorious. (6) And it seems to me that he is not sprung from me but from the angels, and I fear lest in his days a wonder will be wrought on the earth. (7) And now, my father, I am petitioning and imploring you that you go to Enoch, our father, and hear from him the truth, for his dwelling-place is

amongst the angels.' (8) And when Methuselah heard the words of his son, he came to me at the ends of the earth where he had heard that I then was, and he said to me: 'My father, hear my voice and come to me'. And I heard his voice and went to him and said: 'Behold, here I am, my son. Why have you come to me?' (9) And he answered me and said: 'Because of a great distress I have come to you, and because of a dreadful vision I have approached you here. (10) And now, my father, hear me, for a child has been born to my son Lamech whose type and likeness is not like the likeness of a man, and his colour is whiter than snow and redder than the flower of the rose, and the hair of his head is whiter than the whitest wool, and his eyes are like the rays of the sun; and he opened his eyes and lighted up the whole house. (11) And he was taken up from between the hands of the midwife, and opened his mouth and blessed the Lord of heaven. (12) And his father Lamech became afraid and fled to me, and he does not believe that he is from him, but thinks him like one of the angels of heaven; and behold, I have come to you that you may make known to me the truth.' (13) And I, Enoch, answered and said to him: 'Truly the Lord *will make a Promise on the earth*; and according as I was shown, (my) son, and informed you, in the generation of my father Jared *exalted ones of heaven* transgressed the word of the Lord and violated the covenant of heaven. (14) And behold, they committed sin and transgressed the law, and they had intercourse with women and committed sin with them and have married some of them, and from them begotten children, (17a) and they bore children on the earth, the giants, not beings like spirits but like creatures of flesh. (15) And there will be great destruction for one year. (16) And this child which has been born to you shall be left on the earth, and his three sons will be saved with him, when all mankind on earth shall die, (17c) and the earth shall rest and be cleansed of great corruption. (17, see above on 14, 15, 16) (18) And now tell your son Lamech that the one who has been born is truly your son; and call his name Noah, for he shall be your 'remnant', forasmuch as he and his son shall have rest and escape from the corruption of the earth and from all the sins and all the injustice which will be carried out on the earth in his days. (19) And after that there shall be still greater unrighteousness than that which was first committed on the earth; for I know the secrets *of heaven*, for the angels (lit. the holy ones) have shown me and informed me, and I have read them in the heavenly tablets'.

CHAPTER 107

(1) 'And I beheld written in them that generation after generation shall wrong them (the descendants of Noah), and wrong shall continue, until there shall arise generations of righteousness; and evil and godlessness shall come

to an end, and injustice cease from off the earth, and all blessings shall come on the earth upon them. (2) And now, go and make known to Lamech, your son, that this boy is truly and without deception his son'. (3) And when Methuselah had heard the words of his father Enoch—for he had disclosed to him the whole mystery—he returned and acquainted him (his son Lamech) with it, and the name of the child was called Noah, for he will rejoice the earth after all the destruction.

CHAPTER 108

Another Writing of Enoch

(1) Another writing which Enoch composed for his son Methuselah and for those who will come after him and keep the law in the last days. (2) You who did good and have waited expectantly for those days till an end is made of those who work evil, and an end of the power of evil-doers— (3) do you, indeed, wait until sin has passed away, for their names shall be blotted out of the Book of Life and out of the books of the holy ones, and their seed shall be destroyed for ever, and their spirits shall be slain, and they shall cry and lament in a place deserted and void, and in the fire shall they burn; for there is no earth there. (4) And I saw there something like a cloud that was not discernible, for because of its depth I was unable to observe (it); and flames of fire I saw burning brilliantly, and, as it were, brightly shining mountains, circling around and turning hither and thither. (5) And I asked one of the holy angels who was with me and said to him: 'What is this brightness, for it is not heaven but only flames of a blazing fire, and the voices of crying and weeping and lamentation and terrible pain'. (6) And he said to me: 'This place which you see—here will be cast the spirits of sinners and blasphemers, and of those who work wickedness, and of those who pervert everything which the Lord has spoken through the mouths of the prophets about everything that is to happen. (7) For some of them are to be written and inscribed above in heaven, in order that the angels may read of them and know that which will happen to sinners, and to the spirits of the humble, and to those who afflict their bodies and will receive their reward from God, and to those who are abused by evil men. (8) (And what shall befall those) who love God and have not loved either gold or silver or any of the good things that are in the world, but gave over their bodies to torture; (9) and who, since they came into being, longed not after earthly food, but regarded themselves as a passing breath, and lived accordingly (lit. observed this); and the Lord tried them much, but their spirits were found pure so that they might bless his Name. (10) And all their blessings I have recounted in the books. And he has assigned them their reward one by one, because they have been found to be such as loved heaven more than their life (lit. breath) in

the world, and, though they were trodden underfoot of wicked men, and experienced abuse and reviling from them and were put to shame, yet they blessed me (the Lord). (11) 'And now I will summon the spirits of the good who belong to the generation of light, and I will transform those who were born in darkness, who in their flesh were not recompensed with such honour as their faithfulness deserved. (12) And I will bring forth in shining light those who have loved my holy Name, and I will seat each on the throne of his honour. (13) And they shall be resplendent for times without number; for righteousness is the judgement of God; for to the faithful he will show faithfulness in the habitations of upright paths. (14) And they shall see those who were born in darkness cast into darkness, while the righteous shall be resplendent. (5) And sinners shall cry aloud and see them resplendent, and they indeed shall go where days and times are prescribed for them.'

COMMENTARY

CHAPTER 1

(1) There may be an allusion to this verse in Pss. Sol. 4.8: δικαιώσαισαν ὅσιοι τὸ κρίμα τοῦ θεοῦ αὐτῶν ἐν τῷ ἐξαίρεσθαι ἁμαρτωλοὺς ἀπὸ προσώπου δικαίου.

The words of blessing G, Eth. have a coll. sing. for the characteristic semitic plur: in this regular formula beginning a book, cf. Jer. 1.1 etc., Ac. 1.1 (Milik, 143).

The opening v. Dt. 33, '... the blessing wherewith Moses the man of God blessed the children of Israel ...' is immediately followed by the Sinai theophany: Enoch similarly follows this first verse, beginning in the same way, with a similar Sinai theophany, partly modelled on Dt. 33.2. The author has clearly Dt. 33 in mind throughout this passage. Should we perhaps take vv. 8-9 as containing 'the word(s) of the blessing of Enoch'?

righteous elect Eth. reads 'righteous and elect' throughout the Parables, 38.3,4; 39.6,7; 48.1; 58.1,2; 61.13; 62.12,13,15; 70.3 ('elect and righteous'). The emphasis is on the noun 'elect (ones)'.

who are ... to destroy Translators render ἔσονται as παρέσονται 'will be present', a sense which puts an even greater strain on Aram. להון. An alternative is to construe ἔσονται with ἐξᾶραι in the idiomatic semitic construction להון לאעדיה, 'will be (destined) to destroy' (Nöldeke, 216). Cf. Eth. 'who must be ... to destroy' (Knibb, 2, 57). For G, Eth. ἀνάγκη (LXX for צרה Job 15.24, Tg. עיוק) of the distress or tribulation of the end-time, cf. Lk. 21.23, 1 C. 7.26, ThWNT Bd. I, s.v. ἀνάγκη, 349f. (Grundmann).

all the godless The shorter text of Eth$^{tana\,u}$ rasi'ān = רשיעין seems original, cf. 7.6 En^{a} 1 iii 22 (G ἀνόμων), 91.12, En^{g} 1 iv 16. Eth. 'wicked and godless' is a translator's expansion. The nouns רשע, רע in Heb. or Aram. can have the connotation 'enemies (of God)', so that G ἐχθρούς is probably an interpretative rendering of רשיעין. Eth. seems, however, to have read a different Greek text from G [ἀσεβεῖς](?). G καὶ σωθήσονται δίκαιοι seems another translator's gloss.

took up his discourse Eth. 'answered and said'. 'This is a poor attempt at rendering the phrase ἀναλαβὼν τὴν παραβολήν.Ethiopic translators, indeed, found this phrase difficult, and never rendered it literally. Cf. Num. xxiii.7, 18; xxiv.3,15,20,21,23.' (Charles, 1906 2 n. 5) Ullendorff ('An Aramaic "Vorlage"', 266), Knibb 2, 57 argue for an Aram. original ענה חנוך ואמר behind Eth.: it is unlikely, however, that Eth. read any text here different from G; the rendering at v. 3 'I uttered a parable' (Knibb) presupposes a text containing the noun παραβολή. Charles's explanation of the Eth. version of this phrase is more convincing. The Aram. phrase occurs again at 93.3, En^{g} 1 iii 23. For משל (מתלא) 'oracular discourse', see H. N. Snaith, *Leviticus and Numbers* (London, 1967), 284.

(2) **[Oracle of Enoch?] a righteous man ... what I saw** This verse is largely modelled on the Balaam prophecy at Num. 24.3f.: '(3) and he (Balaam) *took up his discourse, and said*:

"The oracle of Balaam the son of Beor,
the oracle of the man *whose eye is opened*,

(4) the oracle of him *who hears the words of God,*
who sees the vision of the Almighty ...'' (RSV);

identical or similar expressions to the words italicised are found in the verse in Enoch. The change from the third to the first person in this verse is awkward, although it is not unparalleled elsewhere (e.g. 12.1-3; 70.1-3; 92.1). The awkwardness, however, would be removed if we could assume that the verse followed its Numbers model by beginning '... And he took up his discourse and said: [Oracle of Enoch] a righteous man ...' If we supplement En^a 1 i 2 by אימר חנוך, 'Oracle of Enoch', we obtain an exact parallel to the Balaam prophecy, and the omission of the words can be readily explained as a scribal error. Cf. Num. 24.3, Tg. אימר בלעם. Eth. 'whose eyes were opened by God' corresponds to the phrases at Num. 24.3, 4 שתם העין, גלוי עינים, for which G has ὅρασις ἐκ θεοῦ αὐτῷ ἀνεῳγμένη|ἦν. Milik reconstructs on the basis of G and renders '(a just man) to whom a vision from God was disclosed', but the translator into Greek misrendered an original עין = 'eye' as ὅρασις (cf. LXX Ezek. 1.22, 8.2). Eth. has preserved a superior text: [οὗ παρὰ τοῦ θεοῦ ὀφθαλμοὶ αὐτοῦ ἀνεῳγμένοι] Eth. [καὶ ὁρῶν ὅρασιν τοῦ ἁγίου] corresponds to Num. 24.4 LXX ὅστις ὅρασιν θεοῦ εἶδεν ...

the Holy One ὁ ἅγιος, קדוש, Isa. 40.25, Job 6.10 (Tg. קדישא), Prov. 9.10. Probably originally a contraction for 'the Holy One of Israel (Jacob)'; see G. W. Schmidt, 'Wo hat die Aussage: Jahweh "der Heilige" ihren Ursprung', ZAW Bd. 74 (1962), 62-66. Enoch prefers the expression 'the great Holy One' (see below, v. 3), reserving קדיש(ין) for 'angels(s)'. (The text 'the Holy One who is in heaven' is to be preferred to G '(the vision of) the Holy One and of heaven').

(3) **the elect** The designation occurs most frequently in the Book of the Parables (40.5; 41.2; 48.1; 51.5; 56.6; 58.3; 62.7,8,11 etc.), but it is also found in the older Book of Enoch at 1.1 ('the righteous elect'), 8; 5.7; 25.5; 93.2 ('the elect of the world'). The term comes from the Old Testament where it is applied to Israel as 'the chosen (of Jahweh)', especially in the phrase 'my(his) chosen ones', e.g., Isa. 65.9,15,22 (LXX 23 οἱ ἐκλεκτοί μου), 1 Chr. 16.13; Ps. 89.3 (LXX 88.4); cf. Dt. 7.6 (14.2). The designation was applied to themselves by the Qumran sect, e.g., IQM 12.1,4 (cf. 10.9), CD 4.3 ('the elect of Israel'). It has established itself in the New Testament as a term for Christians, 1 Pet. 2.9 ('the elect race'); Mt. 24.22, 31; Mk. 13.20, 22, 27; Lk. 18.7; Rom. 8.33; Col. 3.12; 2 Tim. 2.10, etc. On the Elect One see below, 197 Consult ThWNT Bd IV, 186f. ἐκλεκτός (Schrenk).

The great Holy One | G 'my great Holy One' is impossible.[1] Eth. 'the holy and great One' probably stems from an interpretation of G as 'the holy great One' (so understood by Charles [2]).|The original title 'the great Holy One' is found in G at 10.1; 12.3; 14.1; 97.6; 98.6. Occasionally we find ὁ μέγας, 'the Great One' simpliciter, e.g., 14.2; 103.4; 104.1. G 25.3 has ὁ μέγας κύριος, ὁ ἅγιος τῆς δόξης. 14.1,2 is represented in the fragments at En^c 1 vi 10, 11 by קדישא רב]א] and רבא. While the epithet גדול is frequently applied to God in the Old Testament, there does not seem to be any case of הגדול being used simpliciter = 'the Great One'; but cf. Neh. 4.8; 8.6. The addition of the epithet ὁ μέγας, 'the great Holy One' serves to distinguish the term as a title for deity from the common use of ὁ ἅγιος, 'holy one' meaning 'angel' (see above on v. 2).

---

[1] The μου has probably arisen by 'vertical dittography' from (παραβολὴν)μου above.

[2] Eth. generally has the expanded form, but variants occur in the MSS, e.g., 92.2; 98.6 ('great holy One' = G); 104.9 ('the great One' G τοῦ ἁγίου). Sometimes the epithets are reversed, e.g. 10.1, 'the great and holy One', G ὁ μέγας ἅγιος.

The full title 'the great Holy One', occurs again in Aramaic at 1QapGn ii 14, xii.17, discussed by Fitzmyer, *Genesis Apocryphon*, 89, and by Milik, 144. The origin of the title may be traced to Ezr. 5.8, Dan. 2.45, 'the great God' (אלה רב). See also the Schmidt article (above note on v. 2).

3b-9 A poem consisting of tristichs. The discovery of this structure is helpful in the restoration of the text. See especially stanzas 1 and 7 (3b-4, 8c). Was this the original (poetic) structure of Enoch's oracular discourse?

These verses, describing a Sinai theophany, conform to a familiar Old Testament literary pattern, the appearance of Jahweh accompanied or followed by an upheaval in nature. See J. Jeremias, *Theophanie: die Geschichte einer alttestamentlichen Gattung* (*Wissenschaftliche Monographien zum Alten und Neuen Testament*, Bd. 10, Neukirchen-Vluyn, 1965), and J. VanderKam, 'The Theophany of Enoch 1.3b-7, 9' in VT Bd. XXIII.2 (1973), 129-150. En 1.3b-9 has the two-fold Biblical structure in an elaborated form, the appearance of God 3b-4 + 9a, then its sequel in the upheaval of nature (5-7). V. 9 depicts the first as the coming of God to execute a universal judgement, thus supplying an appropriate introduction for the whole Enoch apocalypse.

What is new and original in Enoch is that it is no longer—as in the classic theophanies at Dt. 33.2, Num. 24.1-4, Jg. 5.4-5—an account of a mighty act of God in the past, but a prediction of his future advent in a universal judgement, but couched in the language of the Biblical theophanies.

The passage is of special importance, not only because Jude 14-15 cites v. 9 as scriptural prediction of the advent of God or Christ in judgement, but also as the foundation, in the tradition-history of Hebrew theophanies, for the New Testament doctrine of the Second Advent.

shall come forth from his dwelling G ἐξελεύσεται = Heb. יצא, Aram. נפק, the regular term for Jahweh's appearances; see F. Schnutenhaus, 'Das Kommen und Erscheinen Gottes im Alten Testament', ZAW Bd. LXXVI (1964), 3. The closest parallel to this verse is Mic. 1.3, LXX διότι ἰδοὺ κύριος πορεύεται ἐκ τοῦ τόπου αὐτοῦ καὶ καταβήσεται ... ἐπὶ τὰ ὕψη τῆς γῆς. For the heavenly 'dwelling-place' (מושב) of Jahweh, 1 Kg. 8.30 (2 Chr. 6.21) LXX ἐν τῷ τόπῳ τῆς κατοικήσεώς σου ἐν οὐρανῷ: also Dt. 26.15; Isa. 26.21; 63.15; Ass.Mos. 10.3; a synonymous term is היכל 'palace, temple' of Jahweh, 2 Sam. 22.7, Ps. 18.6. At Dt. 33.2 Jahweh comes (ἥκει) from Sinai: here (v. 4) it is to Sinai he comes from his heavenly abode (see next note).

(4) **And the eternal God ... Sinai** Cf. Mic. 1.3, Eth. [ἐκεῖθεν]. Does this derive from a corruption of ἐπὶ γῆν (or ἐπὶ γῆν from ἐκεῖθεν)? So Knibb 2, 58. G ἐπὶ γῆν could be an echo of Mic. 1.3? A text καὶ ἐκεῖθεν 'and from thence' = ומן תמן could be a misrendering of 'and from Teman'. See below, 106.

The epithet 'the eternal God' occurs at Gen. 21.33 (אל עולם Tg אילה עלמא), Isa. 40.28 (אלהי עולם), Dan. 5.4 LXX (no MT or Theod.) τὸν θεὸν τοῦ αἰῶνος and Rom. 16.26 τοῦ αἰωνίου θεοῦ (as in LXX Gen. 21.33); it is also found at Jub. 13.8; Sib. Or. 3.698; Tob. 14.6 τὸν θεὸν τοῦ αἰῶνος, and for related forms Sir. 36.17 (ὁ θεὸς τῶν αἰώνων), Ass. Mos. 10.7 (deus aeternus). According to Dalman (*Words of Jesus*, Edinburgh, 1902), 164 עולם is here a 'time-concept'; thus מלך עולם means the king who controls infinite time (not 'king of the universe') (see 12.3 for this epithet in Enoch). The 'eternal God' is more usually understood as the 'everlasting God', i.e., the God who is from everlasting to everlasting. En. 84.2 has 'God of the whole world', from which it seems likely that the Ethiopic translator understood the expression as 'God of the universe'. For the parallel phrase 'Lord of the ages' see note on 9.4. Cf. Fitzmyer, *Genesis Apocryphon*, 105 ff., on IQ apGn xix.8.

shall tread ... upon Mount Sinai Since the verb already has a predicate 'upon Mount Sinai' G ἐπὶ γῆν seems redundant. Is it a reminiscence of Mic. 1.3? Does דרך 'tread upon' carry the idea of conquest in this context (so VanderKam, op. cit., 136)?

Sinai, whence the Law was given, is also to be the place of judgement. Cf. Dillmann, 90: '... der Sinai hat fast appellative Bedeutung als der Ort der Gerechtig keits-offenbarung Gottes'. Seir, Edom (=Teman?) and Paran are all also associated with the Sinai theophany (Jg. 5.4; Hab. 3.3.). Cf. J. Jeremias, op. cit., 8.

And reveal himself ... highest heaven Eth. = G φανήσεται = יופע, a verb used, in both Heb. and Aram., for the glorious manifestation of Jahweh in theophanic texts: Ps. 50.2; 79.2 (LXX ἐμφάνηθι): CD 20.25; IQH 4.6,23; 9.31 etc. Eth. = G ἐν τῇ δυνάμει τῆς ἰσχύος αὐτοῦ = En[a] 1 i 6 [בחיל/כוח](?) [נבור]תה, cf. IQH 4.32, בכוח גבורתו. The repeated καὶ φανήσεται in G Eth. suggests either a dittograph or a duplicate rendering or doublet of some kind. The words καὶ φανήσεται ἐκ τῆς παρεμβολῆς αὐτοῦ are bracketed by Charles as an addition (Charles, 6). They are not only against the parallelism but also against the sense: in 3b it is already said that 'the Holy One will come forth from his dwelling', and the writer has gone on to speak of God's advent on Sinai. Eth. appears to have read [ἐν τῇ παρεμβολῇ αὐτοῦ]. Could this be an alternative version of ἐν τῇ δυνάμει (בחיל) τῆς ἰσχύος αὐτοῦ where חיל = 'military forces'? The language is certainly military; VanderKam gives other examples. G 'from the heaven of heavens' seems original; cf. Dt. 10.14, 1 Kg. 8.27 = 2 Chr. 6.18. Was it this Biblical expression which gave rise to the later idea of a plurality of the heavens?. Consult further s.v. οὐρανός ThWNT Bd. V, 496 f. (v. Rad).

(5) **all shall be afraid** Cf. Exod. 19.16 'all the people trembled'. In Enoch, the panic, like the judgement, is to be universal (cf. vv. 7, 9) and cosmic. En. 102.3 'the children of earth shall tremble' seems to be alluding to this verse.

the watchers Watchers (עירין, ἐγρήγοροι G Aq. Symm.), from Heb./Aram. עור 'to be awake', is a term for angelic beings, clearly archangels, peculiar to apocalyptic literature. It appears first in canonical scripture at Dan. 4.10 (13) (LXX ἄγγελος), 14 (17) (Theod. ιρ = עיר), 20 (23)[1], but the origins of the idea are probably to be traced to Ezek. 1 (the eyes of the Cherubim) or to Zech. 4.10 (the 'seven' who are 'the eyes of the Lord'); cf. also Isa. 62.6 ('the guardian (angels)?). J. Teixidor thinks the idea comes from the officers of the Achaemenids called 'the eyes of the king' (JAOS, Vol. 87 (1967), 634). A fuller description occurs at En. 20.1 (Eth. only) 'the holy angels who watch' and at 39.12,13; 61.12; 71.7 'those who do not sleep' (all the latter passages from the Book of the Parables). While this may be held to be a characteristic of all angels (Dillmann), the term comes to be applied specially to a higher caste of heavenly Vigilantes, the archangels, Seraphim, Cherubim, Ophanim, who keep watch by the Throne of God (71.7), or the angel-leaders who have a special area to supervise or role to play (20.1; 82.10 f.). They are for the most part good angels, as the fuller title 'watcher and holy one' indicates (Dan. 4.10 (13), 20 (23)), but since two hundred of them descended from heaven to earth and seduced the daughters of men, the term has come also to refer to these fallen watchers (1.5(?); 10.9,15; 12.4; 13.10 (the 'heavenly watchers'); 14.1 (the 'eternal watchers'), 3; 15.2). At 6.2, 13.8, 14.3 they are called 'sons of heaven' (G υἱοὶ οὐρανοῦ) after Gen. 6.2 בני האלהים (LXX οἱ υἱοὶ τοῦ θεοῦ), itself a general name for angels (e.g., Job 1.6; 2.1; 38.7; Dan. 3.25; En. 69.4,5 71.1; 106.5.) Cf. IQapGn ii 5, 16; IQS 11.8; 'sons of heaven' seems a substitute for

---

[1] For similarities in Enoch's terminology with the Greek Daniel, see Barr, 'Aramaic-Greek Notes', I, 189f.

'sons of God' (see Fitzmyer, *Genesis Apocryphon*, 84). The older book at 15.7 refers to the fallen watchers as 'spirits of heaven' (G πνεύματα τοῦ οὐρανοῦ).

These fallen archangels or 'sons of heaven' play an important part in the theology of Enoch, since it was through their corruption of mankind that evil was believed to have entered the world. Gen. 6.1-4, the story of their fall, is a preface to the Noah story, thus involving heaven, as well as mankind in the universal corruption which led to the Deluge (cf. Test. Naph. 3.5). For the idea that they were astral deities and for parallels with Zoroastrian and Islamic traditions, see Montgomery's *Daniel* (ICC), 232, and Bousset-Gressmann⁴, 322 f.

(5) **the watchers shall quake** Commentators tend to accept the reading of G 'shall believe'; e.g. Milik, 145, thinks the text contained an allusion to the 'Originistic conception of the conversion of the evil spirits'. The superior text is that behind Eth [ἐπισεισθήσονται], of which G is a corruption. Cf. Jl. 4.16; Isa. 24.18, Heb. רעש (Tg. זוע).

seek to hide themselves(?) The sentence καὶ ζητήσουσιν(?) (ms ασωσιν) ... τῆς γῆς is omitted by Eth., probably by hmt; its genuineness is guaranteed by the first frg. at Enᵃ 1 i 7. G is usually read as ᾄσουσιν and translated 'shall sing'. Charles, 6, supports this reading by a reference to the 'singing' of the watchers at 2 En. 18.9 Has this later idea influenced the translator once πιστεύσουσιν had been read? The conjecture of Milik that יענו 'they will be punished' has been mistranslated 'they will sing' is unconvincing, since it involves equating ענו with ἀείδω. VanderKam (op. cit., 142 f.) noted an allusion to this verse at 102.3, 'And all the angels sought to hide themselves from the presence of the Great Glory'. The reading proposed for G ζητήσουσιν ἀπόκρυφα would correspond to Eth. ζητοῦντες ἀποκρυβῆναι at 102.3, and can be readily accounted for as a misreading of יבעון למסתתרה, ζητήσουσιν ἀποκρυβῆναι as ζητήσουσιν ἀπόκρυφα (למסתרתא cf. Dan. 2.22 Theod.). A closer study of 102.3 suggests that there it is not the 'watchers' who seek to hide before the Great Glory, but the heavens themselves and their 'luminaries'. The dislocation of the text there (see below, 311) may have been caused by the idea here of the watchers seeking to conceal themselves. The suggestion of Milik that ἄκρα = קצית 'creatures' ('among all the creatures of the earth') is barely even defensible.

fear ... ends of the earth G τρόμος καὶ φόβος μέγας: so also at 13.3 τρόμος καὶ φόβος. Eth. has the Biblical order 'fear and trembling', Job 4.14; Ps. 2.11; Mk. 5.33; 2 C.7.15; Eph. 6.5.

(6) Vs. 6-7 contain imagery to be found in a number of Old Testament passages. V. 6 is closest to Mic. 1.4, but cf. also Jg. 5.4-5; similar apocalyptic imagery occurs at Nah. 1.5; Ps. 97(96).5; Isa. 40.4; 64.1,3; Hab. 3.6; Jdt. 16.15; Ass. Mos. 10.4; Sir. 16.19; 2 Pet. 3.7-10; cf. also En. 102.2 f. G has a longer text than Eth. which gives a neater *parallelismus verborum*: is G elaborating? G πυρὸς ἐν φλογί = ἐν φλογὶ πυρός LXX Exod. 3.2 לבת־אש.

the high places ... laid low Cf. J. Crenshaw, CBQ Vol. 34 (1972), 43 on Am. 4.13: 'The creator of the mountains is most assuredly their sovereign (hence he strides across the elevated heights *on which are the sacred sanctuaries* as a conqueror tramples on the backs of his victims).'

like wax ... fiery flame While rejecting any idea here of the Essene doctrine of ἐκπύρωσις, Charles nevertheless cites as parallels passages where this idea is implied: Sib. Or. 3.54,60,72,84-87; 4.172 f.; 5.211 f.; 2 Pet. 3.7,10; Life of Adam 49.3 f. The text may well go beyond Mic. 1.4, Nah. 1.5, Ps. 97.5, 104.32, in the direction of this Essene belief, if it does not in fact assume it; cf. IQS 3.29; for a discussion of the latter passage in this connection, Black, *Scrolls and Christian Origins*, 138. n. 1.

(7) **the earth ... sunder** G διασχισθήσεται ... σχίσμα / ῥαγάδι, בקיע ... תתבקע (Nöldeke, 226 f). Cf. Mic. 1.4 'the valleys shall be cleft, יתבקעו'. As Lods noted, the text of Eth II seeks to introduce a reference to the deluge.

a universal judgement A judgement not confined to mankind, as Charles's translation implies 'And there shall be a judgement upon all (men)'. 'All' to be judged and destroyed include the watchers who fell from heaven and their illegitimate offspring as well as mankind (En. 10.6,15).

(8) **the righteous** The word δίκαιος = קשיט, like ἐκλεκτός = בחיר has virtually become a *terminus technicus* in Enoch (e.g., 5.6; 25.4; 39.4; 60.2; 82.4; 95.3; 100.5 etc.). See D. Hill, 'Δίκαιοι as a Quasi-technical Term', in NTS, Vol. XI (1964-1965), 296-302 (Qumran usage, 300 f.). The term, like ἐκλεκτός has passed into the New Testament in this sense. Consult ThWNT, Bd. 2, 189 f. (Schrenk).

make peace Eth. = G εἰρήνην ποιεῖν = 1 Mac. 6.49,58 ποιήσωμεν μετ' αὐτῶν εἰρήνην. The idea is of a state of cessation of the divine anger as at Rom. 5.1,9-11.

he will protect ... upon them Behind G συντήρησις lies an Aram. נטיר (Tg.), cf. ܢܛܘܪܐ. Has an original נטיר been read as a noun by G and as a verb by the Greek version translated by Eth., [συντηρήσει]? And did the parallel line following read 'And he will be merciful (חננא) to them', where the verb was again read as a noun? The translator of G expands by adding καὶ εἰρήνη.

they shall all belong to God Eth. = G καὶ ἔσονται πάντες τοῦ θεοῦ. While the construction is possible in Aram., as in Greek, it seems an unsemitic one. Did the original read 'And they shall all be/become sons of God'? Cf. Dt. 14.1; Ps. 82.6; Pss. Sol. 17.30.

he shall show them favour Charles, 7 (cf. Knibb, 1, 4) suggests that Eth. read [εὐοδίαν δώσει αὐτοῖς] 'he will give them prosperity' becoming 'they shall be prospered'. Cf. Sir. 43.26 for the confusion of εὐδοκία with εὐοδία. Did Eth. read [εὐοδωθήσονται]?

light shall appear Cf. 5.7; 92.4(?); 96.3. 'Light' is the lot of the righteous, 'darkness' the destiny of the wicked (10.5; 92.5; 94.9; 103.8 and passim in the Book of the Parables (e.g., 58.3-6). The Biblical figure of light and darkness for good and evil is a central feature of the Qumran writings.

(9) Cf. Jude 14-15 cited by Pseudo-Cyprian (CSEL, *Cypriani opera omnia* III.3 Appendix, 67), Pseudo-Vigilius (Migne, P. LXII, col. 363), Liber Nativitatis (CSCO *Scriptores Aethiopici*, ed. K. Wendt, Tom. 41,66 = 42,58. See T. Zahn, *Geschichte des nt. Kanons*, II.2, 797-801, Charles, [1906] 5 f.

Behold! he comes Did the original read ארי הא = כי הנה (Tg. Isa. 26.21, Mic. 1.3, LXX διότι ἰδού Tg. ארי הא)? This would account for both variants, deriving from διότι / ὅτι ἰδού(?). G reads οτει which Milik, 186, thinks is ὅτε (כדי), but ει usually = ι. Jude's ἦλθεν points to אתה, a perf. propheticum or perf. confidentiae: 'he will assuredly come ...'; so R. Knopf, *Die Briefe Petri und Juda* (Meyer Kommentar), (Göttingen,1912), 236, and K. H. Schelke, *Der Judasbrief* (Herder) (Freiburg, 1964), 164).

Pseudo-Vigilius *ecce veniet dominus* comes from Jude where the κύριος could be Jesus (Schelke) and the reference to the Parousia. For this verse as the possible source of the μαραναθα formula of imprecation (1 C. 16.22; Did. 10.6; cf. Rev. 22.20), see M. Black, 'The Maranatha Invocation and Jude 14,15 (1 Enoch 1.9)' in *Christ and the Spirit in the New Testament*, ed. B. Lindars et al. (C.U.P., 1973), 189-196.

with ten thousand holy ones En[c] 1 i 15, Eth. = Dt. 33.2 '... myriads of holy ones'(?); see F. M. Cross and D. M. Freedman, 'The Blessing of Moses', in JBL, Vol. LXVII (1048), 198-9, and cf. Milik, 'Deux Documents inédits du Désert de Juda', in Bib., Tom. 38 (1957), 253 and note 2, and Tom. 48 (1967) 573. Cf. also Zech. 14.5, Ps. 68.18 (17). Is the reading ἐν ἁγίαις μυριάσιν at Jude 14 a literary improvement on ἐν μυριάσιν ἁγίων (ἀγγέλων)?

convict all flesh Jude is an abridged adaptation of G: καὶ [ἀπολέσει] πάντας τοὺς ἀσεβεῖς [καὶ] ἐλέγξει(αι) [πᾶσαν σάρκα] Jude καὶ ἐλέγξαι πάντας τοὺς ἀσεβεῖς. The common vocabulary in Jude and G point to a single translator for the original Greek version.

and of all the arrogant ... against him The text of G is confused and repetitious: it probably read originally καὶ περὶ πάντων σκληρῶν λόγων ὧν κατελάλησαν κατ' αὐτοῦ ἁμαρτωλοί. The full expression 'to speak arrogant and hard words' occurs again at 5.4 G καὶ κατελαλήσατε μεγάλους καὶ σκληροὺς λόγους (cf. En[a] 1 ii 13) and 101.3. The shorter form σκληρὰ λαλήσουσιν occurs at 27.2. The longer form seems to be a combination of two expressions דבר קשות. Gen. 42.7,30 LXX σκληρὰ λαλεῖν, and ממלל רברבן Dan. 7.8,11,20, LXX (στόμα) λαλοῦν μεγάλα. Cf. Tob. 13.14 οἳ ἐροῦσιν λόγον σκληρόν 4Q Tob. aram[a] 2 ii 15 די ימ]ללון מלא[קשא (Milik, 184 and 186, L. 17).

CHAPTER 2

Chh. 2-5:1 f. consist of a nature-homily in which the order in the world of nature is contrasted with the disorder in the life of man. The theme is one which is almost a commonplace in ancient, and particularly in Jewish, literature. Charles 8 f. cites parallels from Sir. 43; 16.26-28; Test. Naph. 2.9; 3.2,3; Pss. Sol. 18.12-14. The closest verbal parallel is Test. Naph. 3.2,3 e.g. ὁ ἥλιος καὶ ἡ σελήνη καὶ οἱ ἀστέρες οὐκ ἀλλοιοῦσιν τὴν τάξιν αὐτῶν· οὕτω καὶ ὑμεῖς μὴ ἀλλοιώσητε νόμον θεοῦ ἐν ἀταξίᾳ τῶν πράξεων ὑμῶν. It is possible that the author-redactor of this section of Enoch is drawing on a description of the seasons from a contemporary or earlier astronomical document (see below, 419). Is G ἠλλοίωσαν a gnomic aor. (Blass-Debunner, §333 (171))? For the verb in Daniel and Enoch = שנה, Barr, 'Aramaic-Greek Notes' I 186f.

(1) The Aram. has a longer form of text; the shorter form of G Eth. has arisen by hmt. The introductory formula ἴδετε καὶ διανοήθητε (2.2) is fully preserved at En[c] 1 i 20 חזוא לכון ... ואתבוננא, but, while G retains ἴδετε consistently for חזוא, it employs κατανοήσατε, as here, and καταμάθετε (3.1) for אתבוננא.[1] The order of the verbs is transposed in this verse, and at 3.1, an alteration, which Milik, 148, Ll. 9-11, attributes to the Greek translator, but cf. En[c] 1 i 17-18. A variant formula occurs at 5.1. G διανοήθητε καὶ γνῶτε (see below, 112).

The 'work(s) of heaven (En[c] 1 i 18) are, in this context, the 'works of God' in creation, the heavenly bodies: cf. Job 37.14; Sir. 16.27; Ps. 8.3; 103 (102).22, etc.

in the conjunction ... orbits The term מסורת is best explained, etymologically, as 'bonds' (from אסר 'to bind'), ܡܐܣܘܪܬܐ, and, hence, at 1 QM 3.3, 13 for' '(military) bands, Verbände'. Its astronomical meaning is defined by its use at 1QS 10.4, correctly translated by Driver, *Judaean Scrolls*, 337, as 'their conjunction with one another

---

[1] Here and at 3.1, 5.1 Eth[M] reads 'I considered (ṭayyaqqu)', clearly an inner Eth. variant for ṭayyequ = κατανοήσατε, καταμάθετε. Ullendorff, 'An Aramaic "Vorlage"', 266 (cf. Knibb, 2, 61) reads ṭayyaqu 3rd plur. perf. which he derives directly from an (ambiguous) Aram. חזו. But ṭayyaqa 1.2 renders κατανοεῖν not ἰδεῖν in these verses.

(מסרותם זה לזה)' i.e. of the sun and the moon, producing the 'new moon'.[1] This is the meaning required in this verse, and I suggest reading במסרות [דב]ריהון '(How the (heavenly) luminaries (נהוריא, Gen. 1.16 Tg.) i.e. sun and moon) do not deviate in the conjunction of their orbits'; for דברא in this sense, 79.5 Enastr[b] 26.3 and Jg. 5.21 (Tg. v. 20). Eth. G have omitted במסרות (an unfamiliar noun). The term occurs again at 82.10 Enastr[b] 28.2 Eth. 'their positions', along with 'fixed seasons', 'new moons'.

rises and sets Eth I has ʿaqaba (φυλάσσειν) for ʿaraba (δύνειν). Flemming, 2, explains the variant as having arisen by the introduction of a well-known expression, ʿaqaba šerʿata for ʿaraba. It is also possible, however, that a text once had this phrase, and that šerʿata fell out by ditt. with the following šeruʿ, τεταγμένος, (or that the latter is in fact an accommodation of šerʿata to G): 'How they all rise and keep station ([φυλάσσει τὸ τάγμα]?), each in order (τεταγμένος), in its appointed time'. Heb. שמר is used intransitively, meaning 'to keep station' at 2 Kg. 9.14.

appointed time G ἐν τῷ τεταγμένῳ καιρῷ. Is this a rendering of זמנא (מזמן); cf. 33.4; En[e] 1 xxvii 20?

fixed seasons G ἑορταί, Heb. מועדים Aram. מעדין, 82.9 Enastr[b] 28 למעדיהון לחדשיהון 'the festive times of their new moons'.[2] The reference here at 2.1 is to the 'fixed times' of the solar and lunar year which were festival days, new moon, new year etc.

(2) **observe ... wrought in it** Milik's reconstruction offers much too short a stanza: G περὶ τῶν ἔργων τῶν ἐν αὐτῇ γενομένων could be original: (אתבוננאו) בעבדיא דעבידין בה.

from the first ... is changed Is 'from the first to the last', a 'polar' phrase, meaning 'all of them'? Milik [מן קדמיה לא]חׄרנה; cf. ܡܢ ܩܕܝܡ ܘܠܚܪܬܐ Ps. 139 (138).5; P.Sm. 3491; G ἀπ' ἀρχῆς μεχρὶ τελειώσεως suggests an interpretation 'from the Beginning to the End (of Time)' = συντέλεια (cf. 10.12) G's ὡς εἰσιν φθαρτά reads like 'a Christian gloss' Milik, 147, Ll. 1/2.

all is made manifest to you Does the Ithp. מתחזא perhaps here have the sense of 'are displayed, are spectacular'? Cf. P.Sm., 1234.

(3) G has preserved the opening phrase only, 'Observe the summer and winter' in an already shortened form of text, since there is no doubt that the original at En[a] 1 ii 3 read 'Observe the דגלי [of summer ...]. Eth., which alone preserves most of the verse, is also deficient, since it begins like G, 'Observe the summer and winter', and follows with a brief description which is more applicable to winter than summer and winter. This foreshortening of G Eth. is also clear from the Aram. frgs., as originally read by Milik (the photograph is now somewhat faded), where a lacuna follows 'Observe the דגלי of summer ... upon it'.

A similar tantalisingly fragmentary description of the seasons is preserved in a frg. of Enastr[d] i along with frgs. of astronomical pieces (Enastr[d] ii-iii). See Milik, 274, 296 f. These frgs. are placed by Milik, 296 f., after Eth. 82.10-20, verses which also deal with

---

[1] Cf. Milik, 187 and J. Bowman, 'Is the Samaritan Calendar the Old Zadokite One?' in PEQ 91,33: '... the luminaries (at 1 QS 10.3f.) are the sun and the moon, and they are ... for the Qumran calendar, by divine arrangement, as far as the months are concerned, tied to each other'. For a conspectus of the discussion of this controversial and enigmatic passage, see especially M. Weise, *Kultzeiten und Kultischer Bundesschluss in der 'Ordensregel' vom Toten Meer*, Studia Post Biblica III, Leiden, 1961.

[2] Milik, 187 on 1QS 10.5 translates מועדים by 'the constellations of the Zodiac'. But this 'calendar-hymn' is dealing with the sacred 'seasons' of the new moon, the new year etc., not with the constellations.

the seasons, but contain descriptions of spring (82.15-17) and summer (82.18-20) only: a description of winter is lacking. This gap seems to be filled by Enastr[d] i, supplying the missing piece on winter. Enastr[d] i 4-6 is virtually a duplicate of Ch. 3. (See Appendix, A 418 f.) The pieces on spring and summer at 82.15-20 probably supply the missing portions of the text of Ch. 2.3 (in a fuller recension).

While Enastr[d] i no longer fits precisely after En. 82.20, it clearly belongs to a similar astronomical description of the seasons, which may then have provided the source for En. 82.10-20. Is the author-redactor of the 'nature homily' at En. 2.1 f drawing his material from such an astronomical document?

Observe the signs (דגלי) of summer The word דגל, in Heb. a 'banner, standard' and later a cohort or division of soldiers, is used in the old Aram. for a military detachment or administrative unit (Cowley, *Aramaic Papyri*, passim, Kraeling, *Brooklyn Papyri*, 41 f., Hoftijzer, 55, and especially M. Noël Aime-Giron *Textes Araméens d'Égypte* (Cairo, 1931) 61, and Lesquier *Les Institutions militaires de l'Égypte sous les Lagides* (Paris, 1911), 103. At Ch. 4 Eth. has mawāʿel 'days' for דגלי, which leads Knibb 2, 62 to define דגל as a 'division of time'. Milik, 147, Ll. 2 takes it as σημεῖα 'signs' (a word also used for 'banners', signa), and, in this context, to mean 'natural, regular and cyclical phenomena'. In its meaning 'banners' the word is parallel to Heb. אות (e.g. Num. 2.2) and the context here (and at 82.9) supports the view that, like אות, and σημεῖον, דגל has the meaning 'weather sign', 'token of changes of weather and times', e.g. Jer. 10.2 אותות השמים, LXX σημεῖα τοῦ οὐρανοῦ; cf. Theophrastus, *Enquiry into Plants* ed. Loeb II, 390 f., περὶ σημείων ὑδάτων καὶ πνευμάτων καὶ χειμώνων. See further below, Appendix A, 395.

CHAPTER 3

(1) A substantial frg. closely parallel to this verse has been preserved at Enastr[d] i 4-6 (see Appendix A, 418 f., and Milik, 148, Ll. 4-6, 296). Note especially line 6, where the words 'except fourteen trees ... [whose] leaves remain' are especially close to the text of En. 3. From the opening line, it is clear that it is the 'winter' season (שתוא 'autumn and winter') which is there described.

Consider that all trees G at this point has the longer formula of 2.1, 'Observe and consider', but with the verbs reversed. It is probable, however, that the fragment preserved in G is the beginning of Ch. 5 (cf. Milik, 148, Ll. 9-11). There were two stanzas on the subject of 'trees', each beginning with a version of this formula (Eth. Ch. 3 and 5.1, cf. En[b] 1 ii 4, 9), and this accounts for the omission, by hmt, of Chh. 3 and 4 by G.

The text of En[a] 1 ii seems to be longer than that at En[c] 1 i (cf. En[a] 1 ii 6 with En[c] 1 i 25).

For these fourteen ever-green trees, see Geop. XI. 1 (ed. H. Beckh (Leipzig, 1895), 326), and Milik, 148. Cf. also Jub. 21.12 and Test. Levi, 9.12.

remain without (their leaves) ... years For the meaning of מתקימין in this context, see Milik, 219. Is it rendering (δένδρα) ἀειθαλῆ? Eth. adds 'em-beluy, [ἀπὸ τοῦ παλαιοῦ] lit. '(trees which) remain from the old (foliage?) until the new comes ...' (Knibb, 'with the old (foliage))'. This puzzling reading could be a mistranslation of Aram. מבלי 'without', and a supplement for the lacuna at En[a] 1 ii 6 (following Eth.) read 'without (their leaves) being renewed up to two or three years'.

CHAPTER 4

(1) **the signs of summer** See above on 2.3.

how the heat (of the sun) Did the original perhaps read חמה = 'heat', but used, poetically, for the (hot, summer) sun (e.g. Isa. 24.23; Job 30.28)?

burning and scorching heat En[c] 1 i 26 reads clearly כוייה which I take as the rare word כויה (Heb. Exod. 21.25), LXX κατάκαυμα ܟܘܝܐ P. Sm., 1688. The second word שלקה̊ = θερμός is also rare: cf. late Heb. שלק, 'cook, roast', Syr. ܫܠܩ P. Sm., 4198 coxit, elixavit, ܫܠܩܐ decoctio. Milik, 147, 185, takes both words as verbs: ('Observe ye the signs of summer whereby the sun) burns and glows'. But כוי never appears to be used in the Peal, and שלק is usually transitive.

CHAPTER 5

(1) **in all of them ... covers them** Both G and Eth. are clearly deficient, but the former perhaps offers the least unsatisfactory basis for a defensible reconstruction of the Aram., provided G is construed as 'Consider ... how their leaves are green on them' (ἐν αὐτοῖς = בהן En[a] 1 ii 9). We require a verb in this clause (cf. Milik 'on all of them blossoms their foliage'), and I suggest that an original יעלון lit. 'springs forth, sprouts'[1] has been misread in G as עליהן 'their leaves'. The Aram. would then read, lit. 'Consider all trees [that there springs forth] on them green foliage, and it covers them ...'. For ירוק for 'green foliage', cf. Exod. 10.15, Tg. ירוק באילנא.

in glorious splendour Aram. תשבחה = ܬܫܒܘܚܬܐ, δόξα. For הדר to describe trees cf. Lev. 23.40 פרי עץ הדר 'the fruit of goodly trees' (EVV) i.e. 'ornamental trees'. The noun הדר combines with other nouns, e.g. Ps. 145.5 'the glorious honour of thy majesty (הדר כבוד הודך)' AV. The phrase להדר תשבחה may be taken, therefore, as 'in glorious splendour'.

Examine and consider We have here a variation of the familiar formula: for חקר, 'to search out a subject', cf. Job 5.27, 28.27. G διανοήθητε καὶ γνῶτε reverses the words (γνῶτε = חקרו, cf. Prov 28.11 יחקרנו, LXX καταγνώσεται).

that the God ... these G ὁ θεὸς ζῶν and ζῇ εἰς πάντας τοὺς αἰῶνας are clearly doublets (cf. Knibb, 2, 65). Note the variant reading at En[c] 1 i 30 from En[a] 1 ii 11 viz. 'for all eternity of eternities' cf. G εἰς πάντας τοὺς αἰῶνας.

(2) As compared with G Eth. En[a] 1 ii 11-12 either comes from a shorter recension or has been deliberately abridged: note the careless form of script in comparison with En[c] 1 ii.

The first half-verse is fully preserved in Eth[m] ἐνώπιον γίνεσθαι = היה קדם, היה לפני 'to attend upon' of a servant in the presence of his master. G has lost ἐνώπιον αὐτοῦ by hmt with the following ἐνιαυτοῦ. The phrase היה קדם is synonymous with 'to stand before' (Heb. עמד לפני) and parallel to 'to serve': cf. e.g. 1 Sam. 19.7, 29.8 (LXX εἶναι ἐνώπιον), 2 Sam. 16.19 (par. 'to serve' עבד), 2 Kg. 5.2 etc.

The expression ἀποτελεῖν τὰ ἔργα in the next verse (v. 3) means 'to perform (Eth. faṣṣama) (their) tasks'. The author uses the same noun עובד = ἔργα for 'tasks' as he does for the 'works (of creation)' (En[a] 1 ii 10 עבדיה), viz. '... you have changed your

---

[1] The verb is common in the old as in later Aram. but is attested in Heb. only in this sense.

tasks' (En^{a} 1 ii 12, עבדכון). G πάντα οὕτως καὶ πάντα ὅσα is impossible to construe if ἀποτελοῦσιν is taken as the main verb, and equally impossible if πάντα ὅσα introduces a subordinate clause: πάντα ὅσα may be a ditt. from the earlier πάντα ὅσα ἐποίησεν. Both phrases, however, could be clumsy attempts to render an emphatic semitic double כול, וכול עבדוהי כולהון lit. 'and all his works, all of them' (πάντως (ms παντα ουτως) καὶ παντῆ (ms παντα οσα)?)[1]

Eth. has a shorter text, but with a significant variant, 'and all his works serve him ([δουλεύει αὐτῷ]?)'. This gives the required parallel to 'attend on' of the first clause and seems original: in Aram. יעבדון ליה, gives a word-play with עבדא, ἔργον. G is a scribal error, introducing ἀποτελοῦσιν ... ἔργα from v. 3. I suggest, for the missing line in the two surviving Aram. recensions: וכול עבדוהי כלהון יעבדון ליה ולא ישתנון, 'and all his works wholly serve him, and do not change'. There is also a deliberate paronomasia on שנא 'year' and שני, 'to change'.

all perform his commands G = Eth. could be an expansion of the shorter Aram. text: it could also, however, reproduce a longer Aram. recension. For ממר in the sense of 'command' cf. Dan. 4:14, Ezr. 6.9 מאמר (LXX ῥῆμα). Cf. Barr, 'Aramaic-Greek notes', II, 180f.

(3) This verse, omitted altogether by En^{a} 1 ii, seems original [וחזו די כחדא ימא ונהריא ישלמון ולא ישניון עבדיהון מן ממאמרוהי] Eth. [ὁμοῦ] probably correctly reproduces the original כחדא.

(4) The first clause 'But you have changed your works (עבדכן) has been omitted in G Eth. En^{a} 1 ii seems to have had a shorter recension of the remainder of the verse.

have not been steadfast G = ואנתון לא תקומון (LXX renders קום regularly by ἐμμένειν).

with your impure mouths ... majesty En^{a} 1 ii 13 ביום טמתכן of the ms is clearly a corruption. The addition in G ὅτι κατελαλήσατε ἐν τοῖς ψεύσμασιν ὑμῶν is probably best explained as a translator's gloss on ἐν στόματι ἀκαθαρσίας ὑμῶν.

(you) hard in your hearts ... peace For the expression Ezek. 3.7 (קשי לב LXX σκληροκάρδιοι). Note the word-play in Aram. with קשות ... מלין, 'hard words'.

The expression 'you shall have no peace' is modelled on one from Isa. 48.22, 57.21, 'There is no peace (אין שלום) ... to the wicked' (LXX οὐκ ἔστιν χαίρειν τοῖς ἀσεβέσιν). The phrase occurs a number of times in the denunciation of the wicked (or of the watchers), e.g. 5.5 (G οὐκ ἔσται ὑμῖν ἔλεος καὶ εἰρήνη Eth. om. καὶ εἰρήνη), 12.5, 13.1, 16.4 (of the watchers), 94.6; it occurs again in G^{b} at 98.16, 99.13, 102.3, 103.8 in the form οὐκ ἔστιν ὑμῖν χαίρειν.

(5) **years of your life ... everlasting curse** Eth. 'and the years of your life you will destroy (taḫagguelu)'. This could come from a variant [ἀπολεῖτε] (ms απολιται) = תובדון (Aph. cf. Heb. אבד Piel), 'years of your life you will lose' i.e. your lives will be prematurely cut off; and, in contrasting parallelism, 'the years of your perdition (= אבדנא, in Gehenna) be multiplied'. The wicked will 'curse their days' in this life, but in Gehenna, their years will be prolonged 'under an everlasting curse'.

mercy or peace These are denied to the godless (and the watchers) cf. 12.6. The divine mercy is a frequent theme in the Parables: kings seek it (62.9) but it is also denied to them, 38.6; 50.5. God is great in mercy (61.13) which he shows to the 'righteous and elect', 39.5; 60.5,25; cf. ThWNT s.v. ἔλεος Bd. II especially 477 (Bultmann, who does not mention Enoch).

---

[1] Cf. Ezek. 11.15 and the old Aram. כלכליה, Hoftijzer, 121, and Syr. ܟܠܟܠܗ, ܟܘܠܝ, P. Sm. 1736, 1738.

In his discussion of εἰρήνη Foerster (ThWNT II 408) argues that in I Enoch it refers, not to the opposite of the state of enmity between God and Israel (or mankind), as in rabbinical sources, but simply to the contrasting condition of the destiny of the righteous with that of the godless (and the watchers) at the last judgement. This contrast is certainly emphasised, but 1.7f. implies an act of 'peace-making' by God with the elect, as well as their deliverance at the judgement, by the exercise of the divine mercy. The concept in Enoch is both negative and positive, including both deliverance at the judgement by God's mercy, but also the enjoyment of light and joy and salvation. And here in v. 6 it certainly means much more than that the righteous are to be left in peace (Foerster, ibid., 409.5).

(6) **Then you shall leave (i.e. to posterity) your names as an everlasting curse** Eth. reads [δώσετε] for G ἔσται. As Charles, 12 noted, Eth. alludes to Isa. 65.15 'you shall leave (והנחתם LXX καταλείψετε γάρ) your name for a curse to my chosen ...', where הנחתם has been rendered δώσετε by a confusion of the roots נוח and נתן. I have given preference to the more difficult reading of Eth.

all who curse ... curse by you All who curse will introduce the names of those sinners into their formulas of cursing as instances of persons wholly accursed (Charles, 12). Eth. is so badly corrupted that the original meaning is almost completely lost: the variant [πάντοτε / διὰ παντός], however, could be original, 'by you (by your name) for all time shall all who curse curse'.

and all who are without sin ... execration G has preserved here a long 'addition' unattested by Eth. or the Aram. frgs. Dillmann (SAB, 1092, 1043) explained it as an anticipation of v. 7, presumably as an expanded version. But there seems more to it than this, and it could go back to a longer Aram. recension; note, e.g., the use of λύσις 'remission' (of sins) and κατάλυσις 'dissolution', death', both שריא. These extra lines are important theologically, in view of their ideas and terminology about 'remission of sins' and 'salvation' for the 'sinless', 'dissolution' and 'execration' for the wicked.

καὶ πάντες οἱ ἀναμάρτητοι χαρήσονται. The conjecture of Swete that the αματοι of G stood for ἀναμάρτητοι is confirmed by the occurrence of the word at 99.2 (cf. LXX Dt. 29.18 (19), G = [וכלהון די בלא חטאה יחדון], 'and all without sin shall rejoice'.

καὶ ἔσται αὐτοῖς λύσις ἁμαρτιῶν. For the phrase lit. 'loosing' of sins; שריא דחטיתא, cf. ܫܘܪܝܐ ܕܚܛܗ̈ܐ, Rom. 3.25, in Harclean Syr. for πάρεσις ἁμαρτημάτων, P. Sm. 4316. For the verb שרא in this sense Tg. Jer. II Gen. 22.14. G = ואיתיה להון שריא דחטיתא. Cf. also Dan. 5.12 משרא the 'loosing' of spells, and En. 8.3 λυτήριον (En[b] 1 iii 2).

Καὶ πᾶν ἔλεος καὶ εἰρήνη καὶ ἐπιείκεια. Does ἐπιείκεια here render שלוה? Cf. Dan. 4.24 (27) LXX ἵνα ἐπιείκεια δοθῇ σοι NEB 'peace of mind'? Tg. שליוא 'peace', 'contentedness' G = (?)וכול רחמין ושלם ושלוה 'and all mercy and peace and tranquility of mind'(?).

ἔσται αὐτοῖς σωτηρία φῶς ἀγαθόν G σωτηρία Heb. ישע, ישועה; Syr. uses ܚܝ̈ܐ, 'life'. The nouns ישע, ישועה do not appear to be used in Aram. Whatever word appeared in any original here (and this sentence could be a free expansion of G), σωτηρία is used in its full, classical, eschatological sense, like the verb at 1.1, 99.10. Other passages speak of the denial of 'salvation' to the godless, 98.10,14; 49.1; 102.1; 103.10. In the Parables the 'salvation' of 'the righteous and elect' is 'in the Name of the Lord of spirits', 48.7; 50.3; at 50.3 salvation is defined as deliverance of the righteous at the last judgement, and at 50.2 the way to 'salvation' in this sense is by repentance. Here at 5.6, σωτηρία is defined positively as φῶς ἀγαθόν. See ThWNT s.v. σῴζω, Bd. VII, especially 982f., 984 (Foerster).

καὶ αὐτοὶ κληρονομήσουσιν τὴν γῆν. From v. 7(?) see below.

καὶ πᾶσιν ὑμῖν τοῖς ἁμαρτωλοῖς οὐχ ὑπάρξει σωτηρία ἀλλὰ ἐπὶ πάντας ὑμᾶς κατάλυσις [καὶ] κατάρα. Commentators since Dillmann (SAB, 1892, 1080) have emended the text to read κατάλυσις κατάρα. It seems more likely that καὶ has dropped out by hmt. The noun κατάλυσις gives an effective contrast to λύσις 'remission'. G κατάλυσις, שריא, P. Sm. 4315f. ܫܪܝܐ dissolutio, destructio, ܫܪܝܗ ܕܡܘܬܐ θανάτου κατάλυσις, An. Syr. 32.14. For the righteous remission of their sin: for the wicked dissolution and death. G = ולכולכון רשיעיא ליתה ישועה(?) אלא] 'על כולכון' שריא ולוטא].

(7) If the identification of the frg. at En[a] 1 ii 17 is correct, then the Aram. recension also omitted the long 'addition' in G at v. 6 καὶ πάντες … κατάρα.

light and joy For the 'light' of the elect, cf. 108.11-15. G χάρις 'grace' is equally appropriate and possible as the original reading: χάρις and χαρά are confused elsewhere, e.g. 2 C. 1.15, Phm. 7. Cf. Lods, 102. In Jewish, as in hellenistic usage, χαρά is especially 'festive joy': see ThWNT, s.v. χαίρω Bd. IX, 354.4f. (Conzelmann).

and they will inherit the earth from Ps. 37.11, referring originally to the possession of Canaan. Cf. Jub. 22.14, 32.19: just as Abraham inherited Canaan for ever, so Israel will come to possess (κληρονομεῖν, ירש, 'possess' or 'inherit') the whole earth. See Str.-B. III, 209. The idea is 'spiritualised' to mean 'to possess eternal life' Pss. Sol. 14.10, En. 40.9: cf. ThWNT s.v. κληρονόμος Bd. III 780.20f. (Foerster). Cf. Mt. 5.5.

(8) **Then shall wisdom … elect** As Charles, 13, notes: 'Here φῶς … γῆν is a doublet from v. 7, and τότε … ἐκλεκτοῖς a doublet of the first line of this verse'.

through sinning unwittingly Eth. lit. 'through forgetfulness' (rasi' = res'at, λήθη, Wis. 16.11). I have adopted the view of Dillmann, SAB 1892, 1043 and Lods, 103 that G καταληθειαν is a corruption of κατὰ λήθην, which gives a sense 'parfaitement approprié au contexte' (Lods); cf. Lev. 5.15,18; the reference is to שגגה 'unintentional error', LXX ἄγνοια; cf. 4 Mac. 1.5, λήθης καὶ ἀγνοίας οὐ δεσπόζει, 2.24. The corresponding Aram. is שלו, 'remissness', e.g. Dan. 6.5 (LXX ἄγνοια, Theod. ἀμβλάκημα); cf. also Ezr. 4.22, 6.9. G καὶ οὐ μὴ πλημμελήσουσιν (end of v.) implies 'errors' of this kind.

In an intelligent man … understanding These words, which are found only in G, are usually again explained as an expansion of the Greek translator, or as containing a variant form of the clause found in Eth. only, 'but those who have wisdom shall humble themselves' (see Knibb, 2,66). Again, however, there are features which suggest an Aram. source. The translation adopted assumes that the lines in G follow the Eth. clause 'but those who have wisdom will be humble' (yeganneyu = יתכנעון?) (also an integral part of the original, with its antithetic parallel line 'shall sin no more … from pride').

καὶ ἔσται (sc. σοφία) ἐν ἀνθρώπῳ πεφωτισμένῳ φῶς = [ותהוה באנשא נהירא נוהרא]. For נהיר = 'enlightened, intelligent' = sagax, P. Sm., 2299, cf. Wis. 8.19 ܛܠܝܐ ܗܘܢܢܐ = παῖς εὐφυής. Note the word-play with נוהרא, not so perfectly reproduced in G: for ܢܘܗܪܐ = illuminatio sc. sapientiae P. Sm., 2302.

καὶ ἀνθρώπῳ ἐπιστήμονι νόημα = [ולאנשא שכלא (סכלא?) שכלתנו]. For שכלתנו = νόημα cf. Dan. 5.11,12,14 (Theod. σύνεσις). Did the original perhaps read ולסכלא 'and to a foolish man' in contrast to an 'enlightened man'? Heb. סכל Tg. סכלא LXX ἄφρων, μωρός.

καὶ οὐ μὴ πλημμελήσουσιν [ולא ישליון](?): LXX πλημμελεῖν is used especially of peccata ignorantiae (cf. LXX Num. 5.6,7).

(9) **they shall not be condemned** Behind Eth. [οὐδὲ μὴ κριθήσονται] and G οὐδὲ

μὴ ἁμάρτωσιν lies a single phrase [ולא יתחייבון]. Eth. has a conflate reading. Unlike the wicked the righteous elect will escape condemnation.

the fury of (his) anger G ἐν ὀργῇ θυμοῦ corresponds to אפו חמה, Isa. 42.25, 'the fury of his anger', Tg. חימת רוגזיה.

they shall complete the number of the days of their lives As Lods interprets, men's lives would not be broken off prematurely, but would complete the full complement of their days: cf. 10.17, 25.6. This view seems to be shared by the author of 96.8. There is no question here, or elsewhere in this section of Enoch, of 'everlasting life'.

the times of their festivals(?) G = Eth. 'the years of their joy' is an improbable phrase. Behind τὰ ἔτη τῆς χαρᾶς αὐτῶν I suspect an Aram. עידני חדותהון, 'the times of their festivals': עידן '(definite) time, period' can also = ἔτος 'year' (e.g. Dan. 4.13(16),29(32)). Aram. חדוא like Heb. שמחה is used both for 'festive joy' and the 'festival' itself; cf. Num. 10.10 Pesh. ܚܕܘܬܟܘܢ 'festivitates vestrae' P. Sm., 1199. Cf. Barr, 'Aramaic-Greek Notes', I 197.

CHAPTER 6

(1) The relation of the text of Enoch to Gen. 6.1-4 is discussed below 124f. Vs. 1-7 have been preserved in a somewhat free and abridged quotation in the Chronicle of Michael the Syrian: see S. P. Brock, JThSt 19 (1968), 626-31; cf. J. B. Chabot, *Chronique de Michel le Syrien, Patriarche jacobite d'Antioche* (Brussels reprint, 1963), 1166-1199.

beautiful and comely Milik compares the phrase in Aram. inscriptions לטב ולשנפיר (see H. Donner and W. Röllig, *Kanaanäische und Aramäische Inschriften*, 244.2) and Greek καλὸς καὶ ἀγαθός. The two terms are synonymous, but the repetition is not tautologous; the expression means 'very or surpassingly beautiful'.

(2) **watchers, children of heaven** See the note on 1.5. Eth. = G οἱ ἄγγελοι υἱοὶ οὐρανοῦ, a free rendering of עירין. Lods thinks that 'sons of heaven' replaces 'sons of God' of Gen. 6.2, since the Judaism of the Greek period found the latter objectionable (cf. LXX Gen. 6.2 Ms A ἄγγελοι τοῦ θεοῦ).

lusted after them With Sync. ἀπεπλανήθησαν ὀπίσω αὐτῶν; cf. the use of Aram. טעי בתר = זנה אל, Num. 25.1 AV 'the people began to commit whoredom with the daughters of Moab', NEB '... began to have intercourse with Moabite women'. Cf. also Tg. Am. 7.17.

the daughters of earth I suggest that Sync. Syr. preserve the original (מן) בנת ארעא a phrase corresponding to בני ארעא at 91.14, En^{g} 1 iv 20.

(3) **Semhazah.** For the etymology of the name, see below, 119.

pay the penalty ... sin Eth. G Sync. ὀφειλέτης ἁμαρτίας μεγάλης. Cf. Mt. 6.12 par; Lk. 13.4 and Black, *An Aramaic Approach*[3], 140; ThWNT Bd. 5 s.v. ὀφείλω, 561 and ὀφειλέτης 565 (Hauck). Cf. Tg. Exod. 32.31 '... This people has been guilty (חב) of a great sin (חובא רבא)'. The idea of a penalty to be paid (so Eth.) as well as of guilt incurred is present in the Aramaic word.

(4) **bind ... with imprecations** The original verb is preserved at v. 5 En^{a} 1 iii 5 [ו]ואחרמ. For the construction and a parallel, see P. Sm. 1373 (foot): Sanct. Vit. 14 v ܝܡܐ ܗܘܐ ܘܡܚܪܡ 'he swore with imprecations ...'; ibid. 58r ܐܚܪ̈ܡ ܚܕ ܠܚܕ ܕܠܐ ܢܓܠܐ ܡܕܡ ܕܗܘܐ '... one bound the other with imprecations not to reveal anything that had taken place'. The reading of Syr. in our verse is an abbreviation of the fuller Aram. text. LXX (ἀναθέματι) ἀναθεματίζω renders Hiph. החרים (Aram.

אחרם) 'diris devovere', Dt. 13.15(16), Num. 21.3, and with a personal object Jg. 21.11, CD 9.1 (cf. Lev. 27.29). The oath proposed by their leader was for the watchers to invoke an imprecation calling for mutual annihilation, if any one of them (En[a] 1 iii 2) abandoned this their plan [1]. For Aram. סיר = ἀποστρέφειν Hoftijzer, 191, and cf. Jos. 11.15.

carry it out and do this deed Milik, 151, explains τελέσωμεν αὐτήν in G as a 'classicizing variant' of ποιήσωμεν τὸ πρᾶγμα τοῦτο; Sync. improves it to ἀποτελέσωμεν αὐτήν and omits the original phrase.

(6) **two hundred** According to Origen, Celsus believed the number to be 60 or 70; *Contra Celsum*, translated by H. Chadwick (CUP, 1953), 306 f.

descended ... Jared The connection of Jared etymologically with Heb. ירד = καταβαίνω and the word-play on Hermon and חרם led to the early hypothesis of Halévy of a Heb. original (JA, VI.9 (1867), 356 f.). The discovery of the original in Aramaic would appear to invalidate this hypothesis; cf., however, Knibb (2, 69) who argues for the occasional presence of a Hebrew word in an Aramaic text. More convincing is the view of Milik that such etymologies presume 'a sufficient knowledge of this sacred language on the part of their future readers' and 'cannot be advanced as an argument for the original Hebrew redaction of the Book of Enoch ...' (214). The possibility of a Hebrew as well as an Aramaic Book of Enoch cannot, however, be ruled out altogether.

Origen, in his *Commentary on John*, notes the word-play on Jared, and the observation is repeated in the *Onomastica* (Milik, 152, 214 f.). See Origen on Jn. 1.28, ed. A. E. Brooke (Cambridge, 1896), I, 160: 'Jordan is explained as 'their Descent' (κατάβασις αὐτῶν), and to this is related (so to speak) the name 'Jared' which is also itself interpreted as 'the Descender' (καταβαίνων), since he was the offspring of Mahalaleel, as is written in (the Book of) Enoch (if one favours accepting the book as sacred), in the days of the descent of the sons of God to the daughters of men—a descent which some claim prefigures the descent of souls into bodies, assuming that 'daughters of men' to be figurative language for the earthly vessel (of the body)'. Cf. Philo, *de gigantibus*, II 6-16.

Jub. 4.15 also relates the fall of the angels to the 'days of Jared'. See also Test. Reub. 5.6, and for the development of the legend in Jubilees, P. Grelot, 'La Légende d'Hénoch', 13 f.; in apocalyptic literature, Bousset-Gressmann[4], 332 f. and F. Weber, *Jüdische Theologie* (Leipzig, 1897), 253 f., for later echoes of the myth.

Hermon See Milik, 152, 215. Both Hilary and Jerome comment on this verse: Hilary, in commenting on Ps. 133(132).3 (ed. CSEL, 22 (1891), 689, 9-13, PL, 9, 748-9): 'Hermon autem mons est in Phoenice cuius interpretatio anathema est. Fertur autem id de quo etiam nescio cuius liber exstat, quod angeli concupiscentes filias hominum, cum de coelo descenderent, in hunc montem maxime excelsum convenerint'. Jerome, commenting on the same passage (ed. G. Morin, *Anecdota Maredsolana*, iii 2 (1897), 249, 28-250, 3.17-18; PL 26, 1293): legimus quendam librum apocryphum, eo tempore quo descendebant filii dei ad filias hominum, descendisse illos in montem

---

[1] Cf. Ac. 23.14(21). Paul's enemies inform the chief priests and elders ἀνεθεματίσαμεν ἑαυτοὺς μηδενὸς·γεύσασθαι ἕως οὗ ἀποκτείνωμεν τὸν Παῦλον ... The usual rendering follows the RSV 'we have strictly bound ourselves by an oath ...'. But is this translation forceful enough to convey the distinctive Hebrew idea of the ḥerem? They had vowed total destruction of themselves, like the rebel watchers, i.e., total devotion to death, (in their case) if they partook of food before they had killed the Apostle: '... we have bound ourselves by imprecations to touch no food until we kill Paul ...' Cf. ThWNT s.v. ἀνάθεμα Bd. I, 357 (Behm).

Ermon, et ibi inisse pactum quomodo venirent ad filias hominum, et sibi eas sociarent ... Ermon in lingua nostra interpretatur ἀνάθημα hoc est condemnatio. See also P. de Lagarde, *Onomastica Sacra* 27.5 Ermon anathema sive damnatio; 48.16, Ermon anathema eius vel anathema moeroris. (Milik, 215).

(7) Until Milik's *editio princeps* of the Aram. frgs. we were entirely dependent on the Greek and Eth. versions of the list here and on the Eth. version of the duplicate list in the Book of the Parables at 69.2. Moreover, these versions were extensively corrupted; the varieties of vocalisation of the Eth. names has given rise to a plethora of variations in the manuscripts.

The key in most cases to the decipherment of a name is to be found in the Aram. frgs., but as the list here too has suffered in transmission, it still contains a number of problematic names. In addition, the derivations and explanations of the names hitherto given leave room for further discussion. To simplify an approach to the complex and confused Eth. forms, I have given in the accompanying Table only the consonantal sub-structure of each name; differences in vocalisation which may be relevant for identification etc., and which are not obvious corruptions, are noted in the discussion of each name.

In the Aramaic fragments the names are listed by ordinal numbers following each name, e.g. 'Anan'el, thirteenth to him (i.e. to Šemḥazah, the leader and first name on the list) etc. The archetype of the list in the Greek version was arranged in four columns beginning with the third name, but instead of reading the names from left to right, the scribe of G read from top to bottom, column by column.[1] The list preserved in the citations from George Syncellus (Sync.)[2] lists the names by letters of the Greek alphabet. Eth. 6.7 has the same order as the Aramaic and Eth. 69.2 follows suit. At 8.3f. eight of the dekadarchs are named again and their functions described.

Table of Dekadarchs

	G	*Sync*	*Eth 6.7*	*Eth 69.2*
1 שמיחזה	σεμιαζα	σεμιαζας	smyz Eth^{g} sm'zz	smyz
2 [ארע]ת֯קף	αραθακ κιμβρα (sic)	αταρκουφ (ex αρτακουφ?)	'rkb	'rsṭqf Eth^{g} 'rṭqf
3 רמט֯(ע֯)[אל]	σαμμανη	αρακιηλ	rm'l	'rmn
4 כוכבא[ל]	χωχαριηλ	χωχαβιηλ	kkb'l	kkb'l
5 [אוריאל?]	ταμιηλ (ex 17)	οραμμαμη	ṭm'l	ṭr'l (ex 18)
6 רעמאל	ραμιηλ	ραμιηλ	rm'l	rmyl
7 דניאל	δανειηλ	σαμψιχ (ex 15)	dn'l	dnyl
8 זיקאל	εζεκιηλ	ζακιηλ	'zq'l	nq'l (sic)
9 ברקאל	βαρακιηλ	βαλκιηλ	brq'l	brqyl
10 עסאל	ασεαλ	αζαλζαλ	's'l	'zzl
11 חרמוני	αρεαρως (sic)	φαρμαρος	'rmrs	'rmrs Eth^{u} 'rmns
12 מטראל	βατριηλ	αμαριηλ (sic)	bṭr'l	bṭryl (+ bss'l)

[1] See Milik, 154.

[2] ed. W. Dindorf, *Georgius Syncellus*, Corpus scriptorum historiae Byzantinae, I Bonnae, 1829.

13 ענאל	αναvθα	αναγημας (sic)	'nn'l	'nn'l
14 סִיתוֹאֵ[ל]	ρακειηλ (ex 9?)	θαυσαηλ	zq'l (ex 8?)	ṭryl (ex 18?)
15 שמשי[אל]	σεμιηλ	σαμιηλ	smsp'l	smps'el
16 שהריאל	σαθιηλ	σαρινας	srt'l Eth^{g} str'l	ytr'l Eth^{m} ysr'l
17 [ת]מיאל	θωνιηλ ex 5 ταμιηλ(?)	ευμιηλ	om (but cf 5)	tm'l
18 טוראל	τυριηλ	τυριηλ	ṭr'l	ṭr'l
19 ימי[אל]	ιωμειηλ	ιουμιηλ	ymy'l Eth^{g} ymyl	rm'l (ex 6?)
20 זהר[יאל]	ατριηλ (sic)	σαριηλ	'ṛzyl	'zz'l

(7) **ŠEMḤAZAH** ('Sem seeth'). The yodh here (and in names with a similar formation in the list) is quite certainly nothing more than the familiar yodh compaginis, relic of an old case ending.[1] Noth, 123f., argues that the theophoric element ŠM, usually explained as 'Name' and a substitute for the actual name of the deity (so Lidzbarski, Eph. III, 263), in fact stands for the Phoenician god Esmun and is related to the theophoric names אשמרם, שמשלך from Elephantine, and to שמידע (e.g., Jos. 17.2), אשימה (2 Kg. 17.30, cf. Am. 8.14 אשמת שמרון?) and שמואל (e.g., 1 Sam. 1.20). Noth suspects that אשמן was a name widely known in Palestine and Syria for the youthful god of vegetation (125)[2], but prefers to leave the etymology *in dubio*. Whatever it may have connoted originally, the name probably came to be understood in Judaism as 'the Name (God) seeth'.

AR'TEQIF ('the earth is mighty') Milik, Ar'taqoph 'the earth is power', Knibb, 'the land of the Mighty One', with tqf as theophoric element. The name, however, seems to conform to the pattern of noun + verb (perfect), תקף (Dan. 4.8,17). Although two letters only are clearly visible in Aram. the reading ארעתקף is virtually certain; it is supported by 8.3 where the office of this angel is to teach 'the signs of the earth'[3]. The assortment of corruptions in the versions can be reasonably accounted for by this original reading. At 69.2 Eth^{g} 'Artaqifa is closest to the Aramaic. There is no need to emend to correspond to Sync. Ἀταρκουφ (Charles), since it is now clear that it is the latter which is the corruption. The spelling of most manuscripts at 69.2 'rsṭqf = ארצתקף is a hebraised form of the name.

RAMṬ'EL (RAM'EL) ('burning ashes of 'el'?). Although the decipherment of this name is very problematical, Milik's assumptions of a pan-Semitic root, Arab. rmḏ, Syr. rm', Jewish Aram. rmṣ, is probably the correct explanation, referring appropriately to 'the volcanic activities of the earth's crust' (155). The letters could be read as רמשאל 'evening of God' (Knibb), but this has no support in the versions. Sync. has Ἀρακίηλ, and tells us at 8.3 that the third angel teaches the 'signs' of the earth'; he has evidently confused the third angel with Ar'teqif. Ἀρακίηλ = ארקיאל seems a

[1] See M. Noth, *Die israelitischen Personnennamen im Rahmen der gemeinsemitischen Namengebung* (Stuttgart, 1928), 33f. The omission of the yodh in some names (e.g., Matr'el, Kokab'el, Ziq'el) shows that it was not articulated. Explanations such as '*my* Name has seen' (Milik, Knibb) are to be rejected. Šemḥazah conforms to the 'sentence-form' of proper names, noun + verb in perfect (Noth, 15f.)

[2] Šemḥazah taught 'root-cuttings' (En 8.3).

[3] G σημεῖα = אותות, i.e. meteorological phenomena; see Appendix A, 395 and above, 111.

variant form of ארעתקף (Knibb). Was it a later form substituted for the less well known Arʿteqif? With the form at 69.2 ʾrmn ʾΑρμήν(?) cf. G Σαμμανή, itself probably a corruption of ʿΡαμμανή, or is ʾrmn to be explained as a corruption of Ramiʾel of 6.7 (Knibb)?

KOKABʾEL ('star of el'). According to Sync. 8.3 he taught ἀστρολογία G σημειωτικά, 'signs' (see note). Could it be this angel who is referred to at Isa. 14.13 (כוכבי־אל following הילל בן שחר, 'Lucifer, son of the morning' (AV)?).

[ʾURʾEL] ('fire/light of ʾel). The proposal to read this name must remain conjectural, since the name of the fifth angel has been irretrievably lost in the Aram., and the versions (with the possible exception of Sync.) are probably filling the gap with names drawn from 17 and 18. Dillmann conjectured an original אורם אל behind Sync. ουραμμαμη, a name obviously corrupt in the last two syllables. The suggestion of an original ʾUrʾel (Uriel) might seem to be ruled out by Uriel's role as the second or third archangel in Jewish and Christian tradition (he is also Enoch's angelus interpres in later chapters, scarcely a role for a fallen watcher). Uriel was not, however, the original second archangel: the four archangels of En. 9.1 f. are Michael, *Sariel*, Raphael and Gabriel; G substitutes Uriel for Sariel,[1] and this perhaps reflects a later tradition which had come to assign to Uriel the role of guardian of the nether regions.[2] Uriel may well have had a place in the original list of the twenty dekadarchs and his name may not have disappeared accidentally from that list; some later theologian had another role for the Angel of Fire and Light. Eth. 69.2 reads ṭrʾl which Charles (followed by Knibb) suggests is a corruption of Tamiʾel at 6.7. It seems more likely, however, to be a duplicate of no. 18, probably to replace an unrecognisable name.

RAʿMʾEL ('thunder of ʾel'). At 69.2 the name is generally vocalised as Rumyal, possibly from ʿΡυμιαλ where the ending אל has been read with Aleph as mater lectionis, a spelling which occurs in a number of other names. The angel appears again in the duplicate Greek version of 20.7 as ʿΡεμείηλ, one of the seven archangels ὃν ἔταξεν ὁ θεὸς ἐπὶ τῶν ἀνισταμένων; at 2 Bar. 55.3 he is said to preside over 'true visions'. In Sib. Or. ii 215 f. he is one of the five angels appointed by God to bring the souls of men to judgement. See further Charles, *The Apocalypse of Baruch* (London, 1896), 96.

DANʾEL ('ʾel is judge'). Cf. Noth, 35, 92 (n. 4), 187. Milik identifies with 'the Canaanite hero Danʾel whose wisdom is sung by Ugaritic poems and mentioned in Ezek. 28.3 and 14.14, 20. According to Jub 4.20 Enoch marries the daughter of Danʾel (the Canaanite hero?). Danʾel is again mentioned along with Noah and Job at Ezek. 14.20. Sync. has Σαμψίχ and at 8.3 notes that the 'seventh angel taught the signs of the sun'. This corresponds to the fifteenth name on the Aramaic list where Sync. has Σαμίηλ. We have here at no. 7 in Sync. a duplicate of no. 15. The Eth. Danyal at 69.2 assumes a Greek Δανειαλ from דניאל where Aleph has again been read as vowel.

ZIQʾEL ('fire-ball of ʾel'). G seems to have been influenced by the LXX ʾΕζεκήλ at 1 Chr. 24.16. At 8.3 Ziqʾel teaches ἀστεροσκοπία (G), ἀεροσκοπία (Sync.), both

---

[1] At 10.1 the name of the messenger to Noah is given by Sync. as Ουριηλ but G reads Ισραηλ a form reflected in the Eth. versions; see Yadin, *Scroll of the War*, 328, n. 1. The source of these variants could be an original שריאל.

[2] Similarly IQM 9.14-16 has Sariel as one of the four archangels, in third place. Sariel is included among the seven archangels at 20.1 f. where he appears eighth on the list, with Uriel the first. According to 20.6 Sariel is in charge of 'sinful spirits', and in later cabbalistic sources he supervises the south wind. For his place in the archangelical hierarchy in different sources and periods, consult Yadin, *Scroll of the War*, 237 f., Milik 173.

general terms for the observation of the stars or the heavens. En^{a} 1 iv 3 זיקין is virtually certain, 'fire-balls, lighting flashes, comets?'. At 69.2 Eth^{M} Nuqa'el, Eth^{g} Neqa'el Νηκαηλ(?), where clearly a Zain of the original has been misread as a Nun, ניקאל for זיקאל.

BARAQ'EL ('lightning of 'el'). According to G he teaches ἀστρολογίαι Sync. ἀστροσκοπία (8.3); En^{a} 1 iv 2 [נחשי ברקין] 'signs of lightning' is again a virtually certain reading. The name has been correctly vocalised at 6.7 by Eth^{t} and at 69.2 by Eth$^{g\,t}$. Eth^{tana} at 6.7 reads Baraqyal, the reading of Eth^{M} at 69.2.

'ASA'EL (''el has made') is well attested as an Israelite name.[1] Its morphology varies (Milik, 131), but the two main forms are עש(ס)אל as here and עזאזל at 4Q EnGiants 7 i 6 (exactly as at Lev. 16.8,10,20) and עזזאל at 4Q **180** 1 7-8 (Milik, 313f.). At 69.2 the form in all texts is 'Azaz'el which could be a corruption of 'Asa'el but seems more probably to come from Lev. 16. This fallen watcher or archangel appears again (usually as 'Azaz'el) at 8.1, 9.6, 10,4,8, 13.1 and in the Parables at 54.5, 55.4, 69.2. At 6.7 and 69.2 he is tenth only in the list of the twenty dekadarchs, but he occupies (or comes to occupy) in Enoch and in later tradition a position almost of primacy, next to Šemḥazah, as the chief originator of all evil on earth: he is said at 9.6 not only to have been the archangel who revealed to mankind 'the secrets of heaven', but to have been the one who instructed them in 'all the iniquities on earth' (cf. 10.8 where 'all sin' is attributed to him). In the list of eight watchers singled out for mention at 8.1 f. in connection with their special functions, 'Asa'el comes first, his function being to instruct mankind in weaponry and the arts of seduction (the use of antimony in the beautifying of the eyelids and of all kinds of precious stones for adornment). At 10.4 and 13.1 'Asa'el is the first to be bound and confined in prisons under the earth until finally brought up and condemned to destruction at the last judgement (a legend alluded to at 2 Pet. 2.4 and Jude 6). In later rabbinical tradition it is not surprising to find 'Asa'el linked with Šemḥazah as the two chief fallen angels (Milik, 314, 322).

It seems likely that what we have already in Enoch is a developing role for the tenth fallen archangel in which he comes virtually to challenge and to be placed alongside Šemḥazah as one of the two leaders of the fallen watchers.[2] In view of this it is very strange to find him also cast in later traditions in the role of a beneficent angel. (Milik, 131).

There seems little doubt that 'Azaz'el at Lev. 16 is the same fallen archangel or watcher. Either 'Azaz'el has supplanted 'Asa'el (as he almost appears to have supplanted Šemḥazah (cf. Milik, 31)), or, more probably, עזאזל is a later formation bringing the name into line with a not uncommon form of an Israelite personal name.[3] The occurrence of the Biblical form at 4QEnGiants^{a} 7 i 6 may be explained either as the result of the influence of Lev. 16 (so Knibb, 2, 73) or vice versa as the dependence of Lev. 16 on earlier traditions about the fallen watcher 'Asa'el/'Azaz'el. However the connection is explained, 'Azaz'el has become the arch-demon of the later rabbinical sources and traditions, associated with Metatron in magical texts[4] and apparently also regarded as a kind of desert demon or godling.[5] He appears in fact as the forerunner or prototype in Hebrew demonology of Satan or Belial of later traditions.

---

1 Noth, 254 (no. 1116).

2 Cf. Bousset-Gressmann^{4}, 332, n. 4: 'Es scheint, als wenn hier zwei Traditionen, von denen die eine dem Semjaza, die andere dem Azael die Hauptrolle zuschrieb, verarbeitet sind'.

3 Cf. Noth, 160, 253 (nos. 1046, 1047).

4 See W. H. Rossell, *A Handbook of Aramaic Magical Texts* (1953), 106-107, No. 27, Milik, 129.

5 Cf. M. Gaster, *Chronicle of Jerahmeel*, XXV 13.

ḤERMONI (Milik ḤERMONI adj. "of Hermon', Knibb 'the one from Hermon'). The connection with Mount Hermon is obvious in this context, but the usual explanation that the angel derives his name from the mountain may not be the correct one, and the opposite may be true. חרמן occurs as a nom. pr. at Elephantine (Cowley, *Aramaic Papyri*, 22.4) where the theophoric element is the name of a subordinate deity of the Elephantine pantheon. See Noth, 129, 136, 243, connecting it with the root חרם 'to ban'; Hermon is the name of 'the god of imprecations' (Schwurgott) who punishes oath-breakers with his destroying curse (129). Mount Hermon could have been named after the deity, perhaps as a place with a cultic sanctuary where oaths were taken; it may also have been associated with practices related to the ban-execration such as 'spell-binding' or the practice of the black arts generally. Cf. 8.3 En[a] 1 iv 1, En[b] 1 iii 2, 'Hermoni taught wizardry, the loosing of spells, sorcery and the art of divination', G Ἀρμαρὼς ἐπαοιδῶν λυτήριον Sync. Φαρμαρὸς ἐδίδαξε φαρμακείας, ἐπαοιδίας, σοφίας καὶ ἐπαοιδῶν λυτήρια.

MAṬR'EL ('rain of 'el'). 69.2 has the variant form Batreyal Βατριαλ(?) (Eth[q] Eth[M] Bataryal), where the Mim has been read as Beth and the Aleph as a vowel after yodh compaginis.

After this name 69.2 inserts a name BASASA'EL which has no corresponding name at 6.7. Has it arisen as a corruption of 'Azaz'el? It is clearly an interpolation of a kind, probably a corruption, since it brings the list at 69.2 to twenty one names.

'ANAN'EL ('cloud of 'el'?) Cf. Noth, 35, Babyl. ANANI'EL; ענני and ענניה appear as proper names at Elephantine (Noth, 254, Cowley, *Aramaic Papyri*, 304). Noth connects ענן with Arabic 'anna 'to appear, show oneself' (184). More than one etymology is no doubt possible: in the present context after a name meaning 'rain of 'el', 'cloud of 'el' seems the most natural explanation.

SITHWA'EL(?) ('winter of 'el'?). This name must also remain uncertain, since one letter only of the Aramaic name has survived, and the only clue from the versions is Sync. Θαυσαήλ which, by metathesis, could give Σαυθαήλ, סתאל(?). The alternative suggestion of Knibb to read סתראל ('God has hidden') gives an attested name (Noth, 158 סתריאל), but has no support in the transcriptions. At 6.7 Eth. substitutes Ziqi'el from no. 8 and at 69.2 Tur'el from no. 18.

ŠIMŠ'EL ('sun of 'el'). En[a] 1 iv 4 8.3 'Šimš'el taught signs of the sun', Sync. τὰ σημεῖα τοῦ ἡλίου. The name was sometimes read as Samsape'el (Eth[M] at 6.7, 69,2), Sync. Σαμίηλ cf. no. 7 Σαμψίχ.

SAHR'EL ('moon of 'el'). The corruptions of the versions are extensive, but, if we take the consonants only behind Eth[g] S T R L, the result is סתראל where a He has been originally misread as Tau. Knibb suggests as a possibly alternative reading שחריאל ('dawn of God'), but rightly rejects it in favour of שהריאל.

TAMM'EL(?) ('el has completed'?). Milik Tummi'el (Perfection of God). The name has been omitted at 6.7, but cf. no. 5 in that list.

ṬUR'EL ('mountain of 'el'). This is the only name in the list where all transcriptions are in agreement.

YAMM'EL ('sea of 'el') or **YOM'EL** ('day of 'el'). The transcriptions seem to favour an etymology from יום 'day', but Milik may be right in suspecting an original ימאל ('sea of God') in this context (156). The name is a familiar one; see Noth, 246 and Gen. 46.10, Exod. 6.15 ימואל. LXX Ἰεμουήλ, a mistake for נמואל Num. 26.12, 1 Chr. 4.24?: 69.2 has Ruma'el which Charles (followed by Knibb) thinks is a corruption of Yoma'el = G Ἰωμεήλ, but it could also be a repeat of no. 6.

ZEHOR'EL ('brightness of 'el'), so correctly Knibb, Milik יהדיאל ('God will guide'). For Heb. זהר Tg. זיהור 'brightness, splendour (of the heavens)', cf. Ezek. 8.2,

Dan. 12.3. Aram. זיהרא is also used = סהרא Tg. 'moon', and hence Sync. 8.3. ὁ δὲ εἰκοστὸς ἐδίδαξε τὰ σημεῖα τῆς σελήνης: 69.2 has recourse again to the familiar 'Azaz'el.

The list at 6.8 concludes with a sentence preserved in Aramaic which Milik has translated, 'These are the chiefs of the chiefs of tens' and from this reading of the text goes on to ask (156, L. 13) 'Did the copyist want to say that the twenty angels were 'leaders of leaders because the 200 angels were in turn the chiefs of tens of other demons?' (cf. 152 where the Twenty are described as 'chiefs of decadarchoi'). This translation is based on a misreading of the Aramaic text: En[a] 1 ii 13 reads אלין אנון רבני(ן) (ורבני) עשר[תהן]. The scribe originally wrote 'These are their dekadarchs' and later corrected this by inserting a sublinear Nun after רבני and a supralinear ורבני (as above) to give a text 'These are leaders and leaders of their tens (i.e. their dekadarchs)'. In this way it is made clear that the Twenty were Chief Watchers, presumably archangels, and, at the same time, 'their dekadarchs', the 'chiefs of tens' of the two hundred watchers. Milik has failed to note the sublinear Nun or mistaken it for an insertion mark, and wrongly read the supralinear ורבני as רבני. The reading of En[b] 1 ii 17a has also an interlinear insertion at this point, but the text is too fragmentary to give results. G faithfully reproduces the longer Aramaic text and should now be read as οὗτοί εἰσιν ἀρχαὶ αὐτῶν οἱ δεκαδάρχαι (ms δεκακαι).[1] Eth. I has simply 'These are their chiefs of tens'; Eth II preserves some recollection of the original, 'These are the leaders of the two hundred angels ...'.

Some general observations may now be made on the lists. No. 5 is missing altogether in the Aramaic list and all the transcriptions show corruption. Some uncertainty must attach to 14 and 17, although, in the latter case, no alternative to an initial Tau seems at all plausible. Nos. 2 and 3 are, no doubt, debatable, especially in view of the state of the Aramaic text, but both seem to me to be reliable reconstructions, if the etymology of 3 must still remain dubious. This leaves a total of 15 names, 75% of cases, where the reconstruction or deciphering seems reasonably certain. The proposed etymologies are less certain in a number of cases, whether the form is to be explained as noun + verb (perfect) or construct + noun (e.g. no. 6 'thunder of 'el' or ''el thunders').

The most important observation has already been made by Milik (29), viz. the preponderance of names referring to or describing meteorological, astronomical or geographical phenomena (nos. 2, 3(?), 5(?), 6, 8, 9, 12, 13(?), 14(?), 15, 16, 18, 19(?), 20). Two or three names contain a theophoric element found in Babylonian or Elephantine sources (nos. 1, 11, 13(?)), and at least three have Old Testament counterparts (7, 10, 19).

Knibb has argued that the list at 69.2 is dependent on that at 6.7, explaining differences as 'inner-Ethiopic' variants: the list at 69.2 is an addition to the Ethiopic text of the Parables copied from the Ethiopic version of 6.7 and made during the course of the transmission of that text and not earlier (2, 76). The above examination of the evidence points to a different conclusion, viz., that the variants of 69.2 are best accounted for as coming from a different Greek version of the Aramaic names; and this applies not only to the examples given by Knibb of 'inner-Ethiopic' variants at 69.2 (nos. 3, 8 (Eth[g] Neka'el ex Νηκαηλ), while nos. 5, 19 are borrowings from other names and no. 16 (Eth[m]) and 20 assume two well-known names). Cf. also for additional evidence of Greek variations behind Eth. 69.2, no. 1 (69.2 = G 6.7; 6.7

[1] Swete read οἱ δεκα[δ]άρχαι correctly but omitted the earlier ἀρχαί in his edition, assuming a dittograph.

Eth^{g} = Sync.), 6, 7, 10 (69.2 cf. Sync.). The Book of the Parables is here translated from a Greek source with its own independent tradition; it had its own Greek list of angels transcribed from an Aramaic source, although comparison of the two Ethiopic lists points to the superiority of the transcriptions at 6.7, as a glance down the lists will show. The extra name after no. 12 Basasa'el looks like a corruption, possibly a formation from 'Azaz'el, a Biblical name which appears at nos. 10 and 20. The occasional repetition of a name in both lists (6.7: nos. 3, 6; 69.2 no. 14) points to texts probably so corrupt that translators, in desperation, simply repeat an earlier name.

ADDITIONAL NOTE ON GEN. 6 AND EN. 6

Before the discovery of 4Q En. no one had the slightest doubt that 'The entire myth of the angels and the daughters of men in Enoch springs originally from Gen. 6.1-4 where it is said that 'the sons of God came in to the daughters of men' (Charles, 14). The opposite conclusion is reached by Milik: 'The very close interdependence of En. 6-19 and Gen. 6.1-4 is perfectly obvious; the same phrases and analogous expressions are repeated in the two texts ... The ineluctable solution, it seems to me, is that it is the text of Gen. 6.1-4, which, by its abridged and allusive formulation, deliberately refers back to our Enochic document, two or three phrases of which it quotes verbatim ... If my hypothesis is correct, the work incorporated in En. 6-19 is earlier than the definitive version of the first chapters of Genesis' (31). This last statement makes it clear that 'our Enochic document' to which Gen. 6 refers back is not the 4Q 'Enochic document' reconstructed by Milik, but a putative source document and ancestor of the 4Q Enoch fragments, presumably also an Aram. composition, though it could quite conceivably have been a Heb. document.

The hypothesis has much in its favour. Older Enoch sources have undoubtedly been utilised at En. 6-19; the list of watchers is probably one such source. Moreover, one of the main features of the Enoch legend is the idea of the celestial origin of all knowledge and science, the art of writing, the science of the heavenly bodies, the knowledge of times and seasons, etc. (cf. Jub. 4.17f.); Enoch himself is cast in this role of 'revealer' of these celestial mysteries by authors almost certainly familiar with the Accadian legend, reported by Berossus, of Oannes and the 'seventh antediluvian sage-king'.[1] Similarly, in the watcher legend, it is not the corruption of mankind which is the overriding motif of the legend, but the celestial 'benefaction' of mankind by the revelation of the 'arts of civilisation', in this case with the working of metals and the manufacture of arms as the watchers' chief 'gift' to mankind. This central feature of the legend is nowhere represented in the Genesis narrative, which, by confining itself to the corruption motif, is able in this way to supply a mythological preface (the celestial origin of sin) for the Noah saga. Gen. 6.1-4 reads more like a deliberately brief excerpted notice of the familiar tale, not always felicitous in its abridgment, e.g., the somewhat banal introduction that, as mankind multiplied, daughters 'were born unto them', instead of En's 'surpassingly beautiful daughters were born to them'.

On the other hand, it may be argued that the 'additions' in Enoch to the Genesis narrative are free expansions of the author of Enoch, or, at best, represent old

[1] See P. Grelot, 'La Légende d'Hénoch', 5-26, 181-210; especially 23f.

mythological traditions woven around the Genesis story.[1] A more convincing argument for the priority of Genesis is the change from the expression 'sons of God' to 'children of heaven', like the LXX 'angels of God', to avoid giving offence to Jewish sentiment in the Greek period (see note above on v. 2). Perhaps even more important is the fact that two other fragments, 4Q 180, 181 actually quote verbatim from Gen. 6.1 (Milik, 314).

On the whole the balance of evidence — and literary borrowing one way or the other is obvious — seems to favour the priority of the Enoch tradition, since it is, after all, *the* Enoch legend of the watchers to which Gen. 6 is so briefly alluding. A possible alternative explanation of the inter-relationship of the two narratives is that both, Gen. 6 and En. 6, are descended from a common literary ancestor, which need not have circulated only in an Aram. form.

CHAPTER 7

(1) **These (leaders) and all the rest (of the two hundred watchers)** The shorter text of Eth II G καὶ ἔλαβον ἑαυτοῖς γυναῖκας is ambiguous: it could have as subject either 'the (two hundred) watchers' or simply 'the dekadarchs'. The longer text of Sync. may have been designed to remove this ambiguity. On the other hand, it could be original, and there is space for it in the manuscript of En^{a} 1 iii 13. The rest of the Sync. text is expansion: it is difficult to see how he arrived at the dating of the age of the world as 1170 years from the events of Gen. 6.1-4; the Biblical genealogies at Gen. 5 would give a much larger figure.

took for themselves wives ... from all whom they chose As Milik first noted (156, 132, cf. 'Problèmes de la littérature hénochique à la lumière des fragments araméens de Qumran', HThR, 64 (1971), 349 f) these words are identical with Gen. 6.2; literary interdependence is beyond question. See above, 124f.

to cohabit ... defile themselves with them G εἰσπορεύεσθαι πρὸς αὐτάς = למעל עליהן = Gen. 6.4; cf. En. 9.8-9: בוא אל = עלל על, 'to go into' = coire cum femina. LXX μιαίνειν = Heb. טמא, Aram. סאיב, in sensu sexuali, Gen. 34.5,13,27 etc. (μιαίνεσθαι = לאסתאבה Milik). (The Ēth. variant [καὶ συνεκοιμήθησαν αὐταῖς] may have come from 9.8). The idea in Sync. that the period of illicit relations between the fallen watchers and the daughters of men lasted till the Flood (ἕως τοῦ κατακλυσμοῦ) has probably based on Gen. 6.4,5, where the account of the Flood follows immediately after the story of the corruption of the 'sons of God'.

sorcery and spells... cutting of roots ... herbs The reading חרשה at En^{a} 1 iii 15 is quite certain: cf. ܚܪܫܐ = φάρμακα, Rev. 9.21, P. Sm. 1387. At 8.3 En^{a} 1 iv 1 reads חברא, Heb. חבר, for which LXX has ἐπαοιδή 'spells'. The noun ῥιζοτομία (מגזר שרשין?) does not occur in the LXX, but is found in Theophrastus, *Historia Plantarum*, 6.3.2, 9.8.2; cf. the use of ῥιζοτόμος 'a cutter of roots', 'herbalist', for purposes of medicine or witchcraft (Liddell and Scott, 1570). G βοτάναι = Heb. עשב Aram. עסבא Tg., 'herbs', Eth. ʿeḍaw (Knibb 'trees'). Cf. Josephus's statement about the Essenes in B.J., ii.8.6 (136) (ῥίζαι ἀλεξητήριοι) and Ant. viii 24 (44) (θεραπεία, ἐπῳδαί, ἐξόρκωσις). Healing and magic went together.

(2) **great giants ... to their strength** En^{a} 1 iii 17 has a clause 'were born upon earth',

---

[1] Cf. O. Eissfeldt, *Einleitung in das alte Testament* (Tübingen, 1956), 652.

absent in the versions, unless there are traces to be found in the long addition of Sync. about the γένη τρία of the watchers, the giants, their offspring the Nephilim and their grand-children the Elioud (τοῖς Ναφηλεὶμ ἐγεννήθησαν 'Ελιούδ). This addition is almost certainly a gloss on the 'giants' and Nephilim of Gen. 6.2 (see further below on 16.1), but the clause καὶ ἦσαν αὐξανόμενοι κατὰ τὴν μεγαλειότητα αὐτῶν, as Milik, 150, conjectures, may supply part of the lacuna at En[a] 1 iii 17. If we read די לא before הוו מתילדין we obtain a defensible supplement, 'and there were not born upon earth off-spring (ילודין) which grew to their strength'. The loss or omission of the negative may have given rise to the idea of descendants of the watchers being called Elioud, from ילודין 'offspring', En[a] 1 iii 17? Cf. Barr, 'Aramaic-Greek Notes', I.191.

(3) **fruits of men's labour** For עמל = κόπος, κόποι Ps. 105.44 'fruits of labour'.

unable to sustain them At Sir. 25.22 ἐπιχωρηγεῖν renders Heb. כלכל. Was there a corresponding Aram. Palpel כלכל? With יכילו ἐδυνήθησαν and אכל = κατεσθίειν it would give a characteristic play on words.

(4) **treated them violently** En[a] 1 iii 19 has a clause 'and they began (ושריו correctly read by Knibb) to slay mankind', which I suggest came after the original behind G ἐτόλμησαν ἐπ' αὐτούς. For G τολμᾶν ἐπί a possible Aram. equivalent is ארשיע על; cf. Tg. Exod. 21.14 = Heb. זיד על 'to act presumptuously, insolently against', Dt. 1.43, LXX παραβιάζεσθαι. (The same verb may lie behind ἁμαρτάνειν in v. 5: see note there). An alternative would be אמרחו בהן (P.-Sm., 2222), 'savaged them'. Eth. reads tamayṭu for ἐτόλμησαν and is usually rendered (as by Charles) 'turned against them', i.e. mankind (Flemming-Radermacher 'turned against themselves'). The Eth. verb corresponds to שוב, תוב, ἐπιστρέφειν, which we then require to understand as 'turned against'. At v. 5 the verb is repeated in Eth[q] meaning 'they in turn (tamayṭu) did violence to the birds'. Should the Eth. translator here have read ἐπέστρεψαν καὶ ἐτόλμησαν 'they turned and savaged (them)'?

began to slay mankind G Eth. 'and devoured mankind'. According to Sync. 8.3 the devouring of men is the last crime of the giants. The Aram. has preserved the correct reading, which has been changed by the versions to the more heinous crime still of anthropophagy, no doubt influenced by the later statement at 8.3 Sync. that the giants not only killed, but devoured their human victims. Cf. Knibb, 2, 78.

(5) **began to do violence to ... the birds** G ἁμαρτάνειν seems here to render ארשיע (Aph. רשע 'to harm, do violence to' Eth. 'abbasa I.2 = G). The Greek translator has given the verb the meaning of רשע in Heb. and Syr., 'to act wickedly, sin'. The supplement 'to attack' lit. 'to rise up against' = למקם מן קובל is dependent on the correct identification of [ק]ו̊בל: an alternative would be to supply למקרב לקובל = lit. 'to war against'.

that crawl upon the earth As Milik, 157, Ll. 19-21 (cf. Knibb, 2, 79) notes, En[a] 1 iii 19-21 has a longer text than G Eth.: the supplement '[that crawl on the earth]' is supplied from Gen. 1.26.

devour their flesh ... drinking the blood I have read En[a] 1 iii 21 as בשרהן 'their flesh', i.e. the flesh of beasts, reptiles etc. It was the drinking of the blood which was the breach of divine law (Gen. 9.4, Jub. 7.28,29, En. 98.11, Ac. 15.20). According to G Eth the giants are guilty of cannibalism and drinking their own blood.

(6) **the lawless ones** G ἄνομοι = רשיעין(?). The verse (and chapter) ends in En[b] 1 ii 25 with מתעבד (Plate VII frg k.): Milik supplements '... the earth made ... accusation against the wicked concerning everything which was done upon it'. Nothing corresponding has survived in G Eth.

CHAPTER 8

(1) **Asael** G Ἀζαήλ Eth. ʿazazʾel i.e. Asael, cf. above 6.7, En^a 1 iii 9.

to make swords ... war Sync. is closer to En^b 1 ii 26 than G. Both G and Eth. (and Sync.?) seem to be 'growing texts'. The phrase 'every weapon for war' = Heb. כל כלי מלחמה, cf. Jg. 18.11,16,17, Tg. מאני קרבא, Ec. 9.18 כלי קרב. The addition in G 'the teachings of the angels' looks like a translator's gloss or marginal note indicating the contents of this chapter, viz. what each of the eight selected fallen watchers taught, (cf. Lods, xxxi, n. 1). But Sync. 'every weapon for war' looks original.

metals of the earth G Sync. μέταλλα = מטלון, מטליה? Cf. Levy NCW, III, 89, ܡܛܠܐ P. Sm. The word appears to have been unknown to Eth. which seems to render it as μετʾ αὐτά.

gold ... for women The scribe of En^b 1 ii has inserted above the line at 27a words accidentally omitted in line 27. All the versions shorten the text: the mention of 'silver' by Sync. supports the restoration of דהבה 'gold' at En^b 1 ii 27a ('silver' for 'bracelets' could hardly be mentioned without an earlier reference to gold 'for adornments'(?)). For תיקונין = κόσμος, (women's) ornaments, cf. Jer. 4.30 Tg. תיקונין דדהב = עדי זהב.

dyes ... varieties of adornments G Sync. τὰ βαφικά apparently = τὰ βάμματα = צבעין. Cf. Eth. ṭemʿātāta ḫebr, 'variegated dyes' with βάμματα διάφορα, P. Oxy. 9.14.7 Liddell and Scott, 305. The reading of Eth$^{g\ (tana\ q)}$ 'and the world was changed' is clearly corrupt: Eth. tawlāṭa ʿālam I suggest renders a phrase like (ἀντ)αλλάγματα κόσμου = חליפית תיקונין 'varieties of adornments'. The extraordinary reading of Eth^g. can only be explained as the work of a translator who interprets these innovations of Asael as indeed revolutionary changes.

and the children of men ... transgressed This text of Sync. seems to me to be original since v. 2 begins to refer to the growth of 'godlessness' (ἀσέβεια) among them (cf. Gen. 6.5), and this is only made clear by the Sync. text here and later by the addition of ἐπὶ τῆς γῆς at v. 2. At the same time, the words καὶ ἐπλάνησαν τοὺς ἁγίους seem a moralising addition, perhaps coming from ἀπεπλανήθησαν of v. 2.

For echoes of this verse in Tertullian *de cultu fem.* i. 2, See Charles, 19, especially ii. 10, 'Quodsi iidem angeli qui et materias eiusmodi et illecebras detexerunt, auri dico et lapidum illustrium, et operas eorum tradiderunt, et iam ipsum calliblepharum—tincturas—docuerunt, ut Enoch refert'. Cf also Test. Reub. 5.5,6 and Isa 3.16-24.

(2) As the readings of Eth I show, the Eth. corresponded for the most part to the text of G: Eth II in both readings represents, not variants, but an inner-Eth. scribal rewriting and rephrasing of the verse.

they committed fornication ... astray Milik's reading of frg. p והווא פח[זין] is very probably correct: for פחז in this sense cf. P. Sm. 3080, ܦܚܙܐ, Sir. 23.17 = πόρνος. The line in En^b 1 iii appears to have been shorter than that presupposed by G: could ἀπεπλανήθησαν be an interpretative expansion of ἐπόρνευσαν, understood as 'became idolatrous'?

corrupted their ways The text of Sync. has the exact Biblical locution e.g. Gen. 6.12, השחית ... דרכו, Isa. 1.4, Tg. חבילו אורחתהון.

(3) **spell-binding** Milik, 157 reads חבר[ו] and renders 'spell-binding'. Heb. חבר is rendered by ἐπαοιδή(αί) at Dt. 18.11, Tg. חבורא. In the second half of this verse, however, En^b 1 iii 2 reads חרש where G Sync. have ἐπαοιδαί. The term חרשא Syr. ܚܪܫܐ is a general one for ars magica, embracing 'charms', 'potions' (τὰ φάρμακα) or

'incantations' (P. Sm. 1387). Eth. [ἐπαοιδούς] assumes a masc. חרש, ἐπαοιδός and similarly for [ῥιζοτόμους].

Sync. ἔτι δὲ καὶ ὁ πρώταρχος αὐτῶν looks like a 'classicising' expansion. Flemming-Radermacher emended the scarcely intelligible εἶναι ὀργάς to ἐπαοιδάς and are followed by Charles, 280, although the corruption is transcriptionally unconvincing. Even more difficult is the following κατὰ τοῦ νοός, 'charms against the mind'?: it is probably also an irrecoverable corruption, unless one is prepared to see in it a rendering of על ידועני, περὶ τῶν γνωριστῶν '(Semhazah taught) about sooth-sayers (or sooth-saying?)'. This would imply that the Sync. quotation goes back to a longer recension than that of the Aram. frgs. Sync. has also 'roots of herbs of the earth'. The phrase 'herbs of the earth' is Biblical (Job 5.25, Ps. 72.16); the full expression is Aram. would read שרשי עסב ארעא. If Sync. has preserved the original reading, G Eth. is a 'graecising' of it to the familiar term ῥιζοτομία / ῥιζοτόμος.

In the descriptions of the duties of these watchers, G = Eth. has six names only, Sync. has eight; and the text of Sync. is clearly superior to that of G. The principle behind the selection of the eight watchers, however, is not clear: Sync. cites each correctly, by the cardinal number of its place on his list.

Hermoni ... sophistry On Hermoni, see above on 6.7. Aram. כשפו abstr. is not attested elsewhere; here = Heb. כשף, a word found only in plur. כשפים, 'sorceries'. Aram. חרטמו abstr.is also hitherto unattested, but cf. Heb. חרטם, 'magician': תושיה is attested only in Heb. = 'skill', probably in this context of the 'sophistries' of the magician: cf. Sophocles, *Oed. Tyr.* 501-502. Again Sync. is closest to the Aram: φαρμακείας = כשפו, ἐπαοιδάς = חרטמו, and σοφίας = תושיה.

Baraqel ... lightning While the Aram. frgs. are defective, this reconstruction is quite certain: the same form of words is used in the description of the duties of each of the eight watchers; and in each case the 'auguries' (נחשא, En^{b} 1 iii 3) they teach are related to the etymology of their names. The translator into Greek was obviously unacquainted with many of the terms, and in both G and Sync. general astronomical words, ἀστρολογία, ἀστεροσκοπία, ἀεροσκοπία are substituted for the original terms. Again Eth^{M} renders (א)נחש as a masc. noun ἀστροσκόπους.

The term נחשא Heb. נחש, Syr. ܢܚܫܐ means 'auguries' from נחש 'to practise divination, observe signs' (G σημεῖα), in this context 'doubtless horoscopes and auguries taken from the positions of the stars and from natural phenomena' (Milik, 160, Ll. 2-4).

Zikiel ... fire-balls The Aram. זיקין can mean either 'storms' or 'fire-balls, meteors'; cf. Tg Job 27.20 = סופה ' tempest', Exod. 20.2,3, Tg. Jer. (P-J) 'like fireballs (זיקין) and lightnings (ברקין)'. Cf. also ܙܝܩܐ, P. Sm. 1118. The 'astronomical' context favours the second meaning here.

auguries of earth probably means meteorological or climatic predictions.

secrets to their wives Sync. adds 'and to their children'. The further 'addition' of Sync. μετὰ δὲ ταῦτα ... ἐπὶ τῆς γῆς probably belongs after 7.5. See above, on 7.4 and below on v. 4.

(4) This verse takes up the thread of 7.5f. (Ch. 8.1-3 may even be an interpolation from a separate source into the original Aram. narrative).

The Sync. text introduces extra material at the end of Ch. 8 which links Ch. 9 more closely with 7.5f. viz. μετὰ δὲ ταῦτα ἤρξαντο οἱ γίγαντες κατεσθίειν τὰς σάρκας τῶν ἀνθρώπων καὶ ἤρξαντο οἱ ἄνθρωποι ἐλαττοῦσθαι ἐπὶ τῆς γῆς. The 'addition' was certainly not present in the two Aram. copies which have survived, but it has an authentic ring, and could come from a longer recension. It introduces a contrast to 6.1: instead of mankind 'multiplying', they now, as a result of the cannibalism of the giants, begin to diminish (ἐλαττοῦσθαι = Heb. מעט, Aram. זער). It is unlikely that the frg. at En[a] 1 iv 5 מן ארעא could correspond to ἐπὶ τῆς γῆς (Knibb, 2, 84).

their voice ... Great One Aram. קלא, G βοή, but cf. 22.5 f. for the rendering of קלא by 'voice'. The translation follows the text of Sync[2]. Both Sync[1] and Sync[2] have a much fuller version, but I agree with Charles, xvii, 20, that 'it is natural that the substance of the prayer of men as they were slain by the giants should be given when it is first referred to in 8.4. ... G[s] in Semitic fashion gives the prayer *in extenso* here also (9.3)'. In that case Sync[2] must be rendering a longer recension than that of the two Aram. copies, En[a] 1 iv and En[b] 1 iii since neither has room for the addition in Sync[2]. Contrast Milik, 160, Ll. 5-6, Knibb, 2, 84.

glory of the Great One See below on 14.20. Sync. 'before the Lord of all Lords in greatness' seems a translational expansion.

CHAPTER 9

(1) **Michael, Sariel, Raphael and Gabriel** As Milik, 172, notes '... this is the only place ... where the name of the second archangel (Sariel) is preserved; it occurs three times in Enoch: here, in 10.1 and 20.6'. This is correct so far as the Aram. is concerned, but 'Suryal' is found in many Eth. mss here. At IQM 9. 14-16 the shields of 'the men of the towers' are to be inscribed with the names of these four archangels, one name on each shield, in the order Michael, Gabriel, Sariel (שריאל), Raphael. In all other lists, in the versions and rabbinical sources, while the order of the names varies, Michael, Gabriel and Raphael remain constant, but Sariel is replaced by Uriel (with the sole exception of the list in the Parables where it is replaced by Phanuel).[1] It would seem that 9.1 Aram. preserves an old, probably the oldest, tradition of the names of the four archangels.

It is worth noting that, while G and Sync. here agree with each other, no Eth. ms reproduces exactly the Greek list. What is also surprising in the Eth. is the omission of Raphael in most Eth. mss, and the names of three archangels only in Eth[g (tana)]. No doubt these differences, and the differing order of the names, reflect Eth. traditions. In the occurrence of 'Suryal', the oldest of all the traditions has been faithfully preserved.

the sanctuary in heaven En[a] 1 iv 7 [שמיא] מן קדש, Sync[1, 2] ἐκ τῶν ἁγίων τοῦ οὐρανοῦ. Here τὰ ἅγια = קדש(א) Heb. קדש (cf. Dan. 8.13, LXX τὰ|ἅγια = קדש, 'sanctuary', 'Temple'), presumably refers to the heavenly Temple of God (cf. En. 14.15 f.) (or the heavenly Dwelling-place of God, as at Sib. Or. 3. 307) rather than to heaven as itself God's 'holy place' (cf. Wis. 9.10, ἐξ ἁγίων οὐρανῶν).

the whole earth was full ... against it The Aram. text echoes the language of Gen. 6.11. In the reconstructed text [ה]ל [ין]חטי the טי is quite certain: Milik's [די את]חטי עליה 'so that sin was brought upon it' implies an unattested meaning for Ithp. חטי; the use of חטא ל with an antecedent defining the nature of the 'sin' or 'aberration' is well attested, e.g. Heb. Jer. 33.8 Cf. 106.14 ἁμαρτάνουσι Eth. yegabberu ḫati'ata = [חטי]ן.

(2) **And they went in ... to the angels(?)** The 'four great archangels' (Sync[1, 2] οἱ τέσσαρες μεγάλοι ἀρχάγγελοι) are envisaged as looking out and down (Aram.

[1] See Yadin, *Scroll of the War*, 238. Sariel does appear again at 20.6, but in the list of the seven archangels.

אדיק = Heb. שקף) upon the earth from the heavenly Temple. According to all the versions they then went inside (Sync[1, 2] εἰσελθόντες) for a mutual conferring (εἶπον πρὸς ἀλλήλους as at 6.2). En[a] 1 iv 9, however, reads ואמרו קדמ, which Milik reads as קדמי[הן] rendering 'and said to themselves', but noting (160, Ll. 8-10) that the correspondence of the Aram. and Greek reading 'is not clear to me'. The expression אמר קדם = אמר לפני (e.g. Ezek. 28.9) and means 'to speak, say before, to', dicere coram, a more solemn form of אמר ל: it cannot, on any reconstruction of the suffix, be equated with (καὶ) εἶπον πρὸς ἀλλήλους. The restoration of the latter to πρὸς ἀγγέλους gives a reading supported by the Aram. frg. It also solves the problem of v. 3, where, according to Eth., the angels are addressed by the returning archangels without having been previously mentioned. Cf. Milik, 158, En[a] 1 iv 10, where a clause (without versional support) is introduced to deal with this difficulty. The reading at En[b] 1 iii 10, 10a בני ארעא, is firm (cf. 91.14, En[g] 1 iv 20).

(3) **making their suit ... with groans** = Sync[2], but cf. En. 22.5, En[e] 1 xxii 4.

'Bring our case ... Great One Sync[1] G κρίσιν ('case') = דינא; Sync[2] δέησιν ('request') may be a textual corruption or a deliberate alteration: both nouns make good sense in the context. For the expression 'the glory of the Great One', see above, 104f. The addition of Sync[1] ἐνώπιον τοῦ κυρίου ... τῇ μεγαλωσύνῃ seems an expansion of the translator (from v. 4?)

On the intercession of angels, see Charles, 20 (n. 3), 24 (n. 10) and especially *The Testaments of the Twelve Patriarchs* (London, 1908), 33f. (on Test. Levi 3.5) ThWNT Bd. I, 81.23 s.v. ἄγγελος (Kittel).

the Most High Heb. עליון, Dan. עליא (short for אלהא עליא) and עליונין LXX ὁ ὕψιστος (θεός). Also at En. 10.1; 77.1; 94.8; 97.2; 98.7,11; 99.3,10; 100.4; 101.1,6,9; and in the Parables at 46.7; 60.1,22; 62.7. The full form 'the highest God' is found in Eth. 21.6; 40.10; 61.9 (all in Eth II).

(4) **Then Raphael(?) and Michael ... went in** This reconstructed text is based on the reading at En[b] 1 iii 13 '... and Michael...' together with Sync[2] καὶ προσελθόντες κ.τ.λ. Milik, 174, Ll. 13-16 argues that Raphael and Michael alone 'enter', i.e. the divine throne-room (cf. En. 14.15-25), as two messengers. Sync[2], however, mentions again 'the four archangels'. One would expect Michael not Raphael to come first, but the order in Ch. 10, Sariel, Raphael, Gabriel and Michael last, shows that too much importance cannot be attached to the order of the names.

Lord of the ages Sync[1] (also θεὸς τῶν αἰώνων and cf. 22.14 G κυριεύων τοῦ αἰῶνος). G βασιλεὺς αἰώνων. The titles occur at 12.3; 25.3,5,7; 27.3; 58.4; 81.10; 106.11 (τὸν κύριον τοῦ αἰῶνος). The title is attested at IQapGn 20 (Frg 2) and 21.2 מרה עלמיא; see Fitzmyer, *Genesis Apocryphon*, 77, and as מרא עלמא, in a Palmyrene inscription, c. 150 A.D.; see M. A. Levy, 'Drei palmyrenische Inschriften' in ZDMG, Bd. 15, 616. For βασιλεὺς τῶν αἰώνων cf. 12.3, 25.7 and Fitzmyer, op. cit., 84, E. J. Weisenberg, 'The Liturgical Term 'Melekh Ha-ʿOlam' in JJS, 15, 1964, 1-56.

The Greek and Eth. texts seem to be elaborating on the original expression by drawing on O.T. titles, e.g. from Dt. 10.17 LXX θεὸς τῶν θεῶν, κύριος τῶν κυρίων. For 'king of kings' Ezek. 26.7 (but for the king of Babylon), Dan. 2.37, Ezr. 7.12. See S. A. Cooke, *Glossary of Aramaic Inscriptions*, 77.

our great Lord For the מרא title of God, see Fitzmyer, *A Wendering Aramean*, 87 f.

Thy glorious throne For the expression כסא כבוד, Jer. 17.12 and the concept, ThWNT, Bd. III, s.v. θρόνος especially 161 f. (Schmitz).

Thy Name ... ages G has three adjectives, 'τὸ' ἅγιον καὶ μέγα καὶ εὐλογητόν, Sync two ἅγιον καὶ εὐλογήμενον, Eth. has a conflate text with ἔνοξον for μέγα.

The two additional words in Eth. 'blessed and glorious' probably arose as a marginal gloss to indicate that εὐλογητόν should precede ἔνδοξον. Cf. Flemming, 9 and Zuntz, JThSt 45 (1944) 166, n. 3.

(5) **power (ἐξουσία) over all** Cf. Dan. 4.14 LXX Sir. 10.4 and see ThWNT Bd. II, s.v. ἐξουσία, 562.44 f. (Foerster).

(6) **the eternal mysteries ... heaven** The term μυστήριον(α) in Aram. is רז (רזין, רזיא) a Persian loan-word, Dan. 2.18,19,27,28,29,30,47, meaning lit. 'secret(s)' cf. Heb. נסתרות. It came to be a key eschatological concept in apocalyptic literature, probably originally deriving from its use in Enoch, where it occurs again at 10.7; 16.3; 104.10; 106.19, and in the Parables at 61.5; 65.11; 68.1; 69.16 f.

The motif of a divine revelation of celestial secrets is probably to be traced to Babylonian sources; see especially Grelot, *La Légende d'Hénoch*, 5-26 (conclusion, 23), 181-210. For the development of the idea in Jewish apocalyptic, ThWNT, Bd. IV, s.v. μυστήριον, 815 f. (Bornkamm) and at Qumran, F. Nötscher, *Zur theolog. Terminologie der Qumran-Texte*, 71-75.

The translation follows the longer text of Sync. On a first reading, it seems an expanded paraphrase, but it exhibits features which could be original, e.g. the mention of δόλος (רמיה, רמיו) as one of Asael's legacies to mankind, and the use of ἐπὶ τῆς ξηρᾶς, for 'on the land', parallel to ἐπὶ τῆς γῆς (if this is not a duplicate version of the original על יבשתא). The phrase ἐπὶ τῆς ξηρᾶς is translation Greek.

his abominations ... make Sync. G both have ἐπιτηδεύουσιν, Aram. עתד, יעתדון, Pa. = ἑτοιμάζειν or κατασκευάζειν. Syr. ܥܘܬܕܐ = ἐπιτηδεύματα Hex. Ezek. 8.15 = תועבות 'abominations', cf. P. Sm. 3009 = ἐπιβουλή, insidiae. The reference would appear to be to the mysterious techniques for working in metals, etc., as well as the practice of sorcery at 8.1 f.

Milik, 158, 161, L. 20 suspects that the Greek 'interpreter misread the participle ידעי, 'those who know, experts, artisans', for the perfect ידעו, hence his translation ἐπιτεδεύουσιν ἔγνωσαν ἄνποι'. In his reconstruction (158), he prefers the reading of Sync. υἱοὺς τῶν ἀνθρώπων and reads a text at En[a] 1 iv 20 די להן יעבדון ידעי בני אנשא 'so that the experts among the sons of men should practise them' (the revealed mysteries of heaven). This is an attractive suggestion, accounting for an extremely confused text. The reading of G, however, is probably best explained as a mis-rendering of an Aphel אודע = ἐγνώρισεν as ἔγνωσαν followed by ἀνθρώποις (ms ανποι). But Sync. εἰδέναι τὰ μυστήρια, could well be a mistranslation of ידועי רזין, γνωρισταὶ γνωσταὶ(οἱ) μυστηρίων. For ידועא, cf. P. Sm., 1558. The reading of En[a] 1 iv 20 (Plate IV frg k) is problematical (להן would require to be taken as 'made for themselves'). The translation depends, however, on the text of Sync. ἐπιτηδεύουσιν δὲ τὰ ἐπιτηδεύματα αὐτοῦ: the following εἰδέναι τὰ μυστήρια must be a corruption of ידועי רזין and we should perhaps read ידועי רזין די (בני) אנשא, '(and his abominations) the initiates (lit. the 'knowers of the secrets') among the children of men make for themselves.

(7) **Semhazah instructed ... spell-binding ... spell-binders** Sync. G have a foreshortened text. The Greek version τῶν σὺν αὐτῷ ἅμα ὄντων (common to Sync. and G and behind Eth.) is an obvious mistranslation of חבורין, 'spell-binders', (e.g., Tg. Neoph. Dt. 18.11: see also Levy, CW, I, 237), which has here been confused with חברין = Heb. חברים, 'companions, associates'. Semhazah, who taught mankind spell-binding (8.3) is the fallen watcher put in charge of those who cast spells, i.e. magicians and sorcerers; cf. 40.9 where the archangels are said to be 'over' certain spheres, e.g., Raphael is 'over ... the diseases ... of ... men'.

(8) Cf. 7.1.

they were defiled by the females Sync. cf. Eth. ἐν ταῖς θηλείαις ἐμιάνθησαν may preserve an original reading; cf. Lev. 18.20 Tg. לאסתאבא בה. The addition τῆς γῆς '(the daughters of men) of the earth' seems a tautology or does it represent an original θυγατέρες τῆς γῆς בנת ארעא? See above at 6.2. Eth. has read the Sync. text, but has several corruptions: (1) ḫebura is a ditt. from v. 7; (2) it omits καὶ before ἐν ταῖς θηλείαις, and inserts it before ἐμιάνθησαν, thus making the phrase part of the predicate of συνεκοιμήθησαν; (3) the reading '(all) those sins' may have arisen by inner Eth. confusion of እሎንቱ with ኵሎ. Knibb, 2, 86 follows Ullendorff ('An Aramaic "Vorlage"' 266) in reviving a suggestion of Charles[1906] xxviii, that behind (2) lies an Aram. construction עמהין עם נשיא, but, if this idiom were involved, it would read עמהין נקבות = μετ' αὐτῶν τῶν θηλείων. Moreover, the construction 'to be defiled with (ב)' shows that ἐν ταῖς θηλείαις forms the predicate of ἐμιάνθησαν. Cf. also Milik, 161, L. 22.

hate-charms This clause occurs in Sync. only. I have been unable to find an Aram. or Heb. equivalent for μίσητρα the opposite of φίλτρα), but cf. Heb. לחש, Tg. לחשא, Syr. ܠܚܫܬܐ ἐπῳδή, = criminationes, obtrectationes, Clement of Rome, *Ep. 34.10*, P. Sm., 1933.

(9) For this verse G Eth. have a shortened form which is simply an abbreviated paraphrase of the longer form of Sync. which has all the marks of originality. Its text, however, is clearly itself in some disorder: to construe κίβδηλα with ἐκκέχυται is impossible; κίβδηλα must be an error for κιβδήλους = ממזירין, 'bastards', and be in apposition to γίγαντας (גברין); cf. 10.9 Sync., where γίγαντες are described as κίβδηλοι, υἱοὶ τῆς πορνείας. The last two clauses ἐπὶ τῆς γῆς ... ἐκκέχυται, καὶ ὅλη ἡ γῆ ... ἀδικίας repeat the two parallel clauses at 9.1, the first clause in the form preserved in the Aram. frg. En[a] 1 iv 8; we can, therefore, confidently supply τὸ αἷμα πολὺ in the defective Sync. text. Note that G offers an alternative translation of גברין vid. τιτᾶνες to the more usual γίγαντες.

An allusion to this verse is found in Justin Martyr, *Apol. II.* 4(5) οἱ δ' ἄγγελοι, παραβάντες τήνδε τὴν τάξιν, γυναικῶν μίξεσιν ἡττήθησαν, καὶ παῖδας ἐτέκνωσαν, οἵ εἰσιν οἱ λεγόμενοι δαίμονες ... καὶ εἰς ἀνθρώπους φόνους, πολέμους, μοιχείας, ἀκολασίας καὶ πᾶσαν κακίαν ἔσπειραν.

(10) Cf. Rev. 6.9-11.

cannot escape from the wrongs Most versions read 'it (the groaning of men) cannot go forth (G Sync. Eth. ἐξελθεῖν)'. Charles, 21 f. conjectured that ἐξελθεῖν, למנפק, was corrupt for למפסק 'to cease'. Behind G αἱ ψυχαὶ τῶν τελευτηκότων lies, I suspect, נפשת מיותיא, 'the souls of mortals', misread as נפשת מיתיא, 'the souls of the defunct' (cf. P.Sm. 2057). For ἐξελθεῖν = יצא, נפק, 'escape', Jer. 2.11.

(11) **Thou seest them ... alone** Eth. reads zazi'ahomu = καθ' ἑαυτούς (e.g. Sir. 17.3), but omits ἐᾷς αὐτούς by hmt; G Sync. have lost καθ' ἑαυτούς for the same reason (cf., however, Knibb, 2, 86). An original may have read ושבקת אנון בלחודיהן, 'and thou has left them by themselves' i.e. 'let them alone'.

CHAPTER 10

(1) G περὶ τούτων, עליהון, omitted by Sync. and Eth, may have been introduced as a corruption of עליונין ὁ ὕψιστος.

Sariel See above on 6.7 and 9.1. For the variations in the versions, see Milik, 172: they seem, for the most part, to represent 'a sort of *lectio conflata* of Israel, Sariel and Uriel, written as one or two names'. Theological influences may have led to the replacement of Sariel by Israel: 'Noah ... is warned ... by an angel bearing the name of the eponymous ancestor of the sons of Israel' (ibid., 174).

(2) **Hide yourself** is probably best understood as an injunction to be conveyed to Noah; so Dillmann, 99: Noah is to withdraw to a secret place to receive angelic revelations. Schodde, 75, compares Exod. 3.6. The injunction remains, nevertheless, a curious one, introduced by the solemn 'Say to him in my Name ...'. Has an original Aram. אסתר לך 'I shall protect you' (lit. 'I shall hide you') been misunderstood as אסתתר לך κρύψον σεαυτόν? The text would then refer to God's protection of Noah in the ark.

the End that is approaching G, Sync., Eth. τέλος, סופה in an eschatological sense; cf. 91.19 En[g] 1 ii 21. See ThWNT, s.v. τέλος Bd XIII, 53f. (Delling). Cf. also τελείωσις below, 137f.

the earth ... destroyed Does ἀπόλλυται here represent a prophetic perfect?

a Deluge is about to come For the occurrence of מבול 'Deluge' in Aram. cf. EnGiants[e] 3 i, Milik, 237. Did Eth. read μέλλει ἔρχεσθαι? Cf. Ullendorff 'An Aramaic "Vorlage"', 266, Knibb, 2, 87.

(3) Sync. καὶ τὴν ψυχήν ... συντηρήσει, ויצל נפשו;[1] cf. Isa. 44.20, Ezek. 3.19,21. The longer text of Sync. seems closer to an original. The position of τὸν υἱὸν Λάμεχ points to a second clause parallel to δίδαξον ... ποιήσει, and this is supplied by καὶ τὴν ψυχήν ... δι' αἰῶνος, lit. 'and let him save his life' i.e. 'in order to save his life'. The second half of this parallel clause could, however, go closely with the clause that follows it: 'and he will escape for all time, and from him ... shall be planted'. G = Eth. has 'a drastic abridgement' (Milik, 162): in fact the redactor of G seems to be deliberately replacing the clause φυτευθήσεται ... σταθήσεται by a paraphrase μενεῖ τὸ σπέρμα αὐτοῦ.

Instruct the righteous one For the source of this epithet for Noah, cf. Gen. 6.9, 7.1, Sir. 44.17.

a plant ... established for all generations for ever Cf. Isa 60.21, 61,3 (מטע יהוה), where the figure of speech is applied to the remanent post-exilic Israel. Here the reference is to Noah and his successors, the post-diluvial Israel. For the expression 'plant of righteousness', 10.16 En[e] 1 v 4 נצבת קשטא, 84.6, 93,2, 93.5,10, ('the plant of righteous judgement', 'the plant of eternal righteousness', referring to Abraham); cf. 62.8. The phrase, to describe 'the elect', occurs at IQH 8.10, מטעת אמת, IQS 8.5 מטעת עולם (cf. further IQS 11.8, IQH 6.15, 8.6,9, CD 1.7.

(4) **And to Raphael he said** Here and at vv. 9, 11, Sync. reads εἶπε imper. (unless we emend all three verbs to εἶπεν aor.), thus understanding the phrase as a continuation of the *oratio recta*, so that it is to Uriel (Sariel) the great Holy One gives the commission to instruct the other archangels: 'And to Raphael say ...'. It seems hardly likely, however, that one archangel would convey the instructions of the Holy One to other archangels. The reading of Eth. here may be original, 'the Lord said', as G Eth. v. 9 καὶ τῷ Γαβριὴλ εἶπεν ὁ κύριος.

---

[1] Milik, 161f. has identified traces of vv. 3-4 at En[a] 1 v 4-6, but adds (162) 'only the identification of fragment *l* is relatively certain' (See Plate V). But this is, in fact, very doubtful, since we would require an Aphel infinitive לאצלא (לאנצלא) 'to save (his life)' not the Peal לנצלה. Moreover, l. 4 has a ḥeth not a he (Milik ה[עלמי]) and l. 5 יתה̊, if so read, introduces a rare form of the accus. pronoun. The identification of the frg. must remain uncertain.

into darkness ... cast him in I take the 'darkness' here to be that of the pit which Raphael is to dig for Asael in the desert of Dudael; the καὶ ἐκεῖ καὶ ἄνοιξον clause, that is, is epexegetic: '... cast him into darkness by making an opening ... in Dudael'. Cf. v. 5 and Isa. 42.7. No doubt the origins of the wide semantic usage of σκότος, חשך are traceable to such passages, but it seems unlikely that thoughts of the cosmological 'great darkness' (Milik, 15, 38-40) or even of the darkness of death or hell, much less the 'outer darkness' of Mt. 8.12, had ever crossed the original writer's mind in this verse. According to the text of Sync. the desert was called Dudael. Milik's view of 'an extremely archaic Babylonian feature' in this name (29f., Daddu'el 'the (two) breasts of 'El') is unconvincing. A more likely derivation and connection of the name is with the rabbinical בית חדודי, the wilderness into which the scape-goat for Azazel was driven (Tg. P-J Lev. 16.21-22); cf. Milik, *Bib.* 32 (1951) 395, Knibb, 2, 87, Δουδαήλ, חדודי אל 'the jagged mountains of God', a derivation supported by v. 5. Dillmann, 100, derived from דודא אל 'cauldron of 'El', a description of the desert.

(5) **And place upon him ... rocks** I have adopted the emendation of ὑπόθες of G, Sync. to ἐπίθες (supported by Eth.) proposed by Lods (XXXVI); cf. 54.5 which seems dependent on this verse. On the other hand, both G and Sync. have ὑπόθες 'place under him', and this is a possible meaning: Asael's punishment was to be placed on top of rough and jagged rocks, and then his face was to be covered so that he remained in complete darkness. Sync. λίθους ὀξεῖς (καὶ om. Go) λίθους τραχεῖς could be a doublet of a single Aram. expression אבנין חדדין.

for all time For this use of the expression 'for ever' (עד עולם) for a long, indefinite period of time cf. 1 Mac. 14.41. As v. 6 implies, that period will terminate when, on the great day of judgement, Asael will be thrown 'into the conflagration of fire', the destructive fires of Gehenna (See next verse). This is amplified at 10.12-13 where we are told that this period is to last 'seventy generations', when Semhazah and his band (including Asael) are eventually to be led off and cast into the 'abyss of fire'. Cf. Charles 23.

(6) **the day of the great judgement** Cf. also 22.4, 84.4, 91.15, 94.9, 98.10, 99.15, 104.5: thus 22.4, En^{e} 1 xxii 2, 91.15, En^{g} 1 iv 23, דינא רבא. The 'great judgement' *simpliciter* occurs at 16.1, 19.1, 22.4, 15.4, 91.15. The expression is sometimes 'the great day of judgement', 10.12, 19.1, 22.11; thus 10.12, En^{b} 1 iv 11 יומא רבא. As in Sync. in this verse we also find 'the day of judgement', 22.4 (G, Eth.), 100.4. Other expressions are 'the judgement which is for ever, 10.12, 104,5, and 'that great day', 54.6.

It is clear from 10.12 and 22.4 that both expressions 'great judgement' and 'great day' are found in the Aram. Enoch.[1] Jude 6 prefers 'the great day' (εἰς κρίσιν μεγάλης ἡμέρας, cf. 2 Pet. 2.4 (εἰς κρίσιν), 2.9 (εἰς ἡμέραν κρίσεως).

Other descriptions for the 'day of judgement' are 'the day (of judgement and) consummation' 10.12 (see below 137f.); cf. 16.1 'the day of consummation (of the great judgement)'; 'the (day of) tribulation', 1.1, 96.2; 'the day of tribulation and pain' 55.3; 'the day of tribulation and great shame' 98.10; 'the day of suffering and tribulation' 45.2, 63.8; 'the day of affliction' 48.10, 50.2; 'the day of anguish and affliction' 48.8; 'the day of destruction' 98.10; 'the day of slaughter', 'the day of darkness' 94.9; 'the day of unceasing bloodshed' 99.6; 'the day of unrighteousness' 97.1. Some of these passages refer to the first world judgement by the deluge (e.g.

---

[1] Not surprisingly, Eth. MSS vary between 'great day' and 'great judgement' (e.g., 19.1, 84.4).

10.4,5,12; 54.5,7-10; 91.5; 93.4). Others refer to the final world judgement. See Charles 84.

to the blazing fire Sync. εἰς τὸν ἐμπυρισμὸν τοῦ πυρός, lit. 'to the conflagration of fire' (G Eth. simply εἰς τὸν ἐνπυρισμόν). Sync. corresponds to לשרפת אש of Isa. 64.10 (Tg. ליקידת נורא); for the verb שרף in the old Aram. cf. Hoftijzer, 320.

For the imagery of fire in the Bible in connection with judgement see ThWNT Bd. VI, 935f., 8. s.v. πῦρ (Lang). Within this development, and especially in the emergence of the idea of 'hell-fire', Enoch has a central place: cf. 18.9f. where the 'wandering stars' are punished in a place of fire. Cf. further 19.1f., 21.7-10; in the Parables 54.5-6, 67.4 (cf. 27.1-5); and later at 90.24f., 98.3, 103.8. See further below on v. 13.

(7) **Heal the earth ... healing of the earth** In view of the previous imperatives in the commissioning of the archangels (v. 4 'bind' ... 'cast' ... v. 5 'place upon'), the reading of Sync. Eth. ἴασαι is to be preferred to G ἰαθήσεται, as is also his ἐγρήγοροι, עירין (G Eth. ἄγγελοι: see above 106f.). Sync. G ἴασθαι, רפא; LXX ἴασις, מרפא, cf. 76.4, Enastr[c] 1 ii 2 (Milik, 285) Aram. רפיא 'healing'. The original plays on the name רפאל ('God heals', 'healing of God'). Sync. G δήλωσον here probably = אודע (cf. 10.11, En[b] 1 iv 8, Milik, 175).

that I shall heal its wounds The rendering in Sync. G Eth. by a purpose clause is less satisfactory than to assume a ὅτι, די recitativum. The reading of Eth[b] 'that I shall heal' (= די ארפא) is superior to the reading of Sync. G (ἵνα) ἰάσωνται which requires to be construed as an impersonal plural.

shall not altogether perish Behind G ἵνα μὴ ἀπόλωνται πάντες οἱ υἱοὶ ἀνθρώπων ... ὅλως lies an idiomatic די לא יאבדון כול בני אנשא כולהון. For the idiom cf. Nöldeke, 165, P. Sm. 1736 and above, 113

on account of the mysteries ... children of men The impossible reading of G, Eth. ἐπάταξαν[1] is rightly explained by Knibb, 2, 88 as probably having arisen by 'a confusion of the roots מחא ... and חוא, Aram. מחא, πατάσσειν; אחוי, ὑποδεικνύναι (e.g. 106.19. En[c] 5 ii 26). The confusion would be encouraged if מחתא had just been employed earlier for τὴν πληγήν. Sync. εἶπον looks like a guess.

(8) **devastated by the work ... Asael** G ἀφανισθεῖσα looks like an alternative version of the Aram. verb behind ἠρημώθη (חרב, אתחרב) unless it represents a second complementary verb (חבל, אתחבל). Eth. reverses the words 'works of the teaching' to read 'teaching of the works'; either could be original (יולפן עובדין, עובדי יולפנא). According to G Eth. it is to Asael himself that all sins are to be ascribed; in Sync. it is to his teaching. The works of the teaching of Asael, by which the earth was devastated, consisted of weaponry and metallurgy, cosmetics and female adornments (8.1).

record against him all sins The construction γράφειν ἐπί, כתב על may be understood literally as 'to write down' 'record' against someone, followed by an accus. of what is recorded. It has also the meaning 'to prescribe for someone', 'to decree'. Job 13.26 כי תכתב עלי מרורות may be understood either as 'For thou writest bitter things against me' (RSV) or 'prescribe punishment for me' (NEB). The latter sense here does not seem appropriate with the following πάσας τὰς ἁμαρτίας, unless we are prepared to give ἁμαρτία the meaning of עותא, חטיתא = 'punishment', 'punishments for sins': 'prescribe for him all (manner of) punishments'(?). Milik restores כל עויתא at En[b] 1 iv 5.

---

[1] For some ingenious conjectural emendations of ἐπάταξαν (ἐπέτεσαν, ἐπέταξαν) see U. Bouriant, 'Fragments Grecs du Livre d'Énoch', 105, Lods, 118, Flemming, 11, Charles[1906], 27 n. 6.

(9) **And to Gabriel the Lord said: 'Go, Gabriel, to the Giants"** Milik's restoration of the Aram. frg.y seems convincing. For the use of ὁ κύριος, מָרֵיא(?) simpliciter, see Fitzmyer, 'A Wandering Aramean', 115f. The Sync. text ἐπὶ τοὺς γίγαντας is original, since it yields a word-play with Gabriel ('mighty man of 'El') and γίγαντες, גברין, similar to that on Raphael above, v. 7. G μαζηρέους Eth. manzerān[1] is a transcription of Aram. ממזירין, for which the Greek equivalent of G, Sync. is κίβδηλοι.

This account of the mutual self-destruction of the giants is followed closely by Jub. 5.6-11. The giants slay one another in the presence of their parents, the watchers (5.10, cf. En. 14.6), who are then incarcerated in the abysses of the earth (5.10, cf. 14.5) till the 'great judgement'. According to En. 15.11-16.1 the 'spirits' which emanate from the giants (a kind of third generation progeny), appear to enjoy an immunity in wrong-doing till the last judgement. Did vv. 3-5 in the quotation of Sync. which has no equivalent in G or Eth. (see *Apoc. Hen. Graece*, 37) perhaps belong after v. 10, but consist originally of a longer version of Gabriel's prediction of the doom of the children of the watchers (adapted in the quotation to apply to 'the children of men')? Also appropriate and still referring to the fate of the watchers' off-spring would be after 13.10. The previous vs. about the fate of Mount Hermon are difficult to place. (The Mount is never to be without cold, snow and frost, and dew will not fall on it except as a curse, and at the great judgement it will be consumed by fire).

Muster them (for battle) ... destruction I have adopted the reading of Eth. from waḍ'a II 1 = הוציא, אפיק, ἐξάγειν, since its meaning 'bring out (for battle)' is appropriate in the context; it seems unlikely that a translator would have added on it on his own.

The reading of G ἐν πολέμῳ ἀπωλείας is guaranteed as original by En^b 1 iv 6 בקרב אבדן; Sync. Eth. had evidently difficulty with the expression and paraphrased it. Sync. Eth., however, preserve an original part of the predicate lost in G (by hmt? Lods, 119) in the words εἰς ἀλλήλους ἐξ αὐτῶν εἰς αὐτούς (Eth. = [εἰς ἀλλήλους ἑαυτοὶ καθ' ἑαυτούς]?). The injunction to Gabriel was to send them out in a destructive internecine battle with one another, Aram. אלין לאלין (Dalman, *Gramm.* 114, cf. 89.11, En^d 2 i 24-25, 7.5, En^e 1 ii 25a, G ἀλλήλων). Eth. [ἑαυτοὶ καθ' ἑαυτούς], Sync. ἐξ αὐτῶν εἰς αὐτούς are literal translation equivalents, whereas εἰς ἀλλήλους is the more idiomatic rendering; the two expressions preserved in Sync. and Eth. are clearly literary doublets. Eth. understood the text as 'send them against one another, one with another in war shall perish'.

(10) The traditional verse division seems to me to be wrong: this opening sentence belongs to what follows rather than to the preceding verse. For 'length of days' cf. 13.6, Heb. ארך ימים, e.g. Dt. 30.20, Job. 12.12 etc., LXX μακρότης τῶν ἡμερῶν. The usual translations do not make a coherent sense: the first ὅτι clause, introducing the contents of the petition of the watchers, reads, that 'they (the giants) hope to live an eternal life' (like their parents); the second ὅτι clause then reads, 'and that each of them shall live five hundred years'. The reading of Eth^q seems to me to give only possible sense, viz. that the giants, since, unlike their parents, they are of the earth and flesh, not of spirit and heaven, should *not* be granted to live forever, but be given a life-span of five hundred years. For the expression ἐρώτησις οὐκ ἔσται τοῖς πατράσιν αὐτῶν, 'a petition will not be granted to their fathers', cf. 14.7 ἡ ἐρώτησις ὑμῶν περὶ αὐτῶν οὐκ ἔσται, 'your petition on their behalf shall not be granted'.

---

[1] Like the Aram word, the term in Eth. seems a Heb. loan-word (cf. Dillmann *Lex.*, 191), but it hardly follows that it comes here directly from an Aram. Vorlage; cf. Ullendorff, 'An Aramaic "Vorlage"', 264.

(11) **make it known** Cf. Jub. 4.22. The reading of G, Eth. δήλωσον is probably corroborated by En[b] 1 iv 8 ואודע. Sync. δῆσον is a corruption, as a result of which datives have been changed to accusatives. (Contrast Flemming-Radermacher, 32 n. 11, Charles [1906], 29 n. 2). The command to bind the watchers comes later at v. 12. Michael is commanded first to make known to Semhazah and the rest of the fallen watchers (τοῖς σὺν αὐτῷ, חברוהי, 'his companions') what is to be the fate of their illegitimate offspring, predicted at vv. 9-10. I have adopted the Sync. reading 'the daughters of men' as more primitive than G's 'daughters'.

(12) Jude 6 contains an allusion to this verse, and there may be other reminiscences of vv. 12-14 in Jude (see below v. 14).

when their sons shall slay one another The reading of En[b] 1 iv 10 '[when] their sons shall perish (יבדון)', is not in doubt, but was it the original text translated by G, Sync., Eth. κατασφαγῶσιν, which suggests a verb such as הרג (Hoftijzer, 69)? Have the versions preserved a different recension of the original Aram., or did the Greek translator render יבדון by κατασφαγῶσιν? Cf. Barr, 'Aramaic-Greek Notes' I 197.

seventy generations Milik regards this as the earliest reference to a chronological scheme measured in periods of 'seventy generations', which was to provide a popular chronology of world history, especially in apocalyptic circles, Christian as well as Jewish (248f.). He draws attention to two fragmentary texts (4Q**180**.1, **181**.2) headed by 'Pesher with regard to the Periods (of time) which God created ...', and argues that such a pesher implies the existence of a well-known chronological work on the Seventy Ages which was accorded an authority equal to that given to the prophetic books, the Psalter, etc., 'for which *pesharim* were composed to make them better understood by readers ...' (251). (He refers to this putative work as 'The Book of the Periods'). The text of 4Q**180**.1, **181**.2 is full of lacunae, but it is, undoubtedly, concerned with chronological 'periods' and the lines referring to the watchers (251, lines 4 and 5) mention their illicit liaison with 'the daughters of men' and the birth of the giants in close proximity to the words בשבעים השבוע 'in seventy weeks'. The precise reference in these words, however, is not clear; Milik takes them as referring to the 'seventy ages' of Israel's history from Adam to the End under the dominance of 'Azazel'. If this is correct, then it is to be at the close of the seventy ages that the watchers are to be removed from their imprisonment to face the 'great judgement'.

The Apocalypse of Weeks (93.3-10, 91.11-17) and the Dream-Visions (83-90) are based on the same chronological scheme; the 'seventy hebdomads' is taken up and developed in Dan. 9.24-27, and the scheme reappears in Jub. in the division of the sacred history into 'jubilee' periods from the creation to the revelation on Sinai. See the discussion in Milik (General Index, Chronology, 429) and in Hengel, *Judaism and Hellenism*, Vol. I, 181f., especially 187f.

in valleys of the earth G. Sync. εἰς τὰς νάπας τῆς γῆς, בחלות ארעא(?) cf. 22.2. Eth. has the unusual translation 'under hills/slopes of the earth'. While wagr does once render νάπη (Dt. 3.29), this rendering is strange. The claim of Charles, 24, that νάπη can sometimes render Heb. גבעה 'hill' cannot be substantiated: at Ezek. 6.3 it corresponds to גיאות 'valleys', and not to גבעות (LXX τοῖς βουνοῖς). The second example at Isa. 40.12 is probably a slip: the previous 'mountains' suggests 'valleys' and thus appears where βουνοί not νάπαι is required. Did Eth. arrive at 'under the hills/slopes of the earth' through the influence of the Greek myth of the Titans, each imprisoned beneath a (volcanic) mountain? Cf. Dillmann, 101, Charles, ibid.

the time of the end, until ... the judgement ... absolute G (συν)τελεσθῆναι and its cognates (συν)τελεσμός, τελείωσις can mean simply 'to be finished', 'end' (e.g. at 2.2 'from beginning to end', 10.14 'until the end of this generation'). Here at v. 12, and at

16.1 and 25.4, both nouns and verb are used in connection with the 'great judgement'. The expression ἡμέρα τελειώσεως, here and at 16.1 corresponds to יום קצא of 22.4, En[e] 1 xxii 2, lit. 'the day of the end (of the great judgement)', cf. P. Sm., 3700, Ebed Jesu 143 r ܝܘܡ ܓܡܪܐ dies ultimi iudicii. (The second noun τελεσμοῦ in Sync. is a doublet). Among possible equivalents for the verb (see Dalman, *Words of Jesus*, 155), שלם is employed of a judgement or verdict becoming 'absolute', e.g., Tg. Onk. (cf. Neoph.) Exod. 23.2 'after a majority (of votes) the judgement is absolute (שלם). LXX (συν)τελεῖν frequently renders כלה, 'to be completed', Gen. 2.1 'the heavens were completed' (יכלו). Aram. כלה is used meaning 'to come to an end, cease' (107.1, En[c] 5 ii 28) but for 'to be completed' we require כלל Ishtaph. (Ezr. 4.13,16).[1] Could Aram. כלה also have the meaning 'to be fulfilled' (Heb. Dan. 11.36)? A third possibility is Aram. סוף which, like שלם, can mean 'to be fulfilled' (Dan. 4:30). All three verbs would be appropriate in this context, and possibly also at 16.1, where it is the great aeon which is 'consummated' (τελεσθήσεται). See ThWNT s.v. τελέω, Bd VIII, 59 f., 64 f., 86 (Delling). On the related synonyms πληρόω, πλήρωμα (ܫܘܡܠܝܐ), see ThWNT Bd VI, 297f. (Delling).

(13) **fiery abyss** See above on v. 6 LXX χάος Heb. גי, גיא, גי נורא(?) cf. גי הנם. Cf. Rev. 20.10, 14-15, λίμνη τοῦ πυρός. The idea of 'everlasting incarceration' is found in Josephus, Ant. xviii 1.3 (14) (εἱργμὸς ἀΐδιος) (cf. B.J. ii 8.14 (163). See further, ThWNT s.v. κόλασις, Bd III, 817 f. (Schneider); Str.-B. Bd IV, 1022 f. For ἡ βάσανος, עיקא cf. Tg. עקתא Heb. צוקה, but also עיק. Levy CW II, 214.

they shall be imprisoned Sync. G δεσμωτήριον συγκλείσεως, בית עגון En. 22.3, En[e] 1 xxii 1. Eth. [συγκλεισθήσονται εἰς τὸν αἰῶνα] has preserved the verb in the second clause; ἐν τῇ βασάνῳ, בעיקא 'in torment' Eth. takes (in my opinion correctly) with the first clause. Sync. G is a foreshortened text.

(14) **whoever is consumed by lust and is corrupted** Sync. κατακριθῇ 'whoever is condemned' is almost certainly an 'emendation made by a learned copyist, contextus gratia' (Milik, 190). G κατακαυθῇ is used here with a sexual connotation; cf. πυροῦσθαι 1 C. 7.9 (Pesh. ܠܡܐܩܕ) and Sir. 23.17 (ἄνθρωπος πόρνος ... οὐ μὴ παύσηται ἕως ἂν ἐκκαύσῃ πῦρ, and cf. also *Lyrica Alexandrina Adespota* (ed. J. G. Powell) 8 (c) ὁ ἔρως ἐμέ ... κατεκέκαυκεν (Liddell and Scott, 892). A relic of the original reading may have been preserved at En[c] 1 v 1 [כול די א̊[קד. For יקד in this sense, P. Sm. 1620 and for the form איקד, Dalman, *Gramm.*, 100; for ἀφανίζεσθαι 'to be corrupted', cf. 8.2 Heb. שחת Niph. (Gen. 6.12), Hoftijzer, 295. All who are guilty of the same sin of sexual lust as the watchers will receive the same punishment (cf. Lods, 122).

Could there be a reminiscence of this verse at Jude 6-13 (cf. 2 Pet. 2.2,4,17,18), the sin of ἀσέλγεια?

En[c] 1 v 2 adds a hemistich which has probably been omitted in the versions by hmt: 'until (the end of) all generations'/'for all generations for ever'. For the introduction of the Ist person, 'I shall judge' (אדין cf. Milik) cf. vv. 15,22,11.1.

(15) **I shall destroy ... watchers** En[c] 1 v 2 [אכרת] (cf. v. 16) is ambiguous = ἀπόλεσον G or ἀπολέσω. Dillmann (101 foot) noted that Eth. 'aḫaguelomu imperat. could also be read as a subjunctive 'I will destroy', but he argues that v. 20 καὶ σὺ καθάρισον is against this rendering. But v. 20 may be distinguishing what Michael has to do from the role of the Holy One himself. According to 16.1 the 'spirits of the giants' (15.11) are to continue to plague mankind till the last judgement; the

---

[1] See Black, *An Aramaic Approach*[3], 233f.

imper. implies that they are to be destroyed forthwith by Michael. If we read 'I shall destroy', then their final destruction is to be the work of God, not of the archangel Michael, although Michael is given the task at v. 20 of 'cleansing the earth' i.e. presumably after the Deluge. (The καὶ σύ is emphatic, distinguishing the role of Michael from that of the Holy One). Cf. v. 22 (Eth.). The reading of all the versions '(I shall destroy (or 'Destroy') all the spirits of the bastards) *and the sons of the watchers* (i.e. *the giants*)' has introduced confusion into the narrative (cf. Lods, 123), since the destruction of the giants has already been described at v. 9 (and was the duty of Gabriel, not Michael). Although the original is lost, the text almost certainly read כול רוחות ממזרין ובני עיריא, where the second phrase is in apposition to the first, lit., 'I shall destroy all the spirits of the bastards, the offspring of the watchers'. These are referred to more fully at 15.8-12, 16.1, as spirits which proceed out of the giants: they are noxious spirits, 15.11, ἀδικοῦντα, attacking men (15.12). With this explanation there is no need to assume that the verse is out of place and should have followed v. 10 (so Charles, 25).

The reading of Eth^{m} 'I shall destroy the wiles (πανουργεύματα,[1] ? ערמותא) of all the spirits of the bastards, offspring of the watchers', could be original and lost by hmt with πνεύματα. The Eth. equivalent of κιβδήλων, ተውኔት 'lust' (Knibb) may have arisen as a corruption of the previous ትውልድ (γενεῶν). The Eth. noun renders μουσικά at Dan. 3:10,15, זמרא, and could have come from the work of a translator who misunderstood or was unfamiliar with the foreign word ממזרין.

(16) **I shall destroy all iniquity** En^{c} 1 iii 3 אכרת lit. 'I shall cut off'; cf. Heb. כרת Hiph. = 'destroy' (Mic. 5.10, Zech. 13.2 etc.): כרת is attested here for Aram. for the first time (אכרת = 1st pers. sing. impf. or 2nd pers. sing. imper. Aphel?) (cf. Hoftijzer, 127). I have adopted the reading of G Eth. 'all iniquity': has כולה dropped out of En^{c} 1 iii 3 by hmt עולה כולה?

there shall appear the plant of righteousness ... for ever G has lost a line, 'and it will be a blessing ... deeds of righteousness and truth' by hmt (see textual note). The rendering of the verbs by imperatives instead of futures is a peculiarity of the Greek translation, followed by a number of Eth. mss. As Milik, 191, L. 4, notes, Aram. קושטא is translated by two words in G and Eth. (τὸ φυτὸν) τῆς δικαιοσύνης καὶ τῆς ἀληθείας: 'righteousness' is closer to the original meaning.

(17) **all the righteous shall escape** i.e. Noah, his family and descendants, En^{c} 1 v 5 יפלטון, ἐκφεύξονται cf. 106.18, En^{c} 5 ii 24, יפלט הוא 'he (Noah) will escape (and his sons ...)'.

shall live till they beget thousands 'The picture is a very sensuous one' Charles, 25. For the promise of long life and a large offspring, cf. Isa. 65.20,22; Zech. 8.4; Jer. 3.16, etc. See also En 25.6; 58.3,6; 71.17; 96.8. 'The writer of 1-36 has not risen to the conception of an eternal life of blessedness for the righteous, and so has not advanced a single step beyond ... Isa. 65,66', (Charles, 53, on 24.4).

of your old age G Eth. 'of their sabbaths', a mistranslation of the Aram. text now preserved at En^{c} 1 v 6 but read as שיבתכן; it was first detected by J. Wellhausen in 'Zur apokalyptischen Literatur', *Skizzen und Vorarbeiten*, Berlin, 1899, VI 241, n. 1, 260. (Eth. is dependent on G, but cf. Ullendorff 'An Aramaic "Vorlage"', 264). En^{c} 1 v 6 assumes a change of person 'the righteous shall escape ... all the days of your youth and of your old age you shall fulfill (תשלימון) in peace (שלם).

(18) **all be planted with trees** Trees (especially the kind specified in the next verse) are evidence of prosperity; cf. Isa. 41.19, Ezek. 47.7 and En. 26.1; 27.1; 28.1-3. The

---

[1] Cf. Jdt. 11.8 πανουργεύματα (ṭebabihā) τῆς ψυχῆς σου.

singular in G 'a tree will be planted in it' seems a deliberate interpretation of the translator in order to introduce the Tree of Life of 24.4f. into the new post-diluvial age.

(19) **all luxuriant trees** Eth. [δένδρα τῆς ἀγαλλιάσεως]. G is clearly corrupt, γῆς possibly from a ditt. of τῆς and ἀγαλλιάσονται of ἀγαλλιάσεως (by vertical ditt. from ἔσονται?[1]). The 'trees of joy' ('pleasant trees' Knibb) I take to be a misrendering of an original אילנין רענין; cf. Jer. 17.2 על־עץ רענן, Dt. 12.2 δένδρον δασύ: the adjective is attested for Aram. at Dan. 4.1, but in a derived sense of a 'flourishing man'. The translator appears to have mistaken the adjective for the noun רנה/רננה ἀγαλλίασις(?). Cf. however Ps. 45.7 (LXX 44.8) (Heb. 1.9) שמן ששון, LXX ἔλαιον ἀγαλλιάσεως, which provides a verbal but not a real parallel.

vines or 'vine-yards': כרמיא is ambiguous: ἀμπέλους could mean 'vineyards' and render כרמין: the second ἄμπελος seems, however, to refer to the individual vine (גופנא, גפן).

a thousand measures of wines Lods, 125, notes that the πρόχοος (lit. 'pitcher') was a measure equivalent to the ξέστης (sextarius), about a half-litre or pint; the nearest semitic equivalent would be the log (לג). A thousand log (= 500 litres) would exceed the highest measure, the kor (כור) (466. 56 l.) G is original here; the version in Eth. is a free paraphrase, but whether the work of the Eth. or the Greek translator it is impossible to say.

a thousand seah lit. 'a thousand measures', but μέτρον often renders Heb. סאה in the LXX and the same dry measure is used in Aram.: a seah of wheat = c. 15 litres; 1000 seah = 15,000 litres.

ten baths of oil A seah of olives produces 10 baths of oil = 30 seah, c. 450 litres. See Milik, 192, Ll 9-10, DJD III, 37-41.

Similar 'chiliastic expectations' (Charles, 26) are found at 2 Bar. 29.5, Irenaeus *adv. haer*. v. 33: cf. Isa 5.10 'Five acres of vineyard shall yield only a gallon ...' (NEB). (Note כרם = 'vineyard', and cf. above 'they will plant vineyards/vines'?). Other parallels are at Jer. 31.12; Hos. 2.21f.; Am. 9.13; Zech. 8.12 (Lods, 125).

(20) **And as for you, cleanse the earth** Notice the pronominal καὶ σύ, in contrast to the previous verbs in the first person? The final judgement on the watchers and their off-spring will be the task of the Lord himself; to Michael he now delegates the 'cleansing' of the earth at the Deluge.

I have adopted the reading of Eth. 'remove from the earth': ἀπὸ τῆς γῆς may have dropped out of G because of the previous ἐπὶ τῆς γῆς; it could, also, be a ditt. with ἐπί altered to ἀπό or ἐπὶ τῆς γῆς a ditt.—it seems unlikely that both phrases are original.

(21) For this vision of universal righteousness cf. Apocalypse of Weeks, 91.14. If Milik's identification of יתקשטון is right, we should render either 'will be declared righteous' (Milik, 163, L. 3) or 'will be truly righteous' (see note on 102.10).

all nations ... worship me Cf. 90.30, Isa. 2.2f., 14.2, 60.12. See Charles, 214 (note on 90.30).

(22) **castigation** Aram. מכתש Milik, 163, cf. Tob. 13.16, 4QTob.aram[a] 2 iii 4 = LXX ἐπὶ ... ταῖς μάστιξίν σου.

I shall not again send a Deluge upon it Eth II has preserved the original text. For G οὐκέτι πέμψω we have then to supply an object ('punishment'?). The translator of G (followed by Eth I) may have removed the reference to the deluge, in order to conform

---

[1] U. Bouriant, *Fragments Grecs du Livre d'Énoch*, Plate 40 (5 lines from foot).

the text to his own later eschatology: G ἐπ'αὐτούς also seems an alteration of [ἐπ'αὐτήν] of Eth.

CHAPTER 11

Contextually, 11.1-2 clearly belongs to the previous chapter (cf. Dillmann, 102); the present division is an instance of the arbitrary chapter division adopted by the Eth. editors. For possible Aram. frgs. of Chh. 11, 12, Milik, 162f.

(1) **I shall open the treasures of blessing** The verse is modelled on Dt. 28.8,12: e.g. 12 'May the Lord open the heavens for you, his rich treasure house, to give rain upon your land at the proper time and bless everything to which you turn your hand'. (NEB). The work and labour of the children of men refer to the cultivation of the earth as in Dt. 28.8,12. Eth. has ἐπὶ τῆς γῆς as in Dt. 28.8, from which it may have been introduced by a translator. Or did the original read 'upon the tillage of the ground (על פולחנא די ארעא) cf. 1 Chr. 27.26.

(2) **peace and righteousness** The phrase occurs at 2 Kg. 20.19 שלום ואמת Tg. שלם וקשוט (NEB 'peace and security').

for all generations of eternity G εἰς πάσας τὰς γενεὰς τῶν ἀνθρώπων, is an un-semitic phrase and probably a corruption of τῶν αἰώνων. Cf. 91.13, En[g] 1 iv 18, and 9.4 εἰς πάσας τὰς γενεὰς τοῦ αἰῶνος.

CHAPTER 12

(1) **And before these things Enoch was taken up** 'These things' can refer to the entire sequence of events beginning with the fall of the two hundred watchers (Chs. 6-11), or, more specifically, to the divine commission to the archangels (Ch. 10), or even to the fulfilment of the prediction of the coming Deluge. For λόγος = מלה 'thing', see Barr, 'Aramaic-Greek Notes', II, 190 and cf. below, 321.

G ἐλήμφθη, Heb. נלקח (2 Kg. 2.9), Aram. אתלקח (Hoftijzer, 139 f), Eth. takabta (cf. 71.1,5), refers unmistakably to Gen. 5.24, where לקח means 'to take up' ('God took him up'); cf. P. Grelot 'La Légende d'Hénoch', 11. Eth. renders Gen. 5.24 'God removed him (kabato = LXX μετέθηκεν αὐτόν)'.

(2) **with the watchers, with the holy ones** For 'watchers' see note on 1.5, 6.2. Lods, 127, suggests that the author has Gen. 5.24 in mind, interpreted as 'Enoch walked with God', i.e. consorted with the angels. He may also, however, have been thinking of Gen. 5.22, '... Enoch walked with God ... three hundred years ...', as was evidently the author of Jub. 4.21: 'And he was ... with the angels ... six jubilees of years' (See further below on v. 4).

(3) **was standing** This seems original: 'standing' was the correct posture for prayer: Dalman, *Words of Jesus*, 23. For these titles of deity, see above on 9.4. Eth's addition of 'Enoch the scribe' in this verse '(watchers were calling me) Enoch the scribe' seems a ditt. from v. 4.

the watchers of the great Holy One The fuller text of G seems original: it serves to distinguish the 'good watchers' from the fallen watchers about to be condemned. For the title 'the great Holy One' see above on 1.3.

(3) **King of the ages** See above on 9.4.

(4) **Enoch ... go, declare to the watchers ... who have left the high heaven ...** According to G Eth. Enoch's dealings with the fallen angels takes place after his 'elevation' to paradise (v. 1). A different tradition appears at Jub. 4.22, according to which Enoch's 'testimony' concerning the fallen watchers (along with astronomical revelations) was given during his life-time, in the six jubilees or three hundred years of Gen. 5.22. A third tradition is found at 81.5-6, in the 'astronomical book', where Enoch, having received his revelations from the angels in paradise, is led back to earth, where, for a year, he is to communicate these to Methuselah, and, in the second year, is again 'to be taken up from their midst' (81.6). To this third tradition, but with a feature in common with Jub. 4.22 (Enoch's six jubilees) the Heb. frg. **4Q227** belongs (Milik, 12).[1].

Milik has argued (ibid.) that the 'astronomical book' contains the oldest form of the legend, namely that '... it is during his earthly life, certainly as a creature of flesh and blood, and not as a *redivivus* on leave from his sojourn in Paradise, that the patriarch Enoch explains orally, as father to son or master to disciple ... his revelations ...' With this form of the legend Milik then compares Sumerian and Babylonian parallels.

No evidence from the so-called 'astronomical book' is adduced for this theory. That 'book' (as it has reached us in the Eth. version, supplemented by the Aram. frgs.) is an elaboration of the account of celestial phenomena which Uriel gives to Enoch in the 'Book of the Watchers' (33.3f., cf. 72.1; 74.2; 78.10; 79.2,6; 80.1; 82.7). But there is nothing in these Enoch astronomical 'observations' to suggest that revelations were given to Enoch *during his life-time*. On the contrary, 81.5-6, which is part of the 'astronomical book' (but not cited in this connection by Milik), tells us that Enoch was led back from paradise to earth to communicate his secrets to Methuselah before a second translation to paradise. If anything is characteristic of the 'astronomical book', in this connection, it is this idea of Enoch returning to earth from his sojourn in Paradise, a form of the tradition about the patriarch now attested by **4Q227**.

The only evidence which may be considered for a simple legend of celestial revelations during Enoch's life-time is the place where Enoch communicated with the fallen watchers, which has to be on earth (13.9), together with expressions in 14.2 'I saw in my vision', or 14.8 'the vision was shown to me.' If this was the oldest form of the Enoch legend, then we are obliged to explain 12.1f. (with Charles, 27) as 'an introduction from the final editor', i.e. as a redactional addition, or (with Lods, 127) as referring to Enoch's 'hidden years', consorting with the angels during his life-time. The language of 12.1, however, reflecting Gen. 5.24 (cf. Heb. 11.5) seems decisively against this latter view. Moreover, if the redaction theory is favoured, we then lose from the oldest Enoch traditions the only passage in I Enoch where an explicit reference is made to the central feature of the whole legend, Enoch's miraculous translation to paradise.[2]

On the whole, the evidence seems to me to point to the retention of 12.1 as original to the tradition of the Aram. Enoch. Enoch receives his commission to reprimand the watchers in Paradise to which he has been translated, and, from thence, returns to earth where the fallen watchers are now to be found, in order to carry it out. The variations at 81.5-6, **4Q227** and Jub. 4.22 are later elaborations of the tradition at En. 12.1, the last, Jub. 4.22, as a midrash on Gen. 5.22.

---

[1] After mention of 'six jubilees', Enoch returns(?) 'to the earth amongst the children of men, and he testified with regard to all of them ...' (line 4).

[2] Ch. 71.1 in the Parables is based on this tradition at 12.1.

Enoch, scribe of righteousness So also at 15.1 'righteous man and scribe of righteousness' (G τῆς ἀληθείας; for δικαιοσύνη and ἀλήθεια as translations of קושטא see above 139). At 4Q EnGiants[b] 14 Enoch is further described as 'distinguished scribe' (Milik, 305) and at 93.1 probably 'skilled scribe' (see below, 283). Ezek. 9.2f., the man clad in white linen with an ink-horn by his side, clearly an angelic figure, is the only Biblical parallel to Enoch in his rôle as celestial scribe (cf. Jub. 4.22-3). The idea of a celestial scribe is probably derived from Babylonian sources (Nabu?), and there are parallels in the ancient religion of Egypt. See Charles, 28; Bousset-Gressmann[4] 491; Milik, 237, 262, 305; P. Grelot 'La Légende d'Hénoch', 14f. 'Hénoch le scribe'.

the holy, eternal Sanctuary G τὸ ἁγίασμα τῆς στάσεως is an unparalleled phrase, unless we adopt the suggestion of F. C. Burkitt and equate στάσις with קימא, 'covenant', and render 'the sanctuary of the eternal covenant' (*Jewish and Christian Apocalypses*, London, 1914, 68, Knibb, 2, 92). A more likely equivalent of στάσις, however, is מקום (מקם, Hoftijzer, 165), the Place par excellence where God dwells (Isa. 26.21; Mic. 1.3; 1 Kg. 8.30 = 2 Chr. 6.20). The reading of Eth[tana(q)] seems to me to point to the correct explanation of the phrase: [στάσις τοῦ ἁγίου] stands for מקום הקדש, the Biblical term for the Temple or Tabernacle (Lev. 10.17 LXX τόπος τοῦ ἁγίου). This reading of Eth[tana(q)] can account for the variants, in particular of Eth[M] [τὴν ἅγιαν στάσιν].

children of earth בני ארעא as at 92.14; En[g] 1 iv 20; 92.1; En[g] 1 ii 23 (Milik, 266, 260). Eth. alters to the more familiar 'children of men'.

wrought great destruction G ἀφανίζειν, שחת Aphel, 'corrupt'. It is doubtful if any traces of this verse (or the next) can be detected in the frgs. on Plate V (Milik, 162).

(5) **no peace or forgiveness** See above at 5.4. Eth. 'upon the earth' could be original, but 'forgiveness of sins' seems an Eth. expansion.

(6) **Inasmuch as they delight in their children** Eth. 'esma 'because': I suggest that G περὶ ὧν renders על די 'inasmuch as ...'. Did the original continue the *oratio recta* so that this verse is still part of the message which Enoch is to deliver, 'inasmuch as you rejoice ..., you shall see, etc.' (cf. Knibb, 2, 92)?

they shall have neither mercy nor peace = Eth., and the usual form of this expression. G offers 'a strange construction' (Charles, 288), viz. οὐκ ἔσται αὐτοῖς εἰς ἔλεον καὶ εἰρήνην. Has this arisen from an inner Greek corruption (αυτοις εις), or could it reflect an Aram. לת להון לרחמין ולשלם lit. 'but it (their supplication) will not be to them for mercy and peace'? Cf. Lods, 130.

CHAPTER 13

(1) **You shall have no peace** Enoch delivers the message of the Lord of majesty at 12.5.

Asael Enoch goes first to Asael (G Ἀζαήλ = עשאל, Eth. Azazel), the chief culprit (8.1, 10.4f.). The corruption in G appears to have arisen by the influence of 10.9,11. Cf. v. 3, and see Knibb, 2, 93, Charles[1906] 33, n. 9.

a severe sentence ... issued G ἐξῆλθεν = Aram. נפק, Heb. יצא of the sentence of a judge, e.g. Hab. 1.4, Ps. 17.2.

(3) **fear and trembling** See note on 1.5.

(4) **to draw up ... a memorial and petition** lit. G 'memoranda of a petition' =

דוכרן שאלה(?) i.e. a memorial to serve as the basis for a petition, 'a memorial and petition'.

that I should read Eth. [ἀνάγω] seems almost certainly a corruption of ἀναγνῶ (cf. Charles, 29, Knibb, 2, 93) unless a longer text read both verbs, and one fell out by hmt: ἵνα ἐγω ἀνάγω τὸ ὑπόμνημα τῆς ἐρωτήσεως αὐτῶν καὶ ἀναγνῶ [αὐτὸ] ἐνώπιον κυρίου τοῦ οὐρανοῦ.

(5) **nor to lift up their eyes to heaven for shame** Cf. Lk. 18.13 and for the expression 'to lift up the eyes', Gen. 13.10; 1 Chr. 21.16; Mt. 17.8; Lk. 6.20; 16.23; Jn. 4.35; 6.5.

(6) **with reference to their spirits ... deeds of each one of them** This part of the predicate goes with the main verb, giving the content of Enoch's memorial, i.e. it recorded the worth (or worthlessness) of their 'spirits' and the character of the 'works' i.e. the behaviour, of each one of the fallen watchers. A parallel is IQS 6.17: each candidate for admission to the Qumran sect is to be 'investigated' (דרש) 'with reference to his 'spirit' and 'his deeds''; cf. also IQH 14.11 'according to their spirits he will divide them(?) between good and evil, ... their deeds(?)'. For πνεύματα (רוחות) in this sense, see ThWNT s.v. πνεῦμα, Bd. XI, 388.27-389.12 (Schweizer).

forgiveness and restoration Cf. Charles, 30, Milik, 196, L. 2, Knibb, 2, 94. Eth. nuḫat = μακρότης, but rather than assume that it corresponds to ארך ימים or ארך אפים, I suggest an original ארוכה(א), Heb./Aram. 'restoration', lit. 'healing' but always used figuratively; cf. Isa. 58.8 (of Israel), Jer. 8.22 (LXX ἴασις).

Cf. Irenaeus iv.16.2 'Enoch ... cum esset homo, legatione ad angelos fungebatur, et translatus est et conservatur usque nunc testis iudicii dei, quoniam angeli quidam transgressi deciderunt in terram in iudicium'. Note that Irenaeus is following the tradition at Jub. 4.22, that Enoch's delegation on behalf of the fallen watchers took place in his life-time, although after his translation he was, and is still to this day, a witness to their condemnation by God.

(7) **by the waters of Dan in the land of Dan** The stream known as 'the little Jordan', which may have been a sacred site, or contained a 'sacred place': cf. Josephus, Ant. v. 3.1(178), and especially viii.8.4(226). Like Daniel, Enoch receives this vision by a flowing stream (cf. Dan. 8.2, 10.4). The choice of Dan is no doubt connected with Enoch's mission to judge (דין) the fallen angels (Charles, 31). See further G.W.E. Nickelsburg, JBL 100 (1981), 575-600.

(8) **I lifted up my eyes to the gates of heaven** A text confined to En^{c} 1 vi 4. For שכני עין = 'eye-lids', a poetic expression for 'eye' see Black, *An Aramaic Approach*^{3}, 388, Milik, 196.

wrath and reproof lit. 'the wrath of reproof?' The reading of En^{c} 1 vi 5 is doubtful, but we may compare the phrase in Ezek. 5.15, 25.17 תוכחת חמה, 'furious rebukes' (AV).

speak to the sons of heaven *Oratio recta* seems required by G ἦλθεν φωνὴ λέγουσα. For 'sons of heaven' (בני שמיא) for the watchers, here the 200 fallen angels, see above on 6.2.

(9) **Abel-maim** I have adopted the conjecture of Milik, 196, that behind the confused transcription of G and Eth. there lies the Heb./Aram. place-name of 2 Chr. 16.4: it gives a neat word-play with אבלין = πενθοῦντες, although the name probably means 'wheat-meadow'. G Ἐβελσατά seems to have confused the name with Abel-shittim, Num. 33.49, LXX (Ἀ)βελσαττειν Eth. Ubelseya'el seems hopelessly corrupt, but cf. Knibb, 2, 94.

Senir Another name for Hermon (or part of it) G Σενισηλ Eth. [Σενισηρ] are probably corruptions of Σενειρ, LXX Ezek. 27.5 (B) or Σανιειρ Ca. 4.8.

(10) **And I recounted ... in dreams** En^{c} 1 v 7 seems to agree with Eth.: G καὶ

ἀνήγγειλα appears to have preserved an additional verb and Milik, 197, Ll. 7-8 may be right in assuming an originally longer text, perhaps 'And I recounted before them and explained all the visions etc'.' (cf. also Lods, 134). In that case, En[c] 1 vi 7 = Eth. has preserved a shorter recension; G has retained traces of the longer form of text.

CHAPTER 14

(1) **record ... truth** As Dillmann, 107, noted, maṣḥaf here = 'Schrift' rather than 'book'. The reference is to the following account or record of Enoch's divinely commissioned reprimand of the fallen watchers. There is no justification for Charles's idea that the verse began a new section and should be placed before Ch. 13 (En[c] 1 vi 7-8 has 14.1 following immediately on 13.10). Δικαιοσύνη = קושט means either 'righteousness' or 'truth', and the latter seems more appropriate here. The phrase may, however, be understood as 'words of righteousness', and also as 'words of righteousness and reprimand' taken as a hendiadys = 'words of the righteous reprimand'. On the whole, however, it seems preferable to understand the phrase 'words of truth (and) the reprimand' as 'words of truth concerning the reprimand'.

the eternal watchers See above, the note on 1.5.

the Great Holy One See the note on 1.2. Here again Eth. expands to 'the Holy and Great One'.

in the dream which I saw G Eth. ἐν ταύτῃ τῇ ὁράσει, 'in the following dream': En[c] 1 vi 10 'in the dream which I saw'. V. 2 in G Eth. begins ἐγὼ εἶδον κατὰ τοὺς ὕπνους μου, a similar phrase to that in En[c] 1 vi 10 in v. 1 (cf. Dan. 4.5). It seems likely that there were two such phrases, and that one has been altered in G Eth. to ἐν ταύτῃ τῇ ὁράσει (cf. Knibb, 2, 95). For ὕπνος 'dream', LXX Gen. 31.11, חלום, and see Milik, 'Fragments grec du livre d'Hénoch (P. Oxy. XVII 2069)' in *Chronique d'Égypte*, Tom. 46 (1971), 325, 330, Barr, 'Aramaic-Greek Notes', I, 191f.

(3) **endowed ... created ... men** G = Eth. 'given to men', both taking חלק in its familiar meaning of 'to apportion to, impart to, distribute to'; Milik gives it the meaning 'decree, destine' (198, Ll. 11-12, cf. ܫܠܡ, P. Sm. 1294). This seems to me a less appropriate rendering in this context, and it is better to follow G Eth. 'As he has endowed (חלק), fashioned (עבד) and created (ברא)' is restored to match En[c] 1 vi 12. But cf. Barr, 'Aramaic-Greek Notes, II, 187.

(4) **forasmuch as ... decree** Enoch is warning the watchers that their petition has no chance of success: on the contrary, a final judgement is about to be passed on them by divine decree.

For גזר, and its cognates in this context, see Milik, 197, L. 14, and for שלם in this sense with דין see above, on 10.12. The following vs. 5-6 give the contents of the divine decree. (The words in Eth[M] 'and it shall not be yours' seem a ditt. of the earlier 'it (your petition) shall not be yours (i.e. not be accepted)'; Eth[n ryl2] convert the words into the recurring formula of 1.5, 'you shall not have peace').

(5) **bonds in the earth** G ἐν τοῖς δεσμοῖς τῆς γῆς = באסורי ארעא = 'in imprisonment in the earth' (cf. Ezr. 7.26 אסורין = 'bonds', 'imprisonment'). Charles suggested reading ἐν δεσμοῖς ἐν τῇ γῇ, citing Origen, *Contra Celsum*, V. 52, where the fallen angels 'became evil and were punished by being cast under the earth in chains' (trans. Chadwick) (κολάζεσθαι δεσμοῖς ὑποβληθέντας ἐν γῇ). Cf. Jude 6 (of the same angels) εἰς κρίσιν μεγάλης ἡμέρας δεσμοῖς ἀϊδίοις ὑπὸ ζόφον τετήρηκεν. Charles

makes the further suggestion that the original Aram. may have been misrendered and should have read 'as prisoners of the earth' (כאסירי ארעא), comparing Lam. 3.34 (אסירי ארץ). Milik is inclined to accept Charles's conjecture and to interpret 'the prisoners of the earth' as 'the inhabitants of the underworld', citing in support the use of the word 'prisoners' in this connection in Babylonian and Palmyrene sources (198, Ll. 14-15). But. Lam. 3.34 means 'prisoners in the land', and the reference here seems to be to the decreed fate of the watchers announced at 10.5,12; they are to be confined in the 'valleys of the earth' until the day of the great judgement (cf. 21.7-10). Did the writer think of these 'valleys' as part of the underworld?

(6) **destruction ... total destruction** Both Eth. and G are abridged: 'the destruction of your beloved ones and of (all?) their sons' (i.e. the watchers' grandsons) is clearly read in the Aramaic texts. The decipherment of ... בניהון ובק ... must remain problematical; Milik 'and of (their) possessions [ובק[ניאניא'. For the reading proposed, cf. Jn. 4.12, οἱ υἱοὶ αὐτοῦ καὶ τὰ θρέμματα αὐτοῦ.

you will have no heirs to them i.e. to the sons of the watchers. G evidently read ירותיהון 'their heirs' as יותרניהון, ἡ ὄνησις αὐτῶν. Eth. ṭeryānihomu lit. 'their possessors' (Dillmann, *Lex.* 1221) may have preserved an alternative Greek version of the original reading, perhaps [οἱ κληρονόμοι αὐτῶν] = ירותיהון.

(7) **petitioning** G Eth. κλαίοντες reading בעין as בכין.

will be implemented(?) The text as transmitted in the versions makes very little sense. Dillmann, 108, thought there had been a confusion of λαβεῖν with λαλεῖν, and Charles [1906], 37, n. 11, suggested λαβόντες or λαχόντες for λαλοῦντες. Confusion of מלל λαλεῖν with מלי Pa. 'to fulfil, implement', seems a more likely explanation: וכול מילה לא ממליה, lit. 'and no word will be implemented (pass. ptc. Pa.). Cf. Cyr. *Comm. in Lucam*, 325.26, P. Sm. 2121 for the 'implementing' (ܡܠܐ) of petitions.

(8) **dark clouds were crying out to me** G Eth. ὁμίχλαι lit. 'mists' ('a mist', Charles), 'cloud-mists' (Milik), translations influenced by the most common meaning of ὁμίχλη as 'mist' or 'fog'. It can also mean 'cloud-like darkness', 'gloom' (Liddell and Scott, s.v.), and this is its meaning in Biblical Greek, corresponding to ערפל (Tg. ערפיל), 'deep, dark cloud', e.g. Exod. 20.21 ('in which God dwells'), Jl. 2.2, Zeph. 1.15 where LXX ὁμίχλη (ערפל) is parallel to νεφέλη (ענן).

fire-balls and lightnings G Eth. translate זיקין by the Aristotelian phrase διαδρομαὶ τῶν ἀστέρων, Meteor. 341 a 33. For זיקין with ברקין = 'fire-balls (meteors) and lightning', Tg. (P-J) Exod. 20.2,3.

hastening me on and driving me G κατασπουδάζειν = בהל Pa. (cf. Milik, 194) and θορυβάζειν = דחף (Heb. and Aram.) 'to drive', e.g., 2 Chr. 26.20 (par. בהל), Est. 8.14 (LXX v.l. διωκόμενοι) (Tg. II).

bearing me aloft For the unusual form ἐξεπέτασαν cf. LXX Num. 11.31 (mss) and ἐκπετάζειν at Hos. 9.11, Nah. 3.16. See also Charles[1906], 37 n. 14.

(9) **a wall, built of hailstones** Milik takes τείχους as a collective, 'walls' because of the following κύκλῳ αὐτῶν in G. But Eth. reads αὐτοῦ; the plural αὐτῶν was probably the result of attraction to λίθοις. The reading of Eth. adopted, 'a wall built of ice-crystals' (as in v. 10) presupposes Aram. שורא בנא באבני ברד (בנא ptc. pass. rendered in G by a noun). The idea of a building here (Milik) is out of a place: the 'building' is the house of ice-crystal of the next verse. The writer is thinking of the περίτειχος i.e., the protecting outer wall of heaven, like that of a city (cf. Isa. 26.1, Eth. ṭeqm) Although Aram. אבני ברד had probably also the meaning of κρύσταλλος 'rock-crystal'

(Dillmann, Crystallstein),[1] the primary meaning here is 'hailstones'. The writer's imagination envisages a περίτειχος built of the familiar elements of fire, hail, snow and ice.

it began to terrify me So Eth. but G 'they began to terrify me', the reference being presumably back to the 'tongues of fire'. The phrase seems a trifle lame after the vivid imagery and poetry which proceeds it. Could there be a mistranslation of שורא 'wall' as שרי(ו) 'it began', 'they began'... 'And the wall terrified me' (ושורא מדחל לי)?

(10) **tesselated paving stones** G λιθόπλακες, a neologism for לוחות אבנין, Heb. לוחות אבן Exod. 31.18 LXX πλάκες λίθιναι; cf. also Lucian, *Amores*, 12 λίθων πλαξὶ λείαις ('with smooth stone-facings'). See Milik, 198.

a large house This led into an even larger and more magnificent house where the throne of God itself was located (vs. 15-18). The second House is the traditional (scriptural) Palace or Temple of God in heaven (see below, 148); this 'large house' which led into it was presumably the πρόναος or forecourt, the ante-chamber to the throne-room. The presence of the Cherubim (v. 13) suggests that it was designed to guard the entrance to the throne-room itself. It may well have been suggested by the fore-court of ancient Babylonian temples or palaces.

(11) **upper storeys** G στέγαι = טללין, תטללין(?) (ܬܛܠܠܐ Palm. תטליל), a plur. materiae or plur. majestatis? In view of the following 'in the midst of them' (which can scarcely refer to the subject of the simile 'like meteors and lightning'), 'upper storeys (including the roof)' seems preferable to roof (cf. Liddell and Scott, *P. Petr*. 2, p. 28).

fiery cherubim, celestial watchers(?) The Cherubim are mentioned among these celestial phenomena as at Ps. 18.10 or under the influence of Ezek. 1.13,14 (Lods). Their role, in the upper parts of the house as in the fore-court (above, 146) was no doubt a protective one.

If we are to make sense of the text of G Eth. as it stands 'their heaven was water', we should perhaps read כמיא 'like water', i.e., pellucid and blue (Charles 'their heaven was (clear) as water'); cf. Ezek. 1.22, 'Above the heads of the living creatures was, as it were, a vault glittering like a sheet of ice' (NEB). The thought may also be that of Gen. 1.7 referring to the 'waters above the firmament' and this may supply the clue to the original (cf. 89.2, En[e] 4 i 16 below, 263). Perhaps the original read ושימתיהון מיא 'and reservoirs/ storehouses of waters'. The idea of such 'storehouses' (Heb. אוצרות lit. 'treasuries') of wind, rain and snow is a familiar one in the Old Testament (Dt. 28.12, Job 38.22, Jer. 10.13, etc.) and in Enoch (e.g. 17.3, 18.1, 69.23); Aram. סימתא, ܣܝܡܬܐ, is variously used for 'treasure', 'depository', 'storehouse' (e.g. Isa. 45.3 Pesh. ܣܝܡܬܐ ܕܚܫܘܟܐ = אוצדות חשך).

On the whole, however, it seems more probable that we have to do with a corrupt Greek text from an original ועירי שמיא (cf. 13.10 En[c] 1 vi 8), καὶ [ἐγρηγ]όρους / οὔρους τοῦ οὐρανοῦ: ὕδωρ from מיא[ש]?). Cf. v. 23 and for οὖρος / ἐπίουρος Liddell and Scott, 1274, 649.

(13) **no delights in it** 'Delights, luxuries' G τρυφή (confused with τροφή as at Gen. 49.20 LXX A), Heb. תענוג Tg. פינוקא e.g., Gen. 49.20 = ܦܘܢܩܐ – ܬܦܢܘܩܐ, e.g. Sir. 6.28 vita deliciis abundans, luxuries (P. Sm. 3179, 80). The closest parallel is Sir. 14.16 οὐκ ἔστιν ἐν ᾄδου ζητῆσαι τρυφήν (בשאול לבקש תענוג). Cf also Erub. 54a 'Do good to yourself; for there is no luxury (תענוג) in Sheol'. G = Eth I τρυφὴ ζωῆς and Eth II τρυφὴ καὶ ζωή are free translation equivalents of תענוג (?פינוקא).

horror overwhelmed me lit. 'covered me', G με ἐκάλυψεν. Cf. Ezek. 7.18, Ps. 55.6.

[1] Eth[m] in fact at v. 10 reads 'pearl-crystals'.

(15) **and behold! another house ... opposite me** The text of G (cf. Eth^{tana}) is confused and corrupt. Eth^{M} has preserved the right order and text: the important new vision is of *another house*, larger than and superior to the first, and not of 'another door'[1] (there has been no door hitherto mentioned, only the first house). Eth^{M} has preserved the important words 'and behold! another house greater than that one' i.e. the first house (=[καὶ ἰδοὺ ἄλλος οἶκος μείζων τούτου]) which has survived in G as καὶ ὁ οἶκος μείζων τούτου. The 'door' of the new house is said to be 'opened opposite me'[2], clearly in order to enable the seer to explain how, without going any further, he is able to see and describe the interior of the second 'greater house' (cf. Dillmann, 109). Enoch does not approach this door, much less pass through it, in the same way as he 'entered' the first house (v. 13). But later (v. 25) he is awakened by an angel from his trance and led to the door where he bows low to hear the voice of the Almighty. (There is no question of his going any further and entering the second house into which not even the angels are permitted to enter (v. 21)). The writer is undoubtedly seeking to convey the idea of the distance between the seer and 'the greater house', the heavenly palace or temple (see below); and in order to do so he introduces this 'door' through which he can see and describe but through which he cannot enter the 'greater house'. For the idea of a 'door' (θύρα) into heaven, cf. Rev. 4.1. Cf. also Test. Levi, 5. 1.

another house greater than that one The idea of a House of God in heaven, a Palace or Temple (according as he is conceived as object of worship or as King) is a familiar one in the Old Testament (Isa. 6.1f., 2 Sam. 22.7, Ps. 18.6, 29.9 (Temple of Jahweh), Mic. 1.2, Hab. 2.20; Isa. 63.15 is more general, 'Thy holy habitation'). Both ideas are combined at Ps. 11.4: 'The Lord is in his holy temple, the Lord's throne is in heaven'. En. 14.18 goes on to speak of a 'lofty throne' in this 'greater house', and one naturally tends to think of the throne of God located in the mountain of God in the north-west (also the site of the Paradise of justice) (En. 24,3)[3]. We require to distinguish, however, not only between a celestial and an earthly paradise,[4] but also, in this passage, between a heavenly palace and temple, with its own throne-room, and a throne of God on the mountain of God at En. 24, to which he is to descend for the last judgement.

(17) **its upper chambers** Lods maintained that G τὸ ἀνώτερον was to be understood adverbially, 'at the top of the house'. An original could have read מלעיליה or לעיל מניה 'above it', but Milik's conjecture that a noun עליתה 'upper part' (better 'upper chamber, roof-chamber', Dan. 6.11) was original seems more probable.

(18) For the relation of vs. 18f. to Dan. 7.9,10, see Additional Note, 151f. These verses have been used by the author of 71.5-8, the closing apocalypse of the Book of the Parables. See below, 251.

lofty throne For the scriptural background of the heavenly throne of Jahweh, see 1 Kg. 22.19, Isa. 6.1, and expecially Ezek. 1.26, Dan. 7.9; cf. also Ass. Mos. 4.2, Test. Levi 5.1, and Rev. 4.2f. W. Zimmerli, *Ezekiel 1-24*, (Bib. Komm. I (1969), 16-21)

---

[1] Note the accus. after ἰδού in G, emended by editors to ἄλλη θύρα ...

[2] The kuellu in Eth. may not be redundant (see Nickelsburg 'Enoch 97-104', 149f.), but represent an original תרעא כול פתיח, 'a door completely open' (ganz geöffnet, Flemming-Radermacher).

[3] See P. Grelot, 'La Geographie mythique d'Hénoch au Pays des Aromates', RB, Tom. 65 (1958), 33-69, especially 42f., 46; Milik, 33, 39 and 'Hénoch au Pays des Aromates', RB, Tom. 65 (1958), 70-77, especially 77 and n. 1.

[4] Grelot, op. cit., 42. See further below, 179.

discusses the tradition-history of the throne-vision; see also M. Black, 'The Throne-Theophany Prophetic Commission and the 'Son of Man': a Study in Tradition-History', in *Jews, Greeks and Christians, Religious Cultures in Late Antiquity, Essays in Honor of William David Davies*, ed. by R. Hamerton-Kelly and R. Scroggs (Leiden, 1976), 57-73.

appearance ... ice Cf. Ezek. 1.26, 22 'Above the vault ... a throne', 'a ... vault glittering like a sheet of ice' (NEB) and Rev. 4.6.

wheels ... sun Cf. Dan. 7.9 וגלגלוהי נור דלק Theod. οἱ τροχοὶ αὐτοῦ πῦρ φλέγον (LXX om.). The imagery is drawn in both passages (whatever their relationship) from Ezekiel: the heavenly throne rests on wheels (Ezek. 10.1,2,6,9,11) and is surrounded or accompanied by fire and flame (Ezek. 1.27, 10.6).

watchers(?), Cherubim G καὶ ὅρος Χερουβίν 'and a mountain of Cherubim' is certainly nonsense (Knibb); Charles, 34, suggested a corruption of ὅρασις 'vision', 'and a vision of Cherubim'; others cj. χορὸς Χερουβίν 'a choir of Cherubim', A. Meyer, Flemming-Radermacher. Milik (200) suggests reading ὅρος 'boundary-stone' and retranslates by גדנפוהי 'its sides (were Cherubim), but admits that the retranslation is 'very hypothetical'. A less uncommon word for 'sides' in Aram. is גב , ܓܒܐ (Hoftijzer, 46). Could an original וגבוהי כרובין 'and its sides were Cherubim' have been mistranslated by καὶ ὅρος Χερουβίν through confusion of **גבא** 'side' with Aram. גבא 'high', Heb. גבעה = ὅρος LXX Isa. 31.4, Ezek. 34.6? This would support Milik's idea that we are dealing with two flanking sides of the Throne, carved in sphinxes = Χερουβίν (Milik, ibid.); the Cherubim are then performing their usual role as bearers and guardians of the Throne-chariot, as in Ezek. 1 and 10. Perhaps we have to do with a mistranslation of גב as a preposition = 'beside (it) were Cherubim'? An alternative conjecture would be to posit an original גבא, Heb. גב LXX νῶτος, the 'rim' or 'nave, hub' of a wheel, used, along with **אופנים** (τροχοί), in Ezekiel's description of the throne-chariot at 1.18 (cf. 10.12 where he describes the 'back' of the Cherubim).

Again, however, a corruption at the Greek stage of textual transmission seems preferable, namely a similar corruption as at v. 11 above, [ἐγρηγ]όρους / οὔρους, עירין: 'and (I saw) watchers, Cherubim (ועירין כרובין). Perhaps Eth. preserves an original phrase: 'and (I heard) the sound of Cherubim'? Cf. Ezek. 1.24.

(19) **from underneath ... fire** Cf. Dan. 7.10 נהר די נור נגד ונפק מן קדמוהי lit., 'a river of fire streamed and came forth from before him (from before it (the throne?))'. A common tradition at least lies behind both passages, but note also the identical expressions.

Dan. 7.10, LXX	Enoch
ἐξεπορεύετο ποταμὸς πυρός	ἐξεπορεύοντο ποταμοὶ πυρὸς φλεγόμενοι (Eth. 'streams of blazing fire' = Dan. 7.9 נור דלק Theod. πῦρ φλέγον.

Eth II 'the high (lit. great) throne of the Great One' could imply a variant ὑποκάτω τοῦ θρόνου ὑψηλοῦ. It seems more likely, however, to be an Eth. expansion, perhaps influenced by the phrase 'the lofty throne' in v. 18.

(20) **the glory of the Great One** G ἡ δόξα ἡ μεγάλη is followed by Eth I here and at 102.3 (Eth II in both passages has 'he who is great in glory').[1] At 104.1 G Eth. read

---

[1] A similar expression occurs at Test. Levi 3.4 ἡ μεγάλη δόξα, but the correct reading is probably ἡ μεγάλη δόξα τοῦ θεοῦ.

ἐνώπιον τῆς δόξης τοῦ μεγάλου (Eth. bis); cf. also 9.3 Sync. ἐνώπιον τῆς δόξης τῆς μεγαλωσύνης (G πρὸς τὸν ὕψιστον), 'before the glory of the Majesty'. Of the three forms of the title, the third, 'the glory of the Great One', seems the original; as has been noted above (104f.) רבא is employed simpliciter as a title for deity, and ἡ δόξα τοῦ μεγάλου = יקרא די רבא(?) would correspond in this passage, manifestly influenced by Ezekiel, to כבוד יהוה LXX δόξα κυρίου[1] (used, e.g., to describe the manlike 'appearance' on the heavenly throne at Ezek. 1.28).

brighter ... snow Cf. Dan. 7.9 לבושה כתלג חור.

Dan. 7.9	Enoch G
LXX ἔχων περιβολὴν	καὶ τὸ περιβόλαιον
ὡσεὶ χιόνα	ὡς εἶδος ἡλίου
Theod. καὶ τὸ ἔνδυμα αὐτοῦ	λαμπρότερον καὶ
ὡσεὶ χιὼν λευκόν	λευκότερον πάσης χιόνος

How are these passages related? Are we to explain Enoch as an expanded description based on Daniel or is the account in Daniel an abridged adaptation, in a new composition, of the fuller descriptive imagery in Enoch? Their literary interrelationship cannot be denied.

(21) **by reason of its splendour and glory** Eth. has omitted the preposition διά and read genitives, rendering '... no angel was able to look on the face of the honoured and glorious One'. G has rightly retained the διά phrase; is it a rendering of על הדר יקריה, 'by reason of the majesty of his glory', almost exactly paralleled at EnGiants[a] 9 2, Milik, 316.

(22) **A blazing fire encircled him** Eth I ʾessāt zayenadded = πῦρ φλεγόμενον, נור דלק (see above on v. 19). Eth II ʾessāta ʾessāt zayenadded, lit. 'a fire of blazing fire' seems odd, unless it renders an expression such as לבת־אש Exod. 3.2 LXX φλὸξ πυρός.[2] Could there be a textual corruption, e.g. **በእንተ** διά to **እሳት**, πῦρ 'on account of the blazing fire', a clause going originally with v. 21: '... no flesh was able to look on him on account of the blazing fire surrounding him'?

ten thousand ... before him = Dan. 7.10 רבו רבון קדמוהי יקימון.

Dan. 7.10 LXX = Theod.	Enoch G
μύριαι μυριάδες	μύριαι μυριάδες
παρειστήκεισαν αὐτῷ	ἑστήκασιν ἐνώπιον αὐτοῦ

no need of counsel Charles notes the close parallel at Sir. 42.21 LXX οὐδὲ προσεδεήθη οὐδενὸς συμβούλου (לא צדיך לכל מבין) (cf. also Isa. 40.14) and emends Eth. mekr 'counsel' to makārē 'counsellor' to bring it into line with Sir. 42.21.

every word ... a deed This complementary phrase, omitted in Eth., also echoes Sir. 42.15 (μνησθήσομαι δὴ τὰ ἔργα κυρίου ...) ἐν λόγοις κυρίου τὰ ἔργα αὐτοῦ ((mg מעשיו) באומר אלהים רצונו). God's words are his commands and are immediately acted upon by others, or they are immediately translated into action by himself; cf. Ps. 33.9, '... for he spoke and it came to pass'. A free quotation of the Enoch verse is found in 2 En. 33.4(B) '...and my thought is counsellor, and my word is deed'. Cf.

---

[1] Tg. יקרא דיהוה; for the term applied to God, EnGiants[a] 9.2 (Milik, 316); cf. En. 9.4, En[b] 1 iii 15, 'Thy glorious throne' (Milik, 171).

[2] This seems to be how the phrase has been understood by Dillmann, 'ein Flammenmeer von brennenden Feuer' (Knibb, 'a sea of fire').

Barr, 'Aramaic-Greek Notes' II, 190, who proposes to render 'everything (πᾶς λόγος) is his work.'

(23) **the watchers and holy ones(?)** i.e. the archangels. Eth. 'the holinesses of the holy ones', where qeddesāt, 'holinesses', seems to be a title. Did Eth. read [οἱ ἅγιοι τῶν ἁγίων]? [αἱ ἁγιοτῆτες τῶν ἁγίων] seems unlikely, although [οἱ ἅγιοι τῶν ἁγίων] = קדישי קדישין is possible. Against this, however, is the title given to the archangel Raphael at 22.6, En^{c} 1 xxii 5, 'watcher and holy one' (עירא וקדישא = G Eth. ὁ ἄγγελος ὃς μετ' ἐμοῦ ἦν) and the almost certain reference to the archangels at 93.2 En^{g} 1 iii 21 as 'watchers and holy ones' עירין וקדישין; Eth. 'the holy angels'. The original here was most probably 'the watchers and holy ones'.

24) **I had been prostrate on my face** So G = הוית רמי על אנפי, cf. v. 14 above. Eth. appears to have read a text with περιβεβλημένος, or περικεκαλυμμένος (cf. 13.9) a corruption possibly influenced by passages such as Exod. 33.22, 1 Kg. 19.13. Cf. Knibb, 2, 100.

the Lord For the title, see above, 136 and below, 198.

(25) **raised me** lit., 'raised me up and made me stand'. Charles 'waked me' is impossible, but ἤγειρέν με = אעירני 'roused me, raised me up', is appropriate: Enoch is 'roused' by the angel from his trance (v. 14) and set upon his feet (G ἔστησέν με).

to the door See also on v. 15.

I bowed my face low G = אשפלית אנפי? Eth. appears to have understood ἔκυφον as παρέκυφον, 'I looked (below)'.

ADDITIONAL NOTE: ENOCH 14 AND DANIEL 7

Both passages describe a throne-vision using a common imagery and identical expressions. The imagery is drawn, in both visions, from Ezekiel 1 and 10: the heavenly throne rests on wheels and is surrounded or accompanied by fire (see above on v. 18). Nevertheless, the number of identical expressions not derived from Ezekiel, points to some relationship of interdependence, either of Enoch on Daniel or vice versa, or perhaps on a common archetypal text of this apocalyptic vision in the Book of the Watchers (see especially the notes on vs. 18, 19, 20 and 22). All editions repeat the view of Charles (Charles, 34) that in Enoch 'the writer draws upon Dan. 7' or 'the expression goes back to Dan. 7'. The opposite, however, may have been the case, and this view was, in fact, argued by T. F. Glasson before the discovery of the Aramaic fragments.[1] These fragments have been assigned by Milik, on palaeographical grounds, to the Hasmonaean and Herodian periods (179), and the composition of the Urtext to an even earlier date, c. 250 B.C., almost 100 years before the date usually assigned to Daniel. Comparison of the shorter Daniel apocalypse with the fuller Enoch vision does suggest that Daniel is abridging or adapting phrases or excerpting a whole line (Dan. 7.9 = En. 14.22) from Enoch. At v. 20 the expression in Daniel 'his raiment was white like snow' looks like a shorter form of the fuller line in Enoch

---

[1] *The Second Advent: the Origins of the New Testament Doctrine* (Oxford, 1965), 2-7, and (post Milik) 'The Son of Man Imagery: Enoch 14 and Daniel 7'', in NTS, Vol. XXIII (1977), 82-90. In his Introduction, lii, Charles, with 'complete inconsistency', had also claimed a pre-Maccabaean date for Ch. 14; see H. H. Rowley, *The Relevance of Apocalyptic* (London, 1963), 93f.

(which does not lack poetic quality): 'his raiment was brighter than the sun and whiter than any snow'. Similarly 'the river of fire' in Enoch comes forth 'from under the throne' which seems more natural than Daniel's '... streamed and came forth from before him'. (Was this an editorial attempt to turn attention away from the throne to its occupant?) Perhaps the hypothesis of a common literary archetypal apocalypse is the least unsatisfactory solution of the problem—and this was probably an original Aram. work narrating the saga of the fall of the angels.

CHAPTER 15

(1) The words in G (cf. Eth.) ὁ ἄνθρωπος ... ἤκουσα are a translator's addition from the similar clause after μὴ φοβηθῇς. The readings with the article in this addition, however, ὁ ἄνθρωπος ... ὁ γραμματεύς, could be original, the article here representing vocatives.

righteous man ... righteousness Cf. 1.2 and Gen. 6.9 LXX Νῶε ἄνθρωπος δίκαιος also 92.1, En^{g} 1 ii 23 חכים אנשא 'wisest of men', and for 'scribe of righteousness' (= ספר קושטא?), see above on 12.4.

(2) For the intercession of the angels for mankind, see above on 9.10 and Charles, 21.

(3) **high heaven and the eternal Holy One** G Eth^{tana} read a text usually rendered 'the high, holy and eternal heaven'. The rendering of Eth^{M}, which I have adopted, seems preferable = שמיא עליא וקדישא דלעלם. For 'the Holy One', see above on 1.2.

children of earth For בני ארעא = υἱοὶ τῆς γῆς, see on 91.14, En^{g} 1 iv 20.

(4) **you lusted after the daughters of men** G Eth. 'after the blood of men'. The idea of 'blood-lust' is unsuitable in this context, though a scribe may have been influenced by passages such as 7.5; 9.1,9. I have adopted the conjecture of Knibb, 2, 100, but suggest that the original read בנת אנשא but was misread as בדמא, through the influence of the immediately preceding occurrences of בדם. Are the clauses in the versions in the wrong order, and should this last clause have come first, at the beginning of the verse: 'but you have lusted ... and become unclean etc.'?

(5) **that pregnancy(?) should never fail ... earth** There is serious corruption in the versions even if some meaning can be extracted from them. Cf. Lods, 146, who remarks that it is a strange way of expressing the idea, so familiar in the Old Testament, that descendants should never fail them (cf. Jer. 33.17; 35.19 etc.). I suggest that an Aram. עיבורא Gen. 25.24 Tg(P-J) 'pregnancy' has been misread as עובדא, ἔργον: ... די לא יחסר להון כול עיבורא על ארעא.

(8) **mighty spirits** G πνεύματα ἰσχυρά (= רוחות גברות?) seems original; ἰσχυρός = גבר in malem partem, superbus, tyrannus, Dan. 3.20, Ps. 52.3 (Gesenius, *Thes.*, 263). It is the offspring of these 'mighty spirits' who are to be called 'evil spirits' (v. 9). An Aram. adj. גיבר (Levy, CW, I, 124) would give a word-play with גברים = γίγαντες. The reading of Eth. Sync. is clearly secondary.

(9) Sync. has preserved the longer and more primitive form of text: note πνεύματα ἔσονται 'τὰ πνεύματα' ἐξεληλυθότα = רוחות תהון נפקן(?) cf. ἐξῆλθον G (proph. perf. in a variant text?).

from their bodies Sync. 'from their bodies of flesh' is tautologous, and seems a conflate rendering, from מן בשרהון, ἀπὸ τῶν σωμάτων αὐτῶν and ἀπὸ τῆς σαρκὸς αὐτῶν.

from men they have come G Eth. 'from above', ἀπὸ τῶν ἀνωτέρων crpt ex ἀπὸ τῶν ἀνθρώπων (Sync.), even though it does give an acceptable sense and a parallel line to 'and from the holy watchers is the beginning of their creation'.

and the beginning of their origins. G ἀρχὴ κτίσεως: ריש בריאה(?); cf. Mt. 10.6, Sync. G Eth. ἀρχὴ θεμελίου = ריש יסודא(?) 'the beginning of (their?) foundation', parallel to ἀρχὴ κτίσεως. Could there have been a misrendering of סודא, '(their) assembly, company', (cf. the use of the Niph. of יסד 'to congregate' and סוד, Jer. 6.11, IQS 11.9, 10, P. Sm. 2544 ܣܘܕܐ, coitus, turba (Bar. Heb. *Chron.* 917.5))?

For similar descriptions of the origin of evil spirits or demons, see Justin Martyr, *Apol.*, II 5 (above on 9.9)), Jub. 10, Tertullian, *Apol.*, 22, *Clementine Homilies*, 8.18, Lactantius, *Instit.*, II.15 (ed. Migne, 330); in Lactantius the demons are regarded as wicked angels. Cf. Charles, 36, Lods, 146f.

(10) This verse is bracketed by Charles and Flemming as a translator's expansion, since it is mainly a repetition of phrases found in vv. 7 and 8. Knibb, 2, 101 regards its omission by Sync. as correct. The 'expansion', if such it is, could well have come from an Aram. recension.

(11) **the vicious(?) spirits** The subject in all versions reads simply 'the spirits of the giants' etc., but G repeats the phrase later in a doublet—clearly out of place—which reads πνεύματα σκληρὰ γιγάντων, and which was probably the original subject = רוחות קשות די גברין.

the giants, the Nephilim G νεφέλας is a misreading of Ναφηλείμ: the expression is correctly reproduced by Sync. at 16.1 τῶν γιγάντων Ναφηλείμ = (די) גברין ונפילין. See note on 16.1. Sync. νεμόμενα lit. 'shepherding', is difficult to explain as a corruption of Ναφηλείμ: has it perhaps arisen from a mistranslation of Aram. מרעין = ἀδικοῦντα, misread as from רעה 'to shepherd'? See next note. The 'spirits of the giants, the Nephilim' are, in this context, clearly the 'evil spirits' which issued from the 'bodies of flesh' of the giants. Their destructive role, including 'falling upon, attacking' (ἐμπίπτοντα) supports the derivation of the name of their parents, the giants, from נפל, especially in this sense (below, 154).

they inflict harm ... ground Is this the background to the role of the 'unclean spirits' and demons in the Gospel tradition? Cf. Mt. 1.26; 5.2f.; 9.25 (par. Lk. 9.42, δαιμόνιον). The succession of ptcs. reflect Aram. ptcs. = finite verbs. Sync. G ἀδικοῦντα lit. 'doing wrong to', renders in his context a verb such as רעע Pa., Heb. Hiph. 'injure', or even 'break, smash'.[1]

causing injuries(?) Eth. ḥazan = 'grief' (λύπη), 'pain' (ἄλγημα). Sync. G δρόμους ποιοῦντα has been emended to τρόμους (cf. Knibb, 2, 102), but this hardly accounts for Eth. ḥazan. LXX δρόμος corresponds to Heb. מרוצה from רוץ 'to run'. Has the translator mistranslated רצצין 'injuries' (τραύματα) from רצץ; cf. P. Sm. 3969 ܪܨܨܐ, fractura, contusio, and Jer. 22.17? Cf. Barr, 'Aramaic-Greek Notes', II, 185f.

they eat nothing ... they collapse Sync. G Eth^{tana} προσκόπτοντα lit. 'they stumble': in the LXX the verb most frequently corresponds to Niph. כשל, in the sense of 'stumbling and falling'; cf. Dan. 11.19 (+ נפל). Aram. תקל Ittaph. is similarly used, and cf. Tg. Ps. 109.24 'my knees give way (אתקלו = כשלו) through fasting'. The idea seems to be that those who are possessed by such evil spirits collapse through their abstention from food and drink i.e. through their fasting. Eth. has nothing corre-

---

[1] Cf. Beer in Kautzsch, *Die Apokryphen und Pseudepigraphen*, 247. For the misreading of √רע as ποιμαίνειν, see Rev. 2.27, 12.5, 19.15 = Ps. 2.9. See M. Black, 'Some Greek Words with 'Hebrew' Meanings in the Epistles and Apocalypse', in *Biblical Studies Essays in Honour of William Barclay*, ed. by J. R. McKay and J. F. Miller (London, 1976), 136f.

sponding to ἀλλ' ἀσιτοῦντα and Eth[M] carries the negative forward to διψῶντα καὶ προσκόπτοντα. I have followed the order of words in G μηδὲν ἐσθίοντα ἀλλ' ἀσιτοῦντα καὶ διψῶντα .... As Charles, 37, suspects, Sync. καὶ φαντάσματα ποιοῦντα may be original, and I have taken it into the text.

(12) **will rise against ... women** Sync. G ἐξανιστάναι = Heb./Aram. קום על. Eth. 'against the sons of men and against women' = ἐπὶ τοὺς υἱοὺς ἀνθρώπων καὶ ἐπὶ γυναῖκας = על בני אנשא ועל נשא(?) seems more semitic than Sync. G ἐπὶ/εἰς τοὺς υἱοὺς ἀνθρώπων καὶ γυναικῶν. The ὅτι ἐξ αὐτῶν clause = די ... מנהון, 'from whom they have come forth': the fact that they have come forth from men and women is hardly a reason for their 'rising against' mankind.

CHAPTER 16

(1) **the giants, the Nephilim** Eth. [τῶν γιγάντων]; Sync. τῶν γιγάντων Ναφηλείμ offers a superior text to G Eth. Did Sync. read גברין ונפילין? The expression occurs at EnGiants[c] 2, which Milik, 308, renders 'the giants and the nephilim', assuming that the latter are the offspring of the giants, as in the Sync. gloss at 7.2. But 'the Nephilim' may here be in apposition, 'the giants (namely) the Nephilim', and Nephilim here and at Gen. 6.4. a name for the offspring of the watchers or 'sons of God'. Both at Gen. 6.4, and Num 13.33 (the only two places in the Old Testament where נפילים occurs), the LXX renders by γίγαντες; and Jub. 7.22 calls the sons of the watchers 'Naphidim', although it goes on to assume, in the same verse, the γένη τρία of Sync., giants, Naphil and Elyo. Support for these three 'generations' of the offspring of the watchers is found at 86.4, 88.2, (see Milik, 240, Ll. 20-1), but 'the wild asses, camels and elephants', which are the progeny of the 'fallen stars' which became bulls, represent a quite independent development of the legend. Whatever the source of the Sync. gloss, it seems to have originated through the elaboration of 'the giant Nephilim' into parents and offspring, and their offspring, the Elioud. See above on 7.2. Sync. adds a line, in apposition to τῶν γιγάντων Ναφηλείμ, οἱ ἰσχυροὶ τῆς γῆς, οἱ μεγάλοι ὀνομαστοί, which, if it was ever part of the original,[1] belongs rather to 7.2 where the giants are first introduced. Is Sync. elaborating from Gen. 6.4 LXX ἐκεῖνοι ἦσαν οἱ γίγαντες οἱ ἀπ' αἰῶνος οἱ ἄνθρωποι οἱ ὀνομαστοί.

The word נפלין (Heb. נפילים, in the Old Testament only at Gen. 6.4, Num. 13.33, LXX γίγαντες) is found again at IQapGn 2.1: Lamech suspects that the conception of Noah 'was due to the watchers or ... to the holy ones, or to the Nephil[im]' (i.e. almost certainly the sons of the watchers and not their grandsons). The derivation and meaning of the word is obscure, and a number of ingenious solutions have been proposed (see especially Gesenius, *Thes*., 899, Aram. נפילין, pass ptc. could mean 'fallen ones'; cf. CD 2.18 (the 'watchers of heaven' fall (נפלו) through their 'guilty inclination (יצר אשמה)). The term appears also to be used in Aram. for heavenly bodies, e.g. the constellations (of Orion) or for meteors (Tg. Job 9.9, 38.31, Isa 13.10): but both these uses would more appropriately describe the fallen watchers themselves and not their off-spring, the giants. The ptc. is perhaps best taken in an active sense, the ones who 'fell upon' mankind, with hostile intent (so Aq. ἐπιπίπτοντες, Symm.

---

[1] גברי ארעא רברבין אנשי שמא.

βιαῖοι). See further, E. G. Kraeling, 'The Significance and Origin of Gen. 6.1-4', in JNES 6 (1947), 193f., especially 202f., Fitzmyer, *Genesis Apocryphon*, 81.

from their bodies G ἐκ τῆς ψυχῆς τῆς σαρκὸς αὐτῶν lit. 'from the soul of their flesh' is an impossible phrase; Lev. 17.11 נפש הבשר (LXX ψυχὴ ... σαρκός) means 'the life of the flesh' and is not a real parallel. The more difficult reading of Eth.—and one which can explain the others—is that of Eth[g] 'from the body of their flesh', which repeats the expression used in the same connection at 15.9, G ἀπὸ τοῦ σώματος αὐτῶν, Sync. ἀπὸ τοῦ σώματος τῆς σαρκὸς αὐτῶν, where the longer form is probably a 'conflate' formation (see note above on 15.9). Both Sync. and G read ψυχή, and the reading of Sync. looks like a correction of a scribe who found the combination 'soul of flesh' difficult, rendering '(which came forth) from their soul, as from the flesh'.

day of the end ... the great judgement See above on 10.12.

the great aeon ... completed We have here, it would appear, the cosmological myth of the 'eternal return': see Hengel, *Judaism and Hellenism*, I, 191, II, 128 (for a useful bibliography). For τελεσθήσεται, an appropriate equivalent in this context would be ישתלם (see above on 10.12).

En[g(q t u tana)] has an additional clause, usually bracketed as a gloss, but, as Charles noted (Charles[1906], 46 n. 2) the clause is supported by Sync. ἐφ ἅπαξ ὁμοῦ τελεσθήσεται: ἐφ' ἅπαξ/ὁμοῦ looks like a genuine doublet of כחדא 'together'. The verb שלם in the Ithpa. also means 'to be punished' (Isa. 42.19 Tg. 'the impious (רשיעיא) are about to be punished (לאשתלמא)'. The clause would then read 'together everyone will be punished from the watchers and the impious (rasi'ān)'. Could the latter be an error for 'giants' (ra'ayt)? The Eth. text seems corrupt: perhaps we should read 'together there shall be punished/finished (i.e. destroyed) all the watchers and giants' (kuellomu teguhān wara'ayt). Eth[M] has retained part of the clause and incorporated it within the structure of its reconstituted text, based on the corrupt 'em 'ālam for 'ama 'ālam: thus Knibb '(they will be destroyed) until the day of the great consummation is accomplished upon the great age ('em 'ālam 'abiy), upon the watchers and the impious ones'.

(3) **there was no secret that was not revealed to you** The conjecture of Lods alone gives a suitable sense in the context. The reading behind Eth. οὐκ ἔτι crpt. ex οὐκ ἔστι supports the emendation.

and unspeakable secrets you know G μυστήριον = רזא, here used collectively: there was more than one 'secret' revealed by the watchers. G is obviously a corruption, probably of ἐξουθενημένον, rendered by Eth. mennun, repudiatus, abominabilis. I suggest Aram. רזא מכערא; cf. Tg. Neh. 3.6 and Sir. 13.22 ודברים מכוערים, LXX ἐλάλησεν ἀπόρρητα 'he speaketh things not to be spoken' (AV) (cf. Levy, NHCW, II, 372). Cf. Clement of Alexandria, *Strom*.iii.9 (Migne, II, 24) ὡς οἱ ἄγγελοι ἐκεῖνοι οἱ τὸν ἄνω κλῆρον εἰληχότες, κατολισθήσαντες εἰς ἡδονάς, ἐξεῖπον τὰ ἀπόῤῥητα ταῖς γυναιξὶν, ὅσα τε εἰς γνῶσιν αὐτῶν ἀφῖκτο, κρυπτόντων τῶν ἄλλων ἀγγέλων, μᾶλλον δὲ τηρούντων εἰς τὴν κυρίου παρουσίαν· ἐκεῖθεν ἡ τῆς προνοίας διδασκαλία ἐῤῥύη, καὶ ἡ τῶν μετεώρων ἀποκάλυψις.

CHAPTER 17

This chapter begins a new section of the book, from 17 to 36, dealing with the extra-terrestrial journeys of Enoch. See Introduction, p. 15f.

(1) **I was taken and brought** The fuller text of G seems original (the plurals are impersonal): cf. the Biblical expression at Lam. 3.2, אותי נהג וילך, Tg. יתי דבר ואוביל, LXX παρέλαβέν με καὶ ἀπήγαγεν, 'he led and brought me'.

those who were there These fiery beings which can take human form are not further identified either as angels or demons (they are clearly not the spirit-demons of Ch. 16). Dillmann, 115, suggests a possible connection with the 'will o' the wisps' (Irrlichter) of folklore. For the ability to assume different forms, cf. 19.1 and 2 C. 11.14, and for the phrase 'like blazing fire' cf. Dan. 7.9 נור דלק Theod. πῦρ φλέγον and above, 149.

(2) **a place of dark storm-clouds** Eth. ʿawwelo usually renders γνόφος 'storm', but it can also have the sense of γνόφος 'darkness' (e.g. אפלה Jl. 2.2). It could, therefore, render G ζοφώδη here. It is also possible, however, that the original behind Eth. was γνοφώδη (so Charles, 38, cf. Knibb, 2, 103), meaning '(a place of) storms'. The noun אפלה 'darkness' does not occur in Aram. but ערפלא = Heb. ערפל 'dark clouds' is in common use. Was it 'dark storm-clouds' which Enoch saw in this place where a mountain-top reached up to heaven? And did G perhaps render ערפלא as if it were אפלה, or is ζοφώδη crpt. for γνοφώδη?

summit G ἡ κεφαλή. Eth. appears to have read ἡ κορυφὴ τῆς κεφαλῆς αὐτοῦ. Is this the mountain at 18.8 or at 22.1 or 87.3? See Lods, 154, Dillmann, 115.

(3) **to the uttermost reaches** G εἰς τὰ ἀεροβαθῆ lit. 'to the depths of air' (so Liddell and Scott, but the word occurs here only). We should perhaps read with Eth^{tana} [εἰς τὰ ἄκρα βάθους] = לקצי עומקא. Cf. Charles, 38.

bow of fire ... quivers The lightning-flashes are the arrows of the Almighty, cf. 2 Sam. 22.14-15, Ps. 144.6. G om. 'and a fiery sword', which could be an Eth. expansion.

(4) **to subterranean waters(?)** Lods, 154 suggests an allusion to the ὠκεανός of Greek legend into which the sun sets (*Il.* 18.607f., Hesiod, *Theog.* 131-33, 282, 695f.). Cf. also Charles, 39. But no satisfactory explanation has been offered of the description of this water as 'living water' or 'waters of life'. 'Living waters' normally means 'flowing water', and, while this is a possible meaning here, it seems jejune and inadequate in the context. The other meaning 'waters that impart life' i.e. 'salvation', also familiar in Scripture, is no less unsatisfactory. The proposed conjectured text presupposes an original מיא תחתיא 'the waters underneath (the earth)', i.e. the subterranean waters of Heb. cosmology, mistranslated as מיא חיא 'living waters', an idea all too familiar to the Biblically informed translator. For the 'waters above (the earth)' (מים העליונים) and the 'waters beneath (the earth)' (מים התחתונים) see Jer. Tg.(P-J) Gen. 1.6 and Gen. rabba ad loc. Levy, CW, II, 30. In support of the conjectured text are the descriptions following of rivers and seas under the earth; cf. also 89.2-3, 7-8 and notes. Eth. 'as it is called' is probably a translator's gloss.

the fire of the West Cf. Ch. 23. This is scarcely Gehenna, although the Talmud says that the sun is red in the evening because it passes the gate of Gehenna, and red in the morning when it passes the rose of the Garden of Eden (Baba Bathra 84a Charles, 39). Eth. 'takes hold of' ye'eḫḫez, i.e. the sun 'catches fire' from the fire of the West: G καὶ παρέχον may be a mistake for καὶ κατέχον (note παρεχον πασας and cf. Lods, 155).

If G ἤλθομεν is original, the plural includes Enoch's angelic guides (cf. v. 2); Charles, 294 and Dillmann, SAB 1892, 1045 emend to ηλθο(με)ν on the basis of Eth.

(5) **a river of fire** This is generally identified with πυριφλεγέθων (*Od.* 10. 513) which the author envisages as debouching into ὠκεανός, 'the great ocean towards the West', which encircles the earth. Lods (l.c.) thinks he may also have in mind the Heb.

הים הגדול i.e. the Mediterranean, no doubt as coextensive with the mythical earth-encircling ὠκεανός; cf. Milik, 38f. and 40 (the map of the world of Enoch 1-36). For the admixture of Hebrew ideas and Greek mythology, cf. Sib. Or. 4. 185, 'Gehenna, sister of Styx'.

(6) **the great rivers** presumably the other mythical rivers of the nether regions, Styx, Cocytus and Acheron into which in the Odyssey (10.13) Pyriphlegethon and Cocytus, a tributary of the Styx, flow. The name Acheron was later to become synonymous with Hades, Aen. vi. 295.

the great darkness Dillmann identifies with Hades. Lods, 156 accepts this explanation, unless the author, returning to Heb. thought, means the immense darkness on the other side of the vault of heaven and the terrestrial disk (Job 26.10). Something like this seems to be the view of Milik who envisages a region of darkness encircling the earth beyond 'the great river' (15, 38f.). According to En. 32.2-3, En[e] 1 xxvi 21, it is also encountered by Enoch as he travels 'east of the earth' beyond the Red Sea and before the Paradise of righteousness (232, cf. 291, L. 9). Whether this conception derived from Babylonian sources, as Milik supposes (40), is doubtful. Basically it would seem to be the simple idea of a place in the 'Beyond' where all light ceases (cf. Job 26.10 'the farthest limits of light and darkness' NEB). Cf. further ThWNT Bd. VII s.v. σκότος, 430f. (Conzelmann)

whither no flesh goes Eth. omits the negative, 'where all flesh walks' (Knibb). The translator is thinking of Hades.

(7) **the wintry regions(?) of storm-clouds** Lods prefers the reading of Eth., rendering 'the mountains of black clouds' (Eth. qobārāt = γνόφοι, 'storm-clouds'). Charles suspects an allusion to Jer. 13.16 LXX ὄρη σκοτεινά, הרי נשף 'mountains of the twilight', Tg. טורי קבל. G ἀνέμους = רוחות, and the original behind the translation proposed, presupposes a misunterpretation of the noun רוח = 'region, quarter' as 'winds'; cf. En. 76.14, Enastr[c] 1 ii 14, Milik, 288. Eth. 'adbār, ὅρια, usually rendered 'mountains', is the equivalent of רוחות in this context, 'boundaries' then 'districts, quarters', cf. the use of Lat. fines.

the outpouring ... all the waters Are we to imagine a central spring or source in the 'abyss' from which all the subterranean waters come; cf. the 'fountain of the deep' at Gen. 7.11, 8.2, Prov. 8.28 and En. 89.3,7.

(8) **the mouth ... abyss** Dillmann proposed reading τῆς γῆς after ποταμῶν with Eth., SAB 1892, 1087, and Flemming suggested emending to τὸ στόμα τῆς πηγῆς πάντων τῶν ποταμῶν, 'the mouth of the fountain of all the waters (of the earth)'. See Flemming-Radermacher, 46 n.

CHAPTER 18

(1) **set in order ... earth** G ἐκόσμησεν = Aram. תקן Aph. lit. 'make fast', 'set in order', and hence 'adorn' = κοσμεῖν. The usual translation given here is 'adorn', a meaning which scarcely suits the context: how can the 'winds' or 'the storehouses of the winds' be said 'to adorn' all creation (or better 'all created things' = G: Eth. feṭrat could be collective), and especially 'the foundations of the earth'? Cf. Prov. 3.19, Heb. כונן Tg. אתקין, 'established', 'ordered aright'. The winds are envisaged as 'supporting' the created universe; see v. 3. The 'foundations of the earth' are frequently mentioned in the Old Testament e.g. Isa. 24.18, Jer. 31.37, Ps. 82.5 etc.

Knibb prefers to take 'the foundations of the earth' as object to 'I saw': 'I saw the storehouses of all the winds ... and (I saw) the foundations of the earth'.

(2) **the corner-stone of the earth** Cf. Job 38.6.

the four winds ... the earth The author thinks of heaven and earth being supported by those four winds which blow from the four quarters of the earth (cf. En. 76).

(3) **spread out the heights of heaven** Eth. rababa = Heb. נטה Aram. מתח. Aram. מתח has also an intransitive use (cf. P. Sm. 2247 = se extendit). Could the original have read 'spread out to the heights of heaven'? For the latter, cf. 93.14, En[g] 1 v 23, Ezek 10.19 Tg. לרום שמיא.

and are stationed Eth. yeqawwemu = G ἱστᾶσιν: Heb./Aram. עמד or קום(?).

pillars of heaven a Biblical expression from Job 26.11 (cf. Ps. 75.3, Job. 9.6 'pillars (of earth)'). The winds are the 'pillars of heaven' (עמודי שמיא(?)) cf. v. 11, En[c] 1 viii 28, 29); they reach up to heaven and are stationed (or stand) as pillars between earth and heaven, supporting the firmament of heaven.

(4) **cause ... the orb of the sun to revolve** I have read with G διανεύοντας (= δινεύοντας) τὸν τροχὸν τοῦ ἡλίου = מגללין גלגל שמשא; for גלגל = 'orb, disk' Enastr[b] 6.9, Milik, 284. Cf. Lods, 157f., Dillmann, SAB 1892, 1045 (δινεύοντας), Charles, δύνοντας (with Eth.).

(5) **carrying the clouds** The author is supplying an answer to the question at Job 36.29, 37.16: 'Do you know why the clouds hang, poised overhead?' (NEB, 37.16). As Knibb, 2, 104 notes, βαστάζοντας ἐν νεφέλῃ may come from נשא ב, e.g. Num. 11.17, Job. 7.13.

paths of the angels presumably the paths by which the angels ascend to heaven and descend to earth, employed, for instance, by the rebel watchers. Chh. 34-36 elaborate further on the subject of the 'winds' at the ends of the earth.

(6) **seven mountains** The visions recorded in 18.1-5 include more than what Enoch was shown in the west where his travels first took him (17.14); they appear in fact to have included an 'orbital journey' around the world (cf. Milik, 39). At 18.6, however, we return to the west, or rather to the north-west where the mountain-throne of God is located (v. 18); and towards the south Enoch sees three more fiery mountains, and towards the east another three, making seven in all. The main interest in the subsequent narrative is its climactic vision of the great abyss and 'the forsaken and terrible place' at the ends of the earth (18.2), the place of incarceration for a thousand years of the 'wandering stars' and of the angels who had fallen, the watchers. In the duplicate account of the seven mountains at 24-25 the interest is more directed to the place of paradise, located in the east of those mountains. Cf. further 77.4, and for the paradise in the east, 77,3.

lying towards the south G βάλλοντα (ms βαλλοντας) looks like an attempt at rendering Aram. רמין, lit. 'cast, thrown' and then 'lying, situated'; cf. Dan. 7.9 'thrones were placed' (רמיו, LXX Theod. ἐτέθησαν). Cf. the use of βεβλημένος at Mt. 9.2, Josephus, B.J.i. 32.3 (629); see also Lods, 159.

(7) Cf. Ezek. 28.13, Exod. 28.17f. and 1 Chr. 29.2.

stones of varied hues Lods considers the reference to be to a special stone such as is mentioned in similar terms at 1 Chr. 29.2 'antimony, coloured stones' (RSV), 'stones for mosaic work' (NEB). Gesenius, *Thes.* suggests a variegated type of marble found especially at Petra, and he cites Robinson, *Palestine* III 79, 80.

another of pearls G = Eth. λίθος μαργαρίτης usually rendered by 'pearl' is ambiguous, since the Greek word is used generally for precious stones in the East. It does not occur in the LXX and where ܡܪܓܢܝܬܐ is found in the Peshitta it corresponds to Heb. ברקת at Ezek. 28.13, 'emerald' (RSV), 'carbuncle' (AV): Lods suggests פנינים at Prov. 3.15, Job 28.18, EVV 'rubies'.

stones of the colour of antimony Eth. 'ebna fawwes lit. 'stones of healing': fawwes was originally explained by Dillmann as 'antimony', since it still has that meaning in modern Ethiopic (see Dillmann, *Lex*, 759, 1377, cf. SAB, 1892, 1053). However, since he believed that the latter was mentioned in the next verse (G φουκα), he rejected the explanation in favour of a reading of G's enigmatic ταθεν as a corruption of a transcription of Heb. פטדה 'topaz'. He further suggested the origin of fawwes meaning 'healing' as an attempt to give meaning to ταθεν (ιαθεν, ιατηρ, ιατης); Charles similarly read fawwes as ἰάσεως 'healing' from which he obtained ἰάσπιδος 'jasper'. None of these explanations is transcriptionally probable. Moreover, in Syriac—and no doubt also in older Aram.—'antimony' is ܬܘܬܝܐ (תותיא), P. Sm. 4416, stibium, of which ταθεν is a much closer attempt at transcription. Did Eth. read λίθους στίβι = 'ebna fawwes? Since antimony is a metal not a stone we must presumably render 'stones of the colour of antimony': the Heb. equivalent is אבני פוך Isa. 54.11, 1 Chr. 29.2 ('stones of the brilliant hue of antimony' BDB). For the λίθος φουκ of v. 8, see below.

of carnelians G λίθος πυῤῥοῦ 'fiery stone'; cf. Ezek. 28.14,16 אבני אש LXX (ἐκ μεσοῦ) λίθων πυρίνων. These are the stones known as אדם = σάρδιον Aram. סמקנא the 'red stone', 'carnelian'. This fiery or red stone appropriately symbolises the south. The text implies that all the mountains in the south were made of σάρδιον, unless the versions have abbreviated a fuller original text. The single mountain-throne of God in the north-west, the one 'between them' (v. 8), has two types of stone.

(8) For vv. 8-12 cf. En[c] 1 viii 27-30, Milik, 200.

The idea of mountains composed of precious stones, and in particular, this 'mountain of God', which was his throne (25.3), appears to have been suggested by Ezek. 28.13-14. Certainly Ezek. 1.26 lies behind v. 8: 'Above the vault ... there appeared ... a sapphire in the shape of a throne ...' (NEB). Cf. Charles, 41. The theme of the seven mountains is further elaborated in a parallel account at 24.1-25.6. (Has the author put together two different sources of a traditional apocalyptic vision?).

reached up to the heavens The text of G is defective. Eth. guad'a, lit. 'to strike' here renders ἀφικνεῖσθαι εἰς, נגע ב 'to reach up to, cf. Hoftijzer, 174.

like the throne of God In both Heb. and Aram. (כ)דמות = ὥσπερ, kama, but דמות is a noun in its own right meaning 'likeness, shape'. Could there lie behind the Greek and Eth. in this verse the expression from Ezek. 1.26 דמות כסא 'the shape of a throne' (NEB)?

emeralds G φουκα Heb. פוך 'antimony'. Eth. pēka usually means 'marble', 'alabaster' and this last is adopted by Charles. It may, however, represent the Greek/Heb. φουκα/פוך. Since 'antimony' has already been mentioned in v. 7 for the mountains in the east, I suggest a confusion of פוך with נפך a precious stone translated in the AV and RSV as 'emerald', NEB 'purple garnet'. It is mentioned at Ezek. 27.16, 28.13, and, as one of the gems in the High Priest's breast-plate at Exod. 28.18, 39.11. At Ezek. 28.13 and the two passages from Exodus, it is associated, as in this verse, with sapphires.

(9) **beyond those mountains** The Eth. translator consistently fails to understand the adverb ἐπέκεινα and resolves it into its component parts ἐπί and εκεῖνα rendering here literally 'what was towards these (kuellu is a mistake for 'ellu, so Flemming-Radermacher 47) mountains'. See the full note in Knibb, 2, 105.

(10) **beyond the ends of the earth(?)** The 'great earth' of G = Eth. is an unusual

expression: I suggest a misrendering or corruption, and probably both, of Aram. בריתא דארעא (לעבר) a phrase found in Tg. Prov. 30.4 = אפסי ארץ, LXX ἄκροι τῆς γῆς.

the heavens came to an end The future tense of συντελεσθήσονται is hardly appropriate and has probably been introduced by a translator thinking of the future end of heaven and earth. The Eth. version, 'the waters will be gathered together', seems to have arisen through the influence of Gen. 1.9. Traces of the original may have survived in Eth I ('the heavens' for 'the waters').

(11) Verses 11-16 describe the place of imprisonment of the 'seven stars', the ἄστερες πλανῆται of Jude 13, (see below on v. 13) together with the fallen watchers (19.1). A parallel account of this 'fearful place' is to be found at 21.7-10.

a deep abyss ... fire descending G has lost more than half of the sentence by hmt, viz. τοῦ πυρὸς τοῦ οὐρανοῦ καὶ ἴδον ἐν αὐτῷ στύλους τοῦ πυρός (cf. Charles, 42). Cf. 21.7 where the 'place' (or the 'abyss') is πλήρης στύλων πυρὸς μεγάλου καταφερομένων· οὔτε μέτρον οὔτε πλάτος ἠδυνήθην ἰδεῖν οὐδὲ εἰκάσαι.

(13) **seven stars** The 'seven' is a classical Heb. round number, the perfect number, and is found a number of times in I En. (Dillmann, 119), but here it almost certainly has a further significance. The reference is to the seven planets or 'wandering stars', the ἄστερες πλανῆται of Jude 13, which, from their apparently irregular or 'wandering' course in the heavens (in contrast to the 'fixed stars') became a symbol for apostates (Theophilus, *Ad Autol.* II 15).[1] They are sometimes wrongly identified with the watchers[2] who are also represented in Enoch as 'fallen stars' in the Dream-Visions. Use is made of the 'seven stars' motif to describe apostasy at Rev. 1.20. Cf. Bousset-Gressmann[4], 252, 322f. and below on 88.3 (Milik, 239). As in canonical scripture they are represented as animate beings (cf. Ps. 103.20f., Job 38.7, Sir. 16.27).

(14) Cf. v. 12 above.

a prison Cf. 22.4 En^{e} 1 xxii 1-2, Milik, 229 בית עגון.

the hosts of heaven can refer in the Old Testament to the stars and the angel hosts since the former are considered to be animate angelic beings responsible to God (cf. 1 Kg. 22.19, Isa. 24.21, Lk. 2.13). Jude 13 may be a reminiscence of this verse: cf. Dillmann, 119, Milik, 228, 239 (on 88.3, and see below, 261) G adds ὅτι τόπος ἔξω τοῦ οὐρανοῦ κενός ἐστιν, a displaced gloss on v. 12.

(16) **punishment for their sins** Cf. 21.6. I take ἁμαρτία (ἁμάρτημα) in both passages = חטא, חטאת Aram. חטאה in the sense of 'consequences for sins' and so 'punishment' (cf. Lam. 4.6, par. עון, 'the penalty of Sodom' (NEB)).

CHAPTER 19

(1) There are several major differences in this account of the incarceration of the watchers until the 'great judgement' from that given at 10.4,12,15.8-16, and in the 'duplicate' account at 21.7-10. (1) They are imprisoned in the same 'great abyss' at the ends of the earth as the 'wandering stars': cf. 10.4,12, their prison till the

---

[1] The text of G, usually read as περὶ ὧν πυνθανομένῳ μοι is probably crpt, ex περιπλανώμενα?

[2] e.g. by H. Braun in ThWNT Bd. 6, 240, 32f., 251, 12f.; so also K. H. Schelkle, *Die Petrusbriefe, Der Judasbrief* in Herder's Theol. Komm. zum Neuen Testament, 163.

judgement is in 'valleys of the earth'. (2) According to G (the text adopted), the 'spirits' of the watchers are evidently to roam the earth as the tormentors and corrupters of mankind till the judgement, the rôle at 15.8-16 of the 'evil spirits' of the giants, the bastard offspring of the watchers. Presumably their bodies 'abide' in the 'great abyss', while their spirits are free to haunt mankind, just as it is the 'spirits' of the giants which corrupt mankind after their bodies have been slain (cf. 10.12, 15.8f.). Contrast the text of Eth^{M}: 'The spirits of the angels who were promiscuous with the women will stand here; and they, assuming many forms, made men unclean and will lead men astray ...' (Knibb). Cf. Test, Reub. 5.6: καὶ μετεσχηματίζοντο (sc. οἱ ἐγρήγοροι) εἰς ἀνθρώπους καὶ ἐν τῇ συνουσίᾳ τῶν ἀνδρῶν αὐτῶν συνεφαίνοντο αὐταῖς. See further Lods, 164f.

will abide G στήσονται = Eth. yeqawwemu = Heb./Aram. עמד(?); cf. Exod. 8.18(22) ('dwell'), Hoftijzer, 216.

will harm G λυμαίνεται which Lods claims frequently means 'defiles' (déshonerer), but this is more than doubtful: where it = תעב Pi. it means 'to spoil, corrupt' (Ezek. 16.25, and so also 4 Mac. 18.8). Eth. read μιαίνεται, thus referring to the watchers 'defilement' of mankind (cf. 7.1).

lead them astray ... demons Justin Martyr, Apol. II 4(5).3-4 also attributes the introduction of idolatry to the watchers in a passage reminiscent of this verse: οἱ δ'ἄγγελοι, παραβάντες τήνδε τὴν τάξιν, γυναικῶν μίξεσιν ἡττήθησαν καὶ παῖδας ἐτέκνωσαν, οἵ εἰσιν οἱ λεγόμενοι δαίμονες· καὶ προσέτι λοιπὸν τὸ ἀνθρώπειον γένος ἑαυτοῖς ἐδούλωσαν κ.τ.λ. Note that Justin identifies the demons with the offspring of the watchers. Cf. also Tertullian, *de idol.* iv, Henoch praedicens omnia elementa, omnem mundi censum, quae caelo, quae mare, quae terra continentur, in idolatriam versuros daemones et spiritus desertorum angelorum, ut pro deo adversus deum consecrarentur. Cf. further 99.7, Ps. 106.37, Dt. 32.17. See Bousset-Gressmann[4], 336f.

(2) **and the wives ... transgressed** = G. Did the original perhaps read 'and the women whom the angels seduced' ונשא די מסטין מלאכין; a text which read with סטין would give 'the wives of the angels who transgressed'. (For ܣܛܐ = παραβαίνειν P. Sm. 2594).

shall become sirens(?) I have retained the Greek term. In the LXX σειρήν usually renders יענה NEB 'desert-owl', noted for its plaintive cry (Mic. 1.8, Jer. 27 (50). 39, Isa. 13.21). (The word is also found in the LXX, equivalent to תן, Isa. 34.13, 'jackal', 'wolf' (NEB), also associated with a wailing cry). Could the original have been בנת נעמיין (Tg. Isa. 13.21), lit. 'daughters of loveliness', a term apparently applied to these 'desert-owls' on account of their attractive looks. (So Levy, CW, I, 118, and cf. Gen. 6.2).

(3) **all things that exist** lit. 'the ends of all things', i.e. the totality of all things'. For this use of πέρατα = קצות cf. Milik, 145 (Sir. 16.17 קצות רוחות = universitas spirituum (Zorrell)), 271.

This verse appears to be the source of two passages, one in Clement of Alexandria, the second in Origen. Thus Clement, *Eclog. Proph.*, ed. Dindorf, iii, 453, ὁ Δανιὴλ λέγει ὁμοδοξῶν τῷ Ἑνὼχ τῷ εἰρηκότι· καὶ εἶδον τὰς ὕλας πάσας. Origen, *de Princ.*, IV 35, scriptum namque est in eodem libello dicente Enoch, universas materias perspexi.

CHAPTER 20

(1) Cf. 9.1.f. The earlier list of the four archangels has here been enlarged to seven principal angels, in an order which probably reflects a later, localised angelology: Uriel, Raphael, Raguel, Michael, Sariel (Eth. Saraqiel), Gabriel, Remiel. Their functions are also specified, and in Eth. they are described as 'the holy angels who keep watch' i.e. they are 'watchers' and 'holy ones' (cf. Dan. 4.10,20). (See above 106f.) Does Eth. perhaps go back to a text ταῦτα τὰ ὀνόματα τῶν ἐγρηγόρων καὶ ἁγίων = אלין שמהתהון עירין וקדישין? See Yadin, *Scroll of the War*, 238, Milik, 172f.

It is impossible to trace G ἄγγελοι τῶν δυνάμεων to a common original with Eth. 'the holy angels who keep watch': if anything, the two descriptions could supplement each other and both be original; G simply supplies a title for the list, parallel to the concluding description (only in G[1, 2]) ἀρχαγγέλων ὀνόματα ἑπτά, and probably a virtually synonymous expression. The designation 'angels of powers' occurs once only elsewhere, at 61.10, where the 'angels of powers' are associated with the 'angels of dominion', both expressions recalling the Pauline 'power and dominion' (Eph. 1.21, cf. Col. 1.16) and the 'angels, principalities and powers' of Rom. 8.38. See on 61.10. A parallel of a kind occurs at I QM 8. 11 (cf. I QH 10. 34) **גבורים** and **גבורי כוח** 'mighty ones of power' as a term for angels: see Y. Yadin, *Scroll of the War*, 230; see further M. Black, Πᾶσαι ἐξουσίαι αὐτῷ ὑποταγήσονται in *Paul and Paulinism*, Essays in honour of C. K. Barrett, ed. by M. D. Hooker and S. G. Wilson (London, SPCK, 1982), 74-82.

The angels or archangels in the list which follows are assigned certain duties or areas of responsibility as at Rev. 7.1, 9.11, 14.18, 16.5; cf. 1.20, 2.1f., 3.1f.); cf. also Justin Martyr, Apol. II, 4(5).2 ὁ θεὸς ... τὴν μὲν τῶν ἀνθρώπων καὶ τῶν ὑπὸ τὸν οὐρανὸν πρόνοιαν ἀγγέλοις, οὓς ἐπὶ τούτοις ἔταξε, παρέδωκεν. See further Bousset-Gressmann[4], 320f.

(2) **Uriel ... the one in charge of thunder and earthquake(?)** G ὁ ἐπὶ τοῦ κόσμου 'the one over the world(?)' seems a most improbable original: Milik, 173 (top) suggests reading ἐπὶ τοῦ κόσμου τῶν φωστήρων, but this is the function of Raguel at v. 4, although obviously a most appropriate role for Uriel, the 'angel of light' (cf. 75.3). (There may well have been some confusion of roles in this list as well as corruption of the names as at 6.7f). It is arguable that Eth[M] zaraʿām is a corruption of zaʿālam, but it is 'not entirely convincing' (Knibb, 2, 107). If G τοῦ ταρτάρου is original, it is unique in I En, where the regular equivalent of Gehenna or Sheol is ᾅδης; τάρταρος is also an extremely rare term in the LXX. Charles's suggestion that Eth. 'tremors' has arisen by a corrupt transcription of τάρταρος as τρόμος is transcriptionally unlikely (cf. Knibb, ibid.). Aram. רתיתא, however, = Eth. raʿād, Syr. ܪܬܝܬܐ (plur. = παλματίαι) means 'earthquake', P. Sm. 3992 terrae motus tremuli and this could be original. G τάρταρος may then be explained as a corrupt transcription. Eth. may have come from an alternative Greek version reading τρόμος, tremor (terrae). But cf. Barr, 'Aramaic-Greek Notes', II, 188.

(3) **Raphael ... spirits of men** Raphael's special role, in keeping with his name, is that of healing (10.6). At 22.3,6, however, he has to do with the 'spirits' of men in the nether regions, and this may be referred to here. At 32.6, where he is Enoch's angelus interpres in paradise, he appears to have usurped the role of Gabriel (cf. 20.7).

(4) **Raguel ... tends the hosts of the luminaries** Raguel = רעו אל 'friend of God', Noth, *Die israelitische Personennamen*, 153f., 257 (explained by Milik, 219f. (foot) as 'Delight of God', but understood by the author of Enoch here as 'shepherd of God').

Lods reads the corrupt text of $G^{1,2}$ correctly as ἐκδιώκων as at 23.4 (G), where Raguel explains to Enoch that the 'restless stream of fire', which Enoch has seen in the west, 'follows after' all the heavenly luminaries. Eth. evidently read ἐκδικῶν giving the absurd sense of 'taking vengeance on' the luminaries. Dillmann, 128 suggests that the purpose of this perpetual stream of fire from the west, racing continuously from north to south and south to north, was to provide all the stars, from the remotest in the north or south, with the fire that gives them light. Milik, 219f. pictures Raguel as exercising his role as 'shepherd' and 'following (as a shepherd)' all the luminaries of heaven. Was his role to 'tend' the stars with perpetual light and fire from the west, the source of heavenly fire, since it is there that the sun sets? Lods, 188, notes that the use of κόσμος at v. 2 recalls the usage of the LXX at Gen. 2.1, Dt. 4.19, Isa. 40.26 where it renders צבא 'host' (possibly to be derived from Aram. צבתא, ܨܒܬܐ = κόσμος cf. Heb. צבי(?)). The 'hosts of luminaries' rather than 'the world of luminaries' seems more appropriate in this context.

(5) **put in charge ... the people (of Israel)** Michael, the guardian angel of Israel (Dan. 10.13,21, 12.1 etc.) is appointed to be 'over the benefits (ἀγαθά = טבתא) of the people'. I interpret this verse in the light of 24.6 where it is Michael who instructs Enoch on the blessings in store for the righteous in paradise: but 'benefits' could have a wider connotation and refer to Israel's welfare in the present as well as in the life to come.

(6) **Sariel** = En^{b} 1 iii 7 שריאל Milik, 173. Saraqael, otherwise unknown, seems an inner-Eth. corruption.

which led men astray in the spirit(?) According to Eth. Sariel is in charge of the spirits of men (like Raphael at v. 3) who cause spirits to sin (or who lead spirits astray). The texts of both $G^{1,2}$ and Eth. are manifestly corrupt. Is Eth. a confused rendering of [ὁ ἐπὶ τῶν πνευμάτων ἃ ($G^{1,2}$ οἵτινες) υἱοὺς τῶν ἀνθρώπων ἐπὶ τῷ πνεύματι ἐξαμαρτάνουσιν], 'who is over the spirits which cause the children of men to sin ⌜in the spirit⌝'? Aram. Aph. חטא = Aph. טעי 'to lead astray', πλανᾶν, and the reference here could be to the spirits which cause apostasy; cf. Isa. 29.24 תועי רוח, 'those who err in spirit' (RSV).

(7) **Gabriel** On his changing role, Bousset-Gressmann[4], 328. As the archangel in charge of paradise we would expect him and not Raphael at 32.6.

paradise ... Cherubim The reference to 'serpents' in paradise is usually explained in the light of Gen. 3.1. But there was only one serpent in Eden. The Heb./Aram.(?) שרף = LXX ὄφις (Num. 21.8, Isa. 14.29 ὄφεις πετόμενοι 'flying serpents') is used also for 'the Seraphim' at Isa. 6.2. Dit the writer mean 'the Seraphim and Cherubim'? Only the latter are mentioned in the Genesis account of Eden (Gen. 3.24).

(8) **Remiel** Cf. 6.7 Ῥαμιήλ = רעמאל En^{a} 1 iii 7; and cf. 2 Bar 55.3, 63.6,2 Esd. 4.36, Sib. Or. 2. 215-7, the angel who leads men's souls from the 'dark gloom' of the nether regions to the judgement seat of God. The genuineness of this verse is guaranteed by the number 'seven' (in G^{1} as well as in G^{2}, although G^{1} has only six names). Cf. Charles, 44.

CHAPTER 21

(1) **I went on ... formless void** G ἐφοδεύειν Eth. ʿoda Aram. יבל (Milik, 378). Eth^{M} appears to have read ἕως τόπου ἀκατασκευάστου. $G^{1,2}$, $Eth^{tana\ u}$ ἕως

τῆς ἀκατασκευάστου is explained by Lods, 171, as a contraction of ἕως τῆς ἀκατασκευάστου ὕλης. This may be how the Greek translator understood the expression (cf. the allusion in Origen, *de Princ.*, IV, 35, ambulavi usque ad imperfectum), but the original may have been ἕως γῆς ἀκατασκευάστου, recalling LXX Gen. 1.2, to which the verse is alluding (cf. Dillmann, 123). Eth. renders the adjective here literally by '(to) where there was nothing made', but on its second occurrence at v. 2 by '(a desert place) which was without form ...', the rendering of ἀκατασκεύαστος at Gen. 1.2. Did the original Aram. perhaps read לארעא שהיא, 'to an empty place': cf. ܫܘܐ P. Sm. 4073, EnGiants iii 5 שהוין, Milik, 306.

(2) Cf. 18.12-16. Milik, 228 identified the small frg. En[e] 1 xxi with this verse, but see below, v. 7.

(3) **bound together prostrate on it** G ἐρριμένους = רמין lit. 'cast down', cf. v. 4.

(4) **For what iniquity** G[1, 2] lit. 'for what cause', but αἰτία = עון in LXX (Gen. 4.13; cf. Lods 172). Milik, 229 ועל אידן אשם.

(5) **a leader among them** The same expression in G is used of Michael (24.6). At 72.1, 74.2 Eth. refers to Uriel as 'their (the angel-stars' or angels') leader'. In all three instances all that is probably meant is that Michael or Uriel was an ἀρχάγγελος, i.e. 'a leading angel among them' rather than 'the leader of all the angels'. Cf. the use of רבנין of the 'leaders' of angels (e.g. 6.7, En[c] 1 ii 24). Eth. 'he was guiding me' has misunderstood the text, or read [ἡγεῖτό μου].

do you seek out the truth Charles, 44, cf. Dan. 7.16 אבעא יציבא lit. 'I enquired, sought out, the truth' (LXX Theod. τὴν ἀκρίβειαν ἐζήτουν), Dan. 7.19 LXX ἤθελον ἐξακριβάσασθαι. G ἀλήθειαν φιλοσπουδεῖς seems an attempt to render בעא יציבא, 'to enquire, seek out, the truth'. Eth. 'do you enquire diligently ([ἐξακριβάζει?]) and care ([φιλοσπουδεῖς])' may imply a different text from G or be an attempt to render G (Charles, 44f.). Cf. 25.1 below.

(6) **the Lord** Eth II [κύριος ὕψιστος] a rare nomen dei in 1 En., but cf. Ps. 7.17(18), 13(12).6 etc. and below, 198f.

ten thousand years the sentence imposed on sinners among the Greeks: see Charles, 45, Dieterich, *Nekyia*, 118f., but cf. Rohde, *Psyche*, ii, 179.

the period of their punishment For this use of ἁμαρτία/ἁμάρτημα, see above on 18.16. Eth. renders by 'the number of the time of their sins'; it seems to have read [ἄριθμος τοῦ χρόνου τῶν ἁμαρτημάτων αὐτῶν].

(7) As noted above, Milik regards the identification of En[e] 1 xxi with 21.2-4 as 'fairly certain' (Milik, 228f.). The fragment (Pl. XVIII) certainly belongs to this part of Enoch, since it forms the top lines of a column (xxi) followed by a second column (xxii) containing 22.3-7. Identification is further narrowed by the word נור (πῦρ) which occurs in Chh. 21-22 in two places only, viz. 21.3 and 21.7. In his reconstruction of v. 3 Milik introduces a noun עדבא (σχοινισμός) for which there is no equivalent in G or Eth. I suggest placing the fragment at vv. 6-7 and reading:

[... רבו] שנ֯י[ן ...	21.6	... μύρια ἔτη ...
[... נורא ר]בא הוה [תמן דלק ...]	7	... πῦρ μέγα ἐκεῖ καιόμενον
[...] לשני נור [...]	8	... γλωσσῶν πυρός ...

that place ... the abyss LXX διακοπή renders פרץ, 'a bursting forth, breach', used, e.g. of an outburst of water (2 Sam. 5.20 = 1 Chr. 14.11).[1] The 'breach' in this 'fearful

[1] Another possible equivalent would be חצבא from חצב 'to cut': ܚܨܒܐ has astronomical associations, P. Sm. 1351 = πίθος, πιθίας 'a jar-shaped comet'. 2 Cf. Flemming, 27 and Flemming-Radermacher, 51 n. 20.

place' was evidently an opening through which great columns (or tongues) of fire descended to the abyss. There could be a mistranslation of פרצא, ptc. misunderstood as a noun διακοπή: והוה פרצא לאתרא, 'and it (the fire) was bursting that place, (down to the abyss)'. Should we read 'tongues of fire' with Milik, 228, or perhaps 'tracts of fire', זוני נור (cf. P Sm. 1103 ܙܘܢܐ ܢܘܪܢܝܐ, zona ignea)?

nor to discern Eth[tana] 'āyyeno, inf. = εἰκάσαι (Wis. 9.16). Aram. עיין Syr. ܚܡܐ (cf. Heb. ענין Pol.) is used 'to behold, discern'. The correct reading of G[1, 2] is preserved in Eth[tana] which shares with Eth[M], however, a double rendering of| ἠδυνήθην. The original probably read ולא יכלת למחזא ולא למעיינא '(neither their extent nor their size) was I able to see nor to discern'. An alternative would be to posit two Aram. verbs '(neither their extent nor their size) was I able to see nor could I (ולא כהלת) discern'.

(8) **fearful … terrible** Cf. Dan. 7.7 דחילה ואמתני, Hab. 1.7. Or could δεινός here = זיע? Cf. Bar Hebraeus, *Chron.*, 497 (P. Sm. 1105) ܙܘܥܐ ܘܕܚܠܐ. Did Eth. read [ὀδυνηρός] (cf. v. 9) for δεινός (Charles, 45). Cf. also Dan. 2.31.

(9) **why are you so afraid … spectacle** Eth. has omitted by hmt 'And I replied', giving a text (preserved in Eth[tana]), 'why are you afraid and so terrified on account of this fearful place …' (Eth[M] has rephrased the sentence to read 'why do you have such fear and terror because of this terrible place … '(Knibb)). Eth[tana] preserves the reading 'I am terrified' (dangaḍku) where we require to read (with G) 'why are you (afraid and) so terrified (dangaḍka)'. I suggest that this is more than a scribal error and that 'I am terrified' is part of the reply of Enoch which has dropped out of the text along with καὶ ἀπεκρίθην by hmt; and with it goes the predicate 'on account of this fearful place …'.

(10) **for an eternity** Eth. 'eska la'ālam = εἰς τὸν αἰῶνα. G appears to have a 'doublet' ('dittograph', Charles) μέχρι αἰῶνος εἰς τὸν αἰῶνα, but the longer expression in G could be a corruption of ἀπὸ τοῦ αἰῶνος εἰς τὸν αἰῶνα = מעולם עד עולם(?) The imprisonment of the rebel angels (the 'seven stars' and the watchers) was not to be forever, but for a myriad of years, an eternity'.

CHAPTER 22

This chapter has been the subject of a recent textual and exegetical study by M.-T. Wacker, *Weltordnung und Weltgericht: Studien zu I Henoch 22* (Echter Verlag, Würzburg, 1982).

(1) We should read הובלת, Ophal, with 32.2—Enoch is transported by his angel guides as at 17.1; G and Eth. read ἐφώδευσα 'I proceeded'. See Barr, 'Aramaic-Greek Notes', I, 193f. Does ἔδειξέν μοι render an Ophal אחזית 'I was shown? Cf. Milik 143, L. 3 and below 31.1. Enoch now returns to the fearful place in the west referred to briefly at 17.6 and now described in detail in 22-23, the abodes of the departed, Hades or Sheol, but now no longer, as in primitive Heb. thought, a place of shades beneath the earth, but, as in Egyptian and Greek mythology, located in the west of the earth, in the region of night and darkness beyond the great ὠκεανός itself (see Charles, 46, Dillmann, 124).

a large … mountain … flint-hard rock The lofty mountain no doubt supplies a barrier against the rest of the earth (Dillmann, ibid.), but it also provides the 'hollow places' as promptuaria for departed souls. For 'flint-hard rock' cf. 26.5 and Heb.

צור החלמיש Dt. 8.15 (Tg. טינרא תקיפא), Ps. 114.8, Isa. 50.7, LXX στερεὰ πέτρα. For the redundant ἄλλο here in G see Lods, 174.

(2) **hollow places** G τόποι κοῖλοι: Eth. 'beautiful places' has read κοῖλοι as καλοί. V. 4 below, in En[e] xxii 1, refers to these 'hollow places' as 'pits' (פחתיא), a word which Milik, 230, takes to be the original behind τόποι κοῖλοι or κυκλώματα (v. 8). But 'pits' is a further description of the 'hollow places', and the text should be rendered 'and these (hollow places) are pits (fashioned in this way for their incarceration)'. G τόποι κοῖλοι, κυκλώματα could be terms equivalent to LXX κοιλάδες = גיא, עמק. G κυκλώματα, however, suggests a connection with סחר which seems to have been rendered at 18.15 by the neologism κοιλιόμενοι, usually edited to κυλιόμενοι ('rolling') but more probably a scribal barbarism for κυκλώμενοι (cf. Milik, 228). Did the original here read סחרתא locus circumseptus 'stronghold(s)'? (cf. ܡܣܝܪܬܐ P. Sm. 2593, especially Ephrem's ܡܣܝܪܬܗ ܕܫܝܘܠ, 'the strongholds of Sheol'). Eth. resorts to paraphrase '(things) which roll'.

three of them were dark, and one bright There is no need to emend the text to read 'three' for 'four' and 'two' for 'three' (Charles, 47); see the notes to vv. 8-13. Cf. v. 9 where it is the 'spring of water' which is 'bright' or 'pellucid, limpid' (Eth. has 'living water'). If the reference was to 'pellucid water', 'living water'—clear, flowing water—this would remove the difficulty raised by Enoch's exclamation at v. 2 'How bare are those hollows ... and *dark* to view'. Cf. Charles, 47. All the hollow places were dark, and the text at this point is at fault, applying the adjective 'bright' to the place rather than to the spring.

How bare Charles, 47, suggests that מא (= πῶς) is corrupt for למא '... for what purpose (are these hollow places smooth etc.)'. G λεῖα (Eth. lemuṣ) suggests Heb. חלק (Aram.? cf. Hoftijzer, 89), lit. 'smooth', but 'bare, bald' of mountains (Jos. 11.17 (NEB 'the bare mountain'), 12.7) and here of these 'hollows' in the mountains.

(3) Eth II rewrites 'These beautiful places (are intended for this) that the spirits, the souls of the dead, might be gathered into them: for them (lomu'ellā(o)ntu) they were created ...' (Knibb). The words of Raphael may have read in the original: 'These hollows, for this very purpose they were created that here the souls of all men (cf. En[e] 1 xxii 1) should be assembled.'

(4) **the time of the last day** En[e] 1 xxii 2, lit. 'the time of the day of the end' where קצא is the absolute End; cf. Dan. 8.17, 11.35 'time of the end' (עת קץ) and I QS 4.25 'to the decreed end' (עד קץ נחרצה). The continuation 'of the great judgement' is in apposition, further defining the 'last day' (Milik's 'time of the Day of the End of the Great Judgement' is ambiguous). For a study of קצ(א) see Nötscher, *Zur theol. Terminologie der Qumran-texte*, 47f. G τοῦ διορισμοῦ καί is probably a ditt: it appears to have been read by Eth. and turned into a gloss, 'and the time will be long'.

(5) **the spirit of a dead man** G Eth. have turned the singular into a plural to form a class of spirits; G ἡ φωνὴ αὐτοῦ, however, is evidence of an original singular, as are the following two verses which identify this lamenting spirit with Abel (at Gen. 4.10 the 'blood' of Abel 'cries out' (צעק), here it is his spirit which 'cries out' (זעק) to heaven). The discovery of the original disposes of the theory that this is the first group of the spirits of the departed, viz. the martyrs, and the occupants of the first of the four partitions or compartments of Sheol (so Milik, 231, Ll. 7-8, following Dillmann). The division of Sheol for all who suffer death by violent hands, whether justly or unjustly, is described at v. 12. It is a separate, if related, episode which is here singled out for prominent mention, the first murder of an innocent man (so Lods, 176).

(6) **the watcher and holy one** En[e] 1 xxii 5 עירא וקדישא, rendered freely in G Eth. by ὁ ἄγγελος. See above, 106f.

(7) While the reference is to Abel as the prototype of the innocent victim of violence bringing death, there is probably also here the idea of Abel as typifying the צדיקים, οἱ δίκαιοι, and Cain all the רשעים or οἱ πονηροί of mankind. Cf. ThWNT I 6 Ἄβελ Καΐν (Kuhn).

(8) Cf. above, v. 2. Three 'hollows' for the spirits of the departed, one for the righteous dead. This suggests the schema of δίκαιοι and πονηροί, unless this compartment for the צדיקים, in this special division of Sheol in the west, corresponds to the 'islands of the blest' in Greek legend, and is reserved only for the Hebrew heroes, like Enoch, Elijah, Moses, Noah, Ezra, who, like their Greek counterparts, were believed to have been miraculously transported to Paradise. (See Rohde, *Psyche*, 94f., Dieterich, *Nekyia* (1893) 19f., and Athenaeus, Deipnosophistae (ed. G. Kaibel), Bk. XV, 695b.).

(9) **a spring of pellucid water** Cf. above note on v. 1. For the expression in Syr. *Kalilag und Damnag*, ed. G. Bickell, 14.7 ܒܐܪܐ ܕܡܝܐ ܢܗܝܪܐ 'spring of pellucid water' (P. Sm. 2299). There is no question of this 'hollow of the righteous' being 'bright' or 'illumined', φωτεινός; it is the water which is 'bright' or 'pellucid', 'limpid water' or 'living water' as in Eth. And as Charles, 49, notes: 'In the underworld, souls, according to the Greek Cults, Jewish, Hellenistic, and Christian literature, suffered from thirst: see Dieterich, *Nekyia*, 97f. In the Greek Hades there was a spring of forgetfulness on the left, while on the right was the spring of memory — the cool water — ψυχρὸν ὕδωρ, by the drinking of which consciousness and memory were quickened, the first condition of the full and blessed life'. See Rohde, *Psyche*, ii 2, 310, 390-391'. The author of Enoch seems to envisage all souls or spirits of the dead as 'existing' in four dark canyons in the mountains of the west, held in custody there till the last judgement; only the condition of the righteous seems to have been alleviated by the presence of a stream of fresh water. The ancient concept of Sheol has not been wholly lost.

And yonder one G οὗτος (sc. τόπος) here and in v. 10 is deictic: the angelus interpres points to each partition. This is the first of the four compartments or 'hollows' for the departed.

(10) Cf. 103.6. Both verses speak of 'sinners' who went unjudged (and unpunished) in their lives, in spite of their sinfulness; and on their death they had not suffered the supreme privation of sinners, the deprivation of funeral rites.

(11) **for this great torment** These promptuaria are also themselves to be part of the punishment; cf. 103.7, Lk. 16.23-25.

the accursed Eth. reads καταρωμένων, 'those who curse'.

they shall be bound for ever i.e. for endless time (above, 165) till the judgement. G Eth. δήσει αὐτούς is impersonal singular (Dillmann): Charles 'He shall bind them' seems unlikely in the absence of an antecedent. Eth. adds what appears to be an 'Ethiopic intrusion' (Charles[1906], 58, n. 44) which may be a gloss giving an alternative reading to μέχρις αἰῶνος, viz. ἀπὸ τοῦ αἰῶνος (cf. Flemming-Radermacher, 55). Or did a text read ἀπὸ τοῦ αἰῶνος καὶ μέχρις τοῦ αἰῶνος, cf. LXX Ps. 105(106).48?

that there be retribution on their spirits G (Eth.) ἀνταπόδοσις τῶν πνευμάτων lit. 'retribution of spirits' could mean vengeance for spirits wronged by the sinners. More probably it refers to the retribution to be required from the sinners themselves.

(12) **those who have accusations ... to lay** This third 'compartment' is reserved for men who have suffered death by violence. Both verbs, ἐντυγχάνειν = קבל (here and above at v. 7) and ἐμφανίζειν = Heb. הגיד, Aram. אחוי, have legal connotations: these spirits of men who have been violently slain have, like Abel, accusations to bring and information to lay before the court at the last judgement with regard to their

deaths—accusations and information about their murderers. The spirits in this category are clearly not sinners but the souls of the wretched who have met a violent death at the hands of sinners. For the 'strong sensitiveness' of the ancients to this form of death, see the excellent note in Milik, 231.

(13) The fourth compartment is for the 'not wholly impious(?)', those who collaborated with 'the lawless'.

not wholly lawless ... collaborated The text in both versions is confused and corrupt, and, indeed, almost beyond repair. The main clue to its reconstruction is the expression in Eth^{tana} feṣṣumāna 'abbāseyān = [ὅλοι (G ὅσοι) ἀσεβεῖς] = כולהון רשיען 'wholly impious'. If we remove G = Eth. ⌜ὅσιοι ἀλλὰ ἁμαρτωλοί οἱ⌝, we get 'And that (fourth hollow) has been fashioned for the spirits of men who were not wholly impious, but who were (G ἔσονται = להון ἦσαν) collaborators (μέτοχοι Eth. ὁμοῖοι?) with the lawless. The reference would then be to fellow-Jews who were the accomplices of the Gentile oppressors; their punishment is not so severe as those in the top category of sinners; it would seem to consist of perpetual imprisonment in their hollow place, but not to include any further afflictions at the last judgement.

their spirits will not be punished ... awakened from there En^{d} 1 xi 1 יתנזקון G τιμωρηθήσονται; Aram. נזק lit. 'to cause damage to' (Tg. Jer. 12.14) but also in a legal sense (Hoftijzer, 176). The original of μετεγερθῶσιν Eth. yetnašše'u has been lost: Milik suggests יתקימון '(nor) will they be ... transferred from here', denying any reference to resurrection (Milik, 218f., L. 1), and giving קום Ithpaʿal the meaning 'to remain permanent, to retain (life etc.) lastingly' (how this comes to produce the translation 'nor will they be transferred permanently from there' is not explained). A more likely equivalent seems to me to be יתעירון, עור Ittaphal, 'to be awakened' = ἐγείρεσθαι (μετεγείρεσθαι is a new compound not attested in Liddell and Scott), a verb which is used at Job 14.12 (cf. Tg.) meaning 'to be aroused or awakened from the sleep of death.' Cf. also 91.10, En^{g} 1 ii 13, Milik, 260. Could there be an allusion to Dan. 12.2 (cf. Isa 26.19) '... many ... that sleep in the dust of the earth shall awake, some to everlasting life, and some to ... everlasting contempt.'? By not being 'awakened from there (i.e. from Sheol)' for judgement, such stricken souls would certainly escape the possibility of an even worse fate (cf. Milik, ibid.), but, at the same time, would lose for ever the chance, at the judgement, of 'awakening'—with all righteous and acquitted souls—to 'everlasting life.'[1] For the expression 'righteous judgement', 27.3 (G κρίσις ἀληθινή), 91.12,14 (En^{g} i iv 16,19) 93.5, 60.6.

(14) **Judge of righteousness** G Eth. 'Lord of righteousness' could be a free version of 'Judge of righteousness' דין קושטא. In favour of retaining the full expression 'Lord of glory and righteousness' is the preservation of רבותא 'glory' at En^{d} 1 xi 2. G ὁ κυριεύων τοῦ αἰῶνος = מרא עלמא IQ **20** 2.5 (see Fitzmyer, *Genesis Apocryphon*, 242).

CHAPTER 23

(1) **to the ends of the earth** The reading of G (adopted by Charles) 'to the west

[1] There is no question of 'resurrection' here (or elsewhere in chapters 1-36) in the strict theological sense. That idea probably came from a Christian appropriation of LXX Dan. 12.2. If there is an allusion to Dan. 12.2 here this could imply a later date for the Journeys of Enoch from the Watcher legend; cf. above, Additional Note: Enoch 14 and Daniel 7, 151f.

of the ends of the earth' is less satisfactory than that of Eth., 'to the west, right to (μέχρι = עד) the ends of the earth'. G may have arisen by a scribal slip.

(2) **a blazing fire** For this 'fire in the west', see 17.4.

did not flag Cf. Sir. 16.27 (of celestial bodies) οὐκ ἐξέλιπον ἀπὸ τῶν ἔργων αὐτῶν.

equally ... night Aram. לבך = 'to lay hold on' (Tg. Prov. 31.19); cf. Syr. ܠܒܟ ptc. pass. ܠܒܝܟ = tenens, prehendens, ܠܒܟ ܪܗܛܐ 'to lay hold of a course', P. Sm. 1883, Jacob of Serug, *Hom. Select.*, ed. Bedjan, I 428.9, cursum aggressus est; ἅμα Eth. kamāhu = Heb. יחדיו, Aram. כחדא, pariter.

(4) **This stream of fire** Behind this expression in G Eth. Milik conjectures an original דן דגלה ונורא הוא which he renders '(And Raʿûʾel answered me:) 'This is its function (that this fire ... follows ... the luminaries)''. This Aram. text is reconstructed from frg. c of En[d] 1 (end of lines 3-6), the identification of which can hardly be said to be 'beyond doubt', (Milik, 219, Ll. 5-6). If the frg. has been correctly placed and דגלה is to be read, an alternative rendering would be: 'This fiery sign (follows ... the luminaries)', where דגלה = σημεῖον (see above on 2.3) and the waw in ונורא explanatory. On the meaning of ἐκδιῶκον = רדף in this context 'tend' and hence 'supply (with light)', see above on 20.4 and Milik, 219, Ll. 5-6.

CHAPTER 24

(1) The opening half of v. 1 Eth., 'And from thence ... earth' is omitted in G. Was it introduced by a scribe, drawing on 23.1? See Milik, 220, L. 6. The reading of Eth. ὄρη πυρὸς φλεγομένου is to be preferred to G ὄρη πυρὸς καιόμενα, since the Aram. phrase is נור דלק (Dan. 7.9).

mountains of fire The imagery from this point on represents a further elaboration of the vision of the seven mountains and the mountain-throne of God at 18.6f., and is, in fact, a duplicate version of this Enoch apocalypse (see above, 158). These 'mountains of fire' represent the first great barrier to the mountain-throne of God and the grove of immortality.

(2) Cf. 18.6-8.

whose stones ... fair in form The two parallel clauses, (a) 'whose stones were priceless for their beauty' and (b) 'all valuable and their appearance glorious and fair in form' look like 'doublets' of the same original text; (b) Eth., with the parallel phrases, 'their appearance glorious' and 'fair in form', could be more original (G has simply καὶ πάντα ἔντιμα καὶ ἔνδοξα καὶ εὐειδῆ).

one planted on the other ... no one ... approached any other G ἐστηριγμένα ἐν τῷ ἑνί Eth. 'founded one upon the other' = מתנצבין חד על חד(?). Dillmann, 128, seems to me to interpret rightly that these mountains are envisaged as forming 'staircases' (Stufen) on the east side and on the south, each planted on top of the other and each higher than the other, leading up to the heaven-high mountain-throne of God which is situated 'between them' (v. 3). They thus form a second mountain barrier; and all ways to the mountain-throne and the grove of immortality are blocked, since the mountain gorges are not only deep and crooked but are not connected with one another, thus forming each a natural cul-de-sac.

(3) **a seventh mountain between these (mountains)** G (ἕβδομον ὄρος) ἀνὰ μέσον τούτων = בינתהון, i.e. not 'in the midst of these mountains' but between the three in the east and the three in the south. G καὶ τῷ ὄρει (ms τω ορι) seems a redundant ditt.

Three mss of Eth., but including Eth[b'], the Westenholz ms, preserve the true reading ὑπερεῖχεν αὐτῶν. (cf. Charles, 52, Charles[1906], 61, n. 47). Eth. has 'they were all like the seat of a throne', an unlikely reading. Does it derive from a text 'it surpassed them all in height, like the seat of a throne ...'?

(4) **no tree among them ... like it** Beer and Charles emended αὐτῷ ηὐφράνθη to αὐτοῦ ὠσφράνθη. The text of G still looks odd, οὐδεὶς ἕτερος αὐτοῦ ὠσφράνθη 'no other (person) had smelt it', followed by καὶ οὐδὲν ἕτερον ὅμοιον αὐτῷ 'and no other (tree) was like it'. Eth. may point towards the original text, 'and none of them (i.e. the fragrant trees of v. 3) and no other (tree) was like it'. I suggest, however, that G supplies the missing verb from the first clause (cf. Charles[1906], 61, n. 16): 'and no tree among them flourished (ηὐφράνθη = רענן), and no other tree was like it = ולא חד מנחון רענן ולא אוחרן כותיה. Perhaps the original was even shorter: ולא חד מנהון רענן כותיה, 'and no (tree) among them flourished like it': ηὐφράνθη is best explained as a misrendering of רענן viruit as if from רנן iubilavit; cf. Job 15.32, Dan. 4.1 (Theod. εὐθαλῶν) and above on the 'luxuriant' trees at 10.19. G οἱ δὲ περὶ τὸν καρπόν is clearly corrupt: Flemming-Radermacher emend to ἦν δὲ περιττὸς ὁ καρπός, 'and its fruit was abundant', replacing Eth. 'and its fruit was beautiful'.

its ... wood δένδρον = עץ Aram. עק.

(5) **(how) very lovely to the eye is its fruit** G has simply καὶ τὰ ἄνθη αὐτοῦ ὡραῖα τῇ ὁράσει omitting all reference to the fruit. Eth. has lit. 'its fruit was very delightful to look upon', recalling Gen. 2.9 'every tree which is pleasant to the sight (נחמד למראה LXX ὡραῖον εἰς ὅρασιν)'.

(6) **Michael** Cf. 20.5, Michael, guardian angel of Israel, is in charge of the 'blessings to come' reserved for Israel. Contrast 20.7 where it is Gabriel who is to be 'over paradise'.

CHAPTER 25

(1) **do you marvel at ... tree** G ἐθαύμασας ἐν (τῇ ὀσμῇ τοῦ δένδρου). It seems unlikely that there was any distinction of tense in the Aram.: תשאל 'do you ask' ... תתמה 'do you marvel at' (θαυμάζειν ἐν = תמה ב).

desire to learn Cf. Dan. 7.19 צבית ליצבא lit. 'I desired the truth', LXX ἤθελον ἐξακριβάσασθαι, Theod. ἐζήτουν ἀκριβῶς. Cf. above on 21.5: θέλεις here could also render בעית 'why do you seek to know the truth ...' (cf. Black, *An Aramaic Approach*[3], 188, 244). Eth. had a variant Greek text, possibly closer to the original.

(2) **but most of all** G μάλιστα and σφόδρα seem to be 'doublets', as Eth. shows. The more primitive rendering is probably σφόδρα = שגיא (Dan. 7.28).

(3) **like a throne of the Lord** These words could be a translator's addition from 24.3.

the great Holy One See above on 104f. The Greek text behind Eth., both here and in reading ὁ κύριος τῆς δόξης is to be preferred to G.

the everlasting King Also at vv. 5 and 7; see above on 9.4.

to visit the earth in blessing G ἐπ' ἀγαθῷ = לטוב (Dt. 6.24, Jer. 15.11). As in the theophany at 1.4f. the visitation is in judgement but also in mercy ('for good'), or does the writer wish to distinguish this visitation from that at 1.4f. (Dillmann, 129)? Cf. also 77.1. It seems unlikely that the author has Sinai in mind, as has been argued by Lods, on the grounds that, while Ezek. 28.13 may have contributed to the description of the mountains of precious stones, the passage may also be reminiscent of the ancient turquoise mines on Sinai exploited by the Pharaohs (Lods, 185).

(4) **this fragrant tree** The Tree of Life, cf. 2 En. 8.3 and Bousset-Gressmann[4], 448f. Eth II [δένδρον ὀσμῆς εὐωδίας], has the full Biblical phrase ריח ניחות, LXX ὀσμὴ εὐωδίας (e.g., Lev. 1.9).

allowed to touch it G ἐξουσίαν ἔχει = Aram. שליט ל cf. ܫܠܝܛ ܠ. licet, P. Sm., 4181. Cf. Gen. 3.22-24.

the great judgement See above on 10.12.

a recompense for all ... a consummation forever Eth.$^{g\ t\ (tana)}$ appears to read 'he (God) shall take vengeance on all and bring it (the judgement?) to an absolute end forever'. Cf. 10.12. Did an original perhaps read יתפרע לכול, 'he will recompense all', in which the verb פרע would refer to the rewards of the righteous as well as to the deserts of the wicked.

(5) **Then ... its fruit ... for food** The versions seem overloaded in their description of the 'elect' after the judgement, 'righteous and pious', (Eth. 'righteous and afflicted', teḥḥutān, an unusual rendering of ὅσιοι in G, unless it reflects a variant πραεῖς or ταπεινοί, Heb. ענוים, Tg, עני״). Perhaps we should omit δικαίοις καὶ ὁσίοις as translator's expansions? Charles thinks the author had before him here Ezek. 47.12, LXX ὁ καρπὸς αὐτῶν εἰς βρῶσιν referring to the fruit on the trees beside the mythical stream from the temple to the Dead Sea; he further explains εἰς ζωήν as a mistranslation of Aram. להוא ἔσται as לחיי״א.

it shall be transplanted ... king Eth. reads [εἰς βορρᾶν μεταφυτευθήσεται]. If we read εἰς βοράν we are spared the conundrum posed by the transplanting of this tree 'to the north'—for Dillmann from the south, where he locates this mountain-paradise (with a different 'Tree of Life' from Gen. 3) to Jerusalem, for Charles from the celestial paradise in the north-west (whither it had already been removed from Eden) back to Jerusalem.

There seems little doubt that it is a 'Tree of Life' that is here described; later, in Enoch's further journeyings, he describes 'the Tree of Wisdom' in the 'Paradise of righteousness' far 'towards the east of the earth' (32.2f.). In Gen. 3 both trees are located in Eden. In Enoch, we have a quite different and independent tradition about these mythical trees, the 'Tree of Life' and the 'Tree of Wisdom'.

(6) **and rejoice in the holy place ... in their very bones** Eth$^{g\ q}$, supported by three mss of Eth II, has the variant reading 'and they will bring into it (the holy place) its sweet-smelling odour in their bones', for 'into the holy place they will enter'. Both clauses seem required by the context, unless we render 'they will rejoice in the holy place and bring into it the sweet-smelling odour etc.'. One of the clauses may have been lost by hmt in the original: 'and they will enter (יעולון) the holy place and bring into it (יעלון ליה) its sweet-smelling odour ...'. (The first verb from עלל, 'to enter', the second from עלי, Aph. = Heb. עלה, Hiph., 'to bring as an offering' cf. P. Sm. 2884). 'In their very bones', i.e. in themselves = בגרמיהון, בעצמיהם.

(7) **King of the ages** See above on 9,4.

CHAPTER 26

(1) **a blessed place** Eth. [τόπον ηὐλογημένον πίονα] Charles[1906], 64, n. 21 explains the second adj. as an Eth. gloss. The phrase τόπος πίων, however, in this

connection, is Biblical (LXX Isa. 5.1, 30.23) and is found in all mss. Could it be original? The blessed place here, as at 27.1, is Israel, and the place at the centre of the earth Jerusalem; cf. ThWNT, Bd. III, s.v. Σίων, 323.25-30, 324f. (Lohse).

trees, with saplings ... felled tree The author is seeking to emphasise the extraordinary fertility of this place: even shoots from a dead tree that had been cut down (Heb. עץ Aram. קיסא), will live and spring to fresh life (παραφυάδες μένουσαι = ינקין מתקימין, cf. LXX Ezek. 17.22 παραφυάδες = יונקות, ינקין[1] and En. 2.2, En[a] 1 ii 5, Enastr[d] 1 i 6 for מתקימין of foliage 'remaining' evergreen). For the metaphor cf. Job 14.7-10, Dan. 4.23f. Charles (followed by Milik and Knibb), excise the clause τοῦ δένδρου ἐκκοπέντος, as a 'disturbing gloss' (Charles, [1906], 65, Milik, 221, 'a Christian anti-Jewish addition').[2] Older commentators understand the reference to be to the 'new Israel' ('the new Israel of the time of the Messiah', Dillmann) growing out of the trunk of the old Israel. All that seems to have originally intended, however, was to underline the indestructibility of the inhabitants of this blessed place. There could be a reference to the descendants of Israelites who had been 'cut down' by their Gentile oppressors (Heb. יונקות is especially used of Israel, Hos. 14.7, Ezek. 17.22 and cf. Isa. 53.2). If the Job passage was in the writer's mind, he may even be refuting its contention there that, while 'there is hope of a tree, if it be cut down, that it will sprout again', man has no hope of surviving death, 'But man dieth ... and where is he?' (14.10). Did the writer mean to convey the idea that the new Jerusalem will not only be filled with the righteous living at the last judgement, but that the righteous dead, those who have fallen, 'cut down' by their oppressors, will rise in new life at the judgement? However we interpret it, the clause τοῦ δένδρου ἐκκοπέντος seems an important part of the text.

(2) What follows is an unmistakable geographical outline of Jerusalem and its immediate environs; Lods detects a resemblance to the description of Jerusalem in Josephus, B.J.v.4.1 (136-141). The sacred mount is Mount Sion (Moriah), the Temple mount. The passage has been influenced by Ezek. 47.1-12; cf. also Jl. 4.18, Zech. 14.8 and see further Milik, 222. There is no need to alter δύσιν to ῥύσιν with Eth. (Charles): while δύσις is never used of water 'falling', Aram. נחת is; cf. especially the use of ܢܚܬܐ P. Sm. 2346 and Ezek. 47.1 (המים ירדים). It is Eth. ῥύσις which is the corruption (or correction).

from the east side, a stream ... south Milik, 36, identifies this 'water' which has 'a descent to the south' with the Biblical Gihon which rises from the eastern slopes of Ophel (a continuation of Mount Sion) and 'gives birth to the Cedron, flowing (at first) towards the south' (v. 2). He does not mention the well-known connection of Gihon with the Pool of Siloam (the famous conduit constructed by Hezekiah, 2 Chr. 32.3-4), and the view of commentators since Dillmann that it is the 'water of Siloam' which the author is here describing. The 'descent to the south', I suggest, refers to the 'fall' of this water from Gihon 'underneath the mount' (Ophel and Sion constitute the sacred mount) to the Pool of Siloam. It seems unlikely that the author would omit a stream so renowned in Israel's history (Dillmann, 131).

(3) **another mount higher than the first** i.e. the Mount of Olives. The author then goes on to say that he saw between them a deep, narrow, ravine. Earlier commenta-

---

[1] Usually rendered 'branches' (ענפין Milik), but branches do not 'blossom' or 'burgeon' (βλαστεῖν).

[2] The objection of Knibb, 2, 114 (following Charles), that 'it is difficult to think that both δένδρα ἔχοντα and τοῦ δένδρου ἐκκοπέντος are original' holds for Greek, but not for Aramaic (see textual note).

tors follow Dillmann in identifying this ravine with the Valley of the Cedron (or Jehoshaphat). As Lods, 189, noted, however, v. 4 speaks of another 'lower mount' to the west of this mount: this can hardly be Mount Sion and could only refer to the so-called Mount of Evil Counsel (Gebel Abu Tor). The author must, therefore, be here referring to the lower course of the Cedron through the ravine known as the Wadi en-Nur at the foot of the Mount of Offence (Gebel Batn el-Hawa), itself a continuation of the Mount of Olives. Cf. Milik, 36.

in the midst of them En[d] 1 xii 5 ביניהון G = Eth. ἀνὰ μέσον αὐτῶν (ms αυτου). G gives the meaning of the Aram. correctly in this context: this valley, later identified as the valley of Hinnom, can scarcely be 'between the three mounts' but lying 'among/in the midst of them'. Cf. Ps. 104.10 'among the hills' (בין הרים). It is this description of the valley which identifies it at 27.1 in Enoch's question; see below, on 27.1,2.

alongside this (higher) mount So Eth. [παρὰ τὸ ὄρος]. G seems to combine two readings ὑποκάτω τοῦ ὄρους and ὑπὸ τὸ ὄρος. The first may be ruled out in this context: 'the stream runs (πορεύεται) through the valley under the mount' makes no sense, so that the phrase here may be a scribal error and come from v. 2 G ὑπὸ τὸ ὄρος, on the other hand, means 'at the foot of the (higher) mount' תחות טורא(?) cf. Exod. 24.4, Dt. 4.11. The reading of Eth. is quite independent of G, and either could be original.

(4) **another mount** West of the Mount of Offence lies the Hill of Evil Counsel (Gebel Abu Tor) and it is at its foot and in the midst of these mounts that the deep and parched Vale of Hinnom lies.

other valleys ... at the farthest parts of the three mounts The sing. 'another valley' seems less appropriate in this context where there is more than one ravine at the foot of these three mounts (cf. Milik, 36). G ἐπ' ἄκρων τῶν τριῶν ὀρέων = בסופי/בקצי תלתה טוריא 'at the extremities of the three mounts'; cf. the use of ירכתים in this connection, e.g. Jg. 19.1, 18 'extreme parts of Mount Ephraim'.

(5) **of flint-hard rock** See above on 22.1, Milik, 222, Ll. 7-8.

(6) **I marvelled at the rocky ground ... valley** Milik, 221, renders 'I marvelled at the mountains ...' following his reading of En[d] 1 xii 8. The reading טוריא has much in its favour since it is the 'mountains' and the 'valleys' which are the central objects of interest in this passage. Against this, however, is the reading of Eth. (without variants) clearly presupposing a clause [περὶ τῆς πέτρας], and En[d] 1 xii 8 can be read as על טנרא = על הצור '(I marvelled) at the rocky ground'. What aroused Enoch's wonder was the 'rocky' i.e. inhospitable nature of the surface of these valleys, described at 22.2 as 'bare, bald' (see above, 166).

CHAPTER 27

(1) All versions, except that of Milik, follow G (Eth.) and render (as Charles), 'For what object is this blessed land ... and this accursed valley'. Only traces of a few letters have survived in the Aram. frgs. at En[d] 1 xii 9, but the sentence almost certainly read originally, 'Why (על־מה) is this land blessed ... and this valley accursed ...?'. Eth. adds ἐν μέσῳ αὐτῶν 'in the midst of them', which can hardly have 'trees' as antecedent: the phrase recalls 26.3 where this valley is said to be 'in the midst of them' (ביניהון) i.e. the 'mounts'. Is the Eth. reading a relic of ἐν μέσῳ τῶν ὀρέων?

(2) At 24.6 it is Michael who accompanies Enoch. There has been no indication of

any change in Enoch's accompanying angel, from Michael to Uriel or Raphael since 24.6. Does this inconsistency in the narrative account for the long omission here in G? Or is the scribe of G abridging?

This accursed valley G γῆ κατάρατος '(this) accursed land ...'. Knibb, 2, 115, following Ullendorff, argues that γῆ = גיא 'valley', and, consequently, 'Eth. would hardly be dependent on Gr ... and it is plausible to regard this as a further case where Eth. is directly dependent on a Semitic *Vorlage*'. This is an attractive conjecture, but if the Heb. word גיא is used here—the word is not found in Aram.—why does the Aram. consistently use חלה elsewhere for 'valley'? It seems more likely that G has deliberately abridged this passage, removing the introductory sentence, 'Then Uriel ... to me' and resuming with a description of Gehenna as 'this accursed land' in contrast to 'this blessed land' at v. 1. For Gehenna, see Bousset-Gressmann[4] 233, 278.1, 286, Schürer, History II, 545 n. 108, Str.-B. Bd. IV 1100f., ThWNT s.v. γέεννα Bd. I 655f. (J. Jeremias).

speak hard words against i.e. 'blaspheme'; for the expression see on 1.9 G = Eth. περὶ τῆς δόξης αὐτοῦ = על רבותיה 'against his glory'.

their place of judgement Eth. mekuennānihomu G οἰκητήριον 'habitation'. Eth. = [κριτήριον] 'court' = דינא (בית) cf. Dan. 7.26 'the court (דינא) shall sit'. G appears to have read דירא 'dwelling'; cf. Knibb, 2, 115f., another passage which Knibb claims shows the independence of the Eth. from the Greek and its direct access to an Aram. original. But Eth. probably read a Greek text κριτήριον.

the accursed G = Eth. οἱ κεκατηραμένοι = ליטין but the original could have read לאיטין maledicentes (e.g. Tg. (P-J) Num. 23.8 and cf. P. Sm. 1907).

(3) **In the last days ... righteous** G is corrupt: ἐν ταῖς ἡμέραις τῆς κρίσεως seems a ditt. of the phrase at v. 4, Replacing ἐν ταῖς ἐσχάταις ἡμέραις of Eth. Eth. ὅρασις/δεῖγμα = חזוה, spectaculum. Eth. ἐπ' αὐτῶν = עליהון, 'in them'(?). (For this temporal use of על = ἐπὶ + gen., see Milik, 82).

the pious While G ἀσεβεῖς could be a deliberate alteration, it seems more likely to be a scribal error. Eth. maḥāri usually = οἰκτίρμων, ἐλεήμων, but the translator here is probably rendering εὐσεβεῖς = חסידין (cf. Heb. חסד 'loving-kindness'). There is no need to emend the Eth. text.

This verse is important for the theology of the author: the last great judgement is to take place in Jerusalem and to become a spectacle for the righteous and the pious for all time. Moreover, it would appear that, with the 'Tree of Life' transplanted to a place beside the Temple, Jerusalem itself, or rather the 'new Jerusalem', is to become a second Eden or a paradise on earth.

(4) **in the days ... to them** i.e. to the pious who will bless the Lord for the just judgement that has been meted out to them too—but in their case also a merciful judgement. As Lods, 191, noted, ἐν ἐλέει is to be construed with ἐμέρισεν and not with εὐλογήσουσιν, according to the use of חלק ב e.g. Job 39.17 לא חלק לה בבינה '(God) has not imparted to her understanding'.

CHAPTER 28

(1) As Dillmann first noted (133, cf. Milik, 25f.), this part of Enoch, in particular, is a combination of vision and legend, with an account of real scenes and places which the author himself must have known and visited. It is not always possible to

distinguish the two strands in the author's narrative. The 'theological' motifs in the scenes in 26-27, centring on and around Jerusalem, are certainly immediately recognisable. In the following chapters, 28-32, however, we seem mostly to have details from a real travel-account. Dillmann's attempt (ibid.) to link 28 with Ezek. 47 is not convincing; there seems to be no discernible 'theological' purpose in this chapter.

into the midst of a mountainous part of the desert G Eth. εἰς τὸ μέσον (Eth. + [ὄρους]) Μανδοβαρα (Eth. [Μαδβαρα]). Both versions transcribe Aram. מדברא and make it a proper name here and at 29.1 (G Βαβδηρα). Cf. LXX Jos. 5.6, ἡ Μαβδαρῖτις (see also Lods, 133). That Eth. [εἰς τὸ μέσον ὄρους] has preserved the original text (למציע טורא די מדברא) is confirmed by τοῦ ὄρους τούτου at 29.1; the latter could not possibly still refer to the ὄρος of 27.4 or 27.3. Dillmann suggested Heb. הר־מדבר as the equivalent of Eth. ὄρος / ὄρη Μαδβαρα. A related expression occurs at Ps. 75(74).7, ממדבר הרים LXX ἀπὸ ἐρήμων ὀρέων, lit., 'the desert of the mountains', 'mountainous desert'. Could an original מדבר טורין lie behind the Eth. version? It would account for the plural at 30.1 Eth. κἀπέκεινα τούτων ὀρέων as at 18.9.

and I saw the Arabah(?) G Eth. καὶ ἴδον αὐτὸ ἔρημον '... and I saw it (sc. Madbarâ 'the Desert') deserted' (Milik, 233) can hardly be original, and (except for the ἴδον) could be an explanatory gloss on the proper name Μανδοβαρα. Dillmann, 133, identifies this wilderness with the Arabah, Milik, 233 thinks of the Negeb. Both proper nouns are rendered by ἔρημος in the LXX (Gen. 12.9, Num. 13.22), and this may provide a solution of the apparently tautologous καὶ ἴδον αὐτὸ ἔρημον = וחזית יתה ערבה, et vidi illam Arabah.

a [place] by itself G Eth. καὶ αὐτὸ μόνον after καὶ αὐτὸ ἔρημον is clearly defective; I suggest καὶ [τόπον] μόνον = ואתרא בלחודוהי, 'and a place by itself,' i.e. an isolated spot in this mountainous part of the desert. Support for the emendation comes from 29.1, where, without τόπον here, εἰς ἄλλον τόπον becomes difficult to explain.

(2) **full of trees and plants** G Eth. ἀπὸ τῶν σπερμάτων = מן זרעונין, cf. Dan. 1.12,16, Charles, 57, Milik, 233.

with water gushing forth upon it from above I have adopted the rendering of Eth. [ἀνομβροῦν] with Charles,[1906] 69; contrast Milik, 233 ὕδωρ ἄνομβρον ἄνωθεν '(a stream) of water, not nourished by the rain from above ...'. The adv. ἄνωθεν = מלעיל may have a geographical significance, from a source higher up in this mountainous region.

being carried by a copious aqueduct... north-west G φερόμενον (*Eth.* [φαινόμενον]) (=מתובלין) ὡς ὑδραγωγὸς δαψιλής. G ὑδραγωγός = מרזיבא Tg. 2 Chr. 32.30, P. Sm. 3876. Did the original read במרזיבא רבא, 'by a copious aqueduct' misread (or rephrased) as כמרזיבא רבא, ὡς ὑδραγωγὸς δαψιλής? Milik, 233, identifies very positively with the famous aqueduct from the ʿAin Mousa into Petra. The text certainly gives the impression that the author has a definite place in mind and the details fit the 'mountainous region' of the desert where Petra is situated. The Nabataean aqueduct has been well known since hellenistic times (Milik, ibid.).

taking up water and dew The ʿAin Mousa was not the only source of water feeding the aqueduct; cf. Milik, 233, and Eth.

CHAPTER 29

(1) **in the desert** Eth. '(away) from Madbara'. Eth. = G takes Madbara as a proper name for a place (see above, 175), hence the idea that Enoch went from Madbara.

(2) **juniper trees ... almond trees** G = Eth. κρίσεως δένδρα is obviously corrupt. Commentators tend to favour the conjecture of Prätorius that an original אילני דריחא 'trees of fragrance' has been misread as אילני דדינא (cf. e.g., Knibb, 2, 117). Milik, 233, suggests אילני ברא 'wild trees' rendered first as κτίσεως δένδρα and then corrupted to κρίσεως δένδρα. The translation assumes an original אילני ברותא, i.e. 'juniper trees', mistranslated as אילני בריתא, δένδρα κτίσεως becoming κρίσεως δένδρα. One variety of these trees had a fragrant smell (cf. P. Sm. 607). (Milik, 'Hénoch au Pays des Aromates', 74, identifies the 'aloe trees' of 31.4 with the juniper.) Eth. had a corrupt text which, however, together with G, gives a longer text that makes sense: [καὶ ἐκεῖ ἴδον δένδρα κρίσεως πνέοντα ἀρωμάτων πλείονα κόστου(?) [καὶ] ἀρωμάτων λιβάνων καὶ ζμύρνας] = ותמן חזית אילני ברותא דריח בשמא יתיר מן קשט(?) ⌜ובשמי⌝ לבונה ומור 'And there I saw juniper trees redolent of a fragrance choicer than kostos(?) and ⌜the fragrance of⌝ frankincense and myrrh'. Eth. quasquas, vasa, seems a mistake for quasṭ, quasṭē, a rendering of קשט, an odoriferous root used with frankincense and myrrh in the ritual of the sacred incense; see Löw, *Flora*, IV 99f. There can be scarcely any question here of an actual incense or myrrh, much less of an actual place, as Milik suggests ('Hénoch au Pays des Aromates', 73).

their trees were like almond trees G κάρυα can stand for any nut-bearing tree (Liddell and Scott), but in the LXX it corresponds to Heb. שקד used for either the tree or its fruit (cf. Gen. 43.11, Jer. 1:11). Both G and Eth. read τὰ δένδρα. Cf. Flemming, 35, who notes the proposed emendation of Radermacher to ἀκρόδρυα, 'nuts', which would give 'their nuts are like almonds'. There is no support in the versions for this conjecture.

CHAPTER 30

(1) **And beyond them** the 'trees' of 29.2 or the mountains previously mentioned? Eth. may have read, perhaps correctly, 'beyond these mountains'. Cf. Milik 'And beyond these (mountains) ...'

valleys with perennial streams I have read Eth. qualāta (φάραγγας) (cf. the plur. at v. 3). Edd. read (v. 2) ἐν ᾧ καὶ δένδρον κ.τ.λ. 'and therein there was a tree ...'. For this use of ἀέναος cf. Wis. 11.6 ἀντί μὲν πηγῆς ἀενάου ποταμοῦ Eth. lit. 'water as that which does not fail'; and cf. Tg. Jer. 17.13 מיא דלא פסקין.

(2) **sweet cane** For קניא טביא 'sweet cane' cf. Jer. 6.20 קנה הטוב Tg. קני טובא, LXX κιννάμωμον, Eth. qanānemos šannāy (Dillmann, *Lex.*, 443). Here Eth. appears to be rendering δένδρον ὡραῖον; cf. Milik, 202, Ll. 24-25, who explains G as a corruption of κάλαμοι χρηστοί, which seems transcriptionally improbable. G χροα is impossible (Knibb): Dillmann suggested a corruption of χρεῖα (SAB, 1892, 1051): it seems an unlikely corruption of either χρηστός or χρεῖα (cf. Knibb, 2, 118). Could it be a corruption, by faulty hearing, of ὡραῖον? Note that קניא lit. 'reeds' has been rendered in G = Eth. by δένδρον / δένδρα. The reading 'sweet-smelling tree' (ʿeḍ(a)

maʿāzā cf. En. 24.3, 29.2) has probably come into the text from a gloss on σχοῖνος, understood as the 'mastic tree'.

camel-hay Since we have to do with the family of sweet-smelling reeds (sweet cane) it seems more probable that we should read G as σχοῖνος lit. 'reed, rush', and hence σχοῖνος ἀρωματική 'camel-hay'[1] rather than σχῖνος 'mastic', the resin of the lentisk tree.[2] G at 32.1 gives σχῦνος as the equivalent of Aram. צפר (Eth. 'fragrant trees"), but this is as much a guess as the rendering of Eth; see note on 32.1. The Jewish Aram. equivalent is חילפא דימא, cf ܫܠܦܐ ܕܝܡܐ P. Sm. 1289, Löw, *Flora*, I 695.

(3) **the fragrant cinnamon** En^c1 xii 25 = קונם בשמא = קנמן בשם Exod. 30.23 LXX κιννάμωμον εὐῶδες. See Löw, *Flora*, II 107-113. Cf. Knibb, 2, 118.

CHAPTER 31

(1) **I was shown** En^c 1 xii 26, 27 אחזיאת = ἴδον, τεθέαμαι. See Milik, 202, L. 27.

I saw trees Where Aram. has חזית G reads ἄλση 'groves (of trees)'. It is difficult to find any connection between ἄλση and חזית, but cf. Milik, 202, Ll. 26-27. Has the 'addition' arisen by a ditt. and corruption of ἄλλα? There is nothing corresponding in Eth., an indication that we are probably dealing with a corruption in G.

came forth ... galbanum G σαρραυ, Heb./Aram. צרי, ܨܪܘܐ styrax. See Löw, *Flora*, III 388-395; Eth. sarārā cf. neo-Syr. ܨܪܘܪܬܐ P. Sm. 3439. See further Löw, *Flora*, חלבנה galbanum, III 455-457. On both resins Milik, 'Hénoch au pays des Aromates', 73 (6). The names are by no means synonymous, if this implies two names for the same gum: both styrax and galbanum are ingredients of the holy incense; see Löw, *Flora*, IV 99. Eth. has a doublet 'and there came forth water ...' where the word following 'water' is illegible (see Knibb, 1, 99): I suggest māya lebn = στακτή, giving an alternative rendering of צרי 'styrax' (Dillmann, *Lex*., 42). Milik, 233, suggests an original דמעא lit. 'tear' for νέκταρ; cf. ܕܡܥܐ ? gummi, succus, resina ... P. Sm. 921.

(2) **east of the ends of the earth** India, China, home of the lign-aloes(?).

full of aloes Charles[1906], 71, n. 4, Charles, 59, proposed emending the unintelligible εξαυτης to στακτῆς. En^c 1 xii 28 is unfortunately indecipherable except for the last four words, and Eth. has an equivalent only of G ἐν ὁμοιώματι ἀμυγδάλου plus 'and (it was) hard'.[3] Charles's conjecture may be right (but contrast Milik, 202): Ps. 45(44).8(9) refers to 'aloes', 'powdered aloes' (NEB), אהלות, LXX στακτή, next to myrrh and cassia; they were the three choice aromatics. See Löw, *Flora*, III 411-4.

full of aloes ... the shell of the almond nut It is clear from the next verse that it is the leathery bark of the lign-aloe or 'eagle wood' which resembles the shell of the almond nut, and not the wood itself.

(3) **When an incision is made(?) ... fragrances** En^c 1 xii 29 has a longer text: it is not till the end of the line (of c. 53 letters) that the apparent equivalent of G ὅταν τρίβωσιν is found, viz. כדי מדקן קליפיא אלן (17 letters). In the first half of the line

---

[1] So Milik, 'Hénoch au Pays des Aromates', 73, n. 3, Löw, *Flora*, I 694f. (but cf. Milik, ibid. 72, 32.1 '(nard de choix) de lentisque, de cardamone ...' and Milik, 201, 30.1, 'like mastic').

[2] Löw, *Flora*, I 195; Knibb, 2, 118.

[3] Could this be an original feature of the description? For the hardness of aloe-bark, Löw, *Flora*, III 413, and for the almond nut and its pod, *Flora*, III 142f.

[באי]לניא] En[c] 1 xxvi 16 is certain and [בש]ם ריח a pleasant odour' seems highly probable. At first the clause in Eth. 'and when they take up its fruit' might seem a suitable clause to fill the lacuna, but it seems more likely to have arisen either by scribal error (λάβωσιν for τρίβωσιν), or as a general statement introduced by a learned scribe (cf. Flemming, 36). Milik suggests a suitable parallel to 'is ground' (מדקין impers. plur.), 'When incisions are made', for which קטע or קלף, Pa. 'to cut' or 'to peel', would be appropriate.

CHAPTER 32

Milik's identification of En[e] 1 xxvii frg. e with 32.3 seems to me to be mistaken; it belongs to 33.1 (see below, 180).

(1) **beyond these mountains** Milik's restoration ולהלא מן טוריא אלן seems right; cf. 31.2 and the following 'other mountains' (G Eth. 'seven mountains', is a reading influenced by 24.2). Eth. 'And after these fragrances' again contains a misrendering of ἐπέκεινα (see above on 18.9). The original translated by Eth. may have had simply ἐπέκεινα τούτων as at 30.1, interpreted by Eth. as 'after these fragrances'. Charles[1906], 71, n. 17, explained the omission in G as due to hmt, but this presupposes a Greek text which read ἐπέκεινα τούτων τῶν ἀρωμάτων at 32.1 (31.3 ends with ἀρωμάτων). (Cf. Knibb, 2, 120f.) Eth. is paraphrasing an already corrupt text: 'And after these fragrances, to the north, as I looked over the mountains, I saw seven mountains ...' (Knibb).

choice nard ... pepper En[e] 1 xxvi 18 נרד טב the same formation as at 30.2 קניא טבא. The rare word צפר has as its equivalents, in G σχοῖνον 'camel-hay' and Eth. 'odoriferous trees', the first a guess, the second a vague general equivalent. Aram. צפר has been identified by J. G. Février with the ingredient of the sacred incense צפרן, ὄνυξ, ungues odoratus ('Le Vocabulaire sacrificiel Punique' in JA, Tome 243, 51f.), the onycha or 'aromatic shells' of Exod. 30.34 Heb. שחלת. Cf. Milik 'Hénoch au Pays des Aromates', 74 (6) and Löw, *Flora*, IV 99f. Aram. קרדמן is attested here for the first time = καρδάμωμον, the spice from seed-capsules of Indian plants (not to be confused with κάρδαμον 'cress'), Löw, *Flora*, III 499-500. For 'pepper' פלפלין, see Löw, *Flora*, III 49-62.

(2) **eastwards of all these mountains** The reading למדנח of En[e] 1 xxvi 19 seems certain. It is difficult to find a common ground between G (Eth.) and the firm reading of the Aram.: an explanation from the LXX equivalent of Heb. קדם can hardly account for the connection of ἀρχή with מדנח (Milik, 234, L. 19). We may explain ἀρχάς as an inner Greek corruption of ἀνατολάς although it seems transcriptionally unlikely. Has Eth. preserved a genuine variant [ἐπάνω κορυφῆς πάντων τῶν ὀρέων τούτων], עלא מן רישי טוריא אלן, 'over the summits of all those mountains', where G has given ריש its alternative meaning ἀρχή? Aram. למדנח, πρὸς ἀνατολάς occurs twice in the verse which suggests a scribal error in the En[e] recension. Eth. presupposes a phrase with ἐπάνω (malʿelta), like the two following ἐπάνω phrases in the verse. Cf. Knibb, 2, 122, arguing for a direct rendering of ריש by Eth.; Cf.also Barr, 'Aramaic-Greek Notes', II, 182f.

the Erythraean sea i.e. the Persian Gulf and the Indian ocean; cf. 77.6,7: ימא שמוקה lit. 'red sea,' occurs again at IQapGn 21.17, Fitzmyer, *Genesis Apocryphon*, 153f. See also Abel, *Geog. Pal.*, I 251.

the (great) darkness See note on 17.6. Milik, 234, L. 21, seeks to explain the extraordinary variant of G Eth. 'over the angel Zotiel' as having arisen by corruption of ζόφος (or ζοφώδης); cf. 17.2 εἰς ζοφώδην τόπον. Eth. mal'ak 'angel' could have arisen through its similarity to mal'elto = ἐπάνω αὐτῆς. Dillmann took the reading as genuine and suggested that the angel Zotiel was defending the entrance to Paradise. The name is otherwise unknown, unless we identify it with Sithwa'el, the angel of 'winter' and therefore the 'dark regions' (cf. above at 6.7). A corruption of the text seems more probable.

(3) **beside the Paradise of righteousness** 'beside' = ליד, like ܠܝܕ ܕ (P. Sm. 1549) and cf. Heb. 1 Sam. 19.3. Cf. Barr, 'Aramaic-Greek Notes', II, 179. Paradise is here sited, like the original Eden, in the east (cf. Bousset-Gressmann[4] 283), while at 24.3f., 70.3 it is located in the north-west (at 77.3 in the north). The author seems to be introducing at this point the Eden tradition of the Old Testament in spite of its discrepancy with the situation of paradise in the north(-west) which he gives elsewhere. This is recognised by P. Grelot in his important study 'La Geographie mythique d'Hénoch' in RB, lxv (1958) 63, although at fig. 1 (46) he gives the name 'Jardin de Justice' to the north-west paradise only and describes that in the east simply as Eden. Here, in our text, preserved in the original, however, this paradise in the east is explicitly described as 'the Paradise of righteousness', exactly as the paradise in the north-west is described at 77.3 in the Aram. text. Milik, 'Hénoch au Pays des Aromates', 77, thinks that the author wished 'to unify the two traditional locations', and consequently, he recognises one Paradise of righteousness only,located in the north-west, (40f.). It is impossible, however, to reconcile these two conflicting traditions. They are, no doubt, an inevitable consequence of the 'conflation' of two traditions, the oriental locating paradise in the east and the hellenistic which places Elysium in the west. For the influence of hellenistic traditions on Enoch, see Grelot, ibid. See further Str.-B. IV 1118f., ThWNT, Bd. V 763f. s.v. παράδεισος (J. Jeremias), Bousset-Gressmann[4] 282f., 408f.; Milik, 36f. for the Paradise of righteousness (with the Tree of Wisdom) in the east.

superior to ... trees Eth. kaḥaktihomu la'elketu 'eḍaw = [κρείττονα/ἐπέκεινα τῶν δένδρων τούτων] (for the adverb, Dillmann, *Lex.*, 823, SAB, 1892, 1054) 'Beyond those (former) trees', in a spatial or geographical sense, is difficult, since 'trees' properly have not been mentioned since 31.2, and since then, Enoch has crossed the 'Erythraean sea' and 'the (great) darkness'. I suggest taking 'beyond, above' in the sense of 'superior to', κρείττονα.[1] Did the original read 'superior to all trees' (להלא מן כול אילנין),? altered to 'beyond those trees' when the adverb was understood in a geographical sense?[2]

growing there The corruption in G δυο μεν for φυόμενα has given rise to the speculation that the reference is to the two trees located in paradise, the Tree of Life as well as the Tree of Wisdom (cf. Lods, 95f., Charles[1906], 72, n. 7), but, according to En. 24.2-25, the Tree of Life is located in the paradise in the north-west (cf. Milik, 39).

those who partake G ἅγιοι 'angels (partake)'. Edd. read ἁγίου and construe with καρποῦ '... from whose holy fruit they eat ...' (Charles). The word-order is against this translation. Eth. om. ἁγίοι altogether, and this is probably correct. According to 25.4 it is the saints not the angels who are eventually to partake of the Tree of Life. Perhaps the tenses are wrong and we should render 'of which the saints shall partake and understand great wisdom'?

---

[1] Is G μακρόθεν a corruption of κρείττονα?

[2] Cf. P. Sm. 1009 ܠܗܠ ܡܢ ܟܠ summus, supremus.

CHAPTER 33

(1) The positioning of frg. e of En[e] 1 xxvii is not certain: see Milik, 235, Ll. 1-3. It seems more suitably placed at 33.1 than at 32.3 (see above, 178, 345).

great beasts ... birds Cf. Dillmann, 135, suggesting that Enoch may be here drawing on a legend based on Gen. 2.19-20 that beyond Eden, at the ends of the earth, lay a land with all kinds of wild creatures, beasts and birds.

each differing from the other Cf. Dan. 7.3: '... four great beasts ... different from one another' (שנין דא מן דא).

(2) **whereon the heavens rest** Cf. 18.5.

gates of the heavens i.e. the gates through which the stars emerge and re-enter.

(3-4) (+ 34.1) The identification of these verses at En[e] 1 xxvii 19-21 (Plate XIX) is beyond reasonable doubt, if only three or four words are now legible.

(3) **according to their conjunctions(?)** Eth. darg = conjunctio (Dillmann, *Lex.*, 1097): Knibb 'constellations'. Could darg, ζεῦγμα(?) render מסרות 'conjunctions'(?) See above on 2.1. Eth. menbār = τόπος(?).

(4) **ordinances ... companies** Eth. te'zāz = ἐντολή Aram. פקודא cf. Heb. פקודה. The semitic word can also mean 'command' in the sense of 'sphere commanded, charge, office'; cf. 2 Chr. 23.18; at 1 Chr. 23.11 it denotes a class of officers. Eth I māḫbar = 'company, band', συναγωγή, Heb. קהל Aram. משריתא? Eth II 'their works', 'functions' (Knibb). All the terms here suggested as equivalents of Eth. I have a military connotation which would be appropriate to describe the 'hosts' of heaven.

CHAPTER 34

(1) The Aram. text of En[e] 1 xxvii omits the first sentence of this verse, 'And from thence I went northwards to the ends of the earth', beginning 'And I saw ...'. Is Eth. an elaboration, with an introduction modelled on 33.1?

great and glorious works (of creation) Eth. 'great and glorious wonders' (manker = θαυμάσια LXX Ps. 26.7 נפלאות). I have adopted as the text the noun from En[e] 1 xxvii 21 עבדין 'works (of creation)' — as at 2.1 — but retained the second adjective 'glorious' from Eth., although Eth. 'great and glorious wonders' could be a paraphrase of 'great works (of creation)'. A synonymous expression occurs at 36.4 'great and glorious wonders (ta'āmerāt)'. Cf. Milik, 236, L. 21 '... עובדין 'wonders' just as מעשים in the Hebrew Bible'.

(2-3) For this section 34.2-36.2 on the winds, cf. 76.1-14, Enastr[c] 1 ii 1-10, 13-20, Milik, 284-288. Milik thinks that 34.2-36.2 is a summarised version of the longer section at 76 (203, L. 23): it is also possible that 76.1-14 in the astronomical book is an elaboration of this short section.

(3) **from one gate ... for blessing** Cf. 76.10 where it is the middle gate from which the beneficent winds blow; from the first and third gates come only harmful winds.

CHAPTER 35

(1) **gates opened ... outlets** The number of gates in the west corresponds to the

number of 'outlets', i.e. 'outgoing (gates)', lit. 'outgoings' (muḍā'at, מוצאים, נפקין, ἔξοδοι) in the east. The underlying idea is of the winds or the heavenly bodies (sun, moon and stars) 'coming out' when they rise, and 'going in' when they set or complete their course. The specifying of 'three gates' in Eth. is a mistaken assimilation by the translator of the 'gates' for the winds in the west to those in the north and south at 36.1f. When Enoch 'saw gates of heaven opened, just as I had seen in the east', the reference is to 33.3: these western gates for the winds are like the eastern gates from which the stars emerge (33.3) in so far as the number of their 'ingoings' or 'settings' are the same as their 'outgoings' or 'risings'. The actual number of these gates is not given at 33.3 or in this verse, but this is specified at 36.2 where, in addition to the three gates 'for the winds in the east', there are 'three smaller gates' above them for the stars (presumably because they were considered smaller than the winds). Similarly, but as an independent scheme, there are six gates in the east and six in the west for the sun and moon (72.6). Cf. Milik, 203, Ll. 1-2, who has confused the two schemes, the one for the winds and stars, the other for the sun and moon of 72.6. See Appendix A, The 'Astronomical' Chapters of the Ethiopic Book of Enoch, 386f. How are we to understand חשבוניהון 'their number' at En[c] 1 xiii 24? According to 33.3 the stars appear each to have a number as well as a name. Could this also apply to the winds or does 'their number' refer to the number of the gates?

CHAPTER 36

Cf. 76.7-9 which deals in detail with the various winds that come from the south. See Appendix A The 'Astronomical' Chapters of the Ethiopic Enoch 403f., En[c] 1 xiii 25 has shorter text: '[I was shown three gates open in heaven] for the south wind, for dew and rain ...

(1) **a (hot) wind(?)** Eth. simply wanafās 'and wind' which can hardly be right. If this verse is a summary of the fuller description of the winds at 76.1f. the writer here seems simply to be selecting some of the phenomena associated with the winds from the south: these include 'dew' ('night-mist' or 'mist'?) and rain (76.8-9) and perhaps the nafāsa moq at 76.7. Is this the רוח (ה)קדם (LXX ἄνεμος ὁ καύσων) the 'hot wind' or sirocco (ܪܘܚܐ ܩܕܘܡܐ)? e.g. Hos. 13.15. At 76.7 it is said to 'incline to the east'. Unfortunately the equivalent noun at En[c] 1 xiii 26 is missing. Alternatively, 76.7-9 could be spelling out in fuller detail 36.1.

(3) **which have been shown them** Cf. Milik, 203, 204, L. 27 '(on the paths) which are fitting for them'?

THE PARABLES OF ENOCH

INTRODUCTION

1. The Composition, Character, and Date of the Parables, with an additional note on The Son of Man in the Parables

Since R. H. Charles's monumental works on the apocrypha and pseudepigrapha of the Old Testament, it has been almost universally assumed by scholarly opinion that the 'Book of the Parables of Enoch' at 1 En. 37-71, a section of Enoch described by the author himself as 'The Second Vision ... which Enoch saw' (37.1),[1] was a pre-Christian Jewish apocalypse. And since its 'Son of Man' visions, though presumed to be dependent on Daniel 7, went beyond the Danielic symbolism of 'one like a son of man' for 'the saints of the Most High' by exhibiting a Jewish Son of Man messianism — a heavenly Messiah appearing as God's vice-gerent at the Last Judgement — the Book of the Parables has come to be regarded as supplying the foundation in pre-Christian Judaism for the Son of Man Christology of the New Testament.

This virtually doctrinaire view has received a rude shock in recent years. The discovery and evaluation of Aramaic fragments of 1 Enoch from Qumran Cave 4, some of them substantial, and the total absence of any from the Book of the Parables, has led J. T. Milik,[2] who has been solely responsible for their identification and editing, to a drastic reassessment of the Parables, both with regard to their date and their character and composition (including, of course, their 'messianic' or 'Christological' significance).

As was noted above (9f.), Milik envisages the original Aramaic Enoch as a Pentateuch in which the third Book of Enoch consisted originally, not of the present Book of the Parables, but of a putative Book of the Giants. When this was lost or more probably banned, it was replaced by the Book of the Parables, a Christian composition probably of the third century A.D.

The modern title 'Parables' or 'Similitudes' comes from 1 En. 37.5, but a more accurate designation of this later addition to the Enoch corpus is supplied by the author himself who describes his work as 'the Second Vision (of Enoch)' (37.1). This title contrasts it with the First Vision, even possibly 'with the whole collection of revelations contained in the Aramaic and Greek Enochic Pentateuch' (Milik, 89). In Milik's opinion this First Vision was known to the author of the Parables in its primary Greek Version, substantial portions of which have survived in the Cairo and Chester Beatty Papyri and in the citations of George Syncellus. The longest surviving translation of this primary Greek Enoch — but with the Parables replacing the Book of Giants — is the Ethiopic version.

According to Milik this 'Second Vision of Enoch' was originally composed in Greek, largely based on and drawing its inspiration and motifs from the primary Greek version of the Aramaic Pentateuch. The work, in Milik's view, was not a translation of an Aramaic original (not a single fragment of chs. 37-71 has been found in the rich Enoch material from Cave 4); its affinities are with the Sibylline books: 'Someone well acquainted with the Sibylline Oracles would doubtless have no difficulty in translating the Ethiopic text of the Parables into hexameters and into a pseudo-Homeric dialect; the Greek original of this book was certainly composed in metrical

[1] See further below. Nevertheless, since ('The Book of) the Parables' has established itself in scholarly nomenclature, I find it still convenient to use the term.

[2] Milik, 89-107. Cf. also 'Problèmes de la Littérature Hénochique à la Lumière des Fragments Araméens de Qumran', in HThR Vol. 64, 1971, 333-378. While agreeing in the main with Milik elsewhere, I had always entertained reservations about his views on the Parables, in particular with regard to their messianism, and I gave expression to some of these in an article on 'The 'Parables of Enoch' (1 En. 37-71) and the 'Son of Man' in ET LXXVIII (1976-7), 5-8. At the same time I was prepared in that article to accept the view that the Parables were an original Greek composition but drawing on material from the older book. More recent studies have led me to a further revision of these earlier views.

poetry' (Milik, 92). The book has survived in an Ethiopic version only, all traces of its Greek original having disappeared. Nevertheless, though not a translation and of later date, it is not without value, since the copies of the primary Greek version of Enoch which the author utilized appear to have been, is osme parts (e.g. in the Book of the Watchers), superior to the Cairo and Syncellus fragments and the Greek originals of the Ethiopic (Milik, 91).

Milik's Reconstruction of the Enoch Corpus and Versional Tradition

Original Enoch Pentateuch:
Astronomica, Book of the Watchers, Book of the Giants, Book of Dreams, Epistle of Enoch.

Primary Greek Version
('First Vision of Enoch') (substantial portions surviving in 1 En. 1-32, 89.42-49, 97-107).

Ethiopic Version
1 En. 1-108,

(including Book of the Parables or 'Second Vision of Enoch' (37-71), translated into Ethiopic from a lost Greek original, but the latter based on the Primary Greek Version).

The problem of dating the 'Parables' is a difficult and complicated one. Milik began by attributing its authorship to a Jew or a Jewish Christian of the first or second century A.D.[1] Since the completion of the edition of the 4Q Enoch fragments, however, this date has been moved down to A.D. 250. It is argued that 1 En. 56.5-7 refers to the years of anarchy in the Byzantine empire when the Sassanid Sapor I invaded the West and captured the Emperor Valerian in A.D. 260 (Milik, 95). Along similar lines J. C. Hindley has argued for a date c. A.D. 113 claiming that the references to the Parthians at 56.5-7, usually taken to relate to the Parthian threat to Jerusalem under the Herods in B.C. 40, refer to the confrontation with Trajan, c. 113-117.[2] Moreover, the work draws much of its inspiration from the Gospels, especially in the 'Son of man' and 'Elect One' visions (for the latter, cf. Lk. 23.35), as well as from the 'First Vision of Enoch' (Milik, 91f.). At the same time, it can hardly have been an early Christian work since it is never quoted or referred to until it is attested by Nicephorus of Constantinople (fl. c. 810) and used in the Slavonic Enoch (ninth century?) (Milik, 77 and 92).

This is an impressive array of arguments, the result of which could lead to the rejection of the Parables, in particular their Son of Man visions, as late secondary traditions, inspired by the Gospels rather than the basis of their Son of Man christology. The negative arguments, the silence of Qumran and of versional and patristic tradition, may seem to some decisive for the late origin and composition of the book.[3] On the other hand, the possibility, if Milik is right, that the Second Vision

---

[1] *Ten Years of Discovery in the Wilderness of Judaea* (London: SCM, 1959, Studies in Biblical Theology, 26), 33.

[2] 'Towards a Date for the Similitudes of Enoch', in NTS 14 (1967-68), 551-65.

[3] But cf. A Dupont-Sommer, *The Essene Writings from Qumran*, (Oxford, 1961), 299-300, and M. D. Hooker, *The Son of Man in Mark*, (London: SPCK, 1967), 48. See now J. Coppens, *Le Fils d'Homme*, 147f.

was based partly on good Greek copies of the primary Greek version of Enoch would be of considerable importance for any assessment of the value of the Parables. The Parables are not a completely alien body within the Enoch tradition. Old material from the First Vision could have survived side by side with later traditions. The fact that the Pseudo-Jonathan Targum mentions the names of the wives of Muhammad does not prove that there are no ancient elements in it.

I shall return to this problem of date later, but first we must deal with the question of the composition and character of the Second Vision.

Few have doubted that the basis of the Ethiopic version was a Greek *Vorlage*,[1] or that it was drawing at times heavily—as in the duplicate list of angels at Ch. 69 from Ch. 6—on the First Vision. But Milik's claim that it is not only based on better Greek mss of the older book, but was probably composed like the Sibyllines, in Greek hexameters in a pseudo-Homeric dialect is difficult to substantiate: in fact it is into Hebrew or Aramaic that the Biblical language and allusions, which are found in every other verse, not to mention the peculiar syntax and locutions of the Ethiopic version, most naturally fall.[2]

I do not think Milik would have arrived at this somewhat bizarre conclusion if he had paid more attention to this Biblical element in the Parables, and, more especially, to the source-analysis of the Book and its evidence of translation-Ethiopic i.e. vocabulary which betrays a non-Ge'ez idiom, and is evidence of translation or of mistranslation, the most convincing proof that a document is being translated from another language. Milik assumes, of course, that the Ethiopic renders a Greek *Vorlage*, but he ignores entirely the possibility of a Hebrew/Aramaic *Grundschrift*.

That the author or editor/compiler of the Parables is making use of source material or traditions was the view of all the older students of the Book (Dillmann, Martin, Charles) and is still generally accepted by most scholars, even if Charles's dissection of the sources and their analysis is more confidently precise than can now be accepted. There is a 'Son of Man' source-tradition as well as an 'Elect One' source-tradition combined in the composite messianism of the Parables; there are also Noah apocalypses and a Michael discourse. The traditions, however, are not a haphazard compilation. As Milik has shown,[3] the First Parable imitates fairly closely a series of passages in the Book of the Watchers; the four archangels of Ch. 40 correspond to the four archangels of Chh. 9-10 (the angelology has simply been up-dated to that common in the time of the author). Ch. 54.7-55.2—I would add—describing the Flood, is not simply 'an interpolation' from a Noah apocalypse that does not properly belong in it context: the Flood is introduced at this point immediately after the account of the temporary incarceration of the watchers (on which Ch. 54 is based) simply because the First Vision does the same (10.32f.). The incorporation of such Noah material into an Enoch 'parable' is not always successful, as an occasional illogicality—an artificial 'joining' or 'stitching'—shows up, as at 60.1 and 8, a Noah apocalypse which the writer attributes to Enoch (60.1),[4] but which the same verse, and later verse 8, proves came from a Noah not an Enoch apocalypse.[5] In a word, the Book is not a loose *mixtum compositum* of disparate source traditions, but a work deliberately

---

[1] This has been recently questioned, not only for the Second Vision. See above, 4.

[2] Cf. L. Goldschmidt, *Das Buch Henoch* (Berlin, 1892).

[3] op. cit., 90.

[4] There are no variant readings of 'Noah' for 'Enoch' in this verse.

[5] The speaker at v. 8, who is delivering the vision, mentions the place where 'his great-grandfather', i.e. Enoch, 'was taken up'.

modelled on the First Vision and called by its author/redactor 'The Second Vision (of Enoch)'.

In spite of its character, in places, as a collection of curious oddments, astronomical observations alongside apocalyptic fantasies, the fact that this Second Vision has been composed with the 'First Vision' as a rough framework or model, provides the Book of the Parables with a certain degree of literary unity. Moreover, its apocalyptic eschatology is dominated by the theme of the Last Judgement and of the Elect One or Son of Man at that Judgement; and this gives the book a thematic unity as the Second Vision of Enoch. Though this latter theme may be drawing on different sources and traditions, the final editor has brought these together *as a single theme*, and these two titles refer, for him, to the same central figure.

The view is generally held that the Ethiopic Enoch, including the Second Vision, is a tertiary version i.e. its immediate *Vorlage* was a Greek version, but the *Grundschrift* was a semitic, Hebrew or Aramaic, original. The debate in the Parables has been whether the latter was composed in Hebrew or Aramaic. Charles, following Halévy, opted for Hebrew, Schmidt for Aramaic. Schmidt has been recently followed by Ullendorff and Knibb, with the important difference that the latter argue for a direct rendering of the Ethiopic from an Aramaic *Vorlage* (although they qualify this by adding that the translators could have had a Greek version also available).[1]

I propose to deal with this problem on the basis of the most recent discussion in the Knibb edition (2.41-3).

Ch. 45.1-2 begins the 'second parable' with dire predictions of the fate of sinners who deny the Lord of Spirits. They are to be kept till the Last Judgement, 'the day of affliction and distress'. Verse 3:

'On that day my Elect One shall sit on the Throne of Glory,
And he will choose their works' (yaḫarri megbārihomu)

does not make sense, since the reference (as Sjöberg,[2] Caquot and Geoltrain[3] rightly noted) is to the sinners of vs. 1-2. Charles and Knibb conjecture an Aramaic בחר = 'choose' or 'test, try' behind the Ethiopic verb (Hebrew had also this double meaning, e.g. Isa. 48.10). Caquot and Geoltrain posit a Greek Vorlage ἐκλέξει, as in the sense of LXX Ezek. 20.38. This seems to me to hold the right solution, but if ἐκλέξει is given the same meaning as it has at Ezek. 20.38 'séparer, extraire', then it is translation-Greek; in Ezekiel it renders ברר 'to purge' as well as 'to choose'; we would then require to render 'he will purge' their works. This last conjecture clearly implies a Hebrew *Grundschrift*.

Ch. 52.9 reads:

'And all these things will be *denied* and destroyed from the face of the earth,
when the Elect One appears before the Lord of Spirits'.

'All these things' refers to the seven metal mountains which, according to v. 6, are to be 'like wax before the fire', i.e. to melt away and disappear. To say that 'they will be denied' is, as Knibb comments, hardly appropriate in the context; following a suggestion of Dillmann, he thinks it likely that keḥda is used here with the same

---

[1] Cf. E. Ullendorff, "An Aramaic Vorlage' of the Ethiopic Text of Enoch', *Atti del Convegno Internazionale di Studi Etiopici, Academica Nazionale dei Lincei*, (Rome, 1960), 262.

[2] E. Sjöberg, *Gott und die Sünder im palästinischen Judenthum* (BWANT, Kohlhammer, Stuttgart), iv, 27, 1939, 224, n. 3; *Der Menschensohn im äthiopischen Henochbuch* (Lund: Gleerup, 1946), 75.

[3] A. Caquot et P. Geoltrain, 'Notes sur le Texte Éthiopien des 'Paraboles' d'Hénoch', in *Semitica* 13 (1963), 43f.

meaning as Hebrew or Aramaic כחד 'to wipe out'; later he opts for an original Aramaic יתכחדון; Charles prefers Hebrew יכחד as the original, suggesting at the same time that 'will be denied, will be wiped out and will be destroyed' are duplicate renderings of it. An alternative solution makes these otherwise attractive conjectures less convincing: ἀπαρνηθήσονται would be the Greek equivalent of the Ethiopic verb and this in Hebrew or Aramaic corresponds to the root מאס: 'to deny', but also 'to reject'. Did the original read יתמאסון/ימאסו 'they will be rejected', a verb which is not only attested for the rejection of metals (Jer. 6:30) but is sometimes also confused with the root מסס, Niph. 'to be melted away' (1 Sam. 15.9: cf. Aram. מאס, מסי, מסס)? There could be an intentional pun.

The Noah Apocalypse (Ch. 65) contains a nest of problems. Dillmann was the first to point out that ḫebr in the sense of 'spell', fascinatio, is not found elsewhere in Ethiopic. The word occurs in Aramaic at 1 En. 8.2 (חבר(ה) (Tg. חברא) where it is rendered by ἐπαοιδάς. If this word stood in a Greek Vorlage of v. 6 and was rendered by ḫebr we must assume that the word could have this meaning in Geʿez. Knibb (2.41f.) suggests direct dependence of ḫebr on the Aram.; but the semitic root חבר is also common in Eth; it is easier to assume that ḫebr could mean 'spell-binding'.

The verb yebadder at v. 8 makes little proper sense: 'For lead and tin are not produced from the earth like the previous (metal); there is a spring which produces them and an angel who stands by it, and this angel yebadder'. The root meaning of badara is 'to be light, swift' (= Heb. קלל), and so Flemming-Radermacher render 'and this angel is nimble (behend)'. Charles follows Dillmann in giving the word is secondary meaning of praecellere, 'and that angel is preeminent'. Caquot and Geoltrain connect the root with Syr. ܒܕܪ 'distribute', 'and this angel distributes (the lead and tin)', and this connection is accepted by Knibb but positing an original Aram. בדר. In an earlier discussion I ventured the conjecture that קלל (from קלקל 'to be swift') has the special sense 'to whet, sharpen' of metals (Ec. 10.10) and the adjective קלל = 'burnished' 'and this angel was polishing them'.[1] Beer (*Apocrypha und Pseudepigrapha des alten Testaments*, 273, n. d) suspected a confusion of badara with barada, 'abrada = 'to cool'. Eth. badara = praevertere, 'to prefer', 'to go before, precede' renders בחר at Isa. 56.4 'The eunuchs who choose (בחרו = 'abdara) to do my will' (NEB); cf. Prov. 16.16 '... to gain discernment is better (נבחר) than pure silver' (NEB). Was this the original Heb. verb, but with the sense 'to test, try' = בחן 'to refine, purify'?

Verse 10 contains two 'non-sense readings':

> '... because of their iniquities their judgement (i.e. the judgement of the watchers) will be carried out and *will not be numbered* (or *they will not be numbered*) ...' ('iyetḫuēlaqu).

Charles and Knibb presuppose an original מנע (Heb. or Aram.) '... shall not be withheld by Me for ever' (Charles). This seems clearly the correct solution. The second 'non-sense reading' is:

> 'Because of *the months* which they have invented and learned, the earth will be destroyed and those who dwell on it'.

Halévy and Charles suspect an original Heb. חרשים 'sorceries', misread as חדשים (μῆνες, which Caquot and Goltrain still feel is the original and correct reading). In view of their theory of an original Aramaic directly rendered into Geʿez, Knibb and

---

[1] See 'The Composition, Character and Date of the 'Second Vision of Enoch'' in *Text-Wort-Glaube*, Kurt Aland gewidmet herausgegeben von Martin Brecht, (Berlin, 1980), 26.

Ullendorff argue for an Aram. חדשיא 'months', a word found in Babylonian Aramaic. West Aramaic for 'months', however, is ירחיא. The mistranslation points again to a Hebrew *Grundschrift*.

At 59.1-2, a short 'nature apocalypse', there is conclusive evidence for a Heb. *Grundschrift*. The forces of nature there described, thunder, lightning, are said to be 'for a blessing or a curse' לברכה ולשבועה a phrase paralleled at v. 3 by '(... they will flash) for blessing and for plenty' (לברכה/לשבעה). Clearly, the original Heb. לשב(ו)עה gave the translator of the *Grundschrift* some difficulty, for the same consonants could yield two different meanings, either 'for curse' or 'for plenty'. As with Qumran *pesher* he gives both meanings to ensure that the original intention of the author—whatever it was, 'cursing' or 'plenty'—should not by any chance be missed.[1]

It is difficult to reach a firm conclusion on the basis of such evidence, since most of it—some would say all of it—is ambivalent. Moreover, in view of the variety of traditions in the Parables, there could have been both Heb. and Aram. capitula, a mode of composition familiar in Daniel and Ezra. But there are certainly Hebrew sources behind the variegated source-traditions in the Parables; and the cumulative evidence of Hebraisms and 'Biblicisms' throughout the Book (see Commentary, passim) points to a Hebrew *Grundschrift*.

There remains the problem of date, in particular that of the original *Grundschrift* which I have suggested was, at least in some parts of the Parables, composed in Hebrew.

It can by no means be assumed that the theories of Hindley and Milik represent the last word on the problem. They both raise difficulties and are vulnerable at several points. What is quite certain at 1 En. 56 is the reference to a Parthian invasion or threat of one, menacing Jerusalem, and this points unequivocally to the Roman period, and if my reading of the evidence is correct, to an *earlier* rather than a *later* date in that period. We cannot, in fact, rule out a pre-70 dating for at least some of the oldest traditions in the Book. Thus there does not seem to have been any direct threat to Jerusalem in Trajan's punitive expeditions into Armenia and Persia. The references to the Parthians and Jerusalem are best explained by Sjöberg[2] as referring to the Parthian threat of 40 B.C. when Jerusalem was under the Herods. Ch. 56.7: 'But the city of my righteous shall be a hindrance to their horses', Sjöberg maintains, implies a date prior to the destruction of Jerusalem. To explain 'hindrance' as 'scandalum' and to understand it metaphorically is unconvincing; *Jerusalem was still a defended and defensible city when these words were written.* There are even more formidable difficulties with Milik's theories: to take the references to persecutions at 47.1-4, 62.11 as a 'clear allusion to the first great persecutions of Christians decreed by the emperors Decius, in A.D. 249-259 and Valerian in 257 and 258' (96), is an extraordinary judgement in view of the ambivalence of the evidence, overlooking the many persecutions of Jews as well as Christians, in the first two centuries of the Christian era.

In view of the undoubted fact that the Book as it has reached us is a *mixtum compositum* of sources and traditions, it is only possible to state one confident conclusion, viz. that *there is ancient Hebrew apocalyptic tradition in the Grundschrift used by the final redactor*. On the whole, it seems to me more likely that the *Grundschrift* was itself the *mixtum compositum, and that it was this which was translated*

---

[1] Note the word-play יברקו לברכה paralleled elsewhere in the Parables, e.g. 41.4: '... the closed chambers (in the heavens), אוצרות עצורות.

[2] Sjöberg, *Der Menschensohn im äthiopischen Henochbuch*, 38.

as it stood into Greek. The occurrence throughout of the title 'Lord of spirits' (in Hebrew אדון לרוחות?), a variation of 'Lord of hosts',[1] supports this hypothesis.

In brief, Halévy and Charles were right in proposing a Hebrew *Urschrift* for the Book of the Parables, which I would date to the early Roman period, probably pre-70 A.D.[2]

Additional Note: The Son of Man in the Parables

Milik has argued that the Son of Man visions in the Parables owe their origin and inspiration to the Gospel Son of Man.

It must be noted, in reply, that there is nothing specifically Christian in these chapters; the terms 'son of man' and 'elect one' are well attested in Jewish sources, if not as messianic titles, nevertheless of symbolic or historic figures, the substantive basis for messianism. It is truly remarkable, if the Parables are a Christian composition, that there should be no reference anywhere to the Founder of Christianity. On the contrary, the Son of Man who is to come as the Judge of all mankind is identified, not with Jesus of Nazareth, *but with Enoch himself*.

Some years ago I argued that the final chapter of Enoch, where Enoch is himself designated Son of Man, represented an older and pre-Christian stratum of the Enoch tradition.[3] As H. L. Jansen has pointed out, it was a traditional prophetic *Berufungsszene*, or prophetic 'call' scene preceded by a theophany.[4] I have revived this argument in a recent contribution to the W. D. Davies *Festschrift*.[5] There I have tried to trace the tradition-history of such theophanies from the 9th century, the vision of Micaiah in 1 Kg. 22.19f., of Isaiah (ch. 6) down through Ezekiel 1, Daniel 7 and Enoch 14 and 71. The pattern throughout is strikingly similar, a theophany consisting of a vision of God on his heavenly throne, followed by the commission of the prophet. Enoch 14 is a classic example where Enoch is transported to the Presence of God, then given his commission to condemn the watchers.

This final denouement of the Second Vision at 71 comes as the climax of another Throne-Vision apocalypse which again, like Daniel 7, shows close literary dependence on 1 En. 14-15. It seems undeniable, as many have noted, that there, as in the earlier vision at Ch. 46, the author of the Parables is drawing on Dan. 7.9-13, but no less certain is the fact that the author *is modelling this climactic vision on the Throne-vision of Chh. 14-15*.

'And he translated my spirit into the heaven of heavens, and I saw there as it were a structure built of crystals, and between these crystals tongues of living fire ...'. (71.5

---

[1] See below, 189f.

[2] Since the above was completed, J. A. Fitzmyer's review of the Milik volume has appeared: 'Implications of the New Enoch Literature from Qumran', in TS, Vol. 38, 2 (June 1977), 332-345. I find myself substantially in agreement with his dating and assessment of the Parables (341-345).

[3] See M. Black, 'The Eschatology of the Similitudes of Enoch' above 18, n. 39.

[4] *Die Henochgestalt: Eine vergleichende religionsgeschichtliche Untersuchung* (Oslo, 1939), 114f.

[5] 'The Throne-Theophany Prophetic Commission and the 'Son of Man'', in *Jews, Greeks and Christians: Religious Cultures in Late Antiquity*, ed. by R. Hamerton-Kelly and Robin Scroggs (Leiden: Brill, 1976); see also M. Black, 'Die Apotheose Israel: eine neue Interpretation des danielischen Menschensohns', in *Jesus und der Menschensohn*, Für Anton Vögtle, herausgegeben von Rudolf Pesch und Rudolf Schnackenburg, Freiburg, Basel, Wien (1975), 95f.

trans. Charles) The source is 1 En. 14.18-22; as Charles comments: 'This passage (18-22) is used by the author of 71.5-8'.[1]

What we have is, in fact, another developed Throne-vision apocalypse, but in this case the commissioning of the Prophet Enoch is his designation as 'the Man who is born for righteousness':

> And the angel (Michael) came to me and greeted me ...
> and said:
> You are the Son of Man who is born for righteousness,
> And righteousness abides upon you ...
> And all shall walk in your ways ...

This vision is certainly post-Daniel, and could ante-date our Gospels. It is a purely Jewish apocalypse with an image of a Son of Man figure which has survived, as Hugo Odeberg pointed out, in the Enoch-Metatron speculations where Enoch is translated to a throne next to the throne of God himself (one derivation of *metatron* is *metathronios*, the one seated on the throne after God), and where he can also be described as 'the lesser Jahweh' (יהוה הקטון).[2]

This is undoubtedly the most original feature of the Son of Man figure in the Parables. It is combined, however, with other distinctive features which owe their inspiration to three more obviously Biblical ideals, that of 'the elect one' of Second Isaiah, of the Davidide of Isaiah 11, and of the 'one like a son of man' of Dan. 7.

The two titles 'Elect One' and 'Son of Man' are deliberately brought together by the final author/editor of the Parables.[3] The first is the more common of the two, but both belong to the thought of the author, from whose hand the Book finally came, about the central figure, next to 'the Lord of Spirits', of his apocalypse. This union of the two designations (or 'sources') could, arguably, be the work of a Greek 'translator-editor' of Hebrew sources, but it seems to me more probably the creation of the Hebrew author of the Hebrew original; the use of the Hebrew nomen dei 'Lord of spirits (hosts)' (above, 188) in all parts of the work is a strong indication of the existence of an original Hebrew Book of the Parables.

The term 'the Elect One' points as unequivocally to the elect Servant of Second Isaiah, as does the term Son of Man to Dan. 7.[4] Moreover, Chh. 49.3 and 62.2 apply the prophecy of Isa. 11.1f., about the 'root of Jesse', the Davidide, to the Elect One; and 48.10, 52.4 uses the term 'anointed' of him. The 'messianism' of the Parables thus unites these three strands of Biblical tradition about its central Figure; he is the Elect One, the Isaianic Servant of the Lord, the anointed royal 'son of David', and the Danielic 'son of man'. It must have been a bold mind, perhaps one influenced by hellenistic ideas, which elevated the immortalised patriarch Enoch to virtually angelic status, and invested him with the powers of the manlike one of Daniel's vision, the role of the Isaianic Servant and the destiny of the Davidic King.

2. *Two Unusual Nomina Dei in the Parables*

No single term for deity in the intertestamental literature is as well-known as the

[1] See M. Black, The 'Parables of Enoch' (1 En. 37-71) and the 'Son of Man', (above 182 note 2) 7f.

[2] *3 Enoch or the Hebrew Book of Enoch*, (Cambridge, 1928), Introduction, 63f. That the identification of Enoch with the Son of Man in fact comes from such late Jewish speculations is now argued by M. Jaz in *La Revue Reformée*, 1979, t. xxx, 105f. Cf. M. Black in ET LX.1 (1948), 13.

[3] See above, 185, and J. Coppens, *Le Fils d'Homme*, 128f.

[4] See below, 197, and cf. Coppens, op. cit., 134f.

Ethiopic title 'Lord of spirits'. At the same time, there can be no other single title for deity in Jewish tradition whose origin and significance has received so little attention from Biblical scholars. It is an absolutely unique term in the literature of Judaism, found only in this part of Enoch, where it occurs no fewer than 104 times in 34 chapter. Moreover, it is found in all parts of the Book, including those sections generally regarded as 'interpolations'. It was clearly the preferred and favourite divine name of the final redactor of the Second Vision.[1]

A Biblical parallel adduced as the possible origin of the expression is Num. 16.22, 27.16; in the second passage Moses appeals to 'the Lord, the God of the spirits of all mankind' (NEB) (יהוה אלהי הרוחות לכל בשר) to appoint his successor. This particular epithet of deity is peculiar to these two passages in the Old Testament; it is usually attributed to the advanced ideas of the priestly redactor: while Jahweh was indeed the God of Israel, he was also the God of all mankind on whom all 'flesh', i.e. all human beings, depend (cf. Ps. 104.29). The Hebrew expression here could certainly, formally at least—if there is no better explanation—be a Biblical source of the Ethiopic title.[2] The Targums reproduce the Hebrew literally, with the exception of Neofiti which paraphrases 'the God who rules over (שליט) the souls of all flesh ...'.

The closest parallel hitherto cited occurs at 2 Mac. 3.24, using an expression similar to that of Neofiti. The Seleucid tyrant Antiochus Epiphanes had dispatched his grand-vizier Heliodorus to Jerusalem to ransack the Temple treasury: 'But at the very moment when he arrived with his bodyguard at the treasury, the Ruler of spirits and of all the powers (ὁ τῶν πνευμάτων καὶ πάσης ἐξουσίας δυνάστης) produced a mighty apparition so that all ... were ... stricken with panic at the power of God ...' (NEB). There then appeared an angelic warrior, clad in golden armour and of terrible aspect, accompanied by two young men 'of surpassing strength', who struck down Heliodorus on the spot. (The blows were not fatal for he appears later offering sacrifice and making lavish vows to the Lord.)

The NEB appears to take ἐξουσίας as an abstract for concrete, giving the word the sense of the 'principalities and powers' of the New Testament. This may be arguable, but there is no doubt that we are here in the realm of supernatural powers or celestial spirits of which the mighty apparition was a visible demonstration. The divine title is here designedly selected to match the context of the story: it is the Ruler of celestial spirits and angelic agencies who was responsible for this miraculous deliverance.

The expression 'Father of spirits' at Heb. 12.9 is also quoted in this connection. The writer of Hebrews is using the analogy of respect for earthly fathers as a reason for submitting 'even more readily to our spiritual Father' (NEB) (τῷ πατρὶ τῶν πνευμάτων). The context justifies the rendering of the NEB by 'our spiritual Father'. There do not seem to be any grounds for understanding the expression here as meaning 'Father of celestial spirits',[3] as at 2 Mac. 3.24. The meaning of 'spirits' in this context is much the same as at Num. 16.22, 27.16, i.e. all men as 'spiritual beings'. It seems remotely unlikely that the expression in Hebrew has anything to do with the origin of the 'Lord of spirits'.

In the Qumran scrolls רוח, 'spirit', is used of 'disembodied spirits', 'angels', and at IQH 10.8 God is said to be 'Lord of every spirit' (אדון לכל רוח). In discussing this usage, with particular reference to the angelology of the War Scroll, Professor Yigael

---

[1] Other designations of God found only in the Parables present no serious problem, e.g. 'Lord of the mighty', 'Lord of wisdom', 'Lord of the exalted/rich' all in 63.2.

[2] See F.-M. Abel, *Les Livres de Maccabées*, (Paris, 1949), 324.

[3] Abel, l. c., esprits célestes.

Yadin adds: 'A title of God, frequently found in the Apocrypha and Pseudepigrapha, is 'Lord of Spirits'.[1] As we have noted, the title is, in fact, confined to the Parables of Enoch, unless we include 2 Mac. 3.24 as an example. But the implied suggestion of Yadin that 'spirits' in this title in the Parables is a term for 'disembodied spirits', 'angelic beings' points the way to the probable origin of this enigmatic title for deity. It should be pointed out, however, that the idea that 'spirits' in this title referred to angels is not new; it was made by August Dillmann more than a hundred years ago.[2]

It seems on the whole very unlikely that the Ethiopic translator of the Parables would invent for himself an entirely new name of God, especially since everyone agrees he is drawing on traditional Jewish (or Jewish-Christian) sources. Moreover, when we find again in the scrolls references to the 'hosts of his spirits' (צבא רוחיו) parallel to 'hosts of angels' (צבא מלאכים) (IQM 12.7-8) the possibility emerges that 'Lord of Spirits' is, in fact, something like an 'interpretative transformation' (in a phrase of B. H. Streeter[3]) of the traditional, and indeed probably the most popular, title in the Old Testament, 'Lord of hosts'. This hypothesis becomes increasingly attractive, and indeed receives virtual confirmation when we find the Trisagion from Isa. 6.3 quoted at 39.12 in the following form:

> 'Those who sleep not (i.e. the watchers) bless thee; they stand before thy glory, saying: Holy, holy, holy, is the Lord of spirits: he fills the earth with spirits'.

The motivation for this interpreted version of the traditional title 'Lord of hosts' is evident from the last sentence. The world of the author of the Parables, like that of the Qumran Essenes, was one full of angelic beings and disembodied spirits. This permutation on the traditional title has been dictated by the theology of the circles from which this book ultimately came. The primary purpose of the Parables is to speak to the circumstances and condition of people living, often suffering and dying, in a world whose evil human environment, Gentile oppressors, rich exploiters, the wicked and the hostile, were parts and expressions only of that larger cosmic environment dominated by celestial as well as terrestrial (and sub-terrestrial) 'spirits'. Thus at 41.8 we are told that the Lord who has divided the light from the darkness has also divided the spirits of men, establishing the spirits of the righteous. These words might have been written by the author of the doctrine of the Two Spirits at IQS 3.13f.

The interpretation was certainly one which would readily be made, in view of the cosmic use of the title 'Lord of hosts' in the Hebrew prophets, and the identification of the 'hosts of heaven' not only with the heavenly bodies, stars, planets and constellations (cf. 1 Kg. 22.19), but also with angelic beings (themselves identified with stellar phenomena) (cf. Ps. 103.21, Lk. 2.13).[4] Probably the earliest example of this translation by interpretation is the LXX's κύριος τῶν δυνάμεων 'the Lord of (celestial) powers' (e.g. 3 Kg. 17.1, 18.15, Ps. 45.12: elsewhere the LXX transcribes σαβ(β)αωθ or translates by κύριος παντοκράτωρ). The LXX is followed by the Peshitta and the Ethiopic: (e.g. ܡܪܝܐ ܚܝܠܬܢܐ Jer. 25.28, 32: ʾegziʾa ḫayyālān Ps. 79.5). The meaning is unquestionably angelic powers and agencies or, again, abstract for concrete, the angelic beings themselves. The Eth. 'Lord of spirits' ʾegziʾa manāfest could, in fact, be a translation of κύριος τῶν δυνάμεων.

---

[1] *Scroll of the War*, 231.

[2] *Lex.*, 709.

[3] *The Four Gospels*, (London, 1951), 372.

[4] For modern views of the origin and meaning of the name see O. Eissfeldt, 'Jahweh Zebaoth', *Miscellanea Berolinensia*, II (2), 1950; G. von Rad, *Theology of the Old Testament*, Vol. I, 18f.

Was this interpreted version of the traditional 'Lord of hosts' the creation of the Eth. translator of the Parables or was it already present, in some form, in his *Vorlage*? It may be felt that, since there does not appear to be any trace elsewhere of 'spirit' = רוח, πνεῦμα in this name, its appearance in the Parables is simply a singularity of the Eth. translator. What original expression he was thus interpreting will depend on the view taken of the original language of the *Vorlage* or *Grundschrift*[1]. If the original was in Hebrew and read simply the traditional title יהוה צבאות, the Geʿez version being interpretation, we should perhaps then render by 'Lord of hosts' and explain in a note the unusual and singular 'Lord of spirits'. Or did a Hebrew *Grundschrift* actually read אדון (ל)רוחות, which is not unparalleled in view of IQH 10.8 (above, 190)? A corresponding Aram. would be מרא רוחות / חילותא, and either could have been transmitted to the Eth. translator — the usual explanation of this tertiary version — by way of a Greek κύριος τῶν πνευμάτων / δυνάμεων.[2] 2 Mac. 3.24 may be held to support this last alternative, for the expression there, ὁ τῶν πνευμάτων καὶ ἐξουσίας δυνάστης, has all the appearance of being a similar interpretative paraphrase in Greek of the traditional יהוה צבאות, no doubt a familiar cultic term in Greek as in Hebrew ears. It also has precisely the same meaning as the Eth. title. It seems to me, on the whole, most likely that we have to do with a Greek interpretation of 'Lord of hosts', viz. κύριος τῶν πνευμάτων, taken over by the Eth. translator.

The second epithet occurs in the first Son of Man apocalypse, in a passage which reads like a midrash on Dan. 7.13:

'And there I saw *One who had a head of days*,
And his head was white like wool ...' (46.1f.).

It is generally agreed that the Eth. expression reʾsa mawāʿel corresponds to Daniel's 'Ancient of Days' (Dan. 7.9 עתיק יומיא), but different explanations have been offered of its origin and meaning. Dillmann took the first noun in the expression at 46.1 in its literal sense of 'head' and paraphrased 'betagtes Haupt', which is, for this reason, then described as 'white like wool'. The expression then becomes a nomen dei and appears as such at 47.3, 48.2, 55.1, 60,2, 71.10,12,13,14. Charles took the whole expression to mean 'the sum of days' but always translated 'Head of Days'.

At 46.2, 47.3 Eth. scribes altered the first noun to 'ancient' (beluya) and read as in the Eth. version of Dan. 7.9, 'Ancient of Days' (beluya mawāʿel) (46.2 Eth$^{\mathrm{ryl}^1}$, 47.3 Eth^{ull}). All that this in fact means, however, is that these scribes correctly recognised the title as equivalent to the Danielic 'Ancient of Days', and not that the original reading has been preserved in these two manuscripts of the traditional text (Eth$^{\mathrm{ryl}^2}$ corrects to the familiar reʾsa mawāʿel).

As with the first title 'Lord of spirits' we probably have to do with a translator's coinage which has been arrived at by way of interpretation of a Greek or Hebrew/Aramaic original. The title in Dan. 7.9 is rendered in both the LXX and Theodotion by the literal equivalent παλαιὸς ἡμερῶν, 'old of days', but this is not the only possible rendering. As Montgomery pointed out,[3] the expression appears fairly often in Syriac literature, meaning simply 'old man', e.g. Wis. 2.10 (LXX πρεσβύτης). The regular Greek equivalent of ܥܬܝܩ is ἀρχαῖος so that an alternative rendering of עתיק יומיא would be ἀρχαῖος ἡμερῶν. Has an original ἀρχαῖος been altered to ἀρχή, under

---

[1] See above, 185f., 187.

[2] Tertullian is cited by Tischendorf for the reading at 2 C. 3.18 tanquam a domino, inquit, spirituum, for καθάπερ ἀπὸ κυρίου πνεύματος; von Soden attributes the reading to Marcion.

[3] J. A. Montgomery, *The Book of Daniel*, (I.C.C., Edinburgh, 1927), 297, 300.

the influence of Biblical passages such as Isa. 41.4, Rev. 21.6, 22.13, ἐγώ εἰμι ... ἡ ἀρχὴ καὶ τὸ τέλος to produce ἀρχὴ τῶν ἡμερῶν as a nomen dei?

As noted above, Charles proposed 'sum (ראש) of days' i.e. sum-total of days as a possible meaning of the phrase, = κεφάλαιον τῶν ἡμερῶν. Has this curious title come, not from any rendering of Aramaic, but from a misreading (or (mishearing) by an Eth. scribe of ὁ παλαιὸς τῶν ἡμερῶν as κεφάλαιον τῶν ἡμερῶν? Or was there a Greek corruption which produced the expression κεφάλαιον/κεφαλὴ τῶν ἡμέρων as a nomen dei? The metaphorical use of κεφάλαιον of persons, 'the head or chief' is well attested (Liddell and Scott, s.v.), and, though it seems *prima facie* improbable that the form of a nomen dei could arise by a scribal blunder, this does seem to be the most likely origin of the term. I have, therefore, adopted (τὸ) κεφάλαιον τῶν ἡμερῶν, as the Greek 'original' of Eth. 'Chief/Sum of Days', even though it is something of a philological monstrum.

As Dillmann noted, the title occurs only in the second part of the Parables (from 46.1-60.2) and in the concluding chapter (71.10,12,13). It is relatively infrequent, compared with the more regular 'Lord of Spirits', and appropriately appears in visions inspired by Dan. 7. Although the title appears once only in Aram. at Dan. 7.13, a Heb. עתיק ימים is a possible equivalent.

In some of the passages where the title is used, just as at 46.1, the emphasis is on the thought of the One who is ancient, primordial, at the beginning of all time, eternal.

> 'And at that time that Son of Man was named in the presence of the Lord of Spirits,
> And his name before the Chief of Days,
> Even before the sun and the 'signs' were created,
> Before the stars of heaven were made ...' (48.2-3).

In other passages the accent is on his majesty as the Chief, the First of Days, as at 47.3 (cf. 60.2):

> 'In those days I saw the Chief of Days when he seated himself upon the throne of his glory,
> And the books of the living were opened before him:
> And all his heavenly host which is in heaven above, and his council stood before him ...'.

Dillmann, 156, again rightly, calls attention to the consistent way in which the author, all through this section of the book, as in the rest of I Enoch, employs the divine names in the closest connection with their context.

CHAPTER 37

Verse 1 begins, like En. 1.1, with a verse introducing Enoch in the 3rd pers., followed by the words of Enoch himself, v. 2f. The author appears to be modelling the exordium to his 'Second Vision' on En. 1.1f.

(1) **The Second Vision** Cf. Milik, 89: 'This (title of the Second Vision) contrasts it with the First Vision, that is, with the whole collection of revelations contained in the

Aramaic and Greek Enochic Pentateuch ...' (Milik is assuming that the Parables are based on the primary Greek version of Enoch). Martin, 79, comments, as there is no title 'First Vision' in Chh. 1-36: '... si ces mots sont authentiques, il faut admettre que l'éditeur considère en bloc comme une seule et première vision celles qui sont racontées dans la première partie (XII, XIV, XVII et suivants)'.

a vision of wisdom 'Wisdom' is not usually connected with apocalyptic visions and the expression seems to be unique. In Enoch it is a comprehensive term embracing all the revealed 'knowledge' imparted to Enoch, including knowledge and understanding of the natural world (cf. 93.10-14) as well as the divine mysteries revealed in the Son of Man and related apocalypses. Note now this 'wisdom' is repeatedly emphasised in vs 2 and 3.

son of Jared etc. The genealogy of Enoch taken word for word from Gen. 5. The writer is following established Hebrew practice in introducing a personage of importance, perhaps even supplying what he felt were deficiencies in the First Vision to make up for deficiencies in his own introduction (cf. Dillmann, 139).

(2) **the beginning (the sum) of the words of wisdom** Charles suggests as a possible original ראש חכמה 'sum of wisdom', but notes (and seems to prefer) the expression תחילת חכמה 'beginning of wisdom' (Prov. 9.10). We should probably prefer the (ambivalent) ראשית חכמה of Ps. 111.10 'the chief part of wisdom' (NEB mg.), Prov. 4.7 'the beginning of wisdom'. Could the author have intended 'the *summum* of wisdom', 'the sum of all wisdom'; cf. the expression for the summum bonum in the Syrian mystical writer Bar Sudaili ܪܫܘܬ ܛܒܘܬܐ = ἀγαθαρχία (ed. Frothingham, 97, P. Sm. 3906).

I spoke up and uttered, saying =נשאתי ואדבר לאמר? Cf. 1.2 'Enoch took up (נשב) his parable': here 'anše'a = נשא is used for נשא קול as at Isa. 3.7 (LXX ἀποκριθεὶς ἐρεῖ) 42.2 (LXX ἀνήσει); also at En. 83.5. Is this a Hebrew idiom in Eth. (only this verse and 83.5 are reported in Dillmann, *Lex.*)?

those who dwell on the earth ישבי יבשה(?). Behind the Eth. translation lies the Heb. יבשה or Aram. יבשתא; the latter is used at Dan. 2.10 as a synonym of ארעא. To render by its literal equivalent 'dry ground' (Knibb) makes an odd translation. The expression occurs a number of times, mostly in a pejorative sense, in Revelation: e.g. 3.10, 6.10. 8.13, 11.10 (bis), 13.8,14, 17.8, cf. 14.6.

men of old ... men of later days The expressions in Eth. correspond to Heb. ראשונים (οἱ ἀρχαῖοι) and אחרונים (οἱ ἔσχατοι Ec. 4.16).

Holy One see the note on 1.3.

Lord of spirits Above, 189f.

(4) **wisdom ... received** Cf. 2 En. 47.2.

the lot of eternal life חלק/גורל חיי עולם(?) Cf. Dan. 12.2 חיי עולם.

CHAPTER 38

(1) **congregation of the righteous** Cf. 53.6, 62.8, an expression peculiar to the Parables. It comes from Ps. 1.5 עדת צדיקים (cf. Pss. 74.2, 111.1, 149.1 קהל החסידים, Pss. Sol. 17.18 συναγωγὰς ὁσίων). For Qumran parallels 4QpIsa[d] 1.3 (JBL vol. 77 (1958), 220), 4QpPs. 37.2.5 עדת בחירו 'congregation of his elect ones'; 4QpPs 37.2.10 עדת אביונים 'congregation of the poor', CD 20.2 עדת אנשי תמים הקדש 'congregation of the men perfect in holiness', IQS 5.20 עדת קודש 'congregation of holiness'.

shall be ... driven from the face of the earth Eth. tahawka (hoka III.1) 'emenna probably corresponds to נגרש מן 'to be banished, expelled from'.

(2) **the Righteous One** Although the parallel 'light shall appear to the righteous' would seem to favour the reading of Eth I 'righteousness', Eth II has almost certainly preserved the original reading. The designation — here clearly referring to the head of the 'righteous elect' — occurs only once again, in the Parables, at 53.6, 'the Righteous and Elect One shall appear ...'. The designation 'the Righteous One' (הצדיק) for God appears at Isa. 24.16, but it is from Isa. 53.11, 'my righteous servant' (צדיק עברי) that the term here seems to come. Moreover, as 53.6 shows, the designation goes closely together with the titular 'the Elect One' to describe, also by an Isaianic term for the 'servant of the Lord' (e.g. Isa. 42.1), the Head of the 'righteous elect' (see further below on 39.6). Cf. the similar use of ὁ δίκαιος at Wis. 2.12,18 (at 2.13 he is παῖς θεοῦ); cf. also Ac. 7.52, 22.14. See A. Decent 'Der Gerechte: eine Bezeichnung für den Messias' in ThStKr, Bd. 100 (1928), 439-443, and ThWNT, Bd. II s.v. δίκαιος, 184f.

before the righteous ... Lord of spirits The translation assumes a corruption in the Eth. text in rendering τῶν δικαίων καὶ ἐκλεκτῶν (the correct expression occurs in the next line '(light shall appear) to the righteous and elect'). (For the expression cf. 1.1). An alternative would be to posit a Greek text [ὧν ἐκλεκτὰ τὰ ἔργα αὐτῶν] (די בחירין עבדיהון) where ἐκλεκτός = בחיר (cf. ברור) means 'tried, tested', '... whose deeds have been tested ...'. Dillmann (followed by Knibb) renders 'whose deeds have been weighed (sequl), but the use of this word at 40.5 (46.8) =תלא, תלי LXX ἐπικρεμαννύναι seems preferable: cf. Gesenius, *Thes.* 1504, P. Sm. 4442, Ephrem ed. rom. iii 165 D, ed. Beck, CSCO Tom. 88,2.69: ab eo omnes creaturae dependent (ܬܠܐ).

light shall appear Cf. 1.8 and Isa. 9.2, 60.1.

denied the Name of the Lord of spirits In the Parables this is the offence above all offences 'the very head and front of their (the sinners') offending' (cf. Charles, 71); Dillmann 'ihre Grundsünde' (41.2, 45.1,2, 46.7, 48.10). Its opposite is 'to believe in the Name of the Lord of spirits' (43.4), corresponding to the rabbinical האמין בשם R. Exod. 16.1, R. Dt. 11.10, Schlatter, *Das Evangelium Johannes*, 19. The most likely equivalent of the expression in Heb. is כחש ביהוה Isa. 59.13, Jer. 5.12. According to this last passage the 'denial of God' is the profession of atheism: 'they have denied the Lord, saying, 'He does not exist' (NEB). Cf. also Jos. 24.27 ('renounce God' (NEB)); cf. Jude 4.

It were better for them never to have been born Cf. מוטב היה לו Wajikra R. 26, A. Edersheim, *Life and Times of Jesus the Messiah*, II, 120. The idea was a commonplace in the ancient world: cf. 2 Bar. 10.6; 2 Esd. 4.12; 2 En. 41.2; Mt. 26.24; Job 3.3; Jer. 20.14, etc.

(3) **When their secrets are revealed to the Righteous One he will judge sinners** I have adopted the reading of Eth^{tana} which continues the theme of the role of 'the Righteous One': like the 'Son of Man' he is to be the judge of sinners (69.27). If the majority text is favoured, presumably we should interpret the 'secrets of the righteous' in the light of such passages as 58.5 'the secrets of righteousness, the inheritance of faith'; cf. also 62.7, and Mk. 4.11 par. But Eth^{M} gives poor sense in comparison with Eth^{tana}.

(4) **those who possess the earth** i.e. the foreign oppressors of Israel, a central theme in the Parables; 46.4-8, 48.8-10, 53.5, 62.1-12, 63.

the light of the Lord of spirits Charles emends to 'For the Lord of Spirits has caused his light to appear ...' (reading 'egzi'a man. with Eth$^{g\ q\ u}$ and 'ar'aya berhāno): but 'egzi'a man. is best taken as a nominative absolute (Flemming-Radermacher). (la'egzi'a man. seems to be a scribal error unless it reflects a Heb. לאדון (G-K 143 e).) For the 'light' imagery, cf 1.8.

(5) **mighty kings** Again at 55.4 'powerful kings' (nagašt ḫayyālān), 62.1,3,6,9, 63.1,2,12, 67.8,12. At most occurrences the variant form 'the kings and the mighty' is found. Charles prefers 'the kings and the mighty', but the phrase is identical to מלכים עצומים at Ps. 135.10 (LXX βασιλεῖς κραταιούς). Yadin, *Scroll of the War*, 316, suggests that 'the kings ... the mighty men' at IQM 12.6-7 is a parallel expression. The attempts to identify them ahve not proved convincing. At 46.7 they put their trust in idols, so that they must be foreign rulers of Israel, such as either the Seleucids or the Romans; we can rule out Charles's Sadducaeans and Martin's Hasmonaeans. They are mentioned again at 96.8, the 'mighty', and 104.3f., the 'rulers', in the older Enoch Book, in these passages probably to be identified with the Seleucids.

into the hands ... righteous and the holy For the expression 'to be given into the hands of' 2 Kg. 18.30, and for 'the holy' 'the saints', Dan. 8.24 lit. 'the people of the saints'.

(6) **plead for mercy** Eth. meḥra IV.1 = התחנן ל(?).

CHAPTER 39

(1-2) As the Eth. variants show, the text of these verses has suffered in transmission. In all texts the reference is to the watcher legend; and, as this seems out of place in the present context, the vss. have been bracketed as an interpolation (Charles, 74, Martin, 82): 'Here manifestly 39[2b] 'And mercy shall not be accorded to them, saith the Lord of Spirits' should follow immediately on 38[6]' (Charles, ibid.).

The verses can be defended, however, in their present context if we could assume past tenses, instead of futures (Charles, ibid.). Enoch's speech then ends at 38.6, to resume again at v. 3. The intervening two verses simply record the watcher legend, before Enoch resumes his 'parable' with a description of his translation to heaven. The writer appears to have in mind the sequence of chapter 14.1f., Enoch's reprimand of the watchers, ('the writings of anger and wrath'?), followed by his heavenward ascent to the celestial habitations (14.8f., cf. 39.3f.).

Suspicion of corruption in the text is raised by the expression 'the elect and holy' as, in this context, apparently a description of angels; elsewhere it describes the members of the congregation of the elect and holy, the saints (62,8, cf. 38.4). The phrase 'from the highest heavens' of Eth[M] is also unusual: I suggest a similar mistranslation to that at 106.13, viz. 'some of the highest ones of heaven' [ἐκ τῶν ὑψίστων τοῦ οὐρανοῦ], מן עליוני השמים(?), as the actual subject of the main verb 'will descend'. If we read the text of Eth[m] as 'shall descend to resemble the children of the elect and holy', we have another version of the variation in the watcher legend that the watchers (or their spirits) will assume different forms and continue to plague mankind (see above on 19.1). On this explanation of the text the author of the Parables is providing his own 'watcher legend': at the end of the days there will be fresh assaults of angelic 'watchers' on mankind, and the future tenses of the Eth. text may then stand. Vs. 39.2b and 38.6 are deliberately parallel: the verdict on the 'mighty kings' and on the celestial seducers of men will be the same; they shall receive no mercy. Or are the 'mighty kings' the watchers incarnate? Cf. Test. Reub. 5.6 (above, 161).

(2) **writings of anger ... tumult** Does this refer to the petition on behalf of the watchers which Enoch 'took' (naš'a = לקח) to present to the Lord (13.6f.) or is the

writer here thinking of Enoch receiving (naš'a) a written reprimand from the Lord (cf. 14.1)?

anger and wrath Eth. qenʿāt wamaʿʿāt = עברה / קינה וחמה(?) (cf. Ezek. 5.13).

panic and tumult Eth. guegue'a wahawkᵉ וזעוה (?)מהומה (cf. Zech. 14.13). The panic and tumult of the condemned kings and 'exalted ones' from heaven.

(3) This verse is almost a paraphrase of 14.8f. After completing his commission to the watchers, Enoch recounts his vision, when he was swept up heavenwards by cloud and wind to the heavenly places, where he is confronted first with one house and then with a 'greater house', the latter the palace or temple of the Great One (14.10f.). At 39.4 Enoch sees 'another vision', the 'abodes' or 'resting-places' of the righteous, clearly the author's own elaboration of the First Vision. He seems also to be drawing on 22.3f., the description of the promptuaria of the departed righteous where they are kept till the last judgement.

clouds and a storm-wind snatched me up Eth. ʿawwelo nafās = either סופה or סערה (2 Kg. 2.1). The verb mašaṭa = ἁρπάζειν, חטף(?); cf. 52.1,2C.12.2-4.

(4) **another vision** i.e. in addition to the first vision of the house of the Lord described at 14.10f.?

the resting-places of the holy Eth. meskāb is parallel to māḫdar, and means literally *locus dormiendi*, κοίτων or κοίτη 'bed'; it has also, like κοίτη the meaning of 'place of burial' (Isa. 57.2). At 46.4 meskāb is parallel to manbar 'seat, throne' and there probably means 'couch'. The reference here is to the 'homes' of the holy, where they can rest from their earthly labours, but there may be also an intentional suggestion of κοιμητήριον. An alternative possibility is to posit an original מנוחה, lit. 'resting-place'. Cf. the use of μονή at Jn. 14.2 ('dwelling-place' NEB).

(5) **Righteousness flowed like water** Cf. Am. 5.24.

(6) **the Elect One of righteousness and faithfulness** In view of the clear parallels drawn by the author of the Parables between the 'Elect One' and 'the elect ones', it is not surprising to find the two confused by scribes, as in the alternative reading here, 'the elect ones'. It is arguable that the correction is more likely to have gone from the plur. to the sing., made by a scribe who wanted to bring 'the Elect One' into this vision of the promptuaria of the 'righteous and holy ones'. The context, however, clearly favours the sing.: all Eth. mss read 'The righteous and elect will be innumerable *before him*', which presupposes 'the Elect One' as antecedent (in Eth II 'before him' has no antecedent). Moreover, the words 'of righteousness and faithfulness', alluding to Isa. 11.5, point to an individual rather than to a group.

The title 'Elect One' (בחיר) comes unmistakably from second Isaiah's term for the 'servant of the Lord' (Isa. 41.8,9, 42.1; cf. Lk. 23.35). The writer of second Isaiah is thinking of Israel as 'the elect one'—unless one chooses to interpret his words messianically—but the author of the Parables here takes the thought an important step further. The attributes of his 'Elect One' are clearly inspired by the vision of the first Isaiah's 'shoot from the stock of Jesse' (NEB), the Davidide (Isa. 11.1f.): 'Righteousness shall be the girdle of his loins and faithfulness the girdle of his reins' (v. 5 AV). Like the term 'the Righteous One', 'the Elect One' has come to describe the representative Head of the elect ones, destined to occupy a glorious throne (45.3). This title for the Isaianic servant-messiah occurs only in the Parables at 40.5, 45.3,4 (my Elect One), 49.2,4, 51.3,5, 52.6,9, 53.6, 55.4 (my Elect One), 61.5,8,10, 62.1. Cf. also Lk. 9.35 (ὁ ἐκλελεγμένος) and Jn. 1.34, v. l. ὁ ἐκλεκτός. See ThWNT Bd. IV, s.v. λέγω, 186f. (Schrenk).

(7) **I saw their abode** i.e. of the elect ones; for the variant 'his abode' see above on v. 6.

were radiant like the brightness of fire Cf. 104.2, 108.13. All these passages echo Dan. 12.3 (LXX φανοῦσιν); cf. also Phil. 2.15. The expression berhāna'essāt has parallels at Isa. 4.5 נגה אש להבה Hab. 3.4 נגה כאור Ezek. 1.13 נגה לאש.

wings of the Lord of spirits Cf. Ps. 91.4.

And righteousness ... fail before him Eth^{M} has a double version of this sentence, 'And truth (ret') shall not fail before him'. Was the original subject 'righteousness and truth'?

(8) **There I desired ... my soul longed for that abode** Eth. 'desired' faqada = אוה or הפץ cf. Ps. 132.13,14; 'longed' fatawa = כסף Niph. cf. Gen. 31.30, Ps. 84 (83).2.

There had my lot been assigned before The meaning is not that Enoch's 'portion' or 'lot' had been in this place before his celestial journeyings, but (parallel to the next line) that his heritage (חלק) had been fixed beforehand to come to reside in this heavenly place.

it had been resolved about me Eth. lit 'it had been established' (i.e. by God), ṣan'a = נכון; cf. Gen. 41.32 'God is already resolved to do this' (NEB) (lit. 'the matter has been resolved (נכון) by God').

(9) **established me in blessedness** lit. 'he has set me up, established me in blessedness and honour'; cf. Ps. 7.10 'O let the evil of the wicked come to an end, but establish thou the righteous (תכונן צדיק)'. ברכה is here used with the meaning of 'blessedness', the state of being blessed; 'honour' (sebḥat = δόξα, שבוחה?) repeats the word earlier used and translated by 'praises', but in a different sense. Dillmann (followed by Charles and Knibb), take 'aṣne'a ba = 'to destine' (bestimmen) someone (accus.) for (ba) something', but this requires la not ba, e.g. 65.12 'He has destined thy seed in righteousness (baṣedq) for kingship (lanagašt) and for great honours (lasebḥat 'abiyāt).

(10) **And my eyes looked long** Eth. waguenduya re'yā a'yenteya = ותרבינה עיני להביט (Goldschmidt).

(11) **in his sight there is no end** לפניו אין קץ; 'end', māḫlaqt = קץ Eth. Isa. 9.7.

He knew before ... was created Cf. CD 2.9-10 'He knew ... what was to happen to eternity (הוי עולמים) ... and events (נהיות) until what comes in their times for all the years of eternity'. The reading of Eth. II, 'what the world would become' seems secondary; the parallelism favours the reading of Eth I, 'what was to happen for ever', 'what was to come'.

(12) **Those who sleep not** In the next verse Eth^{q} has the variant (or gloss) teguhān = עירים, ἐγρήγοροι, 'the wakeful ones'. See above, note on 1.5. The full expression occurs again at 39.13, 40.2, 61.12, 71.7.

they stand before thy glory i.e. the כבוד יהוה. See ThWNT Bd. II, p. 235f. s.v. δόξα (von Rad, Kittel).

Lord of spirits An interpretation of יהוה צבאות, here = Isa. 6.3. See above 189f., 'Two Unusual Nomina Dei in the Second Vision of Enoch'.

he fills the earth with spirits A further theological interpretation of the Trisagion. The original may have read 'The earth is full of spirits' (cf. Eth^{tana}). Dillmann describes this version of the Trisagion as 'an alteration of Isa. 6.3 based on the main content (Hauptinhalt) of this second part'. The world of the author of the Parables was one full of angelic beings and disembodied spirits of whom God was supremely Lord. The variant 'the whole earth' Eth^{a} is an accomodation to Isa. 6.3.

(13) **the Name of the Lord** Heb. אדון used simpliciter 'the Lord' occurs once only in the Old Testament at Ps. 114.7 (Mal. 3.1?) (see ThWNT, Bd. III s.v. κύριος 1059.18 (Foerster)). The corresponding absolute use of Aram. מרא is now attested, but, again,

its occurrence appears to be rare (cf. J. A. Fitzmyer, 'The Contribution of Qumran Aramaic to the Study of the New Testament' in NTS Vol. 20, 387f.; and for I En. 9,4, 89.31,33,36, M. Black 'The Christological Use of the Old Testament in the New Testament', in NTS Vol. 18,10; see also above on 9,4, and below on 89.31,33,36). There might seem to be a more frequent use of the expression in this absolute way in the Parables were it not for the presence here and elsewhere of the variant 'Lord of spirits' the regular term for deity in this part of Enoch (e.g., 45.1, 62.1, 65.6, etc.) and in other cases the expression is 'our Lord' (63.7,8). In other cases still where 'the Lord' occurs it is followed by a qualifying phrase, e.g. 41.8 'the Lord who separates the light from the darkness', 63.7 'the Lord of all his works (v. 1. for 'our Lord')': In two instances only is the text firm, 67.10 'the spirit of the Lord' (an Old Testament phrase) and 68.4 (Eth^{M}), in a context following the fuller expression 'the Lord of spirits': '... the Lord of spirits has been angry with them because they act as if they were the Lord'.

The use of 'the Lord' absolutely in this way does seem, therefore, to be almost as rare in the Parables as elsewhere, so that there is much to be said for the variant reading of Eth^{tana} or Eth^{v} 'Blessed is the Name of the everlasting Lord' (lit. 'Lord of eternity'), which may preserve the original reading. On this latter title for deity, see above, note on 9.4.

(14) **my countenance was changed ... unable to look** Cf. Jer. 30.6 (נהפכו כל פנים לירקון): we are to think rather of Enoch's countenance changing for fear and awe than of any transformation (Knibb) or transfiguration; cf. also 1 En. 14.24-25 where Enoch 'fell on his face ...'. The following 'until I was unable to look' seems preferable to 'because I was unable to look'; Enoch's fear and awe were so overwhelming at this vision that he could no longer bear to look. That he was 'blinded by excess of light' is a somewhat free interpretation of Charles (drawing on Dillmann). Heb. ולא יכלתי לראות.

CHAPTER 40

(1) **thousands upon thousands and myriads upon myriads** Cf. 14.22 and Dan. 7.10. The longer form of the expression (En. 14.22 has only 'myriads upon myriads') comes here probably from Dan. 7.10. Cf. En. 1.9 and for the longer Danielic form 60.1, 71.8, Rev. 5.11.

(2) **four presences** Eth. gaṣṣ = פנים 'faces', i.e. the so-called 'angels of the presence'. The term מלאך הפנים comes from Isa. 63.9. When mentioned as a single angel it is sometimes identified with 'the angel of the Lord' (מלאך יהוה). Thus 'an angel of the presence' appears at Jub. 1.27-29, Test. Judah 25.2, but 'angels of the presence' at Jub. 2.18, Test. Levi 3.5. Here in Enoch they are distinguished from the watchers (Cherubim, Seraphim and Ophanim) and identified (v. 9) with Michael, Raphael, Gabriel and Phanuel. For the classification of these archangels and the development of this angelology, see Milik, 174f. and note on 9.1. For a discussion of their respective roles see Bousset-Gressmann^{4}, 325f., H. Bietenhard, *Die himmlische Welt im Urchristentum und Spätjudentum*, (Tübingen, 1951), 103f., and Yadin, *Scroll of the War*, 237f.

the angel that went with me So designated at 43.3, 46.2, 52.3,4, 61.3. This angelus interpres, who acts as Enoch's guide, is the counterpart in the Parables of his accompanying angel in the older Book, Uriel, e.g. 14.25, 21.5. He is sometimes called simply 'the angel' (61.3, 64.2) or 'the angel of peace' (40.8, 52.5, 53.4, 54.4, 56.2, 60.24), a designation usually traced to Isa. 33.7; this angel appears again at Test. Dan 6.5, Test Asher 6.6, Test. Benj. 6.1 where he is the angel of the dead and the guide of souls. See Bousset-Gressmann[4], 327.

all secret things Cf. Dan. 2.29 and IQpHab. 7.5 כול רזי דברי עבדיו הנביאים 'All the secrets of the words of his servants the prophets'.

(3) **the Lord of glory** Although there is a variant 'Lord of spirits' the reading here seems firm: it corresponds to the rare title for deity ὁ κύριος τῆς δόξης at En. 22.14 preserved at En[d] 1 xi 2 [מרה] רבותא (cf. Milik, 218). Cf. also IQapGn. 2.4 מרה רבותא ed. Fitzmyer, *Genesis Apocryphon*, 50, 83, and (by the same author) 'The Contribution of Qumran Aramaic to the Study of the New Testament' in NTS Vol. 20, 387 n. 2. Ps. 29.3 has אל הכבוד 'the God of glory', and the more usual Biblical expression is כבוד יהוה 'glory of the Lord'.

(4) **The first voice** The voices are explained at v. 9, as those of Michael, Raphael, Gabriel and Phanuel. For their different roles, see above note on v. 2.

(5) **the Elect One** See on 39.6.

the elect ones See on 1.3, 39.6.

depend upon See on 38.2.

(7) **the satans** In Job השטן is one of the 'sons of God', a superhuman Adversary: but the Old Testament does not know of superhuman שטנים, although the word is a common one for human adversaries. The satans here have been identified with the 'angels of punishment' (Strafengel, Dillmann, cf. 53.3), although this role is also performed in the Parables, as in the older Book, by the archangels (54.6, cf. 10,4, 90.21). In this passage, the satans are a special class of angels, no doubt subject to Satan (54.6 note), acting as 'accusers', and thus fulfilling the classical role of Satan. At 65.6f., however, they have become simply agents of evil. Cf. Bousset-Gressmann[4], 333 (foot).

(8) **whose words I have ... written down** Here in the Parables as in the older Book (cf. 33.3) Enoch is the 'scribe of righteousness'.

(9) **diseases ... wounds** ḥemām = חלי quesl = מכה(?).

set over ... the powers 'Powers' ḫāyl coll. = δυνάμεις refer, not to the powers of nature, winds, fire, thunder etc. but to astral or angelic 'powers', or rather 'potentates' (abstract for concrete). The term δύναμις is well attested in hellenistic sources in the sense of a 'super-terrestrial being', but in its meaning of an astral or cosmic force or agency, personalised as an 'astral being or angel', it may also owe something to semitic tradition. It is the Aram. or Syr. equivalent of Heb. צבא for the 'host' of heaven, the stars, e.g. Tg. 1 Kg. 22.19; cf. Test. Levi 3.3 αἱ δυνάμεις τῶν παρεμβολῶν 'the hosts of the armies (of angels)' which are ordained for the day of judgement (cf. 3.5 αἱ δυνάμεις τῶν ἀγγέλων). For the Hebrew mind these are the 'forces' of heaven in this quasi-military sense: blended, however, with Greek usage, 'powers' convey the idea of miracle-working agencies or agents.[1]

the repentance (leading) to hope ... eternal life Charles, cf. Ac. 11.18 μετάνοια εἰς ζωήν, 2C.7.10 μετάνοια εἰς σωτηρίαν. For similar ideas, cf. 1 Pet. 1.3, Tit. 3.7. For the expression 'to inherit eternal life' cf. especially Pss. Sol. 14.6: the inheritance of the wicked is ᾅδης καὶ σκότος καὶ ἀπώλεια, but the righteous κληρονομήσουσιν

[1] For the term in the New Testament, see M. Black, Πᾶσαι ἐξουσίαι αὐτῷ ὑποταγήσονται in *Paul and Paulinism*: Essays in honour of C. K. Barrett, ed. by M. D. Hooker and S. G. Wilson (London, 1982), 74-82.

ζωὴν ἐν εὐφροσύνῃ. The rabbinical expression is 'to inherit the life of the world to come'; cf. Volz, *Eschatologie*, 327, 368 and ThWNT Bd. III, s.v. κληρονόμος 780.20 (Foerster).

Phanuel Phanuel is again mentioned at 54.6, 71.8,9,13 as one of the four 'angels of the presence'. He appears as a late-comer among the archangels, replacing Uriel in the list of four at En. 9.1 (Uriel himself appears to have been a successor in this list of Sariel; see Milik, 172f. and above 120; cf. also Yadin, *Scroll of the War*, 238). The name has been variously explained, e.g. as 'presence (lit. face) of 'el' a suitable title for an angel of the presence. The name פנואל itself appears as a place-name at Gen. 32.30, Jg. 8.8,17; cf. also Lk. 2.36 where it is a personal name Φανουήλ, father of Anna the prophetess. The name is a regular formation and was no doubt a familiar Israelite personal name (cf. Noth. *Die israelitischen Personenannamen*, 34 n. 1 and 255). As the name for the fourth archangel, however, Phanuel is extremely rare. It is not, however, confined to the Parables, as Martin maintains (87 n.) but occurs once again at 3 Bar. 2.5 (Φαμαήλ but correctly read as Φανουήλ by the Slavonic version; M. R. James (*Texts and Studies*, Vol. V. 1 *Apocrypha Anecdota* iii, Cambridge, 1897) suspected a corruption of Ῥαμιήλ of 2 Bar. 55.3, En. 20.8 Ῥεμειήλ (cf. 4Q En[a b c] (Milik, 153) רעמאל). See R. H. Charles, *Apocrypha and Pseudepigrapha*, II,534. In further support of this suggestion is the similarity of the role of Ῥεμειήλ at 20.8 'he presides over those who rise (from the dead)' with that of Phanuel of 40.9, who presides over 'repentance (leading) to hope of those who inherit eternal life'. Ramiel has also the further role at 2 Bar. 55.3 of 'presiding over true visions' which coincides with that of the same archangel in 3 Bar. 11.7 where he is 'the interpreter of visions of those who pass through life virtuously'. Cf. also 2 Esd. 4.36 (Hieremihel) and Sib. Or. 2.215-17, the angel who knows all the evil deeds of men and 'from darkness and gloom leads to judgement all the souls of men before the judgement seat of a great God immortal'.

(10) **the Lord of spirits** Eth II 'the Lord Most High' as at Ps. 57.3, 78.56 אלהים עליון (Gen. 14.18 אל עליון). The variant has probably been introduced by a scribe, since the title does not otherwise occur in the Parables and is found once only elsewhere in Enoch (at 98.11).

CHAPTER 41

(1) **a kingdom is divided ... weighed in the balance** This is an enigmatic line which has been variously interpreted. Dillmann took the reference to be to the 'Messianic Kingdom' with its different 'divisions', Charles to the division of the 'kingdom of Heaven' into seven parts, and Schodde, 122, understood the words to refer to the 'kingdom of this world, which is to be divided, i.e. the faithful separated from the sinners, when the deeds of all are weighed in the final judgment'. In view of the 'divisions' of the winds at v. 3, the words might be taken to refer to the partitioning of the natural world and its elements. Is there perhaps, however, an implied allusion to Dan. 5.26-28, predicting the 'division' of the kingdom of Belshazzar, because his deeds had been weighed and found wanting?

'mene: God has numbered the days of your kingdom and brought it to an end;
tekel: you have been weighed in the balance and found wanting;
u-pharsin: and your kingdom has been divided and given to the Medes and

Persians' (NEB). Cf. 38.5 the prediction that 'then shall the mighty kings be destroyed'. An allusion to Dan. 5 would support the view that the historical situation of the writer and his readers was that of the Seleucid or Roman period. For the figure of speech in the second clause cf. 61.8 and in the Old Testament e.g. Job 31.6, Prov. 16.2, 21.2, 24.12, Ps. 62.9, Dan. 5.27, Pss. Sol. 5.6.

deny ... the Lord of spirits See above on 38.2.

(2) **being dragged off** Eth. tasehba = cf. סחר Niph. cf. Jer. 49.20 (50.45).

(3) **secrets of the lightning** etc. Cf. Job 38.22,25-28,34-38. The author of the Parables is following the pattern of the old Book by including such visions of natural phenomena; cf. especially chapter 18f. 'Lightning and thunder' are mentioned several times in the Parables, e.g. 43.1-2,44,59,60.13-15,69.22f.

(4) **closed storehouses** Eth. mazāgebta ʿeṣewāna = אוצרות עצורות with a deliberate word-play? The 'storehouses' (אוצרות) of God for rain, snow, hail, wind, Dt. 28.12, Job 38.22, Jer. 51.16, Ps. 33.7 etc. See also note on 89.3. For the idea cf. 1 Kg. 8.35; these 'storehouses' or 'chambers' are 'shut up' like the sky till their contents are released on the earth.

are distributed Eth. yetkaffalu = יחלקו (Goldschmidt).

storehouse of mist and ... clouds Eth. gimē = ὁμίχλη, עיפה, ערפל, אד (Gen. 2.6); dammanā = ענן, עבה.

clouds from it lit. 'its cloud' i.e. cloud which comes from the 'storehouse of the clouds' and which is perpetually over the earth. The idea is not that of Gen. 2.6 (Beer, Charles) since the mist there rises from the ground, but of the cloud-cover which is always somewhere present over the earth to replenish it with rain (as the gloss on 'its cloud' in Eth^{t} explains 'it produces rain'). Eth. ḫadara lāʿla = שכן על? Job 3.5.

(5) **one is more glorious ... festivals** The text as it stands in Eth^{M} is confused and makes poor sense. Dillmann rightly rejected the meaning 'magnificent' (lit. 'rich') for beʿul and suggested a connection with baʿāl = מועד 'what is fixed, appointed'; he rendered 'ich sahe ihren festbestimmten Lauf'. Eth. meḥwār here = תקופה, the completed revolution or orbit of the sun or moon; cf. Sir. 43.7 συντέλεια, Ps. 19 (LXX 18).7 κατάντημα. Eth. baʿāl = מועד(ים), ἑορταί. For the mention of 'festivals' in this connection cf. IQS 10.2-5 and above, note on 2.1.

they keep faith ... abide Eth. 'keep faith', hāymānota ʿaqaba = שמר אמן, Isa. 26.2, LXX φυλάσσων ἀλήθειαν. Eth. maḥalā = ὅρκος or διαθήκη, ὁρκωμοσία, and nabara 'to abide in', in this context 'to abide by' = ἐμμένειν, e.g. Jer. 38(31).32 LXX οὐκ ἐνέμειναν (Eth. ʾinabaru) ἐν τῇ διαθήκῃ μου. LXX ἐμμένειν = קום Hiph., Jer. 51(44).25. Cf. also En. 5.4. Eth I ḫadaru is a synonymous rendering of ἐμμένειν; ḫabru of Eth^{k} is a corruption; cf. Charles, 81, and Flemming, 46.

(6) **his Name will endure** Eth. yeṣanneʿ which Charles (following Dillmann) renders 'mighty is His name'. Is there an allusion to Ps. 72 (LXX 71).17, LXX πρὸ τοῦ ἡλίου διαμενεῖ (יכון?) τὸ ὄνομα αὐτοῦ? Knibb = Flemming-Radermacher, 'and his name endures ...'.

(7) **hidden and visible path of the moon** i.e. the moon's phases from invisibility to full moon.

one holding a position opposite to the other lit. 'one watches the other'; for the meaning 'stands opposite' see Dillmann, 21 n. 2 (הביט?). The sun and moon are personified or even regarded as conscious beings.

they give thanks ... rest Cf. Ps.148.3. Could ʾakuatēt = αἴνεσις here render מלאכה, as αἴνεσις does at Ps. 72(73).28 (cf. 1 Chr. 9.13, 2 Chr. 24.12 where מלאכה = 'service of the sanctuary'): ʾakuatēt is also used for the liturgy of the eucharist

(Dillmann, *Lex.*, 786). Cf. 69.24 'their thanksgiving (מלאכה?) is their food (מאכל). If we assume an original מלאכה we obtain an oxymoron, lit. 'their work is their rest.'

(8) **for blessing or for curse** Cf. 59.1-3.

made a separation Eth. all mss reads fatara = 'create'; cf. Knibb '(in the name of the Lord) who has created (a division) between light and darkness' = Dillmann 'den eine Trennung schuf ...'. Is faṭara a scribal error for falaṭa? Cf. Gen. 1.4 Eth. falaṭa = הבדיל.

divided the spirits of men Cf. IQS 4. 1f. The division of the 'spirits of men' into two contrasting spheres of light and darkness corresponds exactly to Qumran ideas.

In the Name of basema could be a ditt. from basemu (Eth[1]ana la'egzi'a manā-fest) earlier in the verse and the reading of the Garrett manuscript correct: 'according to his righteousness.'

(9) This verse seems to refer back to the context of 41.2. The reference must be to the Elect/Righteous One as Judge (38.2f., Ac. 17.31).

CHAPTER 42

The short chapter 42 is a fragment of a 'wisdom' poem introduced somewhat abruptly into a largely alien context, except that the writer has been speaking about the heavenly abodes of the righteous (38) and in general about the secrets of the heavens. It is akin to 48, and perhaps belonged originally with this verse in a sapiential piece from which the author of the Second Vision is quarrying his material. But cf. also for wisdom references 49.3 and 5.8, 84.3, 91.10. The praise of wisdom is a favourite Old Testament theme and Charles compares such passages as Job 28.12f., Sir. 24.14f. In several passages Wisdom descends to earth to dwell with men (Prov. 1.20f.,8,9, Sir. 24.7), but here, as at 94.5, Wisdom is rejected by men and returns to heaven. Cf. further 2 Bar. 44.14, etc.

(2) **returned to her place and became established** Eth. ṣan'a III.2, הוכן (Hoph. כון) (cf. Isa. 16.5), lit. 'established herself'(?); the variant of Eth II was read and translated by Dillmann (followed by Charles and Knibb) as nahm ihren Sitz, cf. Dillmann, *Lex.*, 1305f. and 1290. Eth. ṣa'ana III.2 means 'to ride' but does it mean 'to take a seat'?

(3) **chambers** Eth. mazāgebt = אוצרות lit. 'storehouses, treasuries'. Iniquity ('ammaḍā) is, like Wisdom, personified and comes forth from her richly stocked magazines of wickedness. Cf. Prov. 10.2 אוצרות רשע 'treasures of wickedness' (AV).

(Welcome) as rain ... land The contrast is between the rejection and departure of Wisdom from earth to heaven with the greedy reception of Iniquity upon earth.

CHAPTER 43

(1) **called them ... by their names** Cf. 69.21, Isa. 40.26, Ps. 147.4. The subject 'the Lord of spirits' is resumed from 41.8-9.

(2) **how they are weighed in a righteous balance** lit. 'I saw the righteous balance how they are weighed (in it)'. Eth. 'righteous balance' madālewa ṣedq = מאזני־צדק (Job 31.6). Cf. 41.5. The stars and heavenly bodies are hypostatised: they are heavenly

beings, with consciousness and conscience,[1] to be assessed or 'weighed in a balance', like mankind, and to be so judged according to the measure or mass or proportions of light they possess (Dillmann 'ihre Lichtmassen'), but also in accordance with the width of their 'places' or 'areas, spaces', possibly to be understood as the era of the sky they cover or traverse (Dillmann 'die Weite der zu durchlaufenden Räume ...').

how their movement ... angels Eth. miṭat = τροπή, תקופה? For these 'motions' or 'revolutions' of the heavenly bodies, see 72.1f. According to 82.10f. these 'motions' are directed by angels. Does the text mean that every movement of the heavenly bodies is to be directed by a specific number of angels?

the day of their appearing The use of Eth. kunat/kuenat for the 'rising' of a heavenly body (= γένεσις, ἀνατολή, its 'coming into being') is adequately supported (Dillmann, *Lex.*, 864). The difficulty arises with the reading ʿelut, too well supported to be ignored. Flemming suggests that it may represent a careless writing of hellut = hellāwē (τὸ εἶναι). Was the original simply הויתם lit. 'their coming into being' = γένεσις αὐτῶν, translated twice by Eth. as hellut then kunat/kuenat, the surviving readings being a collation and an emendation to ʿelat, 'day'. We should then render 'and their risings, appearances'. Cf. the use of the noun הויה in late Hebrew = ܗܘܝܐ, γένεσις.

(3) **And I asked the angel** See above on 40.2.

(4) **a parable ... for ever and ever** Enoch has not only been given a vision of the heavens and the movements of the heavenly bodies, but these latter, individually named by the Lord of spirits, are somehow to be considered as 'parabolically' representing the names of the faithful upon earth. The text clearly owes much to Dan. 12.3, 'The wise shall be like the stars for ever and ever' (NEB), and cf. 104.2.

CHAPTER 44

(1) **are unable to remain with them** The reference seems to be to 'shooting stars', ἀστέρες διαθέοντες, ἀστέρων διαδρομαί (Arist. *Meteor.* 341a 33, 342b 21). The traditional text (and translations) make little sense. The correct reading is almost certainly that preserved in Eth^{tana} ḫadira of which ḫadiga is a corruption. The shooting stars are understood to be stars which turn into 'lightnings' because they were unable to 'dwell' with the rest of the stars.

CHAPTER 45

(1) **The Second Parable** The First Parable at 38.1-6 (or 38-44?).

who deny ... the Lord of spirits The usual text 'who deny the name of the dwellings of the holy ones' (Knibb) is odd and probably wrong: but it has given rise to the idea that these 'atheists' deny 'the heavenly world' (Dillmann followed by Charles), and that they are to be identified with the Sadducees who deny the life to come (Martin). The text of Eth^{tana} could be a simple correction, but it seems more natural to have the familiar expression 'the Name of the Lord of spirits' coming at the

[1] The stars like mankind are punished for 'transgressing' their appointed orbits 21.6.

beginning of the verse. We are still left, however, with the curious idea of 'denying the dwellings (congregations) of the holy ones'. Could the original have read עדות קדושים 'the testimony of the holy ones (i.e. the angels)' misread as עדת קדושים 'congregation of the holy ones'. For the last expression cf. IQM 12.7 'a congregation of thy holy ones (angels) is in our midst'. The 'testimony' could refer to the solemn divine charge of the Ten Commandments which was given accompanied by the angels (or given by the mediation of angels) on Mount Sinai. The reading māḥdar 'dwellings' seems a corruption of Eth^{tana} māḫbar, συναγωγάς.

day of suffering and tribulation See above on 1.1 and 10.6 and Charles's note here.

(3) **the throne of glory** Cf. Jer. 17.12 for the expression. So also at 51.3, 55.4, 62.3,5, the Elect One sits on the throne, in contexts where he again exercises his role as judge. At 62.5, 69.27,29 it is the Son of Man who is so seated and to whom all judgement is commited (cf. Jn. 5.22,27) and who is placed on the throne by the Lord of spirits (61.8, 62.2). On the other hand, at 47.3 it is the Chief of Days who sits on the judgement throne.

bring their works to the test lit. 'he will choose (yaḫarri) their works'; see above, 185 for this mistranslation. On the whole, the idea of 'testing' or 'trying' seems to fit the context best.

no places of rest The text reads 'their resting-places shall be without number', which can hardly be right; cf. Caquot and Geoltrain 'notes sur le Texte Éthiopien des 'Paraboles' d'Hénoch', *Semitica* 13 (1963), 44. The Greek word was probably μονή: is it possible that an original מנחה = προσφορά was misread as מנוחה μονή, their 'oblations were without number', the idea being that of innumerable offerings to placate the wrath of God? Alternatively, we may explain this line as an interpolation by a scribe (with 39.5f. in mind) who has wrongly understood the reference to be to the elect and not to the sinners. More probably the text is at fault; cf. v. 2 and 38.2.

become heavy Eth. ṣanʿa is interpreted by Charles (following Dillmann) as 'their souls will grow strong' and again wrongly referred to the righteous. Caquot and Geoltrain argue that the word is used here in the sense of 'grow hard, obdurate, perverse' = σκληρύνεσθαι as at Exod. 9.35. A more appropriate original would be תקשה (Niph. קשה; so Sjöberg, *Der Menschensohn im äthiopischen Henochbuch*, 76f.) Cf. Isa. 8.21.

(4) **I will transform the heaven ... the earth** (v. 5) The idea appears to come from Isa. 65.17, 66.22 (and ultimately, no doubt, from Zoroastrian sources). It reappears at En. 72.1 and most notably in the Apocalypse of Weeks at 91.16. See M. Black, 'The New Creation in I Enoch' in *Creation, Christ and Culture, Studies in Honour of T. F. Torrance*, 13-21. As here the idea involves the 'immortal blessedness' of the elect of mankind. Cf. 2 Bar. 32.6, 57.2, 2 Esd. 7.75.

(5) **sinners and evil-doers** lit. 'those who do sin and evil'; for the combination cf. חטאת ופשע Ezek. 33.10.

(6) **I have provided for** For this use of ראה, Dt. 33.21, 1 Sam. 16.1.

satisfied ... with peace Cf. 1.8f.

destroy them ... earth Cf. 69.27.

CHAPTER 46

For 46.1-48.1 see Liber Nativitatis (Lib. Nat.), CSCO, Scriptores Aethiopici, Tom. 41, 54f., Tom. 42, 48f.

(1) This verse, and most of the chapter following, is largely based on the 'Son of Man' vision of Dan. 7. For an analysis and commentary on 46.1-8, see Lars Hartman, *Prophecy Interpreted*, *Conjectanea Biblica*, New Testament Series I (Lund, 1966), 118-126.

a head of days For a discussion of this verse and the title 'Chief of Days', see above 192f.

white like wool Cf. Dan. 7.9.

appearance of a man כמראה אדם Ezek. 1.26, כברנש Dan. 7.13.

full of graciousness Eth. melu' ṣagā = πλήρης χάριτος Jn. 1.14; cf. רב חסד Exod. 34.6, Num. 14.18, Jl. 2.13, Jon. 4.2.

(2) **yonder Son of Man** lit. 'that Son of Man'. There are three expressions in the Parables for 'Son of Man' as a messianic title: (1) 'that/this Son of Man' zeku/zentu walda sab' = ille/hic filius hominis: 46.2,3,4, 48.2. (At 60.10 the expression without the demonstrative adj. is applied to Enoch: 'You, son of man (walda sab') seek to know ...'); (2) the most frequent form is 'that/this son of the offspring of the mother of the living' (i.e. Eve, cf. Gen. 3.20), zeku/we'etu walda 'eguāla 'emma ḥeyāw, the regular Biblical rendering (in both Testaments) of בן־אדם (אדם = 'eguāla 'emma ḥeyāw). It is found, without significant textual variation, at 62.7,9,14; 63.11; 69.26,27; 70.1; 71.17. In two cases only is the title found without the demonstrative adj. viz. 62.7 and 69.27. (At 60.8 the expression occurs, like walda sab' at 60.10, in Eth^{tana} applied to Enoch). (3) The third form is we'etu walda be'si, hic filius viri, at 62.5 (Eth I) 69.29 (bis) and 71.14. Eth II reads 'son of woman' walda be'sit at 62.5 (see below 235f.); 71.14 alone has walda be'si without the demonstrative adj.: 'You are walda be'si ...'.

All three expressions clearly go back to an original בן־האדם (Aram. בר־(א)נשא), ὁ υἱὸς τοῦ ἀνθρώπου. With three exceptions only, 62.7, 69.27 (walda 'eguāla 'emma ḥeyāw) and 71.14 (walda be'si), the term has the preceding demonstrative (zeku, zentu, we'etu) or a qualifying clause following, e.g. 46.3 'That is the Son of Man who has righteousness'. It has been frequently pointed out that the demostrative adj. corresponds to the Greek definite article = ὁ υἱὸς τοῦ ἀνθρώπου. At the same time, the demonstratives are, for the most part, required in Eth. to indicate unequivocally that it is no ordinary 'son of man' that is meant—they mean 'ille homo'.[1]

What applies, however, to languages such as Ethiopic or Latin where there is no definite article does not hold for Greek, Hebrew or Aramaic where there is one, widely used in a number of idiomatic ways. It is a capital error[2] to assume that Heb. בן־האדם or Aram. בר־(א)נשא, '*the* Son of man', could not be used as a designation or with titular force for a particular individual, especially for one of a class 'where usage has elevated into distinctive prominence a particular individual of the class' (A. B. Davidson, *Hebrew Syntax* (Edinburgh, 1894), 25).[3] The designation or title

---

[1] Cf. T. W. Manson in *Studies in the Gospels and Epistles*, (Manchester, 1962) 130f. 'The Son of Man in Daniel, Enoch and the Gospels'. 'Expressions like 'the Elect One' or 'Lord of Spirits' carry their meaning in themselves; but 'the man' will hardly convey any special meaning unless some hint is given that such a meaning is intended. The demonstrative may be a way of giving that hint'.

[2] Cf. Dalman *Words of Jesus*, 239 and contrast G. Vermes in (most recently) 'The Present State of the 'Son of Man' Debate' in JJS Vol. XXIX No 2, 125f. and B. Lindars *Jesus Son of Man* (Cambridge, 1983), 17f.

[3] It is true בר־(א)נשא can be determinate or indeterminate in Jewish Aram., but this does not mean that the former (the emphatic state) can no longer be employed to denote or designate a preeminent individual of a class, e.g. Tg. סטנא, like השטן, at Job 1.6 means '*the* Adversary' and

'*the* Son of man' to refer to the Daniel figure of 'one like a son of man', could as well be used in Aram. or Heb. and with the same force as ὁ υἱὸς τοῦ ἀνθρώπου.

At 62.7, 69.27, 71.14, when the demonstrative is omitted, the familiar Biblical term has, like ὁ υἱὸς τοῦ ἀνθρώπου, already become a title, and at 71.14 'You are the Man' (walda be'si), the context makes the identity of 'the Man' here unmistakable, to be further identified with Enoch himself.

An alternative explanation — and one in some respects even more attractive, since, if the original was Aram., it would remove all possible ambiguity in בר־(א)נשא — is to take we'etu as a self-referring pronoun at 69.26,29, 71.17 as rendering αὐτὸς ὁ υἱὸς ἀνθρώπου (cf. Lietzmann, *Der Menschensohn*, 46) = הוא בר־(א)נשא or הוא בן־האדם, ipse filius hominis, a familiar Semitic locution (see my *Aramaic Approach*[3], 96) especially in referring to an important personage. Cf. 2 Esdr. Syr. 13.1 ܗܘ ܒܪܢܫܐ Lat. ille homo (ed. A. Ceriani, *Monumenta Sacra et Profana*, Vol. V fasc. i (1868), 98), an equivalent phrase occuring in another Daniel-inspired apocalypse. Do the other instances of the Son of Man locution go back ultimately to the same idiomatic הוא בר־(א)נשא or הוא בן־האדם? It is then a solemn designation based on the 'one like a son of man' of Dan. 7, and well on the way to becoming a title.

How are we to account for the use of these three different expressions in the Parables? On (1) and (3), while sab' = ἄνθρωπος be'si = ἄνηρ, we can rule out the idea that there is any difference of meaning between them, since both assume an original ὁ υἱὸς τοῦ ἀνθρώπου, בן־האדם, בר־(א)נשא. What it is reasonable to infer is that we owe these two different forms to the work of two different, probably Ethiopic, translators. The fact that in the majority of cases we have the traditional Biblical expression suggests further that a third translator (or editor) has found the more literal forms (1) and (3) in his manuscript and corrected the majority of these into the regular, familiar Biblical expression. In that case the surviving instances of (1) and (3) represent the oldest stratum of versional tradition. Behind this oldest form must lie ὁ υἱὸς τοῦ ἀνθρώπου, בן־האדם or — if this Daniel inspired messianism was in Aram. בר־(א)נשא, *the* Son of Man par excellence.

Charles, 122, has noted that in ch. 62 'we have a lengthened account of the judgement, particularly of the kings and the mighty. This subject has already been handled shortly, 46.4-8, 48.8-10, 53-54.3; but here the actual scene is portrayed'. It is perhaps worth noting further that the form walda sab' υἱὸς τοῦ ἀνθρώπου is found in the shorter accounts of the judgement of the kings (46.2,3,4, 48.2), whereas the second Eth. version walda be'si occurs in Chh. 62-71 (62.5, 69.29 (Eth I), 71.14). The familiar Eth. Biblical expression is also confined to Chh. 62-71 (62.7,9,14; 63.11; 69.26,27; 70.1; 71.17).

Chief of Days See above, 192f.

(3) **to whom belongs ... righteousness dwells with him** For the 'Anointed One' of

---

so 'Satan'. To restrict בר־(א)נש(א) to a generic sense 'man' or 'a/the man' (or as 'one' with a self-reference) and to deny this κατ' ἐξοχήν usage, is to rob the Aram. language of a mode of expression it shares with Heb. and Greek. Moreover, in the older Aram., the distinction between the absolute and emphatic states (indeterminate and determinate) was by no means completely blurred: Rosenthal, in fact, singles out the use of the absolute state as one of the distinctive features of the early West Aram. dialect. See his *Die Aramäistische Forschung* (Leiden, 1939), 56, and for Qumran Aram. Fitzmyer, *Genesis Apocryphon*, 220. The poetic form too of the locution 'son of man' would also contribute to its development into a term designating a preeminent individual man and so to its titular use. Cf. ThWNT Bd. VIII ὁ υἱὸς τοῦ ἀνθρώπου 405.18 (Colpe). See also M. Black, 'Aramaic *Barnasha* and the "Son of Man"' in ET, 95.7, 200f.

David as the possessor of righteousness, see Isa. 9.6-11, 11.3f., Jer. 23.5, Zech. 9.9, Ps. Sol. 17.21f. etc. For the idea of righteousness 'dwelling' cf. Isa. 32.16 (שכן).

has chosen him Hence 'the Elect One'.

triumphs ... for ever The verb mo'a = 'conquers, overcomes', חזק על Hiph.? cf. Job 18.9, Dan. 11.7. Eth II 'triumphs over all'. I take kefl = גורל in the sense of 'party, followers' and for this meaning see F. Nötscher, *Zur theologischen Terminologie*, 170f. The usual meaning given to the word here 'lot, portion' does not go so naturally with mo'a. Similarly the Anointed One of David is represented traditionally as a victorious figure; cf. Isa 11 and IQSb 20-28. The righteous Son of Man, moreover, overcomes in or by uprightness.

(4) **shall rouse up the kings and the mighty** Cf. Isa. 14.9, i.e. rouse them either to oppose or to pay tribute. For 'the kings and the mighty' = 'mighty kings' see note above on 38.5. Is 'their couches' (משכבותם) a mistake for 'their dwellings' (מושבותם)?

loosen the loins of the powerful Eth. 'loosen the reins ...' leguāmāta ṣenuʿān, 'a strange phrase' Charles[1906], 87 n. 1.5 = מתני עצומים a misreading of מתני = 'loins'. The expression means to demoralise, probably also to disarm the powerful; a parallel is Isa. 45.1 ומתני מלכים אפתח.

break the teeth Ps. 3.7, 58.6.

(5) **cast down the kings** Eth. gafte'a = הפך (cf. Am. 4.11) or הרס = ἀφαιρεῖν (cf. Isa. 22.19, καθαιρεῖν). Cf. Sir. 10.14 θρόνους ἀρχόντων καθεῖλεν ὁ κύριος and Wis. 5.23 ἡ κακοπραγία περιτρέψει θρόνους δυναστῶν and Lk. 1.52 καθεῖλε δυνάστας ἀπὸ θρόνων. Cf. also Dan. 5.20.

with humble gratitude ... bestowed on them Eth. ganaya = התודה = ἐξομολογεῖσθαι(?). There is an implicit allusion to Dan. 4.17,25. Cf. also Wis. 6.2-3: ἐνωτίσασθε οἱ κρατοῦντες πλήθους ... ὅτι ἐδόθη παρὰ τοῦ κυρίου ἡ κράτησις ὑμῖν and Rom. 13.1f.

(6) **faces ... shall colour** Eth[tana] lit. 'will be overturned' yetgafattā' = נהפכו. The expression occurs at Jer. 30.6 (ונכפכו כל פניהם (לירקון 'all faces are turned into paleness' (AV), 'every face changed, all turned pale' (NEB) LXX 37.6 ἐστράφησαν πρόσωπα. The parallel suggests the paleness of shame, but it could also be one of fear. For a similar expression Dan. 3.19, and above, note on 39.14.

worms ... their beds On meskābomu 'their (last) resting-places', see above v. 4. Cf. Isa. 14.11, Job 17.13-15.

no hope of rising An allusion to Dan. 12.3(?) '... those that sleep in the dust (i.e. the righteous) shall arise'.

(7) **who ruled the stars of heaven** Eth. 'judge' in this context makes very poor, if, indeed, any suitable sense; I have preferred the meaning 'rule' for kuannana (Sir. 44.3 LXX κυριεύοντες) here rendering Heb. רדו. Is there a possible allusion to Dan. 8.10 where it is said of Antiochus IV '(And it (the little horn) waxed great even to the host of heaven; and some of the host of heaven and of) the stars it cast down (to the ground and trod upon them)'?

raise their hands against the Most High i.e. 'rebel against' נשא יד על 2 Sam. 18.28, 20.21, Ps. 106.26.

and tread ... earth While the Eth. text again yields some meaning, it is hardly a satisfactory sense in the context, unless we understand 'dwell upon the earth' as virtually 'occupy, possess the earth' (see 235). Charles (89) suggests that there has been a corruption, of a ptc. as a finite verb, יושביה as וישבו viz. '(who) trample on the earth and its inhabitants' אשר ירמסו על ארץ ועל יושביה. Was the original simply '(who) trample on the inhabitants of the earth (על יושבי ארץ)'? The clause continues the allusion to Dan. 8.10.

all their deeds ... unrighteousness An impersonal subject with 'ar'aya = הראה monstrare, ostendere, seems unnatural. Perhaps we should omit kuellu with Eth[b t] and render 'and their deeds of violence (מעשיהם חמס; cf. G-K, 128d, 415) they make a public spectacle of' (lit. 'they cause to be seen'). The reference might then be to the public execution of the Maccabaean martyrs (2 Mac. 7).

their power rests upon their riches The text as it stands makes a possible sense. If a conjecture seems required by the parallelism (cf. Charles, 'their boasting (תהילתם) is in their riches'), then 'their confidence (תוחלתם) is in their riches' gives a closer parallel to 'their faith is in the gods (אלילים?) which they have made with their own hands'. From such a text the mistranslation 'their power' (חילם) would readily arise.

(8) **the faithful who depend upon the Name** See note above on 38.2. Eth[tana] faqada = פקד, ἐπισκέπτεσθαι.

houses of his congregation Cf. 53.6 'the house of his congregation'. mestegubā' = συναγωγή. The full expression corresponds to the rabbinical בית הכנסת Meg. 25b 26a, בי כנישתא Ber. 17a; cf. also with 53.6 IQS 9.6 (cf. 8.5) בית יחד (בית קודש); for כנסת 'assembly' 4QPB 6 (*The Patriarchal Blessings*, ed. by John Allegro 'Further Messianic References in Qumran Literature', in JBL, Vol. LXXV (1956), 175).

CHAPTER 47

(1) **the blood of the righteous** It is usual for editors to treat the Eth. phrase dama ṣādeq (= τὸ αἷμα τοῦ δικαίου) as a collective singular, varying with the parallel phrase with the plural, dama ṣādeqān, 'the blood of the righteous' (v. 2) (cf. Eth[tana m] ḫualqua lasādeq, 'the number of the righteous' (v. 4)). The singular could, however, have a different force, one which is more evident in Greek than in Eth., namely in a singular noun as representing a class; a parallel is the use of ὁ δίκαιος in Wis. 2.12f. where it similarly alternates with the plural (2.12 τον δίκαιον, 2.16, 3.1 δικαίων); the style oscillates between 'the one' and 'the many'. In that case we cannot exclude at the same time a deliberate allusion to 'the Righteous One' par excellence of 38.2, 53.6. Isa 53.11 lies as certainly behind this passage as it does Wis. 2.12-18.

(2) **may not be in vain** Eth. teḏdarrā' 'may not cease'? See Dillmann, *Lex.*, 1329, taḏar'a III.1(f). Cf. בטל, 'cease', ܒܛܠ, P. Sm. 509, irrita, vana.

(3) Cf. for this verse Dan. 7.9-10.

Chief of Days See above, 192f.

books of the living Cf. for the expression 108.3 'book of life' from Ps. 69.28 ספר חיים 'book of life' or 'of the living'. In the apocalypses these books are conceived as records of good and evil deeds on the basis of which every man will be judged and assigned to perdition or to life everlasting; e.g. En. 89.61f., 90.17,20, 104.7, 108.3,7 etc. The books are kept by one of the seven archangels, presumably Michael, En. 89.61 cf. 90.14,22, but at Jub. 4.23, 2 En. 53.2, 64.5, Enoch himself, the 'heavenly scribe', is the keeper of this record of mankind's good or evil deeds. See further Bousset-Gressmann[4], 258.

his council Eth. 'awwed = סוד, e.g. Ps. 89.8.

(4) **the number of the righteous ... reached** Most manuscripts read 'the number of righteousness' which makes poorer sense, unless ḫuelqu 'number' is taken in the sense of μέτρον 'measure' (Knibb). The meaning would then be that the predetermined

measure of righteousness had been reached. The reference, seems rather to be to the fulfilment or completion (lit. the 'coming', qarba or baṣḥa) of the predetermined number of the elect, a familiar idea in the apocalypses; cf. 2 Esd. 4.36, Rev.7.4. See also ThWNT s. v. πλήρωμα Bd. VI, 303 (Delling).

the blood of the righteous ... avenged Eth. tafaqda = בקש Pa. Pu.? Note the coll. sing. = τὸ αἷμα τοῦ δικαίου, but cf. above, n. v. 1.

CHAPTER 48

(1) **the fountain of righteousness** Eth. naq'a ṣedq = מקור צדקה cf. IQS 11.5,6.

all the thirsty drank of them Cf. Isa. 55.1f.

(2) **at that time** (lit. **hour**) This temporal conjunction is here interpreted by Charles, following Dillmann, as referring to the moment when Enoch 'was beholding these visions'. It seems more likely, however, to refer forward to v. 3 which defines it as the time before the creation. The phrase 'at that hour' appears to be unique.

the Son of Man was named Cf. v. 3 'his name was named (יקרא?)'. The naming of a person could have special significance, implying, for instance, ownership; cf. Isa. 43.1 'I have called thee (Israel) by thy name; thou art mine'. It is frequently used in the Old Testament of the 'calling' of an individual or group for a special task, e.g. Isa. 45.3 addressed to Cyrus, Isa. 49.1 to the Servant of the Lord. This latter meaning seems to be here intended; the Son of Man was called or designated, lit. 'named', for his high destiny before the creation. See the discussion in T. W. Manson, *Studies in the Gospels and Epistles*, 135f. and in ThWNT, Bd. V, 252, s.v. ὄνομα (Bietenhard).

(3) **before the sun and the 'signs' were created** Charles (followed by Knibb) translates the second noun by 'constellation' and identifies these with the signs of the Zodiac here as at 72.13,19. But ta'āmer = σημεῖα, אותות in the sense of 'indications', 'signs' in the heavens, e.g. the appearance of certain stars, of the time of year and the corresponding climatic or meteorological changes. See below, note on 72.13. For this meaning of (השמים) אותות as 'signs, tokens of changes of weather or times', Gen. 1.14, Jer. 10.2, Ps. 65.9. See 111 and 395.

(4) **he shall be a staff** Cf. Ps. 23.3 and below 61.3,5, '... that they may lean on the Lord of spirits'. Eth. tamarguaza ba = Heb. שען Niph., Aram. סמך Ithp. + על.

the light of the Gentiles Isa. 42.6, 49.6, Lk. 2.32. The Son of Man is the Servant of the Lord.

the hope ... hearts Isa. 61.1,2 'The Lord hath anointed me to bind up the broken-hearted' (נשברי לב).

(5) **fall down and worship before him** Cf. also 62.9. This is the prediction in second Isaiah about the Servant, Isa. 49.7, 60.10, 'their kings shall serve thee'. The same is foretold of the Davidic king (the Nasi' of Ezekiel) IQS[b] 5.28: 'They (the Gentiles) shall come before thee and worship thee ...'. See Black, *Scrolls and Christian Origins*, 151. Cf. also Dan. 7.14.

celebrate with song Eth. zamara I.2 lasema 'egzi'a manāfest = Ps. 18.49(50) זמר לשם יהוה.

(6) **for this reason** Eth. ba'entaze = למען זאת, e.g., 1 Kg. 11.39.

chosen and hidden The Servant of the Lord is hidden 'in the shadow' of God's hand, Isa. 49.2. The Son of Man is hidden from everlasting to everlasting except among the elect to whom the divine wisdom reveals him (v. 7); cf. Sir. 24.9 (of wisdom): πρὸ τοῦ αἰῶνος ἀπ' ἀρχῆς ἔκτισέν με καὶ ἕως αἰῶνος οὐ μὴ ἐκλίπω.

from everlasting and for ever lit. 'from the beginning of the world ...' Eth[q] 'emqedma 'ālam = מקדם עולם; cf. 1QH 13.1,10 (מקדם עולם) and CD 2.7 'God did not choose them מקדם עולם'. See also Prov. 8.23 מקדמי הארץ and Mt. 24.21 ἀπ' ἀρχῆς κόσμου (Mk. 13.19, 2 Pet. 3.4 ἀπ' ἀρχῆς κτίσεως).

(7) **will preserve ... portion of the righteous** The tenses are perfect when we would expect futures as later in the verse 'in his Name they will be saved'. We have to do almost certainly with prophetic perfects (cf. Dillmann, 24 n. 1). The 'portion' (חלק?) which the Lord of spirits (or the Son of Man) will preserve for the righteous must be their future compensation in the world to come (unless kefl 'lot' = גורל 'the party, company' as above, 208).

loath and despite this world Cf. 108.8f., Gal. 1.4.

in his Name they will be saved A familiar New Testament expression, but here 'the Name' is that of the Lord of spirits.

he will become the avenger ... i.e. the Lord of spirits. Eth. faqādē(u) = ἐκδικητής(?) cf. Dillmann, *Lex.*, 1362. Cf. the use of Heb. פקד 'punish' (Isa. 10.12, Jer. 9.24) or נקם Ps. 8.3 (LXX ἐκδικητής), and of גואל 'avenger' as a name in later Judaism for the Messiah; Knibb 'and he is the one who will require their lives'. Charles adopts the inferior reading bafaqādu and renders 'according to his good pleasure it has been with regard to their lives'.

(8) **downcast in countenance** i.e. in fear and horror at their fate. Eth. lit. 'they became downcast of countenance' = נפלו פניהם(?). Cf. 62.15 Eth. lit. '(they ceased) to make their countenance downcast' ('atḥata gaṣṣa = הפיל פנים?).

on account of the deeds ... hands Charles takes with the previous clause: 'the strong who possess the land because of the works of their hands'.

anguish and affliction Cf. the Biblical צרה וצוקה Isa. 30.6; צר ומצוק Ps. 119.143.

save themselves Cf. the Biblical expression הציל נפש, Ezek. 14.14,20, Isa. 47.14. Is there an allusion to the latter verse in the following metaphor of 'stubble of fire' (v. 9)?

(9) **give them ... into the hands of** See above on 38.5.

As stubble in the fire Exod. 15.7,10, Isa. 5.24, Ob. 18, Mal. 4.1.

so shall they burn before the face of the holy Cf. 27.2,3, 90.26,27. The reference can only be to the fires of Gehenna. See Bousset-Gressmann[4], 286 and ThWNT s.v. γέεννα Bd. I, 655f. (Jeremias). At 27.2,3 and 90.26,27 the torments of the wicked appear to be an ever present spectacle for the righteous, whereas in the Parables it appears to be a temporary spectacle only; the wicked are to vanish forever in Gehenna from the sight of the righteous; 48.9, 62.12,13; cf. Rev. 20.14 and 2 Esd. 7.36 'apparebit lacus tormenti, et contra illum erit locus requietionis; et clibanus gehennae ostendetur, et contra eum jocundatis paradisus'.[1]

lead in ... water Exod. 15.10.

(10) **the day of their affliction** Cf. 45.2, 50.2.

before them (i.e. my elect ones (v. 9)) **they shall fall ... rise again** The variant 'before him'—God is the speaker at v. 9—has been introduced by a scribe who understood the text to mean 'fall down and worship', but here it is the 'downfall' of the kings of the earth in the presence of the righteous which is meant. The writer may have Ps. 36.13 in mind: 'There are the workers of iniquity fallen (נפלו): they are cast down (דחו), and shall not be able to rise'. In the words 'they shall not rise again' there may be a deliberate allusion to the resurrection which the wicked kings will be denied.

take them by his hands For this Biblical expression Isa. 51.18 ואין מחזיק בידה. Cf.

[1] ed. G. H. Box, *The Ezra Apocalypse* (London, 1912), 336.

Isa. 41.10(9) LXX ἀντιλαμβάνεσθαι: Israel, on the other hand, God's Servant, is 'taken by the hand', i.e. helped by God.

denied ... and his Anointed One For the denial of the Lord of spirits see above, 195. This description of the Son of Man as 'Anointed One' occurs here and at 52.4 and nowhere else in the Parables. It might seem as if the writer deliberately goes out of his way to avoid the use of such a term in connection with his Son of Man or Elect One, presented by him in the role of the humiliated servant of the Lord. Perhaps it was already becoming a term especially associated with the Davidide in a political sense. Or it may be he reserves the term to the climax of this vision, and is sparing in its use simply because it is a special term of majesty with royal as well as high-priestly and prophetic associations. It is used here, however, in a verse which clearly seems reminiscent of Ps. 2.2, and, after an account of the rôle of the Son of Man as the 'Righteous and Elect One', with features derived not only from Isa. 42, 49 and Dan. 7, but from the description at Isa. 11.1 of the 'root of Jesse', one soon to figure prominently at 49.3 f. At 52.4 it seems mainly to be the universal dominion of the Son of Man which the author has in mind, going back again to Dan. 7 (see below on 52.4). See ThWNT, Bd. V, s.v. χρίω, 505f. (de Jonge).

blessed be the Name For these doxologies concluding a vision see the note above on 22.14.

CHAPTER 49

For 49.1 — 51.5 see Lib. Nat., Tom. 41,57f., 42, 50f.

(1) **before him** Could refer to the Lord of spirits or 'his Anointed One', the Son of Man (48.10) or Elect One (49.2). If we are dealing with different sources the author has woven them skilfully together.

(2) **mighty ... righteousness** = בכל רזי צדק (ḥāyāl) כביר(?) (cf. Job 36.5 כביר רוח לב).

disappear as a shadow ... no continuance ... the Elect One standeth As Charles noted, the phraseology is borrowed from Job 14.2: '(Man born of woman) ... fleeth also as a shadow and continueth not' (lit. 'does not stand') (ויברח כצל ולא יעמוד). The Elect One standeth (עמד): 'unrighteousness will have no standing ground because the Elect One standeth' (Charles). The play on words is only possible in Heb. For the phrase 'to stand before, attend upon, be a servant of', עמד לפני, 1 Kg. 17.1, 18.15, above, 113. In this context, however, the verb also carries the idea of the continuance and the permanence of the Elect One in his position before the Lord of spirits.

(3) For this verse cf. Isa. 11.2: Eth. reproduces practically verbatim the clauses in this verse: רוח חכמה ובינה, רוח עצה וגבורה. Here, as at 62.2, Isaiah's 'messianic oracles' concerning the Davidide are expressly applied to the Elect One; and this follows closely on the author's own rare use of the term 'Anointed One' at 48.10 for the same Elect One figure. See below 218 and J. Muilenburg, 'The Son of Man in Daniel and the Ethiopic Apocalypse of Enoch', in JBL LXXIX (1960), 204f.

who sleep in righteousness After this otherwise almost word for word translation of Isa. 11.2, Eth. substitutes this surprising clause for 'the spirit of knowledge and of the fear of the Lord'. Eth. ʾella nōmu clearly here means 'those who sleep in death', thus introducing the idea of the spirits of the faithful dead. It seems very unlikely, however, that this is what stood in the original. Is the text corrupt? See Textual Notes, 351.

(4) **the secret things** This seems to be an interpretation of Isa. 11.3: '... he shall not judge after the sight of his eyes etc.'

an idle word Cf. 62.3, 67.9. The expression occurs at Dt. 32.47 דבר רק, LXX λόγος κενός '(no) useless things or word'; cf. Ps. 4.3. where ריק 'emptiness, vanity' is parallel to כזב 'deceit'. Cf. also Mt. 12.36 in a similar context of judgement. The expression means a great deal more than 'idle chatter' or 'careless talk': it includes deceptive speech, including unfulfilled promises or covenants, false witness, all the evil that the tongue can create.

CHAPTER 50

(1) **a change shall take place** Eth. miṭat 'directio rerum (fatum) ...' (סבה), Dillmann, *Lex.*, s.v. Cf. 1 Kg. 12.15, 2 Chr. 10.15 נסיבה ('this turn to the affair' NEB). What is meant is a reversal of fortunes for both the righteous and sinners.

the light of days(?) shall remain upon them The expression 'light of days' seems unusual. It is explained by Dillmann, 163f, from 58.5,6: the change for the righteous is like coming from the dominion of sinners out of night into sun-light. Eth[t] preserves a remarkable variant, 'the Ancient of Days (shall remain upon them)', which, if it is not the true reading, must be regarded as an extremely bold scribal theological intrusion, from the Biblical idea of Jahweh dwelling with Israel (Zech. 2.10, Jl. 3.17,21 etc.). It seems more likely that a corruption has come into the text, possibly in the word 'days' (mawāʿel) from the beginning of the verse. Did the original perhaps read 'the light of the peoples (אור עמים, misread as אור ימים)'; cf. Isa. 51.4 where the divine judgement is so described. In support of this conjecture is the reference in v. 2 to the repentance of the Gentiles.

(2) **evil shall have been heaped up** The locution is Biblical, e.g. Hos. 13.12, צפונה חטאתו. The text should perhaps be understood in the light of Job 21.19 'God reserves punishment for' (צפן און ל ...); יצפן און לחטאים 'punishment(?) will be reserved for sinners'. Cf. Rom. 2.5.

the Gentiles lit. 'the others' = אחרים, but in the sense of 'the aliens, the foreigners'; Ps. 109.8, Job 31.8,10. Dillmann, 164, die Nichtisraeliten. The idea of the conversion of the Gentiles as at 90.30,33,34, 91.14. Cf. οἱ λοιποὶ τῶν ἀνθρώπων, Rev. 9.20 and Black, *An Aram. Approach*[3], 176.

(3) **his compassion is great** Eth. meḥratu, cf. Isa. 54.7.

(4) **unrighteousness** Eth. waʿāmadā = גם רשעה לא תקום(?) lit. 'unrighteousness assuredly shall not stand'. Cf. Job. 18.5 for this use of גם = profecto.

(5) **I shall have no mercy** Cf. 60.5, 2 Esd. 7.33, 2 Bar. 85.12.

CHAPTER 51

(1) **that which has been entrusted ... committed** There are several possible Heb. equivalents for Eth. māḫḏant: (a) cf. Isa. 48.6 נצורות, lit. 'things guarded' (EVV 'hidden things'); (b) Lev. 5.21,23 (MT) פקדון, NEB 'deposit', LXX παραθήκη 1 Tim.

6.20 (so Goldschmidt); (c) 1 Sam. 22.23 משמרת, lit. 'a thing (or person) protected'. The variant of Eth[g] may be explained as a paraphrase of māḫḍant, but it could also be closer to an original = τὰ τεθησαυρισμένα ἐν αὐτῇ, אשר שפונות בה; cf. Dt. 33.19 'the hidden wealth (שפוני) of the sand' (NEB). Eth. tamaṭṭawa III 2 = ... נסמר ל παραδοθῆναι(?) 'to be delivered up', especially of captives (cf. 95.7). This verse (with v. 2) and 61.5 represent the author's view of the resurrection. No doctrine or belief has been more widely debated or more extensively documented: see especially Str.-B. Bd. IV, 1166f.; 1167 deals with Enoch, 1168c (last sentence) implies that for the author of the Parables non-Israelistes were excluded from the general resurrection, but this does not agree with the evidence of 50.2 which allows repentance to Gentiles and therefore, presumably, a share in the destiny of the righteous. For an up-to-date bibliography, Schürer, *History* (1979), II, 539f.

Abaddon Eth. ḥagul = ἀπώλεια, אבדון, אבדנא. See above, 282 n., 113.

A clear allusion to this v. occurs in the Pseudo-Philo's *Liber Bib. Antiquitatum* 3.10 'et reddet infernus debitum suum et perditio restituet paratecam suam'. Note especially the phraseology 'debitum suum' and 'paratecam suam' (= פקדון). While there are many other parallels with Enoch (cf. *Pseudo-Philon Les Antiquités Bibliques*, ed. C. Perrot and P.-M. Bogaert, Sources Chrétiennes, 230, Tom. II, 291), none reproduces the actual terminology so closely as the quotation in Pseudo-Philo. This may be important for the dating of the Parables, since the *Liber Bib. Antiquitatum* is now being placed c. A.D. 100 (See D. J. Harrington, 'The Biblical Text of Pseudo-Philo's *Liber Antiquitatum Biblicarum*' in CBQ 33, 1 (1971), 16). Cf. also 2 Esd. 7.32 'The earth shall give up those who sleep in it, and the dust those who rest there in silence; and the storehouses shall give back the souls entrusted to them' (NEB).

(2) **he shall choose** The Lord of spirits or the Elect One of v. 3?

(3) **on his throne** Eth I 'on my throne'. An original על כסאו could be read either way in a manuscript where no distinction was made between Yodh and Waw. At 47.3 the Chief of Days sits upon his throne, but at 45.3, 62.5 it is the Elect One who sits upon his throne.

all the secrets ... mouth Knibb renders 'all the secrets of wisdom will flow out from the counsel of his mouth', and this is the reading of most manuscripts. Charles, however, noted that a number of manuscripts put 'all the secrets' in the accusative and some manuscripts read 'and from the counsel of his mouth': he proposed to read 'afuhu yāwaḍḍe' 'his mouth shall pour forth all the secrets of wisdom and counsel'. Cf. Job 37.2 והגה מפיו יצא.

appointed him Eth. wahaba = נתן, e.g. 1 Sam. 12.13; etc.; cf. Knibb, 2, 136.

(4) **all will become angels in heaven** The objection of Charles, 101, that the text of Eth II is 'wanting in sense' can no longer be sustained in the light of IQH 3.19-23, 6.13, Lk. 20.35f., and cf. Str.-B. Bd. I, 891. See further Black, *Scrolls and Christian Origins*, 139f.

(5) **shall arise** Eth. tanše'a = יקום.

go and walk thereon This need not be a ditt. as Charles thinks. Does Eth. yānsosewu (from sosawa) = יתהלכון as at Gen. 13.17? The elect are to go through the length and breadth of the land as did the old Israel when it entered into possession of the land of promise. The prophecy about 'inheriting, possessing' the land is now to be fulfilled for the elect of the new Israel; cf. Ps. 37.3,9,11,29,34, Mt. 5.5.

CHAPTER 52

Chapters 52-57 form a new section in the Parables corresponding to Enoch's journeys in the older Book, especially chapters 17-36. Certain visionary scenes or objects in these earlier chapters are singled out for special mention or development, e.g. the mountains at 18.6, the 'hollow places' of 22 become the 'deep valleys', the places of punishment, of 53,54.

(1) **carried off ... a whirlwind** cf. 39.3, 14.8.

towards the west Dillmann suggests that the writer has in mind stories of fabled mountains of metal in the west inspired by tales about the Phoenician mines in Spain, similar to the tales of the mountains of precious stones in the south at chapter 24.

(2) **which are ... to happen** Eth. zayekawwen hallo = ἃ μέλλει/δεῖ γενέσθαι LXX Dan. 2.30 מה די להוה.

a mountain of iron ... lead The allegory of the six metal mountains clearly has its source of inspiration at Dan. 2.31f., especially v. 39f., where the metals of the great image in Nebuchadnezzar's vision represent the empires which successively subdued Israel. And just as Daniel's empires are broken in pieces (2.44), so Enoch's kingdoms of iron, copper, silver, gold etc. melt in the fire before the Elect One (52.6), and are destroyed from off the face of the earth (52.7-9). The symbolism of the world-empires as 'mountains of metal' may have been suggested by Dan. 2.35, where the stone that smote the image—the new Israel or her Messiah—becomes a great mountain. For 'metals' in classical mythology, Origen, *Contra Celsum*, VI.22 (cf. Charles, 102f.) and as symbolising the different ages of the world, see Hengel, *Judaism and Hellenism*, I, 182, II, 129, n. 559.

Charles argues that there ought to be seven not six mountains mentioned, and this is certainly what we would expect if the writer is using the material at 18.6. He suggests that the metal omitted in this list is supplied by 67.4 which, while mentioning five metals only, nevertheless includes 'tin' (nā'k), a metal which he considers was not included in 52.2. But there are difficulties with this proposal. The 'soft metal' naṭabṭāb is not defined by Charles, but, as Dillmann argues, in view of 65.7,8 it can only be 'tin' at 52.2.[1] We are still left then with six metals only. However, naṭabṭāb 'Tropfmetall' is Greek σταγών which is used for other kinds of 'soft metal' such as stannum, an alloy of lead and silver. If it referred to a soft metal other than 'tin' at 52.2 and nā'k 'tin' was originally included in the list, we would then obtain seven metal mountains.

(3) **What are those things which I have seen in secret**? The 'things in secret' which Enoch has seen are the mountains of metal, their 'secret' being that they typify the world-empires which have successively oppressed Israel.

(4) **All these things ... shall serve the dominion of his Anointed One** The world-empires shall become subservient to the rule of the Elect One. The author has clearly in mind Dan. 7.18, the universal dominion of the Son of Man over all the empires of the world. The variant of Eth$^{q\ u\ tana}$ 'shall be in the dominion of his Anointed One' is also a possible original. For the term 'Anointed One', see on 48.10. Eth. kona la = היה ל could also mean 'shall belong to (the dominion of his Anointed One)'.

(5) **the angel of peace** See above, 199f. The description of the angel as 'angel of peace' is appropriate in a context which envisages the cessation of warfare (v. 8).

which the Lord of spirits has prepared(?) Translators differ in their choice of verb. Dillmann rejected the meaning 'decreed' for takala, although admitting it as a

---

[1] At 65.7,8 naṭabṭāb is a general term which can describe either 'tin' or 'lead' and since 'lead' is specified at 52.2, it must then be 'tin'.

possible sense, and opted for 'which the Lord of spirits planted'. (If we can speak of mountains having 'roots' they can also be said to be 'planted', 168). Knibb renders by 'established'. Charles and Flemming adopt the reading of Eth I, which Charles emends and renders 'all the secret things which surround (kallala) the Lord of spirits', a translation which cannot be faulted linguistically, but which makes an odd sense in this context. The same is true of Flemming-Radermachers' 'which the Lord of spirits holds encompassed' (umschlossen hält). The exact equivalent of kallala is Heb. עטר, 'to surround, encompass', which looks like a mistake for עתד, 'to prepare'; cf. the use of ἐπιτηδεύειν at 9.6 for the 'preparing' of τὰ μυστήρια τοῦ αἰῶνος.

(6) **all these ... Elect One ... wax before the fire ... water which streams down from above** The whole verse with its Biblical imagery recalls 1.6 with the significant difference that the theophany is of the Elect One not of God (ὁ θεός). The language and imagery is borrowed from Ps. 97.5, Nah. 1.5 and virtually verbatim from Mic. 1.4, כדונג מפני האש כמים מגרים במורד ...

become powerless under his feet The Elect One, like God in the earlier theophany, shall tread upon the mountains and find them soft and yielding (Eth. dekumān = אמלל(?) Nah. 1.4 'languishing' AV) before or under his feet. Cf. also Isa. 16.8.

(7) **Either by gold or by silver** Cf. the close parallel at Zeph. 1.18 'neither their silver nor their gold shall avail to save them' (NEB): (גם־כספם גם־זהבם לא יוכל להצילם); cf. also Isa. 13.17, Ezek. 7.19, Ps. 49.7-10, Jer. 4.30. In the allegory here the gold and silver mountains typify the wealth of the world-empires which subdued Israel: that wealth will not save them on the 'day of the Lord'.

(8) Cf. Hos. 2.18, Isa. 2.4, Zech. 9.10, Ps. 46.9.

(9) **rejected and destroyed** Eth. yetkaḥḥadu 'denied (and destroyed)'. Dillmann suggests giving a Heb. meaning to takeḥda = 'be wiped out' (Knibb adopts the suggestion but proposes an Aram. יתכחדון). Charles considered 'denied' and 'destroyed' alternative versions of an original Heb. יכחד, coming from a Greek version, ἀπαρνηθήσονται and ἀφανισθήσονται. An original ימאסו 'they will be rejected' (cf. Jer. 6.30, of the rejection of silver) could have given rise to ἀπαρνηθήσονται. (Was there a pun on מסס 'to be melted away?) See above, 186.

CHAPTER 53

(1) **a deep valley** For the expression cf. Isa. 30.33 LXX φάραγγα βαθεῖαν (העמק הרחב?).[1] As at 22.1f. the description of the 'hollow places' follows that of the seven/six mountains. The former describes Sheol and the underworld: the 'deep valley' of the Parables here has been identified with the valley of Jehoshaphat in Jerusalem, where, according to Jl. 3.2,12, the Gentiles were to be assembled for the last judgement (Dillmann, Charles, Martin, Schodde). Against this identification is the apparent location of this valley 'in the west' ('There my eyes saw etc.' referring back to 52.1,2 where Enoch journeys westwards). Nothing is, in fact, said, in these verses, about the judgement of the Gentiles, only about this assembly of all mankind with its offerings and tributes. It is true, in v. 3, this 'deep valley' is described as a place where 'the

---

[1] Some Eth. mss appear to have read a plural 'valleys' but this seems secondary tradition, the result perhaps of reading qualāta (sing. Eth^{tana}) as plur. (cf. Dillmann, *Lex.*, 410).

angels of punishment prepare their instruments of torture for the 'kings and the mighty',' but it does not follow that this was the place of torment itself. In fact, 54.1-6, which follows the appearance of the Elect One (53.6-7) goes on to describe another 'part of the earth' with another 'deep valley' into which the 'kings and the mighty' are cast: the two 'deep valleys' are clearly distinguished, and it is this second which we are to identify with the 'valley of Jehosaphat and Gehinnom'. Dillmann, 169, thought of this first valley, with its appeasement offerings, as lying in the neighbourhood of the metal mountains, and destined also, like them, to be totally destroyed. According to the Onom. Sacra (ed. Lagarde, 300) the two 'valleys' are identical: φάραγξ Ἐννόμ... παρακεῖται δὲ τῇ Ἱερουσαλήμ · λέγεται δὲ ἔτι νῦν φάραγξ Ἰωσαφάτ. On the valley of Jehoshaphat see especially A. S. Kapelrud, *Joel Studies*, (Uppsala, 1948), 147, referring to 'an ancient Biblical 'valley' tradition with sinister associations' (Isa. 22.1-8,12-14, Ezek. 39.11-29, Zech. 14.1-15); this valley, however, 'cannot be located on any map ... (but) belongs to the sphere of mythology ... perhaps conceived as ... in the vicinity of Jerusalem'.

land and sea and the islands i.e. the whole inhabited world. 'Islands' (איים) is a comprehensive term for the islands and coastlands, usually of the Mediterranean. The Gentiles bring their offerings and gifts to him, i.e. the Elect One rather than to it, the deep valley. The writer is drawing on Ps. 72.10: 'The kings of Tarshish and the islands shall bring gifts (מנחה), the kings of Sheba and Seba shall present their tribute (אשכר), and all kings shall pay him homage, all nations shall serve him' (NEB). The gifts and possibly tribute are designed to win the favour of the Elect One: but the valley is so wide and deep that it cannot receive enough gifts and tribute—it can never be filled with them. The idea would seem to be that no amount of tribute or offerings would be sufficient to 'buy off' the Elect One.

(2) **Their hands fashion idols(?) ... swallowed up** The traditional text, as it stands, can only be understood as it is literally rendered by Knibb (= Dillmann, Flemming-Radermacher): 'And their hands commit evil, and everything at which (the righteous) toil, the sinners evilly devour'. But the resultant sense is poor, and suspicion of textual corruption seems justified (cf. Dillmann, 169). Charles, 104, emended the verb yeṣāmewu, '(at which) (the righteous) toil' to yeṣṣāmawu, rendering '(the sinners devour all whom) they (lawlessly) oppress'. He supports the emendation by noting the reference at v. 7 to 'the oppression (ṣāmā) of sinners'. The basic meaning of gēgāy (Knibb, 'evil') is 'iniquity, error' (ἀνομία, πλάνη) (Dillmann, *Lex*., 1200, top) so that it is not surprising to find Eth[l] reading ṭā'ot 'idolatry, idols' as an explanatory gloss. Does this supply the clue to the difficult text? I suggest that gēgāya gabra = עשה און, where און = 'idols' (abstract for concrete cf. Isa. 41.29, 66.3): '(their hands) fashion idols'; and zayeṣāmewu lagēgāya = פעלי און 'the workers of iniquity'. For this last phrase, cf. Wis. 15.8 κακόμοχθος = zayeṣāmu la'ekuy. The parallelismus membrorum requires a pass. verb: has an original Pu. יבלעו lit. 'shall be swallowed up', 'shall be destroyed' (cf. Isa. 9.15) been misread as Qal '(sinners) shall swallow up'. The reading 'sinners' (ḫāṭe'ān) has probably come into the text either as a gloss or as a consequence of misreading a passive as an active verb.

Who shall not abide ... be exterminated for ever and ever Eth. yetqawwamu (Eth[tana] yeqawwemu) is taken by Dillmann as = tanše'a, in the sense of 'be stirred up' and so 'frightened off' 'aufgescheucht werden' ('from the face of the earth'); Flemming-Radermacher 'hinweggetrieben werden' (hence Knibb = Charles 'banisghed'). The next phrase wa'iyaḫālequ lit. 'and not cease' is taken as part of the predicate by Dillmann (= Flemming-Radermacher, Knibb): thus Knibb 'will be banished from the face of his earth, unceasingly, for ever and ever'. Charles, however, read wayaḫallequ '(from off the face of His earth) and they shall perish (for ever and ever)', a reading now found in Eth[tana]. I have adopted this text, but reading the negative with the previous verb wa 'iyetqawwamu ('iyeqawwemu) in the same sense of the phrase as at 41.2, 'and they could not abide' = ולא יקימו (perhaps with the added sense 'they could not withstand, stand up to, resist'). For yaḫālequ the equivalent Heb. verb would be יכלו, lit. 'they will be finished, exterminated'.

(3) **angels of punishment** See note at 40.7 on the 'satans'; cf. further Isa. 37.36, Ps. 78.49 מלאכים רעים 'messengers of evil'. There does not seem, however, to be an obvious Biblical equivalent of Eth. malā'ekta maqšaft; the variant malā'ekta ma''atu at 63.1 οἱ ἄγγελοι τῆς ὀργῆς αὐτοῦ suggests an original מלאכי עברה, ἄγγελοι ὀργῆς. Does it come from Job 40.6 (11) LXX ἀγγέλους ὀργῇ, ἐν ὀργῇ, τῆς ὀργῆς (Eth)? Further at 56.1; 62.11; 63.1.

continually preparing We seem to have here a case of the Hebrew auxiliary use of הלך G-K § 113u.

(iron) fetters of Satan These are presumably the chains (cf. 54.4) by which the wicked are bound in Gehenna, the place where they are also forged (54.3-5). The kings and potentates are first bound by them and then cast into Gehenna.

(5) **the kings and mighty** Cf. 38.5, 46.4 and note below, 235.

(6) **The Righteous and Elect One ... Lord of spirits** 'The Righteous and Elect One here only. I have adopted the reading of Eth[tana] and taken 'and his congregations (coll. sing.) as subject of the second verb, lit. shall not be prevented, i.e. probably legally prohibited by interdict. Most editors follow Eth[M] taking yāstare''i as transitive '(the Righteous and Elect One) shall cause the house of his congregation to appear' (Charles), but re'ya IV 1 = ראה Niphal. Is the writer thinking of the Temple (cf. Dillmann, 90.29 and Ps. 74.8) or of synagogues of the faithful? See above, 209.

(7) V. 7 reverts to the theme of 52.1-9 and could have been displaced from this section, perhaps after v. 6 or v. 9.

shall become level land(?) Most manuscripts read 'will not be as the earth' which is interpreted to mean that the mountains will not be firmly based like the earth, which is established forever (cf. Ps. 78.69, Ec. 1.4). Alternatively, we may omit the negative with Eth[tana, b] 'will be ... as the earth' i.e. flat and level (so Dillmann). The manuscripts further vary between 'before his righteousness' (baqedma ṣedqu) and 'before his face' (baqedma gaṣṣu). Eth[m] has the unusual reading 'as his righteousness (as the earth)'. The suggestion behind the translation adopted (which would then agree with Dillmann's interpretation) is that the original Heb. was כארץ המישור ὡς γῆ πεδινή like the level land' (for the expression, Jer. 48.21, Ps. 143.10, Dt. 4.43). The variants in Eth. could then have arisen by a failure of a translator to recognise this expression and render part of it (ה)מישור by ṣedq 'righteousness', the meaning of the word (e.g. Isa. 11.4) with which he was most familiar. The unusual reading of Eth[m] would then be closest to the original. The phrase 'before his face' may not be original but have arisen out of the difficult 'before his righteousness'.

hills ... like a fountain of water Cf. J. C. Hindley, 'Towards a Date for the Similitudes of Enoch' NTS, 14, 1967-8, 562f.; Hindley considers that there is 'an extraordinarily close historical parallel' to the account in Dio Cassius (LXVIII 25§6) of the earthquake in Antioch in A.D. 115, where the collapse of hills was accompanied by streams of water.

CHAPTER 54

(1) **a deep valley with burning fire** i.e. Gehenna. See note on 53.1 and cf. 56.3-4. Their torment is to become a spectacle for the righteous (48.9, 62.12). For reference to the place of final torment in Enoch elsewhere see 10.6,13, 18.11, 19.1-3, 21.7-10, 90.24, and for their relationship to the concept in the Parables, Dillmann, 284.

(2-3) Cf. Isa. 24.21,22 and note on 90.25.

their fetters i.e. 'fetters were being fashioned for them'. Eth. mabā'elat, mā'bal, כלי ברזל, σκεῦος σιδηροῦν at 3 Kg. 6.7; cf. also Ps. 149.8 בכבלי ברזל LXX ἐν χειροπέδαις σιδηραῖς, Nah. 3.10 בזיקים, χειροπέδαις, Dan. 4.23 באסור די פרזל, Theod. ἐν δεσμῷ σιδήρῳ; 'of incalculable weight' = אין משקל 1 Chr. 22.3, 14 ('more bronze than could be weighed' NEB).

(5) **the host of Azazel** Cf. 10.8f., on which the author is basing his text.

the depths of hell Eth. mathetta ⌜kuellu⌝ dayn, lit. 'the lowest part of ⌜all⌝ = שאול תחתית Dt. 32.22. LXX ἕως ᾅδου κάτω. Eth. 'all hell' is a curious expression; Charles rendered 'into the abyss of complete condemnation' (dayn). Could ኵሉ be a corruption of ሲኦል 'Sheol', 'to the lowest part of Sheol', and dayn, infernum, represent first a gloss and later replace the corruption?

cover over them The correct text is found in the Garrett manuscript (Eth^{a^1}) 'cover over them (mal'eltihomu = ἐπάνω αὐτῶν) with rough stones', corresponding to 10.5 'place upon him (ἐπίθες αὐτῷ (lā'lēhu) λίθους τραχεῖς καὶ ὀξεῖς). The traditional text 'cover their jaws' (malātehihomu) is an obvious corruption, yielding a virtual non-sense.

(6) **And Michael etc.** On the four archangels in the Parables, see above, 199, cf. also 9.1 n.

in becoming subject to Satan The idea that the watchers were the subjects of Satan is peculiar to the Parables, reflecting a later demonology. Satan is no doubt conceived of as the Head of 'the satans'. Cf. Charles, 66.

the burning furnace The expression occurs at Zech. 12.6 כיור אש 'an oven of fire' and Dan. 3.6,11, אתון נורא יק(י)דתא 'the burning fiery furnace' of Nebuchadnezzar. At En. 98.3 it is used of the fire of Gehenna. Cf. Str.-B. Bd. I, 673 on Mt. 13.42. The rabbis use the term תנורא for the 'furnace' of Gehenna, based on Isa. 31.9, Mal. 3.19.

(7) The view that 54.7-55.2 (Charles, Martin) is an 'interpolated' Noah apocalypse overlooks the evident dependence, here as elsewhere, of the Parables, the 'Second Vision', on the older 'First Vision': in this passage the destruction of the watchers corresponds to the account at En. 8.1-9.11 of the condemnation of the watchers, followed at 10.2 by an account of the deluge, The Parables follow the same pattern.

punishment ... go forth Cf. the expression at Jer. 4. 4., 21.12 יצא הקצף.

waters ... above the heavens ... beneath the earth Tradition knows only of the two 'chambers' or 'treasuries' of waters, those above the heavens and those beneath the earth, e.g. Jer. Tg. and Gen.r. on Gen. 1.6 מים העליונים; מים התחתונים (Levy, CW, II, 30).

(8) **the waters ... above ... masculine, ... beneath ... feminine** Charles traces these ideas to Babylonian mythology. The expressions are found in the Jerusalem Talmud, Berach ix 2 המים העליונים זכרים והתחתונים נקבות

(9) **under the ends of heaven** Cf. Gen. 6.17: the Deluge is to destroy all flesh 'from under heaven' (מתחת השמים); for the expression here used, Dt. 30.4 = Neh. 1.9 בקצה השמים 'the uttermost parts of heaven'.

(10) **inasmuch as they acknowledged** Eth. waba'enta za'a'marewwo = וכאשר ידעו; cf. Jer. 3.13, 14.20, Isa. 59.12. The reading of Eth^{tana} 'inasmuch as they did not acknowledge' could be original.

CHAPTER 55

(1) **In vain have I destroyed** Cf. Gen. 8.21. There is nothing, however, in the account of the Deluge in Genesis to suggest that it was all to no purpose (baka = ריקם, הנם, לשוא?), unless it is the statement 'for the imagination of man's heart is evil from his youth', i.e. man is incurably evil, and, therefore, the punishment of the deluge was to no purpose.

(2) **a pledge of good faith** Eth. simply hāymānot 'faithfulness' = אמונה(?). Gen. 9.13 has אות ברית 'a token of a covenant'. Could hāymānot here = אמנה, Neh. 10.1, 'covenant'.

so long as the heavens are above the earth Cf. Dt. 11.21 'As the days of the heaven above the earth' i.e. so long as heaven endures above (or resting on) the earth'. See S. R. Driver, *Deuteronomy*, ICC, 121.

(3) **my command [with regard to the host of Azazel]** The first four or five words of the text of this verse are confused and corrupt, with some part of the text missing (cf. Charles[1906], 99). Charles reads with Eth[g t] 'this is in accordance with my commandment', and takes it to be the last sentence of 55.2 referring to the 'pledge of faithfulness' as the divine command. All other editors follow the traditional verse division and take it as the opening words of 55.3. The verse itself clearly refers back to 54.6f., the fate of Azazel and his host, and is usually explained as the resumption of 54.1-6 after the 'Noah interpolation', thus originally following 54.6. But we are then left with the problem of an utterance of the Lord of spirits in the first person for which there is no preparation at 54.6, whereas it follows naturally the first person in 55.2. The suggested emendation of an obviously deficient text and the consequent translation assumes that a phrase referring to Azazel and his host, lateʿyenta 'azāze'ēl has been lost by hmt: 'And this is my command [with regard to the host of Azazel] when I am pleased to seize them by the hand of the angels'. The divine compassion announced in 55.2 does not include the fallen watchers, Azazel and his crew.

cause my anger ... to abide on them For the equivalent expression in Heb. הניח חמה, Ezek. 5.13, 16.42, 24.13.

(4) **who occupy the earth** See below, 235.

throne of glory The reading of Eth II 'my glory' can hardly be right. The Elect One (Son of Man) sits on his own throne and not on the throne of the Lord of spirits. The reading 'on the right hand of my glorious throne' is theologically more correct and this may account for the variant.

CHAPTER 56

(1) **they went, holding scourges and fetters of iron and bronze** The reading 'eḫuzān of Eth[M] is preferable to ye'eḫḫezu, since it corresponds to Heb. אחזים = ἐχόμενοι; cf. Ca. 3.8 אחזי חרב 'holding swords'. See also Dillmann, 173: Aram. אחידין. Eth. mašger usually renders παγίς (פח) 'snare', but is generally given the meaning here of 'bond', δεσμός (cf. Dillmann, *Lex.*, 267). It may, however, be a scribal error for ma'asert, the more usual word for δεσμός as at 54.3. I have adopted the reading of Eth[g (m u)] 'scourges and fetters of iron and bronze', since v. 2 repeats the phrase 'those who hold the scourges' (again Eth[g]; Eth[M] omits the object and leaves an impossible text, 'those who are holding?').

(3) **elect ... beloved ones** Eth II 'each to their elect ones...' identifies the 'beloved ones' with the children of the angels of punishment. Dillmann interprets, in the light of 90.26,27, of the 'blinded sheep', the apostate Israel, and Schodde thinks of the 'kings and the mighty'. Charles notes, however, that the term 'beloved one' is especially used at 10.12, 14.6 to refer to the children of the watchers (cf. En[c] 1 vi 16, EnGiants[a] 7 i 7), whom he describes as 'demons'. Could the reference, in this context, be to the children of Azazel (Asael) and his crew, i.e. the watchers, even if, in the oldest form of the legend, the 'beloved ones' of the watchers destroyed themselves in the presence of their parents. (9.9,12)?

(4) **their leading astray** i.e. of mankind. Eth. seḥtat = πλάνη, טועה, שגה (e.g. Isa. 32.6 LXX πλάνησις).

shall not ... be reckoned i.e. there will be no more days to count.

(5) **the angels ... send forth their chiefs** The angels assemble and decide to send forth their 'chiefs', presumably the archangels, who are the 'angels of punishment', to intervene on behalf of the righteous by stirring up the Parthians and Medes to send their armies to attack the land of the elect (v. 6). See above on 54.6 for the archangels as 'angels of punishment'. (Cf. Dan. 10.13-21 for the idea of angelic intervention). Eth. 'and eject (yewaddeyu) their chiefs' seems strange; Dillmann renders 'ihre Haüpter richten'. Does Eth. wadaya here translate ἐκβάλλειν, which occasionally renders הוציא in the LXX meaning 'to send forth', especially on a mission or campaign? (e.g. 2 Chr. 23.14). Alternatively, yewaddeyu may be a corruption of yawaḍḍe'u = יוציאו, 'they sent out their chiefs'. (Eth[g] has 'they cast' yewaddequ and Charles (followed by Knibb) renders 'hurled themselves ('ar'estihomu) towards the east'). Their mission was not to destroy the Parthians and Medes, but to incite them to invade 'the land of the elect ones', and no doubt confront the Romans who occupied it.

to the Parthians and the Medes The usual Biblical description is 'Medes and Persians' (Dan. 6.8). Here, however, the 'Parthians'[1] are the successors of the Persians and the order of names 'Parthians and Medes' for 'Medes and Persians' shows that the Parthians were now the dominant partner. The order is the same at Ac. 2.9, Πάρθοι καὶ Μῆδοι, but cf. 1 Mac. 14.2, ὁ βασιλεὺς τῆς Περσίδος καὶ Μηδίας. We are some centuries from the Persian period into hellenistic times; the Parthians emerged as a world power in the Roman period. It was early suggested (by Laurence, Hofmann and others) that the reference in these verses was to the Parthians' incursion into Palestine c. B.C. 40 in the attempts of the Hasmonaean Antigonus to oust, with their help, Hyrcanus II and Herod from their position under Rome as the rulers of the country. The attempt almost succeeded but in the end it was Rome and Herod who prevailed. (Josephus, *Ant.* xiv.13,14 (324-481), B.J. i.13 (248-273). Dillmann rejected this theory and argued that the events predicted were all in the future: the Parthians were the people 'from the North' and their role that of Gog and Magog in the prophecies of Ezek. 38,39 and in passages such as Jl. 2, Zech. 12,14 about a last final battle with the Gentiles before Jerusalem. The earlier view was revived by E. Sjöberg (*Der Menschensohn im äthiopischen Henochbuch*, 38), who also argued that 56.7, 'The city of my righteous ones shall be an obstacle to their horses', required a date of composition before the destruction of Jerusalem. More recently J. C. Hindley[2] has raised several objections to this view: he recalls, for instance, the views of Krieger and Lücke (see Dillmann) that if the reference was to the events of B.C. 40 we would

---

[1] Eth. sab'a parṭē = אנשי פרס (Heb. has no separate word for Parthian); Syr. ܦܪܬܝܐ.

[2] Above, 182f., 187.

require to assume that, while vs. 5-6 were based on knowledge of these events, v. 7 could only be prophecy, when the writer was looking for a miraculous deliverance, like that from Sennacherib. 'It is surprising, in that case, that the crisis and the certainty of deliverance form so small a part of the book as a whole.' (553) On the other hand, Hindley agrees that the reference to the invasions of Palestine or Syria by the Parthians does imply a date in the Roman period: he opts for the Parthian penetration to Antioch in the time of Trajan (110 A.D.). But on that occasion there was no threat to Jerusalem and no invasion of the 'land of the elect': the only serious Parthian menace so far as Palestine and Jerusalem were concerned were the events of B.C. 40, and v. 6 agrees well with Josephus's description of the ravages of the 'invasion' at B.J. i.9 (269). The reference to the city of the righteous as a 'hindrance' or 'obstacle' to the invader is a clear allusion to Zech. 12.3, probably interpreted here with reference to the Parthian siege of Jerusalem in B.C. 40 (see note on text v.). Even though Parthian troops did penetrate to Herod's Palace and the Temple area, it is unlikely that they did so with a body of cavalry; Jerusalem probably did prove a very real obstacle to their horses. Moreover, the idea that the city could still be defended and prove an obstacle to an invader, clearly does imply that the city was still standing; and this is important for the dating of the Parables. See above, 187.

from their dens(?) Eth. 'among their flocks'. If this text is sound, we require to interpret the pronoun as referring to the angels (and similarly at v. 6 if we read 'their elect ones'). (So Dillmann, 174). It is unusual, however, for an antecedent to be so far removed from its pronoun, and, in any case, we require a parallel to 'from their lairs' and this is not provided by 'among their flocks'. Could there be a mistranslation of מערה (מערותיהם) = 'their dens' read as מרעה (מרעותיהם), 'their flocks'; cf. Nah. 2.11 NEB 'the *cave* (מערה) where the lions' cubs lurked'. MT מרעה = LXX νομή 'pasture'. Cf. also Isa. 32.14. We would then require to read 'from the midst (ʾemmāʾkala) of their dens', parallel to 'from their lairs'.

(6) **they shall come up** Eth. yaʿarregu = יעלו, ἀναβήσονται i.e. 'set out for war' (Jg. 12.3, 1 Sam. 7.7). The idea is also that of 'coming up' to and against Jerusalem.

my elect ones In view of the firm reading at v. 7 'the city of my righteous ones', I have adopted (with Knibb, 2, 140) the reading of Eth[tana] 'my elect ones'. 'Their elect ones' could only refer back to the angels of v. 5, again a somewhat remote antecedent. The alternative reading 'his elect ones', taken to refer to God, makes tolerable sense, but an antecedent is desirable. Has the phrase 'says the Lord of spirits' accidentally dropped out of the text (it appears, e.g. at 55.3), or does Enoch's accompanying angelus interpres speak in loco dei?

a trampling-place and a highway Eth. mekyāda waʾasara = מרמס ומסילה(?) or mekyād = גרן 'threshing-floor' (Charles); Dillmann detected an allusion to Isa. 21.10.

(7) **shall prove an obstacle to their horses** An allusion to Zech. 12.2,3, where Jerusalem is to be אבן מעמסה Tg. אבן תקלא 'a stone of stumbling'. The Parthians were noted for their cavalry: did the original perhaps read פרסה (פרסותיהם 'their horses') (cf. Exod. 10.26)? For the view that this verse implied that Jerusalem was a defended city when the Parables were written, see above, 187.

incite to warring Eth. našʾa (II.1) qatla cf. Heb. Dan. 11.25 יתגרה למלחמה.

and fail to keep faith with one another This line is usually rendered 'And their right hand shall be strong against themselves'. This is practically meaningless, although Martin understands it of the right wing of the Parthian army (leur droite deploiera sa force contre eux). No attention is paid by translators or commentators to the variant of Eth[q] hāymānotomu and its use after the verb ṣanʿa II.1 meaning 'to confirm, ratify an agreement', e.g. 69.10, which maintains that written documents are not necessary 'to confirm an agreement'. A similar Heb. expression occurs at Dan. 9.27, הגביר ברית, Theod. δυναμώσει διαθήκην, Jer. 9.2 לא לאמונה גברו, LXX οὐ πίστις ἐνίσχυσεν. Here I suggest [[οὐκ] ἐνίσχυσεν ἡ πίστις αὐτῶν ἐν αὐτοῖς], lit. 'their faithfulness is not firm among themselves', i.e. 'they fail to keep faith with one another'. The conjecture is supported by the parallel reading of Eth^{n} adopted for the next line 'so that a man shall not trust his brother ...'. This reading would also support the identification of the occasion with the Parthian invasion of B.C. 40—the Parthians were notoriously perfidious (cf. Josephus, *Ant.*, xiv.13.4 (341), or does the verse refer to the outbreak of civil strife in Jerusalem itself at this time? Cf. also 41.5, 43.2 for the virtue of trustworthiness.

shall not trust his brother For this idea of internecine and family strife cf. Ezek. 38.21, Hag. 2.22. These passages may well have been in the writer's mind, possibly even conceived as being fulfilled in the events he is alluding to; for at the time of the Parthian menace, there was 'division' among the Jews themselves, some of them siding with the party of Antigonus, others with the faction of Herod (Josephus, ibid.).

of their men(?) The two forms of the text transmitted give poor sense: 'corpses' either 'innumerable' or 'a number of corpses' 'from their deaths' ('emmotomu) is an unlikely text. We should emend to 'emmetomu, ἐκ τῶν ἀνδρῶν αὐτῶν (cf. Eth^{tana}) Heb. מן מתיהם, from מת 'male', a word used in the Old Testament in similar contexts for the male members of the population put to the sword (Dt. 2.34, 3.6 etc.).

(8) **shall open her mouth** Cf. Isa. 5.14 פערה פיה ... שאול.

be engulfed therein Cf. Ps. 91.16 טבעו גוים בשחת.

their destruction Sheol shall not remit(?) The text is extensively corrupted. Flemming-Radermacher follow Eth^{g} and emend the first verb to taḫadga/(pass.) and render 'ihrer Vernichtung ist freier Lauf gelassen'. The translation adopted follows Eth^{q} reading 'iteḫaddeg 'Their destruction Sheol shall not remit ...'

CHAPTER 57

(1) **a host besides of chariots** i.e. besides the 'hosts' mentioned at 56.1,5. The events now described in Enoch's vision follow on immediately after the defeat of the Parthians. These events are usually interpreted as referring to the return of the dispersion; cf. 90.33, Isa. 27.13, 43.5, 49.12,22,23, Tob. 13.12, 2 Esd. 13.17f., 2 Mac. 2.18. Other commentators (e.g. Schodde) think of the convergence of the Gentile nations against Israel. They come on 'wagons', perhaps 'chariots', with the speed of the wind. (Charles suggests reading 'like the wind' כרוח misread as ברוח 'with the wind').

from the east ... until the middle of the day Eth. lit. 'to mid-day' 'eska manfaqa 'elat. All commentators assume that this is a somewhat inept rendering of μεσημβρία in its geographical sense (Knibb 'a not entirely satisfactory rendering', Dillmann '... μεσημβρία ... nicht ganz passend übersetzt ...'). (Another term for 'mid-day' qatr can also = plaga meridionalis.) It is possible to take the phrase quite literally = 'until mid-day (as Neh. 8.3 עד מחצת היום): the whole operation is envisaged as taking place within a single day (v. 2) so that this first part, the ingathering of the diaspora (or the assembling of the Gentiles) was to be completed by the first half of the day, mid-day. The words 'from the east and from the west' are 'polar' expressions, intended to convey the meaning 'from the whole inhabited earth'. This alternative understanding of the text would solve the problem why no mention is made then of the north or of the south. Jerusalem, on which they are to converge as their assembly point, was thought to be at the centre of the world, not 'in the south'.

(2) **the noise... was heard** Eth. demḍ = המון(?). Jer. 47.3, of the 'noise' of chariot wheels. The repeated 'and the sound thereof was heard' may be a ditt.

the pillars ... moved The Biblical language of Hag. 2.6,7, Jl. 3.16.

(3) **fall down and worship** As at Isa. 27.13.

CHAPTER 58

(2) **glorious shall be your lot** Their lot is preserved for them by the Elect One, 48.7.

(3) **light of eternal life** Cf. Job 33.30, Ps. 56.14 אור החיים 'light of life'. Although this phrase has a Biblical antecedent, the Eth. text here could be corrupt (baberhāna a ditt.?) and Eth[q] original: 'And the elect shall possess eternal life'.

(4) **eternal Lord** See above note on 9.4.

(5) **seek out in his Name**(?) However the texts are construed, as they stand the result makes unsatisfactory sense. Eth II reads (Knibb): '... it will be said to the holy that they should seek in heaven the secrets of righteousness, the lot of faith; for it has become bright as the sun upon the dry ground ...'. Eth I (Charles): '... it shall be said

to the holy in heaven
That they should seek out the secrets of righteousness, the heritage of faith.
For it has become bright as the sun upon earth ...'.

The text seems to be at fault. The suggestion of mistranslation of basemu በስሙ by basamāy በሰማይ goes give meaning to the v. Note basema at the end of the previous verse.

it shall arise ... pass away Eth. šaraqa, ḫalafa = זרח, עבר(?), both prophetic perfects and cf. Mal. 4.2: 'For unto you that fear my name shall the sun of righteousness arise' (זרח).

(6) **light inexhaustible ... limit of days** Eth. lit. 'light that cannot be counted' (see above on 48.1). The phrase 'a limit of days' (ḫuelqua mawāʿel) could render מספר ימים, 'an (allotted) number of days' or ḫuelqua could be a mistake for ḫelqata mawāʿel = סוף ימים, τέλος τῶν ἡμερῶν, קצת ימים (cf. Dan. 4.31).

light of uprightness established forever Eth. ṣanʿa = Heb. נכון(?).

CHAPTER 59

(1) In this chapter on natural phenomena the writer is drawing heavily on Job 36.31; 37.5,13; 38.24-27. Cf. 17.3f. and 41.3, a similar passage, except that here the natural phenomena are related more fully to the life of mankind, as blessings or curses, a source of plenty or its opposite.

and ... the luminaries Presumably all the luminaries of the heavens, sun, moon and stars as well as the lightnings.

their ordinances Eth. kuennanēhomu = חקותיהם or משפטיהם; cf. Job 28.26, Ps. 148.6.

for a blessing or a curse Cf. Job 36.31f. Note the word-play, in Eth. yebarrequ labarakat = יברקו לברכה. See above, 187 n. 1.

(2) **the sound thereof is heard among the habitations of earth** The text is badly corrupted and has been variously emended (cf. Knibb, 2, 141f.). The translation proposed rests on a simple emendation of the text from 'and the habitations of earth' (wamāḫdarāta yabs), which makes no sense in this context, to '(... the sound thereof is heard) in the habitations of earth' (bamāḫdarāta yabs).

(3) **for blessing and for plenty** This whole verse reads like an alternative version of v. 1, except that 'plenty' Eth. ṣegāb = שבעה(?) replaces 'curse' (margam שבועה(?)). Was the translator in doubt about how to read an original לשבעה, whether as שבעה 'plenty' or שבועה 'curse'? See above 187, and the significance of the observation for the question of a Heb. original (Aram. 'curse' is לוט(א), ܠܘܛܬܐ).

CHAPTER 60

(1) **In the year five hundred** Since Dillmann it has been argued that chapter 60 is a later interpolated 'Noah apocalypse'. Certainly v. 8, which refers to the visionary's great-grandfather, the seventh from Adam, i.e. Enoch, could only refer to Noah. It is not surprising then to find the dating 'in the year five hundred', since this is a date mentioned at Gen. 5.32 in the life of Noah; it could hardly apply to Enoch, whose life-span was three hundred and sixty five years (Gen. 5.23). The 'interpolator' appears to have adapted a 'Noah Apocalypse' (similar perhaps to that identified at DSDI, 84f.) to the Book of the Parables. If we substitute 'Noah' for 'Enoch' we restore something of the original Noah apocalypse. The 'interpolator' has, however, clearly intended his readers to take this chapter as an *Enoch* vision, even though his original source obtrudes at v. 8. There is no variant 'Noah' for 'Enoch' in the texts of this verse. Eth^{tana} reads 'a son of man' for 'my (your) great-grandfather', a further attempt to assign the apocalypse to Enoch. Perhaps the 'adaptation' was made by the original author of the 'Second Vision'? See above, 8.

The verses on the two monsters, Leviathan and Behemoth (7-10, 23-24), represent a development of the flood legend (cf. Dillmann, 182f.), their role being to devour the victims of the flood, Leviathan in the depths of the ocean (where he traditionally belongs), Behemoth lying in the desert to the east of paradise. See further below on vs. 23-24. The judgement therefore, of which v. 6 speaks would then be originally the Noachic judgement of the deluge, not the last judgement. As adapted by the 'interpolator', of course, the apocalypse would be understood to refer to the last judgement; and the Leviathan-Behemoth legend was probably also abbreviated to become, as in later tradition, associated with the eschata after the last judgement. Verses 11-21, describing various meteorological phenomena are perhaps less intrusive than they first appear, since the flood involves an upheaval in the world of nature.

in the seventh month i.e. the festival of Tabernacles.

In that Parable I saw This is an unusual use of the term 'parable', which normally refers to an utterance of Enoch describing a vision; here the word practically means 'vision'. It is 'a clumsy attempt to connect this (interpolated) chapter with the main context', betraying the hand of the 'interpolator' (Charles). It appears to refer back to 58.1 and especially 57.2.

heavens to quake Cf. 1.6,7.

host of the Most High Cf. 1.9, 40.1, 71.8,13 and Dan. 7.10. For the title for God, see above, 130.

(2) **Chief of Days** See above, 192f.

sat ... glory Cf. 47.3 and Dan. 7.9f.

the angels and the righteous ones That 'the righteous' should surround the heavenly Throne along with the angels in what seems to be a judgement scene (cf. v. 6) is a difficult, though possible, conception, and could be the work of an ignorant interpolator (Charles, 114). Textual corruption cannot be ruled out, ṣādeqān for qeddusān, (cf. v. 4), or that a translator is rendering ἅγιοι by ṣādeqān (cf. 1 Tim. 5.10). Cf. further, however, 70.4, which mentions, after the patriarchs ('the first fathers'), 'the righteous who from everlasting dwell in that place' (i.e. paradise). Has the author the idea of a celestial company of 'righteous ones' dwelling from time immemorial with the angels in paradise and so being present at the last judgement?

(3) For this verse cf. 14.24, 71.11.

a great trembling ... fear took hold of me Cf. En. 1.5 λήμψεται αὐτοὺς (Eth. yenašše'omu) τρόμος καὶ φόβος μέγας: 13.3 ἔλαβεν αὐτοὺς (Eth. naš'omu) τρόμος καὶ φόβος. Heb. אחז, Isa. 33.14 LXX λήμψεται τρόμος τοὺς ἀσεβεῖς.

my loins gave way Eth I ḥaquēya tafatḥa = ויפתחו מתני Isa. 45.1 and cf. 46.4 (note). Eth II has a double rendering of יפתחו, taqaṣ´a watafatḥa; the first verb is used in a similar connection in the Eth. version of Nah. 2.10 (see Dillmann, *Lex.*, 476)

dissolved were my reins Eth. lit. 'my reins melted' i.e. with fear = ונמסו כליותי(?) usually of the 'heart' but 'reins and heart' go together (Jer. 11.20). Cf. also 1 Mac. 2.24. The variant in Eth. 'my whole being' (Knibb) kuellantāya is obviously a corruption of kuelyāteya, 'reins'.

(4) **Michael ... the holy ones** Michael is the chief archangel (40.4,9) and sends to Enoch his accompanying 'angel of peace' as angelus interpres (60.24).

my spirit returned Eth. manfaseya gab'at = שב רוחי Jg. 15.19.

not been able to endure Eth. 'ikehelku ta´āgešo = לא יכלתי לכלכל cf. Mal. 3.2: a characteristic Heb. word-play.

(6) **the righteous Judge** (bis) The majority text 'who worship the righteous judgement' is obviously wrong. Knibb's conjecture that kuennanē comes from דין 'Judge' misread as 'judgement' offers a reasonable solution (the mistranslation could have come from either an Aram. or a Heb. original). The noun דין as a term for God is attested at 22.14 En[d] 1 xi 2 דין קושטא 'righteous Judge' (wrongly translated by Milik as 'Judgement of Righteousness'). The further expression 'those who deny the righteous judgement' is possible, but the same mistake may have been made here: 'to deny' the righteous Judge is an even worse sin than to withhold the worship which is his due. Knibb accepts the majority reading 'those who worship the righteous judgement', arguing that a contrast is intended between those who 'worship' and those who 'deny', although he qualifies this by adding that possibly the reading 'who do not worship' should be adopted. Punishment, however, is prepared only for those who do *not* worship, but deny the righteous Judge. For this passage cf. 2 Esd. 7.37: 'Look and understand who it is you have denied and refused to serve, and whose commandment you have despised' (NEB).

take his Name in vain Eth. 'ella yenašše'u semo baka = אשר ישאו שמו לשוא (cf. Exod. 20.7 = Dt. 5.11).

a visitation i.e. punishment; ḥatatā = פקדה, ἐπισκοπή.

(7) **two monsters ... Leviathan ... Behemoth** (8) Eth. ʿanābert = תנינים, κήτη. See Bousset-Gressmann[4], 285f. for the mythological development in apocalyptic and rabbinical traditions of the תנינים of Gen. 1.21, Isa. 27.1 etc.; for apocalyptic 2 Esd. 6.49-52 and 2 Bar. 29.4 are especially important since they seem to be drawing on these vs. from the Parables. See further Charles[1906], 105.

separated from one another Although paired, male and female, they are separated each to go to his or her own sphere, Leviathan to the ocean depths, Behemoth to the empty wilderness.

over the fountains of the waters Cf. Gen. 7.11, Job. 38.16 etc. i.e. the sources of the oceans. According to Dillmann Leviathan's task was to control these 'springs' to prevent a future Flood, but there is nothing to suggest this in the text. V. 25 gives a more plausible reason for their presence in their special stations, namely to complete the destruction of mankind drowned by the flood — they are to be devoured by these monsters of the sea and the dry land. The idea that the monsters themselves are to supply the food for the banquet of the righteous in the messianic age is a later development. See further on vs. 23-24.

(8) **covers with his belly an empty wilderness** Eth. zaye'eẖḫez = אשר אחז, Piel 'overlay, cover' (Job 26.9); 'engedeʿā = גחון Gen. 3.14. Eth. babadw za'iyāstare''i = ἐν ἐρήμῳ ἀοράτῳ lit. 'in a desert that cannot be seen', but ἀόρατος renders תהו in the LXX (Gen. 1.2); the reference is to the emptiness of the desert. Cf. Dt. 32.10. See Dillmann, 184, and below, 323f.

Duidain (or Dendayn) The etymology suggested by Dillmann דין דין ('Judgement of the Judge') seems improbable. The form given by Eth[u] Duidain is closer to the Δουδαήλ of 10.4 with which a connection has been suspected (see above note on 10.4, 134). Is it the Daneben of Pseudo-Philo 48.1?[1] Charles connected it with Gen. 4.16 'the land of Nod to the east of Eden'.

east of the Garden ... righteous dwell On Eden or Paradise and its location see above note on 32.3 and Charles on this verse: it is only in the Parables that Paradise is inhabited by 'the elect and rightous'.

my great-grandfather was taken up The reference is plainly to Enoch (and the scribe of Eth[ull] spells it out with a gloss 'Enoch'), and the text implies that the speaker is Noah (cf. 65.2,9; 68.1 where Noah, in these Noah apocalypses, refers to Enoch as 'my great-grandfather'). En. 60 is to be regarded as originally a 'Noah apocalypse', in spite of the references at v. 1 to Enoch. The variant of Eth[tana] looks like a scribe's attempt to remove the implication that the speaker can only be Noah. If Enoch is the narrator, the reference in Eth[tana] is to himself. The verb tamaṭṭawa is impersonal sing. 'he (one) received', unless it is explained as a (rare) passive 'was handed over' (Dillmann, 184); Eth[tana] tamayṭa 'was changed, transformed' could be original, perhaps = נהפך (cf. 1 Sam. 10.6).

seventh from Adam Cf. 93.3 and Jude 14, Jub. 7.39.

(9) **the other angel** i.e. the subordinate angel of vs. 4 and 11.

the might ... how they were separated Presumably by their 'might' is meant their capacity to consume the victims of the Flood, and the reason ('how' = 'why') male and female are separated is that one devours those drowned in the seas, the other those destroyed by the Flood on the 'dry land'.

(10) **Son of man** A form of address peculiar to Ezekiel; here only in the Parables.

---

[1] See M. Black, 'The 'Two Witnesses' of Rev. 11.3f. in Jewish and Christian Apocalyptic Tradition' in *Donum Gentilicium*, New Testament Studies in Honour of David Daube, 231f.

V. 11f. where one would expect an immediate answer to Enoch's question in v. 9, begins a long section (11-23) on the 'mysteries' of nature. It is not till vs. 24-25 that Enoch's question receives an answer. These verses need not, however, be an arbitrary inserted 'interpolation', and there is no justification for rearranging the text for vs. 24-25 to follow v. 10: they are an expansion in the style of the passages on nature in the older Book, perhaps of 11.1f. 'And in those days I shall open the treasures of blessing in the heavens ...' which follows immediately on the description of the flood.

(11) **things first and last** Dillmann understands the words to refer to those things which Enoch saw 'first and last', i.e. a 'polar' expression meaning 'everything he saw'. But the phrase has a profounder meaning in this context: Eth. zaqadāmi wazadaḫāri = τὰ ἀρχαῖα καὶ τὰ ἔσχατα = אשר בראשון ובאחרון(?), (cf. also the expression 'the First and the Last' as an epithet for God, e.g. Isa. 44.6, Rev. 1.17 etc.).

foundations of the heavens The reading of Eth[ex] 'foundations of earth' continues the contrast between heaven and earth. Cf. Ps. 104.5.

(12) **storehouses of the winds** Eth. mazāgebta manfasāt, lit. 'magazines of the winds' = אוצרות רוחות; cf. Jer. 10.13 (= 51.16); Ps. 135.7; Job 38.22 etc. and above, 202, 263.

the wind-spirits are divided As appears from v. 15 onwards the various forces of nature described, winds, thunder, lightnings, seas etc. are under the control of 'spirits' or 'angels', a peculiar feature of this chapter. (Cf. 18.14-16 where natural phenomena or powers are personified or actually regarded as intelligent beings). For similar ideas, Jub. 2.2 'angels of the spirit of fire', 'angels of hail', 'angels of thunder'; so also Rev. 7.1,2, 14.18, 19.17. The same word for 'spirit' רוח is also the word for 'wind'. I have rendered by 'wind-spirit' to convey the intention of the text which regards the 'winds' as 'spirits'. The division of the 'wind-spirits' probably refers to the different areas or spheres where each respectively operates.

weighed ... gates(?) of the winds are numbered For the idea of the winds being 'weighed', cf. Job 28.25. I have adopted the conjecture of Flemming-Radermacher, 78, and read ʿanāqeḍ 'gates' for ʿanqeʿt 'springs', since the latter seems inappropriate to describe winds (or their sources) and the idea of 12 'gates' of the winds is familiar from 76.1f.

the intensity (lit. the 'powers') **of the phases of the moon** Cf. Dillmann, Knibb for this interpretation: Eth. berhān is collective (but note the variant berhānāta).

according to its right strength I emend wakama to babakama ḫayla ṣedq parallel to the previous babaḫayla manfas 'each according to the powers of the wind'.

(13) **the thunders ... they roll** (lit. 'fall') Eth. naguadguād bamudāqātihomu lit. 'thunders in their falling': mudāq is a rare word occuring again once only at 100.6 (mudāqa ḫāṭiatomu 'the overthrow of their sins' Charles). Dillmann's interpretation of this v. 13 has been followed by all subsequent translators as 'places where they fall' (Charles), the idea being of thunder 'falling') like an object ('Gegenstand' Dillmann) down an incline until it reaches 'places of rest' (v. 14 meʿrāfāt). It is perhaps simpler to understand naguadguād 'thunders' as reproducing קולות in its absolute use, e.g. Exod. 9.23 ('thunderings' were the 'voices' of God); see S. R. Driver on 1 Sam. 12.17. The 'voice of God' 'falls' from heaven, Dan. 4.28.

(14) **pauses** Eth. meʿrāfāt = Heb. מנחות lit. 'resting-places'(?).

by the operation ... proceed inseparably On this understanding of the text, both thunder and lightning are controlled by the same 'spirit'.

(15) **utters its voice** Eth. qālo yehub = Heb. יתן קולו.

makes (the peal) cease ... between them Eth. lit. 'the spirit makes it (the peal) cease at its appointed time (of duration)' (bagizēhā = בזמנו cf. Est. 9.27). The 'equal division' between them, i.e. lightning and thunder, has been interpreted to mean that the peals of thunder are not to be stronger than the lightning flashes require (so Dillmann, Knibb). The 'equal division' seems more likely, however, to refer simply to the simultaneous timing of lightning flashes and thunder-peals, with an equal time-lapse between each lightning-flash and thunder-clap. Cf. the use of a similar expression at 1 Sam. 30.24 יחדו יחלקו 'they shall share and share alike' NEB).

the storehouses of their occurrences ... regions of the earth Eth. mazgaba gizēyātihomu, lit. 'the store-houses of their times (i.e. of thunder-claps and lightning-flashes)'. The 'store-houses' or 'treasuries' (mazāgebt = אוצרות, see above, 228) are as full of thunder and lightning storms as the sea-shore is of sand. The figure then changes to that of a rider and his horse, reining in his steed, turning it back, then driving it forward—all this in order to cover the entire earth with thunder-storms.

(16) **upon all the shores of the land** Eth. bakuellu 'adbāra medr translated usually as 'amid all the mountains of the earth' (Charles), with the consequent problem of how the sea comes to be scattered among all the mountains of the earth. (Dillmann thinks of the tides penetrating under the earth into the mountainous regions and so nourishing the springs). Beer in Kautzsch, *Die Apokryphen und Pseudepigraphen*, 270, suggested that ὄρος should be read as ὅρος = 'boundary'. Eth. dabr in fact does render ὅρια = fines = גבול (Gen. 10.19, Mic. 5.5. etc.), a word which is used, like its synonym קצה, for the 'bounds' of the seas (Ps. 104.9 cf. Jer. 5.22, Num. 22.36 of the 'bounds' of a stream). For קצה = 'sea-shore' Num. 34.3.

(17) **an angel of ill omen**(?) The text as it has been transmitted makes little coherent sense: Charles renders 'is his own angel', i.e. the hoar-frost has a special angel of its own. Is Eth. mal'aka zi'ahu a corruption of mal'ak za'ekuy 'an evil angel', the opposite of the following 'good angel'? Cf. Ps. 78.49 for the expression מלאכי רעים 'angels of evils' i.e. 'that bring evil, disaster etc.'. Eth. 'asḥatyā = כפור(?) Job 38.29, Ps. 147.16. For these angels of natural phenomena Str.-B. III, 818f. At 2 En. 5.1 the angels who guard the treasure of cold and snow are 'terrible angels' (Charles. *Apocrypha and Pseudepigrapha*, II, 433.

a good angel According to rabbinical tradition it was this angel who offered to quench the fiery furnance in Dan. 3.19f. Str.-B. ibid.

(18) **never fails** Eth^{tana} ḫadga (leg. ḫadgat) 'albo, ἀνεκλιπές ἐστιν, לא יעזב. Cf. Jer. 18.14: 'Will the snow cease to fall (יעזב) on the rocky slopes of Lebanon?' (NEB).

a spirit all to itself Eth. bāḥtitu = καθ' ἑαυτό = לבדו.

(19) **the storm-cloud** Eth. gimē corresponds most frequently in the Old Testament to ערפל or ענן which go closely together. (The translation 'mist' owes more to Greek usage than to its Heb. original).

its course has the Glory in it So Eth I meḥwāra zi'ahu sebḥata bo(tu). Cf. Flemming-Radermacher 'sein Lauf zeigt Klarheit(?)'. The clause gives the main reason for the 'spirit of cloud' having a separate 'storehouse' of its own. It is the dwelling of 'the Glory (כבוד = LXX δόξα); cf. Exod. 20.21, God dwells in the 'storm-cloud' (ערפל). Knibb suggests that the phrase 'in glory' is a gloss on 'in light'. But the words are so firmly attested that I prefer to take them as an original part of the text. The connection of 'the Glory' (כבוד) with the 'weight' of the storm-cloud is ancient Hebrew 'theology' (cf., e.g. ענן כבד of Exod. 19.16) and eventually the 'cloud' came to be thought of as the 'envelope' of the Glory (1 Kg. 8.10f.). See ThWNT, Bd. II, 242f. s.v. δόξα (Kittel).

and in its storehouse is an angel This should probably be taken along with the idea of the storm-cloud as containing 'the Glory'; it is this angel which is the agent that brings the Glory into the storm-cloud. The alternative reading conveys the further idea of the light shining through the storm-cloud; cf. Job 37.11 for the 'courses' of the clouds and 37.15 for the light shining through them. Eth[t] obelizes the 'angel' in its text and places 'light' in the margin, but this looks like a gloss on the original text 'angel'; 'light and an angel' seems a conflate text.

(20) **the spirit of the dew ... the ends of the heaven** The author is drawing on 34.1,2, 36.1 and cf. 75.5. The 'dew of heaven' (טל השמים) is the gift of God (Gen. 27.28); cf. Zech, 8.12, Dt. 33.28 (the heavens distil it).

(21) For vs. 21, 22 see especially Job 38.25-27, 33-38.

unites with the waters on the earth Cf. 54.8; presumably the idea is that the rain from heaven joins up with 'the waters beneath the earth', i.e. spring-waters or the waters of lakes and rivers on the earth, to produce inundations (cf. Dillmann, 190). The verse is an elaboration of 41.1-6. The 'addition' of Eth[M] 'and if it always (bakuellu gizē (בכל עת)) united with the waters on the earth', looks like a clumsy ditt., but could, as Flemming, 67, suggests, be the protasis of a sentence where the apodosis has been lost. The scribes of Eth[tana (q)] take the first clause of v. 22 as apodosis: 'and if it always united with the waters on the earth, water would be plentiful(?) for the inhabitants of the earth'; ἱκανοῦσθαι = Heb. רב, but meaning 'abundant' rather than 'enough'; Eth. kona la Dillmann, *Lex.*, 864; cf. 3 Kg. 12.28 ἱκανούσθω ὑμῖν, רב לכם, 'it is too much for you'. The meaning would then be that the union of the two sources of water, the heavens and the 'springs' on earth 'at all times', would then produce the needed inundations which earth's inhabitants required. The clause is omitted by Eth[q] and I have followed this text, but with some hesitation. The protasis 'And if it always united ...' could have been lost by hmt.

Lib. Nat. Tom. 43, 79.7, 44, 69 has a free quotation: 'And when the spirits move out in winter-time (lagizē keramt), angels lead them out, and they bring the rains as the Most High wills; they are sent from God', as Enoch says.

nourishment ... from heaven Eth II 'from the Most High from heaven', corrected by Eth I to 'in heaven'. The original almost certainly read 'they are nourishment from the Most High, from heaven.'

(22) **and the angels measure it**(?) Cf. Isa. 40.12, Job 28.25. The angels do the work of the Almighty. The reading 'received it' (yetmēṭawewwo) is that of all texts: Charles interprets freely and renders 'take it in charge', Knibb 'comprehend it' (meaning?). A textual error 'measure it' (yemēṭenewwo) restores a more intelligible text, and one which agrees with the Old Testament sources about the measuring of the waters by God. Since the rain is the gift of God to nourish the earth for the good of its inhabitants, it must be supplied by angelic agents in proper measure, not too much, not too little.

(23) **Garden of the righteous** Eth. gannata ṣādeqān, גן צדיקים the equivalent in the Parables of παράδεισος τῆς δικαιοσύνης, 32.3, 77.3, En[e] 1 xxvi 21, Enastr[b] 23.9 פרדס קשטא, but 77.3 gannata ṣedq 'Garden of Righteousness'.

(24) **to be feasted** The passive yessēsayu (tasēsaya = Polpal כלכל 'to be supplied with food' 1 Kg. 20.27?) is firmly attested. Commentators have emended to an active yesēseyu 'will provide food' to bring the text into line with the tradition that Leviathan and Behemoth were created in order to provide food for the righteous at the messianic feast (see Bousset-Gressmann[4], 285). Dillmann rightly insisted that this was a later development of the legend: originally the role of the two monsters, was to devour the victims of the Flood. The tradition in 2 Esd. 6.49-53 is ambiguous, but obviously dependent on these verses in the Parables (see above on v. 7). The deficient state of the text in these verses may have been the result of attempts either to remove

parts of the text which conflicted with this later tradition or to leave a text which could be interpreted in harmony with it. For the confused state ot the mss, see Knibb, 2, 168, Flemming, 67, Charles[1906], 108.

(25) **punishment ... fall upon them** Eth. ʿarafa II.1 diba = Heb. נוח על, e.g. Ps. 125.3, 'the scepter of wickedness shall not rest (ינוח) upon the land allotted to the righteous' (RSV); cf. Isa. 30.32 '... the chastisement which the Lord inflicts (יניח) on her' (NEB). There is a deliberate word-play on the name Noah (נוח). See further on v. 25 and on 106.18, Milik, 213f.

falls (ceases) upon all The text is that of Eth^{tana} (omitting a long dittograph in Eth^{M}). The verb 'aʿrafa = נוח LXX ἐπαναπαύεσθαι lit. 'rest upon' has also the nuance 'come to rest' i.e. 'to cease' = καταπαύειν, giving a further word-play on Noah (cf. 106.17). The reference in the apodosis is to the post-diluvial covenant of mercy. In both v. 24 and 25 the text of Eth^{M} and particularly Eth I has been marred by haplography and dittography (cf. Knibb 1, 170).

CHAPTER 61

two angels So Eth^{q}, Eth^{M} 'those angels'. The identity of the angels to which the majority text here refers is not immediately obvious. It can hardly refer back to the myriads of 60.1. Even more puzzling, at first reading, is the variant of Eth^{q} 'two angels'. Charles's characteristic solution is that the angels here referred to may have been named in some preceding part now lost. The angel interrogated by Enoch in v. 2 can only be Enoch's own angelus interpres, since in v. 2 he has seen these angels flying off towards the north. The solution of the problem is perhaps to be sought in the vision of Zech. 2.1ff. of 'the man with the measuring-rod in his hand'. He is described in v. 3 as 'an angel' and is joined (v. 3) by 'another angel'. Is the author of the Parables assuming an acquaintance with the story of the two angels with the measuring-rod measuring off Jerusalem for the returning exiles? The Enoch vision is, in some respects, a midrashic treatment of that story; here, however, it is the 'portions' of the elect which are to be measured in the new land of promise. In that case we ought perhaps to read 'two angels' with Eth^{q}.

took to themselves wings Milik, 97, notes that 'with the exception of the Seraphim and Cherubim, early Jewish literature is not familiar with any winged angels', and claims this as a Christian element in the Parables. As is noted below on v. 6, however, this passage is alluding to Ezek. 1; the 'winged angels' here could be the Cherubim and Seraphim.

towards the north i.e. more precisely to the north-west where the Garden of Righteousness is located (cf. 70.3 and 77.3). In Dillmann's interpretation the angels are to bring back from the paradise in the north measurements for the new earthly Jerusalem which the righteous are to inhabit. But although v. 4 does refer to the 'elect dwelling with the elect', the 'measures of the righteous' which the angels are to bring back from paradise, is a general term referring to the heritage of the righteous, perhaps their allotted share of the new Jerusalem, no doubt in a transformed heaven and earth (cf. 39.4f., and Dillmann, 192).

(3) **the lots of the righteous** Eth. lit. 'amṭān = מדות i.e. 'measured and allotted portions' as at Jer. 13.25. The reference is to the future heritage of the righteous, the 'measured portion' of each one. The author may well have in mind the allocation of

the promised land in paradise as corresponding to the dividing out of the allotted portions of the land of Canaan at the time of the conquest, (Num. 33.54, Jos. 13.6, 19.1f.).

the lines of the righteous Eth. ma'āser lit. 'ropes' (so rendered by Charles, Knibb, etc.) corresponds to Heb. חבלים as a synonym of מדות; cf. Ps. 16.6 'my lines (חבלים) have fallen in pleasant places'. I have retained the archaic 'lines' (cf. NEB at Ps. 16.6): 'allotments, lots, apportionments' give the required sense.

that they may stay themselves Eth. tamarguaza diba = נשען על lit. 'lean upon', Mic. 3.11, Isa. 10.20, of 'leaning upon, trusting' God. Cf. above, 210.

(4) **faithfulness ... righteousness** Eth. hāymānot = אמן, אמונה, אמנים, 'fidelity, faithfulness'. The 'measures' are the rewards of loyalty, abstract for concrete 'the loyal ones'? I suspect a mistranslation of לאמונים read as [τῇ πίστει] when it should have been rendered by [τοῖς πιστοῖς]; cf. Ps. 12.2 'the faithful (AV), LXX αἱ ἀλήθειαι, 2 Sam. 20.19. Cf. Isa. 26.2. The sentence is further usually construed as meaning that 'they' (the measures) will 'strengthen righteousness', but this is a most improbable original (Eth II 'the word of righteousness' is no improvement). The phrase ṣanʿa II.1 with ṣedq suggests Heb. החזיק בצדק (Job 27.6, Sir. 49.3) lit. 'to hold fast to righteousness'. Has the Eth. text been foreshortened from 'and to those who hold fast to righteousness'? For the Eth phrase, cf. 69.10. A scribe has further expanded to read 'shall strengthen the word of righteousness'.

(5) **these cord-measures ... depths of the earth** Eth. (ʿellu) ʾamṭānāt = חבלים or מדות, חבלי מדה(?). The 'depths of the earth' (ʿemaqa medr), in this context, would seem to refer to the underworld of Sheol; cf. Isa. 14.15 māʿmeqtihā lamedr = ירכתי בור (par. שאול). Such depths are plumbed by these חבלים; perhaps the Psalmist's חבלי שאול 'the pangs of Sheol' (Ps. 18.6) was in the writer's mind, or is he thinking of Zechariah's חבלים 'staffs, binders' (Zech. 11.7,14), 'Grace' and 'Union' (RSV), the 'bands' of a united people? In view of the community of the elect of v. 4, now to be united with the risen dead (here those who died a violent death), the allusion is not impossible. Is the reference to the scattered people among the elect ones who lost their lives in flight from the Gentile oppressors, in the deserts or on the seas?

by wild beasts If Flemming's conjecture 'monsters' is correct, could this imply the 'resurrection' of those 'devoured' by the twin monsters, Leviathan and Behemoth at the flood? Note that the author emphasises the universal character of the 'return to life' of all from destruction: 'For none shall be destroyed ... none can be destroyed'.

stay themselves on the day of the Elect One 'rely on the day of the Chosen One' (Knibb). If the reference is to the 'return to life' or 'resurrection' of the righteous, this would presumably follow 'the day of the Elect One' i.e. the day of judgement. Is the text sound? Should we perhaps emend to 'on the Name of the Elect One' as at v. 3? Cf. also 48.4.

(6) **one illumination like to fire** Cf. Ps. 78.14 אור אש 'fiery light', 'a glowing fire' (NEB), Isa. 50.11 באור אשכם '(walk) in the flame of your fire'. For 'fire' as a manifestation of angelic beings and their presence cf. Ezek. 1.13 נגה לאש 'a fiery light'. Martin interprets of the light of illumination by which they will see the Elect One, just as their power and voice is given to praise what they see.

(7) A fuller doxology occurs at v. 11.

the First of voices Eth. maqdema qāl (coll.) = [ἀρχὴ τῶν φωνῶν](?) ראש קולות, lit. 'the chief of voices'(?). The expression has given difficulty to translators and interpreters. Dillmann, 193, rejected a literal and grammatically possible 'and that first word they praised' referring to the command the angels had just received, as hardly suitable, and preferred to interpret the phrase as = 'before the word' i.e. before any other word, and so 'before all things' (Knibb 'before everything'). (In his *Lex.*, 464, Dillmann had earlier suggested that an original maqdema kuellu = ante omnia). Flemming-Radermacher render 'they praised him with the first sound (mit dem ersten Laut)'. At Ezek. 1.24 the sound of the wings of the Cherubim is 'as the voice of the Almighty, the voice of speech' (AV) (LXX^A φωνὴ τοῦ λόγου), and this is followed at v. 25, by 'And there was a voice from the firmament that was over their heads ...'. Does the author of Enoch have this passage in mind? The angelic host is endowed, inter alia, with a single voice, which they employ with wisdom and skill (a 'voice of speech'?) to praise 'the First of voices' (the voice from above?), the Primal Voice. Cf. Tg. to Ezek. 1.24 where the sound of the wings of the cherubim, again 'the sound of their speech', is also represented as praising and blessing the King of the ages.

with the spirit of the living creatures Dillmann understands the words to refer to the spirit of eternal life and takes them with the verb 'they were wise'. But Eth. manfasa ḥeywat πνεῦμα τῆς ζωῆς, appears in this context to be a misreading of רוח החיה, πνεῦμα τοῦ ζώου 'the spirit of a living creature' (or coll. 'living creatures'), i.e. the 'living creatures' of Ezekiel; cf. especially Ezek. 1.21. Cf. also Ezek. 10.15,17,20 for these 'living beings', and below, 251.

(8) Cf. 45.3; Ps. 110.1.

the holy ones in heaven above It seems strange that in this judgement scene only angels seem to be judged, good and bad alike. Dillmann thought that 'the holy ones in heaven above', qeddusān bamalʿelta samāy, might go back to an original 'the saints of the Most High' of Dan. 7, קדישי עליונין. (The title 'the Most High' occurs in the next v. in Eth. II). Or is the writer (or editor) of this verse extending the powers of the Elect One to include a judgement of celestial beings, 'the holy ones in heaven above'? Cf. 1 Cor. 6.3.

(9) **when he shall lift up his countenance to judge** The Biblical expression נשא פנים, when used of God or man, can mean either 'to be gracious to' or 'to show partiality towards'. The implication here is that the Elect One will graciously proceeds to judge the 'holy ones' or 'saints', since their unseen ways will all be shown to be ways of righteousness.

in the Name he utters Translations which render literally 'according to the word of the name of the Lord of spirits' suggest that the Elect One is about to judge the 'holy ones' (or 'saints') in accordance with the commands of the Name of the Lord of spirits, i.e. the divine ordinances. What is meant, however, as Dillmann, 194, noted, is that the judgement is to be 'in the name of the Lord of spirits which the Elect One takes on his lips' (banagara semu laʾegziʾa manāfest).

The verse is especially appropriate if it is the 'saints of the Most High' who are being judged. They will alone have cause to celebrate, on such an occasion, and glorify God.

(10) **all the hosts ... shall call out** Eth II 'And he shall summon (yeṣēweʿ) all the hosts ...'. Dillmann boldly emends the accusative kuello of all manuscripts to a nominative kuellu thus making 'all the hosts' the subject of the verb, on the grounds that it was illogical for God (or the Elect One) to 'summon' the angelic host which was already present at the judgement scene. He was followed by Flemming-Radermacher; Charles and Knibb adopt the reading of Eth II and read 'And he shall summon all the hosts ...'. The nominative kuellu is attested at the second occurrence of (wa)kuello in all other texts by four of the best manuscripts, thus supporting Dillmann's emendation, and by Eth II at 'all the angels of power and all the angels of dominions ...' The verb ṣawweʿa I.2 can be transitive or intransitive, in the latter case = קרא e.g. Isa. 44.7, 50.2.

Cherubin, Seraphin, and Ophannin On these classes of angels see Bousset-Gressmann[4], 326. Here they seem to have taken the place of the archangels.

angels of power ... angels of dominions These words are still regarded by commentators and others as our main ancient authority for these New Testament terms, e.g. 'the principalities and powers' of Rom. 8.38. Cf. Bousset-Gressmann[4], 326. Sanday and Headlam, in their commentary in Romans, quote the entire verse from the 1893 translation of Charles: 'And He will call on all the host, the heavens and all the holy ones above, and the host of God, the Cherubim, the Seraphim, and Ophanim, and all the angels of power, and all the angels of principalities, and the Elect One, and the other powers on the earth, over the water, on that day ...'. In Charles, 121, it is stated: 'These are exactly St. Paul's 'principalities and powers'.' If we are to be guided, however, by the Ethiopic version of Col. 1.16 and Rom. 8.38, the second expression malā'ekta 'agā'ezt means 'angels of dominions', the latter word corresponding to κυριότητες and not to ἀρχαί, 'principalities'. Moreover, in both places the text reads 'angels of power' and 'angels of dominions' and not simply ḫayl as at Rom. 8.38 (coll. for δυνάμεις) and 'agā'ezt at Col. 1.16 (κυριότητες). Eth. may be simply translating δυνάμεις and κυριότητες.[1]

the Elect One See above, 189. The Elect One is here included among the higher echelons of the angelic hierarchy (the angels of the Presence are at the head), in the praise of God: but he comes after the 'angels of power' and 'angels of dominions' but before the 'other powers of the earth and water'. Is the author thinking of his Elect One as enjoying angelic status but one subordinate to all celestial 'powers'? Was it this 'subordination' of the Elect One to these higher heavenly powers which led to the protestations of Eph. 1.21 (cf. Col. 1.16) which place these angelic powers under the authority of Christ and maintain that they were even created by him?

powers on the earth ... over the water These are polar expressions meaning over the whole earth: the reference is to lesser angels in charge of the phenomena of nature; cf. 60.12f.

(11) Cf. 49.3 and note that there are seven virtues, אמונה, 'faithfulness', חכמה, 'wisdom', ארך אפים, 'patience', רחמים, 'mercy', צדקה, 'righteousness', שלום, 'peace', טוב, 'goodness'. Cf. IQS 4.2f. and Gal. 5.22f.

For the doxology cf. 39.10.

(12) **who sleep not** i.e. the watchers (עירים); so also at 71.7. See above on 1.5 and Dillmann on 12.2f.

spirit of light Cf. IQS 3.25 רוחות אור וחשך 'spirits of light and darkness'.

Garden of Life See above note on 32.3, 60.8,23. Eth. gannata ḥeywat = Heb. גן החיים, a term found here only in 1 En. but occurring in later tradition, e.g. in Jellinek, *Beth ha-Midrash*, Seder gan Eden, 2.52f., Str.-B. IV, 1133.

exceedingly Eth. fadfāda'emḫayl lit. 'more than power': cf. 90.7 fadfāda waḫayyāla, vehementer (Dillmann, *Lex.*, 609) and the expression παρ δύναμιν Il. 13.787 and ܠܒܪ ܡܢ ܚܝܠܐ B.O. iii 1 598, 600 (P. Sm. 1259a). But we should perhaps emend to fadfāda waḫayyāla = מאד מאד(?).

(13) **mighty deeds** Eth. ḫaylo collective sing. = גבורות(?) Dt. 3.24, Isa. 63.15, Ps. 145.4,12, LXX δυναστεῖαι; cf. the New Testament use of δυνάμεις.

---

[1] See M. Black, Πᾶσαι ἐξουσίαι αὐτῷ ὑποταγήσονται, in *Paul and Paulinism*, Essays in Honour of C. K. Barrett, 79.

CHAPTER 62

(1) **the kings and the mighty,** See 196.

those who possess the earth lit. 'those who inhabit the earth/land'. As Dillmann noted, this expression cannot here refer to the earth's or the land's inhabitants in general, but to those that 'possess the earth or land', i.e. the Gentile oppressors: it varies with the expression 'those who possess the land' (ye'eḫḫezewwā layabs, 62.3,6, 63.1,12, 67.12 and cf. Lib. Nat. Tom. 41,59) and 'those who rule the earth' (yemallekewwā layabs, 62.9) and 'those who tread the earth underfoot and dwell upon it', 38.4, 46.7). The problem remains how a translator could use ḫadara = 'dwell' in a sense virtually synonymous with 'possess'. It is tempting to suggest that an original ירשי הארץ, 'the possessors of the land' has been misread as ישבי הארץ, 'the inhabitants of the land'. More probably, ḫadara medr corresponds to שכן ארץ, 'to occupy, possess the land' (cf. Gesenius, *Thes.*, 1408b, secure possedit terram), in most contexts in Enoch of the occupation or possession of land by alien inhabitants. In this verse the phrase could mean 'those who occupy/possess the earth'.

lift up your horns The figure conveys the idea of pride or arrogance: 'lift up your (proud) horns', like a lordly animal. Cf. Ps. 75.4-5: 'To the boastful I say, 'Boast no more', and to the wicked, 'Do not toss your proud horns ... against high heaven nor speak arrogantly against your Creator' (NEB). The imperative is ironical: the mighty are asked to do the impossible, to open their eyes and recognise, and, in their arrogance, acknowledge the Elect One. If they were able to recognise him they would be 'downcast of countenance' (cf. 48.8, Dillmann, 196).

(2) **the Elect One sat** Eth[all mss] 'the Lord of spirits sat'. Charles emends to read 'anbaro 'seated him' as at 61.8. The context requires either this emendation or the variant 'the Elect One sat' in the Lib. Nat., 41, 59, which has both readings.

spirit of righteousness ... all the sinners The description of the expected 'root of Jesse' at Isa. 11.4f. is here applied to the Elect One, as it is also earlier at 49.3 (cf. 52.3 where he is the Lord's Anointed). These allusions and quotations from Isa. 11 are of great importance in considering the messianism of the Parables. See above, 189.

(3) **there shall stand up** i.e. in respect, Heb. קום, cf. Gen. 23.7. Eth[t] justice shall not fail before him' could be original.

no lying word See note on 49.4.

(4) For Biblical and other parallels to this verse, Isa. 13.8, 21.3, 26.17, IQH 3.8f.

pain come upon them Eth = ובאו עליהם חבלים(?).

enters the mouth of the womb ... in bringing forth Other parallels to this common figure of speech at 2 Kg. 19.3 = Isa. 37.3, Gen. 35.17 ('Rachel had hard labour (ותקש בלדתה)').

(5) Cf. Isa. 13.8 'And they shall be afraid (ונבהלו): pangs and sorrows shall take hold of them (חבלים יאחזון); they shall be in pain as a woman that travaileth: they shall be amazed one at another; their faces shall be as flames' (AV) (NEB '... shall burn with shame'). Cf. also Wis. 5.1ff.

downcast of countenance Eth. lit. 'they will cast down their faces' (yātēḥetu gaṣṣomu = יפילו פניהם).

that child of woman I have opted for the more difficult reading of Eth II against Eth I, which reads 'Son of Man' (walda be'si) as at 69.29 (bis), 71.14 (see above, 206). Whereas the formal 'child of the offspring of the mother of the living' of the Ethiopic Bible appears in the next vs. 7 and 9, the form here in Eth II 'child of woman' is, in my opinion, a survival from an original Heb. alternative to the earlier walda be'si = בן־האדם, in this case walda be'sit = ילוד אשה, an expression found at Job. 15.14, 25.4 (cf. 14.1), i.e. 'child of woman' and particularly appropriate in this context. In both expressions the writer is drawing attention to the humanity of his Elect One, but in this case deliberately pointing a contrast between the mighty potentates who are confronted and put to shame by this 'child of woman'.

(6) **him who reigns over all, the One who hides himself** The text can be rendered in several ways. The least satisfactory is 'who rules over everything which is hidden' (Flemming-Radermacher, Knibb); Dillmann renders correctly 'der verborgen war' ('who was hidden'). What is being stressed is the reign of the Elect One 'over all' the mighty kings. The clause 'who rules over all' recalls παντοκράτωρ in Greek, a term reserved in the LXX to render the 'Lord of hosts' (κύριος παντοκράτωρ or 'the Almighty' (שדי). The phrase 'the one who is hidden (in hiding)' recalls Isa. 45.15 אל מסתתר 'a God who conceals himself', but is here applied to the concealment of the Elect One or Son of Man (cf. 48.6,7).

(7) **in the presence of his (heavenly) host** Eth. baqedma ḫāylu = לפני צבאו i.e. under the surveillance of his heavenly host(?) Eth[u] 'by his power' is a possible alternative.

(8) **congregation of the elect and holy** (or **holy and elect**) Cf. 38.1, 53.6. For the figure of speech Isa. 40.24. The idea of the 'congregation' of the elect as a 'planting' is found at 10.16. It is possible that we have here a 'translation-variant' from an original נצב (cf. also יצב) parallel to קום = lit. 'to stand' 'to be established'(?); נצב is also used = 'to plant', and so Eth. at 10.16,18,19, En[c] 1 v 4, 7, 8 (10.18 καταφυτευθήσεται).

(9) **set their hope on** Eth. safawa III.2 = קוי especially of 'looking to', 'planting hope or trust in' God; Isa. 40.31 'They that wait upon the Lord ...'; Ps. 37.9.

that Son of Man See above on 46.1 = בן־האדם הלזה 'yon son of man'(?); cf. Gen. 37.19, 'yon dreamer' i.e. Joseph.

(10) **affright them** Eth. guague'a II.2 = 'incite to haste', e.g. 2 Chr. 26.20 (LXX κατέσπευσαν) Hiph. הבהיל; and this is the meaning adopted by the translators 'drängen', pressera, press'. But Heb. הבהיל means also 'to terrify', e.g. Job 23.16, 'the Almighty ... fills me with fear (הבהילני LXX ἐσπούδαζέν με)' (NEB). This is a much more appropriate sense in this context; Eth. is translation language.

hastily go forth Eth. yāfṭenu wayeḍā'u = ומהרו ויצאו.

their faces ... filled with shame The Old Testament speaks of 'shame' (חרפה) 'covering' the face (כסה: Jer. 51.51).

darkness shall grow deeper on their faces Cf. Nah. 2.10, Jl. 2.6: '... all faces shall gather blackness' (AV) which Charles thinks is the source of Eth. Cf. also 46.6.

(11) **angels of punishment** See above on 53.3, 56.1.

(12) **drunk from them** Eth. lit. 'is drunk (tesakker) from them'. For the figure of speech and expression Isa. 34.5,7, Jer. 46.10.

wrath ... rests upon them Eth. = ינוח עליהם הרון (Goldschmidt) Ec. 7.9 for a parallel expression.

(13 **saved** Cf. 48.7.

never ... see the face Eth. = לא יראו מעתה or מעתה לא יראו עוד (Goldschmidt).

(14) **shall abide by them** Eth. = ישב על, cf. 1 Sam. 30.24, Ps. 104.12. Or could the phrase have the force of ישב = 'preside, rule over'? Parallelism supports the meaning of 'abide, reside', and cf. the use of שכן and ישב of the Lord 'abiding' with his people, e.g. Jl. 3.17,21; for שכן על = 'abide by', Ps. 104.12. See above on 62.1.

shall eat and lie down The expression is used to describe the security of the Remnant at the Return by Zeph. 3.13: '... for they shall feed and lie down (... ירעו ורבצו)'.

(15) **be raised up** Commentaries since Dillmann seek to deny any reference in these verses to the idea of resurrection: this verse signifies only 'that all the humiliations of the righteous are at an end' (Charles). But the reference to 'the garments of glory' in v. 16, interpreted by Charles of 'the spiritual bodies that await the righteous' clearly supports a reference to resurrection. The author may well have Dan. 12.2 in mind: 'And many that sleep in the dust shall awake ...'.

cease ... downcast countenance Eth. lit. 'ceased to cast down their countenance'. Cf. Job 9.27 אעזב פני lit. 'I will relax my countenance'; NEB freely 'I will show a cheerful face and smile'.

clothed with garments of glory Eth. lebsa sebḥat; cf. Sir. 6.29,31 בגדי כבוד LXX στολὴν δόξης applied to 'wisdom', 'a gorgeous robe' (NEB). The further description of these garments in v. 16 as 'garments of (eternal) life' points unequivocally to the idea of 'garments of immortality'; cf. 2 Esd. 2.39,45, 2 C. 5.3-5, Rev. 3.4,5,18, 4.4, 6.11, 7.9,13,14.

For this verse cf. Isa 61.10.

(16) **garments of life** Presumably 'garments of eternal life'; cf. also בגדי ישע LXX ἱμάτιον σωτηρίου 'garments of salvation' at Isa. 61.10.

shall not grow old For the expressions Dt. 8.4, 29.5.

nor ... pass away Eth. ʾiyaḥalleq = לא יכלה and cf. Isa. 21.16 of 'the glory of Kedar' 'failing' (וכלה כל־כבוד קדר). Flemming, 74, suggests that the continuation of this chapter is to be found at 69.26-29.

CHAPTER 63

(1) Charles renders 'In those days shall the mighty and the kings implore (Him) to grant them a little respite from His angels of punishment ...'. To construe, however, as Charles does, strains the syntax of the sentence: 'angels of punishment' goes more naturally with 'implore' than with 'respite' as a predicate.

angels of punishment See note above on 53.3.

a brief respite Eth. ʿeraft = מנוחה (1 Kg. 8.56 (cf. 5.18) 'respite' (from enemies), Jer. 45.3 'relief' (from sorrow)). The phrase ʿeraft nestit lit. 'a little respite' occurs again at v. 6. Could the original have read נחת זרועו demissio bracchi eius (in castigando) misread as נחת זעיר? This could then imply not only a request for a respite from their punishment but for a cessation of it.

confess their sins Cf. Num. 5.7.

(2) **Lord of kings** etc. Cf. 9.4 for a similar doxology. Eth[e h v ull] give the Biblical 'Lord of Lords' (Dt. 10.17, Ps. 136.3) as a variant to 'Lord of kings'; there are no exact Biblical parallels to the other expressions, 'Lord of the exalted' (or its variant 'Lord of the rich'), 'Lord of glory' and 'Lord of wisdom'. The appellation 'King of kings', applied in the Old Testament to the Babylonian kings (Ezek. 26.7, Dan. 2.37) is used of God at 9.4 Sync. Was it avoided in the Old Testament because of its associations with an alien culture?

Lord of the exalted The majority reading 'Lord of the rich' is unusual as well as un-Biblical, whereas 'the exalted' (leʿulān) occurs several times in these chapters (62.1,3,9).

(3) For this verse in its context, cf. Job. 12.22.

every secret thing ... brought to light Eth II yebarreh kuellu ḫebuʾ = יאור כל

נסתר(?). Dillmann, *Lex.*, 499, explains yebarreh as a quasi-passive of yābarreh = 'will be brought to light'. The variants in the manuscripts show that scribes had difficulty with this verb.

Thy righteousness is beyond reckoning We should perhaps take Eth. ṣedq as a collective singular = צדקות, 'deeds of righteousness'.

(5) **Would that ... respite** For the Heb. construction מי יתן G-K p 151 a, d. For 'respite' see note on v. 1 above.

give thanks Eth. 'amana. At v. 1 this verb is used in the stem III.1 = 'to confess (sins)'. It can have the meaning 'to make confession' (= ὁμολογεῖν) also in the simple stem, and is so rendered by Knibb. In a context, however, where the emphasis is on glorifying and praising God, the meaning 'to praise, give thanks' = ἐξομολογεῖν (Dillmann, *Lex.*, 735) seems more appropriate both here and at v. 8. Cf. Heb. ידה (Hiph.) Ps. 6.5. Dillmann, 35, 199 prefers the meaning credere in both verses.

(6) **we long for** Eth. manaya III.2 = Heb. חכה Pi., e.g. Job 3.21, LXX ἱμείρεσθαι.

We pursue it Eth[g] nesadded = διώκομεν = נרדף (Charles καταδιώκομεν). Eth II nessaddad 'we are driven off' (Dillmann, Flemming-Radermacher, Martin, Knibb).

(8) **our Lord is true** The 'mighty kings' identify themselves with 'believers'. Cf. Dt. 7.9 'the faithful God' (האל הנאמן).

(10) For this verse cf. Ps. 49.16,17.

Then will they say Eth. lit. 'they will say to them' = יאמרו להם an ethic dative, 'they will them say'(?).

ill-gotten gains lit. 'wealth of unrighteousness'. Cf. Lk. 16.9 μαμωνᾶ τῆς ἀδικίας.

they will not keep us ... pit(?) of Sheol The reading of Eth[q] 'will not be able ('itekel) for 'will not keep' ('itekalle') is probably a corruption, unless the original read 'will not be able to keep' (לא יוכל יכלא). Eth. kebad, 'weight', τὸ βάρος is given the meaning vehementia by Dillmann in this verse, and translated by Pein (Knibb 'torments'), (Heb. כבד is used for the violence of battle (Isa. 21.15) or of the divine rage (Isa. 30.27)). But the 'flames of the weight of Sheol' is an un-Biblical expression. Eth[tana] reads 'from the weight of Sheol', ἀπὸ τοῦ βάρους/βάρεως ᾄδου where, I suggest, βάρεως stood originally for בירה 'stronghold', a mistake for בורה 'pit', 'the pit of Sheol'. The addition 'the flames of' may have originated as an explanatory gloss of the rare kebad. (For the confusion of βαρύς and βᾶρις by Eth. scribes, see Dillmann, *Lex.*, 851). Did the original perhaps read '... they will not keep us from going down into the pit of Sheol'? The variant 'from their midst' (adopted by Flemming-Radermacher) refers back to ill-gotten gains': 'They will not keep us from going down (to the grave) from their midst to the pains of hell'. If the proposed emendation is adopted then the original contained no reference to hell-fire (cf. Dillmann, 199).

(11) Cf. 62.10 with which this verse has much in common.

shall be driven ... in their midst Eth. yessaddadu = יגרשו or ירדפו.

CHAPTER 64

In Charles's atomistic treatment of the Parables, chapters 64-69.25, containing a fresh vision to Noah about the watchers, their evil deeds and eventual condemnation, are 'out of place'. The judgement of the watchers has already been described at 55.3-4.

In this earlier passage, however, that judgement is related in order to provide a spectacular warning to the 'mighty kings', whose own judgement follows at 62.1f. (cf. also 67.12). But it is in the 'patchwork' style of the Parables to attach a fresh vision, in this case one ascribed to Noah, of the 'figures' or 'faces' of the watchers which are seen 'in that place', i.e. Sheol (63.10), presumably suffering their punishment along with the 'mighty kings'. The final author of the Parables is no doubt incorporating material from a separate apocalypse which may have had an independent origin and literary existence. Probably some significance is to be attached to the fact that this judgement of the 'angels' (מלאכים v. 2) follows the judgement of the 'kings' (מלכים): but connection of pericopae by Stichworte is a familiar literary device in the growth of traditions. (Cf. 67.13 where Charles thinks we should read 'kings' instead of 'angels').

(1) **other presences** Dillmann (and Knibb) equate gaṣṣ/gaṣṣāt with πρόσωπον(α) and cf. 40.2 (the archangels as 'presences'). In both cases an original פנים seems probable, but perhaps here with the additional connotation of ὄψις 'appearance', perhaps even 'apparition'(?).

CHAPTER 65

The narrative in vv. 1-3 is in the 1st pers. throughout: the true reading in v. 1 has been preserved only in Eth[t]. If we read 3rd pers. with Eth[M], the translation to 1st pers. at vs. 4-5, where Noah is the speaker is awkward and confusing, since there is no preparation for it. The 1st pers. has been preserved in the best texts at v. 3 'And I said to him ...'.

(1) **brought low** Dillmann noted that the verb 'aṣnana, although usually transitive, can mean 'to bow oneself down': he suggested שקע, נשקע (Am. 8.8, 9.5) as a Heb. equivalent, meaning 'to sink down'; and hence Charles 'had sunk down'; Knibb renders 'had tilted'. Eth. 'aṣnana renders κλίνειν (Jn. 19.30), and an alternative equivalent more appropriare to the context is כרע lit. 'to bow down', but figuratively of being 'brought down' (of enemies in death Jg. 5.27; cf. Job 4.4 of 'tottering knees'). As Dillmann noted here (see below) and at v. 3 the 'earth' is being personified, 'is brought low' = 'falls into a decline': but since it is the earth, its 'decline' means that it is 'shaken' or 'totters' (cf. again v. 3 and note).

(2) **set out** Eth. lit. 'lifted up his feet' 'anše'a 'egara נשא רגלים (cf. Gen. 29.1).

to the ends of the earth i.e. to the entrance of heaven where Noah could meet his great-grandfather Enoch; cf. 106.8.

with an embittered voice Cf. Gen. 27.34, Est. 4.1.

(3) **so afflicted and shaken** Eth. sarḥa = חלה, כשל (Prov. 24.16 נכשל = ἀσθενεῖν)(?): כשל has also the sense of 'stumbling, tottering'. Eth. 'shaken up' 'anqalqala = התקלקל (Jer. 4.24)(?).

(4) **thereupon at once** Eth. badeḫra we'etu gizē. The phrase 'at once' we'etu gizē = εὐθύς. Eth. badeḫra = postea is very rare (Dillmann, *Lex.*, 1110): has it perhaps arisen her by a misreading of באחת 'at once' (Prov. 28.18) as a form of אחר 'after', thus giving a doublet of we'etu gizē? Perhaps the original reading is preserved in Eth[u]: 'And in its (the earth's) fall, at once there was a great commotion ...(?)'. Cf. Ps. 56(55).13, דחי LXX ὀλίσθημα.

fell on my face Cf. 60.3.

(6) **the satans** See above on 40.7.

the power of spells Eth. ḫebrāt 'spells': ḫebr is not attested elsewhere in this sense (Dillmann, *Lex.*, 598 and above, 186); here = חבר (Dt. 18.11, Ps. 58.6, Isa. 47.9,12); cf. En. 8.3 En[a] 1 iv 1 [חבר[ו Tg. חברא. The context suggests חֹבְרִים '(the powers of) spell-binders' rather than 'spells' (חֲבָרִים); ספאתרהפ 'the powers of the binders of spells' (חברי חבר).

those who make molten images Eth. = נסכי/יצרי פסל cf. Isa. 40.19, 44.10 (cf. Hab. 2.18).

every created thing Or 'of every creature'. So Eth[v]. The usual reading 'images of the whole earth' is usually rendered by translators (following Dillmann) as 'images for the whole earth'.

(7) **silver comes forth ... earth** Cf. Job 28.1-2.

soft metal See above on 52.6.

(8) **lead and tin** Cf. Num. 31.22.

a mine ... produces them Eth. naq' is usually rendered 'spring', but באר is a more general word; and lead and tin come from 'pits' dug in the earth, i.e. from mines.

purifies (them) See above, 186.

(9) **took hold of ... hand** Cf. חזק ביד Gen. 19.16.

(10) **fully carried out** Eth. tafaṣṣama = כלה proph. perf.?

withheld before me Eth. 'iyetḫuēlaqu, 'will not be counted' = ימנה Niph.: read with Charles Heb. ימנע Niph., 'withhold, be held back'. Cf. Knibb, reading Aram. יתמנע. See above, 186.

because of ... sorceries Halévy and Charles suspect an original Heb. חרשים 'sorceries' has been misread as חדשים (μῆνες), which Caquot and Geoltrain still feel is the original and correct reading. In view of their theory of an original Aramaic Knibb and Ullendorff argue for an Aram. חדשיא 'months', a word found in Babylonian Aram. West Aram. for 'months' however, is ירחיא. See above, 186.

(11) **return** (to heaven) Eth. megbā'a = תשובה 'return' (1 sam. 7.17). 'They have shown them what is hidden': the subject is the 'angels' of 64.2, i.e. the fallen watchers (so Flemming-Radermacher) and the object 'the children of men' to whom these guilty secrets of heaven have been disclosed. As at 14.5 the penalty for the watchers includes exclusion from the heaven from which they came.

this reproach ... secrets Cf. 67.1 Eth. ḫis = חרפה, כלימה. This is based on Gen. 6.9 as at 67.1; Noah had no part in the wickedness which the watchers had communicated to mankind.

(12) **established your name** Cf. 'established your seed' below: 'aṣne'o lasemeka/lazar'eka, and cf. Isa. 66.22 'so shall your race (זרעכם) and your name (שמכם) endure (יעמד)' (NEB). Eth. =אעמיד שמך/זרעך(?). So Dillmann 'er hat befestigt deinen Namen'. (This rulès out Charles's 'destined thy name', a possible rendering of 'aṣne'a by itself.)

to be among the holy ones Does this mean that Noah is to be one of the 'saints' or that he is to find an established place among the 'holy ones' of heaven, i.e. the angels.

a fountain Cf. Dt. 33.28, Ps. 68.26.

CHAPTER 66

(1) **angels of punishment** See the note on 53.3 above, 218.

(2) **not show their power** Eth. lit. 'not to raise (their hands)' 'iyānše'u 'edawa =

לא ישאו יד i.e. in a display of power by releasing the waters beneath the earth. For the expression Ps. 10.12, Ezek. 20.6.

keep watch Eth. ye'qabu = ישמרו, or 'store them up' (Gen. 41.35), 'keep within bounds' Exod. 21.29).

angels were over ... waters Cf. Rev. 16.5 ὁ ἄγγελος τῶν ὑδάτων.

CHAPTER 67

(1) **your lot has come up before me** Eth. kefl = חלק, μοῖρα, perhaps 'way of life'; 'arga ḥabēya = עלה לפני (Jon. 1.2 for the idiom).

love and uprightness Eth[tana] feqr (feqra?) ret' = 'love of uprightness' אהבת צדק. 'Love and uprightness' is an usual, and certainly an un-Biblical, combination. Was the original perhaps חלק אהב הצדק 'the lot of a lover of uprightness'?

(2) **a wooden (vessel)** Eth. simply 'eḍaw = עצים (cf. Exod. 7.19 'articles of wood'); the ark, according to Gen. 6.14, was made of 'gopher wood' (עצי־גפר). At 89.1 Noah himself constructs the ark; this is the only place where it is the angels who build it.

completed this work Charles and Flemming read (wasoba) wad'u (lawe'etu mal'ekt[1]), '(... and when) they have completed (that task)' (Charles), both emending Eth[g q u] wad'a to wad'u. But this is an unattested use of wad'a in this sense of 'to complete' (cf. Dillmann, *Lex.*, 932): Eth[tana] tawaḍ'a 'is completed' is perhaps possible, but we would then require to read we'etu mal'ekt, 'when the work is completed'. Eth II waḍ'u (Eth[o] maṣ'u) can only be rendered 'and when they come forth'. Dillmann understands 'when they have come forth for that task', i.e. not for the building of the ark but for the release of the waters of 61.1,2. Knibb follows, but takes mal'ekt as subject and renders '(when the angels) come out for that (task)'. But mal'ekt is 'task' and not malā'ekt 'angels', and here = מלאכה 'work' (Ezek. 15.3 LXX ἐργασία). Perhaps Eth II should read 'awḍe'u = הוציאו meaning 'to produce', 'when they have produced this work'; cf. Isa. 54.16.

seed of life Cf. Gen. 7.3.

transformation ... inhabitants Behind tawlāṭa yebā' lies a phrase with בוא = 'to come to pass' and some form of הפך (הֵ֫פֶךְ?) in the sense of 'change, transformation' (cf. 89.36 En[c]4 10 אתהפך) (Heb. סיבה, נסיבה 'turn of events' is also possible). Eth. 'iyenbar yabs 'erāqo suggests לא תבוק יבשה (Isa. 24.3, Nah. 2.3).

(3) **I will spread abroad** Eth. zarawa = זרה Pi., with a word-play on the previous זרע 'seed, offspring'.

it shall not be barren Eth. 'iyemakker 'it will not take counsel' (sic). For this nonsense reading Charles suggested emending to 'iyemakken 'it will not be fruitful'. Martin (followed by Knibb and Caquot and Geoltrain, *Semitica* 13 (1963), 51) render 'I will not put (them) to the test', a meaning attested for makara II.2 but not for I.1; they suggest an allusion to Gen. 8.21 and 9.11. On the whole, Charles's conjecture seems preferable.

multiply upon the earth Eth[M] 'in front of the earth' qedma yabs = על־פני יבשה. The translator has misrendered על־פני which in this context means 'on the face of, upon' in its sense of 'in front of'. (A correct rendering appears earlier in the v.).

---

[1] Charles[1906] malā'ekt.

(4) The author at this point is elaborating on the fate of the watchers (and the 'satans'[1]) on the basis of the tradition of the Book of the Watchers (10.12f.). Here their imprisonment is is a 'burning valley' located 'in the west' among the mountains of metal described at 54.1f. (although these are there said to lie 'in another part of the earth'). From the following verses it would seem that the writer has in mind not only the place of temporary punishment of the watchers but Gehenna itself. He envisages a place beneath which fire and streams of 'water' pour forth. Dillmann finds in this passage a vision of the vale of Gehinnom ('west' means Jerusalem) but extended down to the Dead Sea. It seems likely that some kind of volcanic or earthquake activity in this area was in the writer's mind (see below on v. 6).

those angels who revealed iniquity (to mankind) i.e. the watchers.

(5) **a great convulsion ... waters** Eth. hawk = רעש usually of earthquakes (Am. 1.1). The 'waters' of the 'burning valley' are evidently from the same underground source (66.1) as the waters which brought the judgement of the flood: they also are to be waters of judgement, since they are to be changed into the fires of destruction (v. 13).

(6) **a smell of sulphur** Eth. ṣēnā tay = ריח גפרית. Noah's vision here suggests that the author had some acquaintance with volcanic or earthquake phenomena, such as molten lava with its sulphurous odours. The fiery 'molten metal' (or 'soft metal and fire') is in some way conceived as connected with or proceeding from the 'metal mountains' of the earlier vision (52.1f.), now located here; this fiery molten metal (lava) joins up with the subterranean waters which, along with the waters from above in the reservoirs of heaven, produced the flood. This 'fire and brimstone' (including the 'waters') are to be the instruments of the final punishment in Gehenna (v. 13).

(7) **through its ravines** i.e. the 'valleys' of the land. If 'that land' is the 'burning valley' of v. 4, perhaps a different word was used in the original to refer to the 'ravines' (נחלים) (גיא, עמק = 'valley' cf. גיא חנום 'Valley of Gehenna'). Does the idea of the πυριφλεγέθων underlie the thought of the passage?

(8) **who occupy the earth** See note on 62.1.

healing of the body ... punishment of the spirit Eth. fawweso šegā, 'healing of the flesh' cf. Prov. 4.22. The reference is clearly to the hot springs, e.g., at Kallirhoe to the east of the Dead Sea (Josephus, *Ant.* xvii.6.5 (171-172)) to which Herod the Great resorted, or those at Machaerus (B.J., vii 6.3 (186-189), in an area which contained sulphur mines (Charles, 134).

Since these waters come from the prison of the watchers and will end by changing into everlasting fire (v. 13) they can be said to be for 'the punishment of the spirit'. This use of these hot springs by 'kings' may imply a knowledge of Herod's visits, giving some confirmation for the dating of the Parables c. 40 B.C.

full of lust Eth. tawnēt = תאוה 'desire, lust'. The thought seems also, however, to be that, while these hot springs provide healing for the bodies of the 'kings and mighty', they are also a mode of punishment 'for their spirits'. These spirits are so full of lust that their bodies are 'punished' in these waters by the diseases they seek to cure, and they daily witness this punishment. Such are the consequences of 'denying' the Name of the Lord of spirits.

see their punishment daily Perhaps see the instruments of their final punishment i.e. the waters of the hot springs which come from the fires of Gehenna.

(9) **burning of their bodies** (with lust) Dillmann, followed by Charles, interprets of

---

[1] See on 40.7 and on 69.3f.

the literal burning of their bodies in Gehenna, leading to regret, sorrow and repentance: 'The punishment will work repentance in the kings which will be unavailing' (Charles). The context, however, is referring to the sin of lust of the kings and powerful (vs. 8 and 10); for the figure of speech of 'burning' for lust, Hos. 7.4, Sir. 23.17, 1 C. 7 and cf. Ezek. 16.26 where the Egyptians are גדלי בשר 'great of flesh', i.e. 'lustful' (NEB). See above on 10.14.

the price they will pay Eth. ṭawlāṭ = τὸ ἄλλαγμα Dt. 23.18(19), מהיר ἀντάλλαγμα Job 28.15 = תמורה (Job 15.31)?

a lying word Cf. 49.4, 62.3. Does the expression here mean: 'Before the Lord of spirits this is, indeed, true'?

(10) **deny the spirit of the Lord** The expression is unique and confined to Enoch, as Dillmann and Charles noted. The usual phrase is 'to deny the Name of the Lord of spirits' (e.g. 41.2, 46.7, 48.10). Has a theologically minded scribe altered the usual expression, so that the original read 'deny the Lord of spirits' as at 48.10?

shall change their temperature The watchers and the satans are incarcerated in 'that burning valley' (v. 4) which is here envisaged as filled with underground water heated to be a punishment for them during their temporary imprisonment. When they ascend from these waters to be taken off to the place of their final destruction (cf. 90.24) the waters become cold. They are then, however, to become hot again and to be transformed into fire for the punishment of the kings and the mighty.

(12) **a warning** lit. 'testimony'. Eth. samāʿt = Heb. עדות.

(13) **body of the kings** Eth. šegāhomu lamalāʾekt 'bodies of the angels'. Clearly 'kings' מלכים is meant. Flemming maintains (Flemming-Radermacher, 86) that malʾak = 'ruler' as well as 'angel' so that is no need to assume a mistranslation. But nagašt is the usual word for 'kings'. It seems more probable that the translation 'angels' is a slip of the translator. Or is it possible that the author of the original Heb. text is deliberately punning throughout these two chapters 67 and 68 on the similar sounding words for 'kings' and 'angels', and under the guise of retelling the story of the fall and punishment of the angels, developing the story to enable him to bring in the fate of the kings for which that of the watchers is a terrible warning?

CHAPTER 68

(1) I have adopted the text of Eth$^{tana(g)}$ against Eth^{M}, according to which it is Enoch who imparts this instruction to Noah. The order of words in Eth^{M} (the subject 'my great-grandfather Enoch' is in an odd position) favours the reading of Eth$^{tana(g)}$; the subject 'Michael' is supplied from the context. Should we then assume a change of subject in the last sentence, 'and he (Enoch) had composed them for me ...'?

Book of the Parables This Noah apocalypse seems here to assume the existence of a 'Book of the Parables of Enoch', and this is taken as a clear indication that these chapters are a later composition incorporated in the 'Second Vision'. (So Dillmann, Charles etc.).

composed them Eth. dammaromu, συνετάξατο; cf. the similar Heb. idiom at Job 16.4, חבר Hiph. 'to join words together'.

(2) **power of the spirit seizes me** The text is not in doubt, but this 'strange expression' (Charles) is. A clue to the original may be provided by the parallel verb yāmeʿʿāni which almost certainly renders ירגיזני 'made me tremble' (so Halévy,

JA, vi.9 (1867), 379 f., cf. Knibb, 2, 158). The verse recalls Exod. 15.14 'Nations heard and trembled (ירגזון), agony seized (חיל אחז) the dwellers in Phoenicia' (NEB). Heb. חיל(ה) means 'agony, anguish', and the parallelism at Exod. 15.14 shows that it is the 'agony' of fear which is meant there. Could the original in Enoch have read 'the anguish of fear seizes me' (חיל יאחזני)', where 'anguish' has been misread as חיל, δύναμις, and the sentence expanded to 'the power of the spirit seizes me'?

the sentence regarding the secrets kuennanē 'enta ḫebu'āt, [ἡ κρίσις τῶν κρυπτῶν]. As the following 'the judgement of the angels' indicates, what is referred to here is the punishment of the watchers for revealing 'celestial mysteries' to mankind; 'the punishment regarding the secrets' or, perhaps, 'the punishment of the secrets' is a compendious phrase for punishment for revealing such celestial secrets. At the same time, the writer may well be using the homonym מלכים/מלאכים to point to the impending punishment of the 'kings' (cf. above on 67.13).

and not be filled with fear wa'iyetmassaw (cf. Eth$^{b^2}$) lit. 'and not melt', καὶ οὐ τακήσεται, i.e. with fear. Cf. Heb. מוג (Am. 9.5) and מסס (Nah. 2.11).

which is to be executed Eth. tagabrat, proph. perf. (Dillmann, vollzogen wird).

(3) **whose heart ... faint for fear at it (the punishment) ... reins do not quiver with fear** Eth II lit. 'who is there who does not soften (his) heart' (za'iyāraḫarreḫ lebbo) = Heb. מי הוא אשר לא ירך לבו. (Is the Eth. verb here, like the Heb., intransitive, [οὗ οὐκ ἐκλύεται ἡ καρδία]: Eth$^{t^2}$ corrects to a pass.); cf. Dt. 20.3 אל־ירך לבבכם, LXX μὴ ἐκλυέσθω ἡ καρδία ὑμῶν, Jer. 51.46, Isa. 7.4. Eth I za'iyerasseḥ, 'dessen Herz sich dabei nicht schuldig fühlte' (Flemming-Radermacher) is an inner Eth. scribal 'improvement'. Eth. 'iyethawwak kuelayātihu = לא ינועו כליותיו, cf. Isa. 7.2 וינע לבבו, LXX καὶ ἐξέστη ἡ ψυχὴ αὐτοῦ.

from those ... afflicted(?) Dillmann understands 'emennēhomu 'from them' as virtually repeating dibēhomu 'against them' and renders: 'Ein Gericht ist ausgegangen über sie, über die die sie also hinausgeführt haben', i.e. 'over those whom they (impersonal plur.) have led forth' from their temporary incarceration to their final place of punishment, i.e. the fallen angels. Charles renders 'because of those who have thus led them out', the reference in this case being to the punishment of 'the satans who are rigorously punished because they seduced the angels into sin'. The suggestion behind the translation proposed is that an original ענה Pi. 'oppress' appears to have been misread as ענה = 'reply', the reading of Eth^{tana}. The reference throughout is really to the 'kings' not to the watchers whose secrets had led mankind astray. The 'kings' are to be punished 'in like manner' kamaze = οὕτως as they had 'afflicted' (ענו, ἐκάκωσαν) the righteous.

(4) **And it shall come to pass ... stand** It makes better sense to take this clause as part of the oratio recta of Michael, and to emend wakona, καὶ ἐγένετο to wayekawwen, καὶ γενήσεται and read qomu, ἑστᾶσιν, with Eth^{tana}.

no mercy in the sight of the Lord The reading of Eth$^{e\ v\ h^2}$ alone makes good sense. Dillmann 'ich werde nicht für sie sein unter dem Auge des Herrn' is followed by all successive translators; 'I will not take their part' (Charles, Knibb). Martin notes and translates the variant in the margin: 'Il n'y aura pas pour eux de miséricorde.' Eth. bawesta 'ayna = לנגד עיני (cf. Job 4.16, Ps. 5.6).

they have fashioned images of the Lord Translators render: '... they act as if ('amsāla = tanquam) they were the Lord'. If this was original, it could be a reminiscence of Isa. 14.14, where the King of Babylon declares 'I will be like the Most High' (אדמה לעליון). In favour of the alternative 'they have fashioned images of the Lord' is the mention of 'idols' at v. 5.

(5) **the sentence ... secrets** See above on v. 2.

neither idol nor man shall be accorded his portion The reading of Eth^{tana} ʾamsāla = ὁμοίωμα, εἰκών, εἴδωλον (read ʾiʾamsāla 'neither an image (idol)') follows closely on v. 4, alluding, by implication, to Exod. 20.4: neither idol no human being shall be given what belongs only to the Lord ('his portion' i.e. worship and service). Did the original perhaps read 'neither idol nor angel nor man'? This could be a reference not only to idolatry, but also to that form of it which became Emperor-worship in the Graeco-Roman world. For 'his portion' Eth II reads 'their portion'. (The singular has been changed to a plural in the Rylands manuscript (Knibb, 1, 198). This understanding and form of the text may have given rise to the last sentence in its present form: 'neither idol nor man will receive their portion (or punishment), but they alone will receive a sentence for ever and ever.' Was bāḫtitomu 'they alone will receive' introduced by a scribe, the original reading being 'but they will receive their sentence for ever and ever'?

CHAPTER 69

(1) **the sentence (upon them) ... occupy the land** The verse, though made the beginning of a new chapter, follows on 68.5. This fearful judgement 'will terrify them and make them tremble (cf. 68.2)', presumably the מלאכים, the 'angels', but again almost certainly a cryptic homonym for the 'kings', מלכים. The last clause is difficult, and Dillmann, 210, regarded the text as corrupt. If we give the expression 'those who occupy the land', however, the meaning it has in earlier usage (cf. the note on 62.1), the clause can yield an appropriate sense. I take the verb ʾarʾayu as an impersonal plural (Michael and Raphael could be the subject, but this seems unlikely); 'inasmuch as this (fearful sentence) has been shown to those who occupy the land'. The 'kings' have been apprised of their fate and are consequently in terror and trembling.

(2) **and these are their names** These words are taken from 6.7. The opening clause (of which the words from 6.7 are a duplicate) has been thought to have formed the introduction to v. 4f. where the names of five evil angels are given. Most commentators accordingly bracket vs. 2-3 as an interpolation from 6.7: Knibb considers these verses to be a 'secondary insertion ... copied from the Ethiopic version of 6.7'. It was introduced 'at a very late stage in the transmission of the text of Enoch' (2,139). Knibb goes on to argue that variations in the names at 69.2 are the result of 'mistakes of copyists'. See above, 123.

(2-3) For notes on these names, see above on 6.7.

(3) This verse corresponds to 6.8 and seems to be an elaboration of the text of Eth II 'These are the leaders of the 200 angels and of all the others with them'. But the author of the Parables has a new use for the verse: it has been taken over and adapted to enable him to introduce the archangel rulers of the 200 watchers, the five following demon-satans, a higher angelic echelon than the watchers, and here described as 'centurions, quinquegenarii and dekadarchs'.

(4f.) The five named angels, Yeqon, ʾAsbeʾel, Gadreʾel, Penemuʿe, Kasdeyaʾ, who are to undergo this fearful judgement, are generally identified with the satans (see above on 40.7, and cf. Dillmann, 211, Charles, 136f.). Just as at 54.6 the watchers are said to have become the subjects of Satan, similarly here Yeqon and ʾAsbeʾel are said to have led the watchers astray. They belong to a higher echelon in the 'angelic' hierarchy than the watchers. The writer is evidently accommodating the older tradition of the fallen angels, the watchers, as 'evil angels' within a later demonology where Satan and his host, the 'satans', had come to occupy the highest place in the hierarchy of evil angels. Two of these satans even take over the role of the earlier 'watchers': 'Asbe'el (or Kasbe'el) has assumed the role of Semhazah (vv. 5, 13) and Gadre'el that of Asael as the angel in charge of weaponry (v. 6). For a detailed study of the individual names, see especially L. Gry, 'Mystique Gnostique (juive et chrétienne) en finale des Paraboles d'Hénoch' in Le Museon, 52, 1939, 337-377.

Yeqon A derivation from יקום 'rebel' (cf. Dillmann) or 'he will arise' (Knibb) is unconvincing. Gry, op. cit., 340, suggests that the name is a corruption of יקרון, described as an 'angel' in Mandaean sources. Could it be Greek εἰκών (LXX Isa. 40.19 'idol', Tg. איקון, ܐܝܩܘܢܐ יקונא Levy, NHCW, II, 260[1])? Yeqon 'led astray (the watchers)' = Heb. טעה Hiph.: טעות = 'idol'. Levy, CW I, 312.

sons of the angels Also at v. 5 and 71.1; cf. 39.1, 106.5. The identity of these 'sons of heaven' (En. 6.2) is not in doubt: they are the watchers, the 'sons of God' of Gen. 6.2. Both En. 6.2 (G ἄγγελοι υἱοὶ θεοῦ) and Gen. 6.2 LXX (ἄγγελοι τοῦ θεοῦ) avoid the straightforward translation 'sons of God'. (See above on En. 6.2). The Eth. circumlocutory 'sons of the (holy) angels' is to be similarly explained (it also succeeds in retaining part of the LXX rendering of Gen. 6.2). (There is no need to assume a 'false translation' or mistranslation of בני אלהיא as בני מלאכיא (Schmidt, Knibb).)

(5) **'Asbe'el** Dillmann = עזביאל deum deserens, Schmidt חשביאל, 'thought of 'El', corresponding to the role of this angel as 'planning evil' (cf. Ezek. 38.10). The same angel at v. 13 is named Kasbe'el and perhaps this was the original form = כזבאל 'lie of 'el'(?) or כשפאל 'sorcery of 'el' (Knibb, cf. Gry, 340f.). Cf. also חשבנה Neh. 10.25 and Noth, *Die Israel. Personnenamen*, 189.

defiled their bodies Cf. 7.1 G μιαίνεσθαι ἐν αὐταῖς and Heb. טמה Niph. (Lev. 18.24), LXX μιαίνεσθαι.

(6) **Gādre'el** None of the derivations proposed is convincing. The derivation from עדר, עזר 'God is my helper' (Charles, Knibb) seems inappropriate. Dillmann took the name to be synonymous with עזביאל, deum deserens (עדר Niph. 'to be lacking, to fail'?). A derivation from גזר 'to cut' used of the sword (1. Kg. 3.25) and then 'to destroy, exterminate' (Hab. 3.17) is perhaps less speculative than Kuhn's otherwise appropriate הרגיאל 'Mordengel' (ZAW, Bd. 39 (1921), 270). In the Book of the Watchers it was Asael who introduced mankind to weaponry. The name is found in Aramaic incantation texts (Gry, 340, Montgomery, *Aramaic Incantation Texts from Nippur*, (Philadelphia, 1913), No 14 (183).

weapons of death Eth. = כלי מות, Ps. 7.14.

(7) **by him they have gone forth upon** Eth. 'emenna 'edēhu lit. 'out of the power of', i.e. 'through his agency', 'through the power of Gadre'el' = מן ידו (בידו). Eth[tana q] may be original through them' = בידיהם i.e. through 'the weapons of death'. Translators assume 'the weapons of death' to be the subject of the verb waḍ'a, lit. 'from his hand they (newāya mot, coll. sing.) have gone forth upon those who dwell on earth'. The subject could, however, be 'death': 'from him/through his agency death has gone forth upon those who dwell on earth'; and this would go well with the reading 'through them'. If the plur of Eth[a] is more than a scribal correction of the sing., it could be taken as an impersonal plur. 'they have gone forth (to war)', an

[1] A suggestion first made by Schmidt, 'Original Language', 344.

idiomatic use of waḍ'a = יצא: 'through him (them) war has been waged on those who dwell on earth for this time and for ever and ever'.

(8) **Ṗēnēmu'e** (variants Tenemu'e, Tuni'el). The etymology of the name has baffled commentators and conjectures have been wholly unconvincing (e.g. Kuhn, op. cit., 270 suggests that it is a shortened form of פנה אמונה 'he has done away with trust'). Heb. פנמו means 'inside, within', but how this explains the role of this angel is not clear. The name may be imported from another culture (Accadian or Phoenician?). Thus Panammu is attested in cuneiform sources and Aramaic inscriptions, but its derivation is obscure. See D. W. Thomas, *Documents from Old Testament Times* (Edinburgh, 1958), 54, H. Donner—W. Röllig, *Kananäische und Aramäische Inschriften* (Wiesbaden, 1962), 216, Cooke, *North Semitic Inscriptions* (Oxford 1903), comparing the name πανaμύης found in Caria.

bitter was sweet and sweet bitter So Eth^{n} with what seems to be the original text, alluding to Isa. 5.20: 'Woe unto them that call evil good and good evil; that put darkness for light, and light for darkness; that put bitter for sweet, and sweet for bitter'. The majority of manuscripts read 'he taught bitter and sweet', an abridged text. Cf. Kuhn, op. cit., 270.

(9) **many ... have gone astray** Cf. 99.2, 104.9 for the sins here mentioned.

(11) **touched them** Eth. = נגע 'to touch' i.e., 'to harm'.

(12) **Kasdeya'** Dillmann, 212, suggested a possible חסדיה, an attested name (1 Chr. 3.20 cf. Noth, *Die Israel, Personennamen*, 183), but one, as usually explained ('God hat Liebe gezeigt', Noth) which is hardly suitable for a satan. Could it be derived from חסד 'reproach, shame'? Kuhn (cf. Caquot and Geoltrain, 'Notes sur le Texte Éthiopien des 'Paraboles' d'Hénoch' (above 185, n. 3), 52, Knibb) suggested כשדיא 'Chaldaeans' (ZAW 39 (1921), 270). Cf. Gry, *op. cit.*, 342f.

which come through the noontide heat Cf. Ps. 91.5,6 which Kuhn (ibid.) suggests lies behind the entire verse. Cf. Dt. 32.24.

Tabā'et Could this proper name be connected with ܛܒܥܐ daemonissa quae infantes et mulieres strangulat; P. Sm. 4383.

The list of satans breaks off abruptly at v. 12. What now follows is 'an independent section dealing with the divine oath' (Knibb). A new chapter might well have been made at this point.

(13) **And it was he ... in glory** Eth^{q} alone gives an intelligible text. For attempts to make sense of Eth^{M} see Charles and Knibb. The introductory clause 'And this was he who' occurs at vv. 4, 6 (zewe'etu za-), 12, 13 (zentu we'etu za ...), introducing the particular evils which were the specialty of each of the five satans. The concluding task of the fifth satan, Kasdeya, according to Eth^{q}, was to count the Gematria value of the divine Name, lit. 'to count the Chief of Days', and to do so for the benefit of Kasbe'el, who revealed the mystery to the angels, in this case, since Kasbe'el is here the same as 'Asbe'el and fulfilling the role of Semhazah, to the watchers who corrupted mankind. That the difficult and remarkable text of Eth^{q} is original is confirmed by the gematria suspected by Dillmann in these verses. (See note on BIQA and 'AKA').

the sum of the oath Dillmann 'Das Haupt des Schwures', Beer 'Das Hauptschwur', described in detail in vv. 15f. It seems more likely to mean the numerical sum of the letters in the 'hidden Name' contained in the oath (cf. v. 14).

BIQA Dillmann suspected Gematria here and for 'AKA' in v. 15. The latter = אכע, 91, would correspond to יהוה אדני which appears in a similar Gematria claimed for an alleged acrostic at IQS 10.1-8 (see Driver, *Judaean Scrolls*, 339). BIQA may be a corruption of 'AKA' (Dillmann, Flemming) or an independent Gematria: read as

BIQAH = ביקה its number is 117 corresponding to the number of יהוה האלהים. The secret of the powerful oath which Kasbe'el revealed would then be the knowledge of the numerical value of these nomina sacra.

(14) **This (satan) ... before that Name and oath** The subject could be either the fifth satan Kasdeya or the second satan Kasbe'el. Since it was the second, in the role of Semhazah, who had to do with the solemn oath of the watchers at 6.2 (although it was the watchers themselves not their leader who proposed swearing this oath), it seems more likely that Kasbe'el is the subject. Does the writer intend to suggest that Kasbe'el tricked Michael into revealing the Name on the pretext that he would thereby terrify the watchers and so deter them from their plan to seduce the 'daughters of men'?

the hidden Name The 'ineffable Name', שם המפורש, is clearly meant, but this seems to be the only place where it is described as 'the hidden, secret Name'. See ThWNT, Bd. V, 265f. s.v. ὄνομα (Bietenhard): as part of the 'oath' it becomes 'eine geheimnisvolle mächtige Potenz' (Bietenhard); cf. Jub. 36.7. See further M. Gaster, 'The Sword of Moses' in JRAS, 1896, 155f.

The reading of Eth$^{m(u)}$ 'that evil **(እኩይ)** ... Name' is more than a scribal error: the text copied must have contained a version of the consonants אכע the Gematria for יהוה אדני. The original possibly had '(that he might pronounce) the hidden Name אכע'. (Eth^{tana} reads 'this evil oath' for ʾAKAʿ at v. 15.) With Eth. zakara cf. זכר Hiph. memorare ore et lingua (Gesenius). It can mean 'to call on the Name of the Lord' (Isa. 28.13), but here simply to pronounce the Name.

(15) **he had placed** The verb can be read as an impersonal singular 'and one placed' the equivalent of a passive verb 'there was placed' (cf. Knibb). The antecedent 'God' is not explicitly mentioned in the context but is implicit in a context with Gematria for 'the Name'.

ʾAKAʿ See note above on v. 13.

(16) Vs. 16-21, 26-29 are composed in poetic form, with parallelismus membrorum and, for 16-21, with a recurring concluding refrain 'from the creation of the world and unto eternity' or 'from eternity to eternity.'. Has an older poetic source on the Oath, an 'Oath poem', been incorporated by the author within his description of the satans?

The text of v. 16 cannot be right. Flemming-Radermacher suspect a lacuna after '(These are the secrets of) this oath' and suggest supplementing with '(Everything has been created) and is firmly founded (ṣenuʿ) through his oath'. Eth$^{tana\ d}$ reads 'his firmament' (ṣenʿu); perhaps we should read 'And the firmament (ṣenʿ) through his oath and the heavens were suspended'? Cf. Dillmann, 39, n. 2 and Dillmann, *Aeth.*, 43, Annotationes, 19. Or is this text an expansion of 'And the firmament of heaven (ṣenʿu lasamāy) through his oath was suspended .'?

(17) **the earth was founded upon the waters** Cf. Ps. 24.2, 136.6.

sweet water Cf. Ps. 104.10,13. Eth^{M} lit. 'beautiful waters'. Was the original perhaps מים חיים 'living waters'? Cf. Eth. Gen. 21.19 (LXX ὕδατος ζῶντος) māyā teʿum, aqua dulcis. Cf. also the curious variant of Eth$^{u\ 3\ mss}$ laḫeyāwān māyāt lit. 'for the living, waters'. (Dillmann read a conflate text, schöne Wasser für die Lebendigen).

from the creation of the world Cf. 71.15 and ἀπὸ κτίσεως κόσμου Rom. 1.20; ἀπὸ καταβολῆς κόσμου Mt. 25.34, 13.35 (v.l.) Lk. 11.50, Heb. 4.3, Rev. 13.8. Above, 211.

(18) This verse has been corrupted in all mss. Eth^{q} has preserved a satisfactory text on two important respects, but it has also obvious corruptions (e.g. 'barrier' for 'wrath'). It is not, however, as badly corrupted as Eth^{M} which has lost two nouns

necessary for the sense of the passage ('barrier' and 'boundary'), both fortunately preserved in Eth[q].

he placed for it the sand as a barrier Cf. Jer. 5.22, Job 26.10, Ps. 104.9 etc.

pass beyond its boundary Eth. ḫalafa = עבר in its primary meaning: cf. En. 2.1 En[a] 1 ii 1. For the meaning 'to pass beyond' and so 'to transgress', see below at v. 20 and at 18.15, 21.6, 41.5, and cf. Prov. 8.29, Num. 22.18.

(19) For this verse cf. Prov. 8.28.

(20) **deviate not from their ordinance** See above note on v. 18.

(21) **he calls the by their names** Cf. 43.1.

Verses 22-24 are treated by Charles as an interpolation. They could be an expansion of the poem on the oath to take account of later ideas about the spirits of nature.

(22) The text of v. 22 may be crpt. (cf. Dillmann, 213 foot) or be simply a compendious summary. The waters, like the heavenly bodies, know their place: without deviating from their ordinances, their spirits complete their tasks, and all the winds likewise from all the regions and directions of the wind-rose from which they blow ('from a' the airts the winds can bla'). Eth. ḫebrāt = חבל, 'regions' (cf. Zeph. 2.5) (Flemming, Charles) rather than 'bands', 'groups' (Knibb).

(23) **And there are kept** All texts read 'there' where we would expect 'by it' (botu) i.e. by the oath: 'there' suggests a description of the heavenly regions as at 60.16f. We should perhaps emend the first 'there' (baheyya) to 'by it' (botu). In its second occurrence, however, Eth[q] reads 'by this (oath)' instead of 'there'.

the storm-cloud See on 60.19.

(24) **give thanks** See above on 63.5.

their sustenance ... thanksgiving This is a difficult text and the suggested parallel at 41.7 does not explain the difficulty. Was there a mistranslation of מלאכה, λειτουργία as βρῶσις/βρῶμα: ומלאכתם בכול תודה הית, 'and their service was in all manner of thanksgiving'(?). See above, 202f.

(25) **is binding upon them** Eth. dibēhomu ṣan'a lit. 'is strong over them', 'prevails over them', perhaps = חזק עליהם (cf. 1 Chr. 21.4).

is not spoiled Eth. māsana = שחת Pu. (cf. Prov. 25.26 of the spoiling of a 'spring', מקור).

Verses 26-29 take up the theme of the revelation of the Son of Man, last appearing at 63.11. It seems unlikely that these verses are at home in their present context and that the subject of the verbs in v. 26 'they blessed and glorified etc.' is the 'spirits' (of nature) of 69.24. There is a formal connection with the doxologies and v. 28 does refer to the watchers, the subject, along with the satans of the last chapter, but this may be no more than coincidence or a recensional linking. Where, if at all, they may have appeared originally in the Son of Man sections is a matter of debate, (perhaps at the end of 61 and the beginning of 62?). Charles thinks of vs. 26-29 as sypplying the conclusion of the Third Parable, part of which, like Dillmann, he suspects has been lost and replaced by the 'interpolations' of the previous chapters.

(26) **revealed to them** Cf. 48.6, 62.7 the motif of the Son of Man absconditus yet revealed (through Scripture and its interpretation?) to the elect.

(27) **the sum of judgement** i.e. the whole judgement; cf. Ps. 139.17, and especially Jn. 5.22,27 τὴν κρίσιν πᾶσαν δέδωκεν τῷ υἱῷ, claimed by Charles as a direct quotation.

those who have led the world astray A reference to the watchers (cf. v. 28). This should almost certainly be read as subject to the verb in v. 28 yet'assaru, in spite of the verse division which places it at the end of v. 27 (so also Knibb).

(28) **with chains ... be bound ... be imprisoned** Eth. basanāsel yet'assaru = יאסרו בנחשתים (cf. Eth. 2 Kg. 25.7): yet'āṣawu = יעצרו. Note the word-play in Heb.

(29) **that Son of Man** Note the distinctive version walda be'si. Eth[n] walda be'sit 'son of woman' has arisen by a dittograph from be'si tare'ya, but note Eth[g] 2° walda be'sit. Cf. above on 46.2.

shall be strong Eth. ṣana'a = קום 'be established'(?).

CHAPTER 70

Except for the common expression 'Son of Man' Chapter 70 has little in common with chapter 69 and reads like a quite independent piece of text. Do chapters 70-71 come from a different source, perhaps an even older tradition? Cf. Coppens, *Le Fils d'Homme*, 134f.

the name of a son of man was raised up Since Dillmann Eth[M] has been rendered (as by Charles): 'And it came to pass after this that his name during his lifetime was raised aloft to that Son of Man and to the Lord of spirits ...'. (The reference it is assumed, is to Enoch, mentioned at 69.29). The awkward expression, however, 'his name alive' (semu ḥeyāw) looks like an inner Eth. corruption of the variant preserved in Eth[v] 'the name of a son of man' semu lawalda 'eguāla ['emma] ḥeyāw). If the original reading was 'the name of a son of man' (i.e. Enoch)[1] was raised up' (the context is all about Enoch's translation to heaven), a scribe, probably a Christian scribe, may well have taken exception to the use of this now theologically loaded term 'son of man' of Enoch, and redrafted the v. to read 'his name alive was raised up *to the Son of Man* ...'. For a similar solution based on the reading of Eth[u] see A. Caquot 'Remarques sur les Chapitres 70 et 71 du Livre Éthiopien d'Hénoch' in *Apocalypses et théologie l'espérance*, Paris Ed. du Cerf, 1977, 113.

(2) **on a chariot of the spirit** Cf. 2 Kg. 2.11: Elijah is transported in a 'chariot of fire', Enoch in a 'chariot of the spirit'.

his name was bruited abroad among them i.e. among the dwellers on earth. Eth. waḍ'a = יצא promulgari and sem = שם in the sense of 'fame'.

(3) **between two regions** Eth. lit. 'winds' = רוחות cf. Ezek. 42.16 and below note on 76.1. The reference is to the account of the measuring by the angels of paradise at 61.1f.

(4) **the first fathers, and the righteous** i.e. the patriarchs. Are 'the righteous' here a separate category of dwellers in Paradise from the earliest times (lit. 'from everlasting')? See above, 167. Or should the 'and' be taken as epexegetic, 'the first righteous fathers'?

---

[1] The term would then be an oblique reference to Enoch himself. For this use of 'son of man' as virtually a pronoun = 'one' but with a reference to the speaker (in this case the seer), see my *Aramaic Approach*[3], Appendix E, by Geza Vermes, 'The Use of ברנשא/ברנש in Jewish Aramaic'. The term 'name' itself could be construed as referring to the person of Enoch (so Dillmann, 216). The implication of adopting this text and interpretation would be that these verses originally referred to the 'ascension' of Enoch in terms recalling that of Elijah.

CHAPTER 71

As Dillmann, 218, noted the vision of Enoch's translation in this chapter resembles closely the account at chapter 14 of his celestial journey to the 'places' and Throne of God. Indeed, as Charles, 34, pointed out, 14.18 is being used by the author of 71.5-8. It seems, in fact, that chapter 14 has provided a model for chapter 71. See further M. Black, 'The Throne-Theophany Prophetic Commission and the 'Son of Man': a Study in Tradition-History', in *Jews, Greeks and Christians: Religious Culture in Late Antiquity, Essays in Honor of William David Davies*, edited by Robert Hamerton-Kelly and Robin Scroggs (Leiden, 1976), 167.

(1) **was translated** See above on 12.1.

sons of the holy angels Cf. above at 69.5.

Their garments were white Cf. Dan. 7.9, Mk. 9.3.

(2) Cf. v. 6 below, 14.19 and Dan. 7.10. The reference is not necessarily to the streams of fire that issued from beneath the divine throne. The reason for two streams of fire is not clear.

hyacinth Eth. yāknet = ὑάκινθος = ספיר ('sapphire') (Ezek. 28.13) (LXX mostly for תכלת 'blue'; Rev. 21.20 'turquoise', 'aquamarine'(?)).

I fell on my face Cf. 14.25.

(4) **in the presence of the holy ones** Did the original perhaps read 'in the presence (lit. before the face of (לפני)) the Holy One'?

(5) **And he translated my spirit** i.e. Enoch was translated to the heavenly realm by Michael. The text has been altered in Eth II, perhaps under the influence of Ezek. 11.1 ('A spirit lifted me up ...' NEB), to 'And a spirit translated Enoch ...'. The reading of Eth I adopted is parallel to v. 1 'my spirit was translated ...'.

in the midst of those luminaries Eth. berhān collective refers to the luminaries of v. 4.

a house as it were built of hailstones Eth. kama botu lit. 'as in it what was built of crystals'. I suggest that kama botu is a mistake for kama bēta (accus.) = כבית 'a house as it were etc.'. (cf. v. 6, where the house is mentioned.)

tongues of fire of the living creatures(?) Eth. 'essāt ḥeyāw translated 'living fire' yields an un-Semitic expression (Dillmann, *Lex.*, compares Latin ignis vivit). A possibility is that a verb ζωπυρεῖν lies behind the phrase. The alternative adopted is to take the original as לשני אש חיה = γλῶσσαι πυρός ζῴου, the 'living creature', i.e. the חיה (LXX ζῷον) of Ezek. 1.13ff. I take ḥeyāw here as a collective = ḥeyāwān the reading adopted at v. 6 below. Cf. 14.11 καὶ μεταξὺ αὐτῶν Χερουβὶν πύρινα and Ezek. 10.2,6,7.

(7) Cf. 61.10, 39.13 and 40.2.

(8) This verse is modelled on 14.22 (cf. also 40.1).

Phanuel See above on 40.9.

go in and out A new feature: there is nothing corresponding at 14.22f.

(10) Cf. above on 46.1 and Dan. 7.9.

indescribable Eth. za'iyettaragguam; cf. ܠܐ ܡܬܡܠܠܢܐ P. Sm. 4496 inerrabilis.

(11) **my whole body became weak from fear** Eth. tamaswa lit. '(my whole body) melted'; cf. 60.3 kuelyāteya tamaswa lit. 'my reins melted (with fear)'. The text here seems to be a freer version of the same underlying expression (לבי) ונמסו כליותי(?).

spirit of power Cf. Isa. 11.2 רוח ... גבורה.

(12) Vs. 12-17 are cited in Lib. Nat., Tom. 41, 62f., 42, 55.

(13) **that angel (Michael) came to me** Eth[g m t] 'he (the Chief of Days) came to me', a possible original.

(14) **You are the Son of Man** Here walda be'si as at 46.3; cf. also 60.10.

born for righteousness For the construction Job 5.7 '... man is born for trouble ...'. Note the variant 'in righteousness'. Dalman, (*Words of Jesus*, 249f.) detected an allusion to the צמח צדקה, the Branch 'of righteousness' of Jer. 23.5, 33.15; in that case we have another Davidic 'messianic' feature attached to the person of Enoch.

(15) **He proclaims ... peace** Eth. = קרא שלום (Jg. 21.13, Zech. 9.10 (דבר שלום).

world to come Charles points out that this is apparently the earliest use of the rabbinical expression העולם הבא. See Dalman, *Words of Jesus*, 148.

since the creation of the world See above on 69.17.

(16) **with you ... their dwelling-places** Cf. 39.4,7, presumably Paradise is meant.

CHAPTER 80.2-8

THE PERVERSION OF THE HEAVENLY BODIES

As noted elsewhere (Appendix A, 411) Chapters 80.2-82.3 is 'an intrusion of non-astronomical material' into the astronomical section. Charles, 170, regards it as 'an interpolation'. It has, however, been 'adapted' to its astronomical context (cf. especially v. 5). It takes up again the theme of chapter 5, viz. the perversion of nature and the heavenly bodies, probably to be regarded as a consequence of the sin of men (v. 2 it is in the days of sinners these perversions take place). Cf. 2 Esd. 5.1-13.

(2) **all work on the earth** In its context the reference is to agricultural work.

the heavens shall be shut up Cf. Jer. 3.3, 5.25, and, especially also for the terminology, Gen. 8.2, Hag. 1.10. Eth. II is usually rendered (as by Charles) 'and the heaven shall withhold (it)' i.e. the rain. But the use of a transitive verb 'shall withhold', tāqawwem, without an object, is difficult (Eth[tana etc.] 'the heavens shall stand still (te(ye)qawwem)' is probably an inner Eth. change to meet the difficulty). Moreover, the Biblical expression, in this connection, is 'the heavens shall be shut up' (3 Kg 8.35 (2 Chr. 6.26) בהעצר השמים , LXX ἐν τῷ συσχεθῆναι τὸν οὐρανόν). We should, therefore, probably read a passive verb, 'the heavens shall be shut up' (συσχεθήσεται ὁ οὐρανός yetqāwam samāy). Or should we read ἀνέξει οὐρανός [ἀπὸ δρόσου] c. LXX Hag. 1.10?

(4) Cf. Jl. 2.10, Am. 8.9, 2 Esd. 5.4.

(5) Charles (171, cf. Charles[1906], 153 n. 4) brackets the first two lines of this verse, which he regards as 'very corrupt', as 'an interpolation', and, building on some conjectures of Halévy, offers an unconvincing reconstruction in Aram. Certainly the mention of 'drought' is quite out of place in the context of a verse dealing with the aberrations of the moon, and clearly belongs (as Knibb notes) to the end of v. 2, where I have placed it.

the remotest regions of the Great Waggon to the west Eth. ṣenf, ἄκρον, קצה, lit. 'extremities, borders'. The 'Great Waggon' ṣaragalāt ʿabiy, Ἅμαξα, the 'Waggon' (Il. 18.487, Od. 5.275) or Great Bear Ἄρκτος μεγάλη (Strabo, 2.5.35, 36, Cicero, *de Natura deorum*, 2.41.105). See Liddell and Scott, 76, 242. The reference is to the remotest parts of the north-west, the least likely places to suffer drought.

(6) **stray from the commandments (of God)** Knibb (following Dillmann and Flemming-Radermacher renders '... heads of stars in command ('ar'estihomu lakawākebt te'zāz)', but, as Charles, and before him Flemming and Beer had noted (Charles, 172, Charles^{1906}, 153 n., 7, Flemming, 108, Kautzsch, *Die Apokryphen und Pseudepigraphen* II, 285 n. b) the noun te'zaz, ἐντολή is part of the predicate, although normally we should have read πλανηθήσονται ... ἐκ τῆς ἐντολῆς ('emte'zāz). The accus. te'zāza is read by the corrector in Eth^{q} and this is arguably possible (cf. Flemming, ibid.). More probably, τῆς ἐντολῆς alone stood in the Greek text, to be construed with the verb (cf. 3 Mac. 4.16), but has been wrongly connected with τῶν ἀστέρων by the Eth. translator. For the expression, cf. Ps. 118(119).110, LXX ἐκ τῶν ἐντολῶν σου οὐκ ἐπλανήθην.

leaders of the stars Cf. 82.10f.

CHAPTER 81

THE HEAVENLY TABLETS

(1) **heavenly tablets** See below, 313.

(3) Cf. 22.14 for a similar doxology.

the great Lord, the King of glory See above, 104, 168.

I wept ... sons of Adam So Eth^{tana}. That Enoch should bless God (Eth^{M}) because of the 'sons of Adam' (a unique phrase replaced in Eth^{M} by 'sons of eternity') seems an unlikely text. Eth^{tana} agrees with 90.40f. where Enoch again weeps after blessing the Lord.

(4) The reading of Eth$^{t^2}$ (which I have adopted) has all the appearance of being a conflate text; and it probably was for the corrector of Eth^{t}. It nevertheless could be the original text: Eth$^{tana^a\ g\ u\ (m)}$ (adopted by Charles) is clearly an inferior form of text.

(5) Eth II has 'three holy ones'. I have adopted the full number 'seven holy ones' of Eth I, but either reading could be original. Cf. 87.2, 90.21,22, and 20.

no flesh is righteous in the sight of the Lord Cf. Job. 9.2, Ps. 143.2, Gal. 2.16, Rom. 3.19.

As Charles, 173, notes, vs. 5 and 6 serve as an introduction to 91-104(8). Enoch is miraculously transported by seven archangels back to his own house and instructed to spend a year with his son imparting all his secret lore. At 82.1f. these revelations are to be committed to writings which Methuselah is to pass on to 'the generations of the world'. This scheme then provides a framework for the revelation of the Dream-Visions (cf. 83.1,9, 85.2), the 'Epistle of Enoch' (92,1), and for the so-called Noah 'apocalypse' (106.8, 107.3, 108.1).

(6) **until you again shall give them your last charges** Cf. Charles, 172. Eth. 'azzaza I.2 is the equivalent of Heb. צוה Pi., Aram. פקד Pa., used in its special sense of the giving of the last instructions of a father to his sons; cf. Gen. 49.33, 2 Kg. 20.1, Isa. 38.1. Knibb (following Dillmann, Flemming-Radermacher) 'until you have re-gained your strength'. Charles rejects Eth. kā'eba 'again' on the grounds that it has arisen by a ditt. of עד 'until', עוד 'again'. But it is not meaningless: the writer may be thinking that this will be the second 'elevation' of Enoch, and presumably he gave last charges to this family before his first disappearance. Alternatively, the word may have

been inserted by a scribe understanding ʾazzaza as 'become strong'. The variant reading of Eth[q], 'comfort, exhort him' seems an inner Eth. change.

(9) **be destroyed** Behind Eth. ḫabʾa III.1 I suspect ἀφανισθήσονται rendering Aram. כחד 'to conceal, efface', Ithp. 'to be destroyed'. See above, 185f.

(10) **Lord of eternity** See above on 9.4.

CHAPTER 82.1-3

(1) Tertullian, *de cultu fem.* i 3 alludes to this verse: cum Enoch filio suo Matusalae nihil aliud mandaverit quam ut notitiam eorum posteris suis traderit.

(2) **to you and to your children** Cf. Ps. 78.5,6.

and they shall celebrate all the wise ... slumber This fuller text, preserved only in Eth[tana], seems original.

(3) **better ... than rich food** Cf. Ps. 19.10

CHAPTER 83

FIRST DREAM-VISION

(2) **before I took a wife** i.e. before he was 65 (Gen. 5.21): her name was Edna (85.3 and Eth[e]), the Aram. equivalent of Eden, 'paradise'. See Milik, 42. Jub. 4.20 gives her name as Edni, and Edni is there Enoch's daughter-in-law (Jub. 4.27). 'We must not look in this for an 'ascetic' tendency, but rather an allusion to the rites of incubation which demanded temporary continence' (Milik, l. c.). But the writer may well have associated visions with an ascetic life.

when I was learning to write Cf. Jub. 11.16 (Abraham is taught by his father to write at the age of 14). In the light of this parallel Dillmann thinks the reference is to Enoch receiving his first visions in very early youth. According to Jub. 4.17 Enoch was the first to learn writing. (Cf. P. Grelot, *La Légende d'Hénoch*, 14.). Its invention is attributed at 69.9 to Penemue, one of the satans. Does the text imply that the art of writing was disclosed to Enoch in a vision?

(3) **in the house of ... Malalel** Heb. מהללאל Mahalaleel Gen. 5.15 who later (v. 6) expounds Enoch's vision.

tottered and shook Eth. yetnaḍḍāḫ wayethayyad: naḍḫa III.1 cf. Heb. נוד Hithpalel Isa. 24.20 (of the earth tottering); Syr. ܢܘܕ Ethpalpal P. Sm., 2309: hēda III.1, cf. Aram. טרף Ithp. 'be shaken, tossed about', σαλευθῆναι: cf. ܛܪܦ Ethp. Ac. 27.27 = διαφερομένων, 'as we were tossed about' (by the sea). Has Eth. read ἡρπασμένον giving טרף its meaning in Heb.?

(4) **the earth was swallowed up** For the figure of speech, Exod. 15.12.

mountains crashed down on(?) mountains Eth. yessaqqalu lit. 'were suspended' (Charles): saqala here = תקל II (Heb. כשל Jer. 46.12), Levy, CW II, 551, LXX προσκόπτειν: impingunt et corrunt (יתקלון) montes super montes(?).

hills sank down on hills Eth. saṭama III.1 = טבע; cf. 88.5 (En[e] 4 i 19), Prov. 8.25.

torn up from their roots Eth. yetgazzamu 'emguendātihomu = יתעקרון (cf. 91.11 En^{g} 1 iv 14, gazama = עקר).

and sank Eth. yessaṭṭamu = יטבעון or ישקעון (for שקע 'sink down', 88.5, En^{e} 4 i 19).

(5) **came into my mouth** Eth. lit. 'a word fell into my mouth'. Dillmann thought that the expression was intended to convey the involuntary nature of Enoch's outcry. It may owe something, however, to the idiom at Isa. 9.8 '(The Lord has sent forth his word against Israel) and it shall fall on Israel' (NEB) (ונפל בישראל). Cf. also Dan. 4.28. The author is emphasising not only that Enoch's cry was involuntary, but that it was given to him as an oracle from above.

I lifted up (my voice) See the note at 37.2. For נשא קול Syr. has ܐܪܝܡ ܩܠܐ but ܐܪܝܡ without ܩܠܐ does not occur. Was the Heb. idiom shared with the older Aramaic? (נשא 'lift up' does occur: see Hoftijzer, 186) Eth^{tn} read 'I began ('aḫazku) to cry aloud', an inner Eth. variant for the rarer 'anšā'ku, 'I lifted up (my voice)'.

The earth is destroyed Eth. = אבד ארעא. Is this a prophetic perfect, 'The earth is to be destroyed'?

(6) **What ails you ... cry so** Eth^{tana} has retained the Heb./Aram. idiom (cf., e.g., Isa. 22.1); Eth^{M} is a translator's simplification.

(7) **A terrible thing ... and horrible indeed is your dream-vision** The two words here used in Eth. ṣenu' 'strong' and ḫayyala I.2 'to be powerful' probably correspond to Aram. חסין and תקיף. Cf. Isa. 2.13 Tg. '(the kings of the nations) strong and powerful' (תקיפיא וחסיניא); Dan. 2.37 חסנא ותקפא 'power and strength'. Both words are used in varying contexts with different shades of meaning, e.g. Isa. 28.2 Tg. מחן תקיפן וחסינן 'strong i.e. fierce and terrible plagues', Isa. 21.1 Tg. 'the land in which terrible things (חסינן) have happened'; Tg. 2 Sam. 2.17 קרב תקיף 'a strong i.e. fierce or terrible battle'.

secrets of the sins Does this mean that Enoch's vision had to do with the secrets concerning the terrible fate awaiting sinners (abstract for concrete) upon earth?

(8) **the Lord of glory** See above note on 22.14.

remnant may remain Eth. kama yetref terāf: tarafa 'to be left over'; cf. Heb. נוח, Hiph. 'to let remain'. It seems likely that the root נוח in the old Aramaic could also have this meaning. There is a deliberate play in the etymology of the name Noah (נוח): see further below on 257, 322.

(10) **wrote down my prayer** Is this the prayer at 84.2-6?

(11) **And when I had gone forth below** i.e. presumably to the door of the house where he had received the vision in his sleep. Does Eth. tāḥta = תחיתה '(to) the lower part of the house' i.e. to the atrium or courtyard? (For the noun Kraeling, *Brooklyn Papyri* 9.15; P. Sm. 4425, ܬܚܬܝܬܐ contrasted with ܥܠܝܬܐ 'upper rooms'). Enoch 'went out' (waḍ'a) into the atrium to gaze aloft at the heavens with thanksgiving that all was as it always had been, with no sign of the heavens collapsing.

stars on the wane(?) Eth. weḫudāt(a) kawākebt is usually rendered by the banal 'some stars' (Knibb), 'a few stars' (Charles). I suggest emending to waweḫudāta kawākebt, [καὶ τὴν ἔκλειψιν τῶν ἀστέρων], וחסירות כוכביא, 'and the waning of the stars.' Cf. Enastr^{b} 26.3, Milik, 294, ܚܣܝܪ P. Sm. 1341.

he had known it in the beginning Eth. za'a'mara diba qadāmi. Flemming-Radermacher suggested a misreading of ἐποίησεν 'as he had made it' as ἐνόησεν. Charles[1906] recalls the conjecture of Margoliouth that הכין 'had established' had been misread as הבין 'had known'. An alternative would be to assume a confusion by a translator of Aram. כון Pa. in the sense of 'he established' with בין = 'consider' (cf. En. 5.1, En^{a} 1 i 10 אתבוננו). (Cf. Levy, CW I, 358, where כון has the meaning 'die Gedanken auf etwas lenken'.)

Lord of judgement A unique title but one which is entirely appropriate in a context where the world stands condemned to destruction by the deluge. The nearest parallel occurs at 22.14 (En[d] 1 xi 2) where God is described as 'Lord of glory' (as above at v. 8) and then 'Judge of righteousness' (דין קושטא). The titles 'lord' (מרא) and 'judge (דין) occur together in the Brooklyn and Elephantine papyri (see Hoftijzer, 167), but apparently for different officials of the judicature.

windows of the east Cf. 72.3,7, 75.7 where 'gate' (fenot) is the term used for the 'exits' of the sun or moon, 'windows' (maskot: cf. Dan. 6.10,11 כוין = θυρίδες) are for the emergence of the stars (see note on 72.3), not, as here, for the sun and moon.

the path which has been shown to it The identical phrase occurs at 36.3; Milik, 204 suggests that there has been a misunderstanding of 'on the paths appropriate to them (חזיאן להון)'.

CHAPTER 84

(1) **I lifted up my hands** Cf. Ps. 28.2 נשא יד (= ארים יד(?)) 'to raise the hands in prayer'; cf. also Neh. 8.6.

the great Holy One See above on 1.3.

breath of my mouth ... for the children of men These expressions are borrowed from 14.2 (En[c] 1 vi 10-11). The phrase 'the flesh of men' for 'mankind' (which occurs again at vs. 4 and 5) is found at Job 12.10 בשר איש Tg. בסר גבר lit. 'the flesh of a man'. The variants here 'children of the flesh of man' or 'children of the men of flesh' seem to be different attempts at rendering the original expression. This phraseology for the more usual 'children of men' is confined to this verse, so that it is possible that Eth[u] is original, reading simply 'the children of men', the longer expression having arisen either by the accidental repetition of šegā from the previous lesāna šegā or under the influence of the phrase 'flesh of men' for 'mankind' at vs. 4 and 5.

And he gave them breath ... therewith This reads like a repetitious expansion.

(2) For similar prayers of Enoch, cf. 9.4-11, 63.2-4, and EnGiants[a] 9, 10, (Milik, 317), with which there may be a connection. See Knibb, 2, 193, and note on next verse.

Lord of the whole creation Cf. 82.7 and 58.4.

God of the whole world Cf. 1.4, A variation of 'eternal God'.

all the heavens are thy throne From Isa. 66.1.

(3) **nothing is too hard for thee** Cf. Jer. 32.17,27, and Gen. 18.14. A similar expression occurs in Aram. at EnGiants[a] 9.4 וכול צבו לא יתקפתכה (Milik, 316) within a doxology which has been suspected as lying behind this verse (Knibb, 2,193f.).

from the seat of thy throne The Eth. 'emmanbartā has given difficulty since manbart is only attested meaning vita, victus (Flemming-Radermacher, 108, Lebensgrund, Charles[1906], 162, n. 13, 'state, condition, life, food'). Knibb (2, 194), following Dillmann, 254, treats 'emmanbartā as a gloss and reads simply 'from thy throne' (cf. Eth[q]). Charles emended to 'emmenbārāta menbārika which he took to be a rendering of ἀπὸ τοῦ τόπου τοῦ θρόνου σου = ממכון כסאך, Ps. 89.15. But we should then have read a sing. 'emmenbāra (menbārāt is plur.). The text should be read as 'emnebrata manbārika [ἀπὸ τῆς καθέδρας τοῦ θρόνου σου] from nebrat sessio καθέδρα Ps. 139(138).2; cf. also 2 Chr. 9.18 LXX. Charles boldly rearranged the text to give a neater parallelism: 'Wisdom departs not from the place of Thy throne, Nor turns away from Thy presence.'

(4) **angels of thy heaven are doing wrong** Cf. En. 106.14 (En^d 5 ii 18, Milik, 209).

(5) **great King** Also at 91.13; cf. 9.4.

leave me a posterity upon earth Eth. tarafa II.1 = Heb. הניח Aram. אניח (with a word-play on Noah): see note above on 83.8 and below, 322. Eth. daḫārit = אחרית in the sense of 'posterity', Ezek. 23.25 (LXX τοὺς καταλοίπους σου), Dan. 11.4 ('descendants', NEB).

make the earth without inhabitant Eth. ʿarqa II.1 = שכל Pi., תכל Pa. fig. orbare; cf. Ezek. 14.15 (of the earth), 2 Kg. 2.19 (Tg. תכל).

eternal destruction Eth. ḥaguel laʿālam = סוף אבדנא(?) as at 91.19 En^g 1 ii 21, Milik, 260.

(6) **my Lord** See note above, 198f.

as a seed-bearing plant Eth. takla zarʾ lit. 'a plant of seed'; cf. Gen. 1.11 'whose seed is in itself' (אשר זרעו בו). For this figure of speech see the note on 10.16 and cf. 93.1 En^g 1 iii 18f. (Milik, 263). Knibb suggests a connection with $EnGiants^a$ 10 (Milik, 316).

CHAPTER 85

THE SECOND DREAM-VISION

(1) **after this** Eth. ʾemdeḫraze = מן בתריה 91.12, En^g 1 iv 15, 19, Milik, 266.

I saw in another dream Eth. ḥelma reʾiku = חזית חלם, lit. 'I beheld a dream', Dan. 2.26, 4.2,6,15, 7.1. Cf. 89.7, En^e 4 ii 1.

I will show the whole dream to you, my son Cf. 79.1 $Enastr^b$ 26.6, Milik, 294.

(2) **lifted up (his voice)** See note on 37.2, 83.5.

incline your ear Eth. ʾaṣnen ʾeznaka = Heb. הטה אזנך Aram. אצלי אדנך. Is this a Biblical expression only? (It is the only example cited for Ethiopic by Dillmann, *Lex.*)

dream-vision Eth. reʾya ḥelm = חזות חלם cf. Dan. 4.6 חזוי חלמי 'my dream-vision'.

(3) **vision on my bed** Cf. Dan. 4.7.

on the earth Eth^M 'from the earth' suggests Gen. 2.7.

that bull was white ...the other red The white bull is Adam, the heifer Eve and the two bull-calves Cain and Abel. 'White' symbolises 'purity', and 'red', in this context, is intended to symbolise the murder of Abel, 'black' Cain's wickedness. (But the 'white', 'red' and 'black' of 89.9 probably refers to pigmentary and racial or ethnic differences. See note, 264).

two bull-calves Eth. ṭāʿwā = vitulus: cf. 89.12 En^e 4 ii 12, Milik, 241, עגלה 'bull-calf'.

(4) **gored** Eth. guadʾa = נגח (Exod. 21.28).

(5) **grew up** Eth. lehqa = רבה 'grow to maturity'.

the black bull-calf ... had intercourse with a heifer $Eth^{g\ m\ q}$ maṣʾa ἦλθεν, בא, על is the more difficult reading. I suggest reading meslēhā (for meslēhu of the mss), lit. 'he went in to her (namely) that heifer', corresponding to the parallel phrase at v. 7, maṣʾa ḫabēhā, and = על לות תורתא דא, εἰσῆλθεν πρὸς ἐκείνην δάμαλιν, where על לות, Heb. בא אל means coire cum femina (Dt. 25.5, 2 Sam. 12.24 etc.). The majority reading 'that heifer went in to him' is possible, but a rarer usage (cf. 2 Sam. 11.4). The words 'that heifer' (תורתא דא) is a reference, not to the previously mentioned heifer at v. 3, Eve, but to 'that (well-known) heifer', Cain's wife, who, according to Jub. 4.1,9 was Cain's sister and named Awan.

(6) **that first cow** Eve in search of Abel. Cf. Jub. 4.7 where Adam and Eve mourned for Abel twenty-eight years. The idea of Eve's mourning (אבל) for Abel (הבל) is probably a midrash on the proper name in spite of the difference in spelling.

(7) **had intercourse with her** The same idiom as at v. 5.

(8) **another white bull** Seth, father of the Sethites (v. 9 'many white bulls, and they resembled him'). The variant reading of Eth[tana g q u] 'two white oxen' could be original and refer to Seth and his sister who became his wife; cf. Jub. 4.8,11.

many black bulls and cows If we follow the Eth. word order we should render 'many bulls and black cows', but as Dillmann, 256, notes, 'black' apparently goes also with 'bulls' as well as with 'cows'. This would be clear from the Aram. where the adj. referring to both 'bulls' and 'cows' would be masc.: תוריא שגיאיא ותורות אכומיא, lit. 'many bulls and cows black'. The reference appears to be Gen. 5.4: these sons and daughters of Adam are described as 'black' like Cain and the Cainites, i.e. (probably) they were not included in the righteous progeny of Adam.

(9) **in my sleep** Eth. = ἐν ὕπνῳ μου = בחלמי see note, 145.

(10) Some Greek fragments of the Dream-Vision (and from the astronomical book[1]) have been preserved in P. Oxen. XVII 2069 G^{2069}, ed. A. S. Hunt, *The Oxyrhynchus Papyri*, Part XVII, 1927, 6-8, No. 2069. The fragments have been reconstructed by Milik in the light of the Ethiopic and the surviving Aram. fragments, 'Fragments Grecs du livre d'Henoch (P. Oxy. XVII 2069) in *Chronique d'Egypte*, 46 (1971), 321-343. For the fragment 85.10-86.2 see 323: 86.1-3 is fragmentarily preserved at En^f 1, Milik, 244f. The identification of this fragment in the original Aram. is quite certain (cf. Knibb, 2, 196 note) and where Greek, Aram. and Eth. are extant, in whatever fragmentary or corrupt condition, some attempt at reconstruction is called for. The lines of the Greek fragment vary in length from a minimum of 25 to a maximum of 30 letters per line: Frg 1 recto and Frg 2 recto belong to the same leaf and are fragments of vs. 1 and 2 respectively.

CHAPTER 86

(1) **I saw with my eyes** The fuller expression, 'I lifted up my eyes to heaven and saw' occurs at 87.2 and 89.2, and should perhaps be read here. (So Milik, 244, supplement to En^f 1.1-2). The expression 'I lifted up my eyes' is a familiar Biblical one (cf. Dan. 4.31, Isa. 37.23 (= 2 Kg. 19.22)). G^{2609} here and at 87.2 has ὼν ἀναβλέψας = הוית חזה (cf. En^e 4 i 16).

a single star What follows is the 'zoomorphic' version of the fall of the watchers. The 'single star', the first to fall, is Asael, not Semhazah, the leader of the 200 watchers, since at 88.1 this first star is seized and cast into the abyss by one of the four archangels, and this corresponds to the action of Raphael at 10.4f. on seizing Asael and casting him into the abyss. In rabbinical tradition Semhazai and Azael descended to

---

[1] See notes on 77 and 78 in Appendix A 'The 'Astronomical' Chapters of the Ethiopic Enoch', 408, 410 and in Appendix B Apocalypsis Henochi Graece: Corrigenda et Addenda, 421.

earth together, but Azael alone was responsible for the seduction of the 'daughters of men'. See Charles, 187, and Milik, 321f.

became transformed (into a bull) The verb here yetlēʿāl is usually rendered 'it arose', e.g., '(... a star fell from heaven), and it arose and ate and pastured ...' (Knibb). Dillmann, 256, interprets this as meaning that this fallen star rose up after its fall in a new form resembling an animal ('in einer neuen (thierähnlichen) Gestalt'). This is certainly what the context requires: at v. 3 the other stars who followed 'became bulls amongst those cattle' (i.e. the bulls, heifers and oxen of 85). The text seems deficient and corrupt: yetlēʿāl could have arisen through the influence of the previous malʿelta and be a corruption of yetʿallaw or yetwēlaṭ = Aram. הפך Ithp., Heb. הפך Niph.; cf. 89.36, En^{c}4 10, Milik, 205 '... that sheep was changed (אתהפך) and become a man'. A predicate seems required to obtain a parallel with v. 3 'and they (the many stars) became bulls': an original perhaps read: והוא מתהפך לתורא, 'and was changed into a bull'. In view of the poor condition of the Aram. fragment (see Milik, Plate XXI), and the uncertainty about the length of the lines, the proposed supplement could be accommodated without difficulty in En^{f}1.2-3.

(2) **destroyed their stalls ... and their calves** Milik, 245 reads 'pastures and stalls'. Eth. wallaṭa = ἀλλοιοῦν Aram. שני Pa. 'to change' (trans.), but with the meaning, in certain contexts, 'to change for the worse'; Ezr. 6.12 it is used with חבל (AV 'to alter and destroy', LXX ἀλλάξαι ἢ ἀφανίσαι (τὸν οἶκον τοῦ θεοῦ). It is even possible here that an original in fact read חבלו 'destroyed' and was misread as חלפו, a synonym of שני and the verb selected here by Milik. Aram. חבל would be particularly appropriate no less for the 'destroying, corrupting' of the calves of the black oxen than for their stalls and pastures. The great white oxen are the Sethites and the black oxen the Cainites.

to butt(?) one another The majority reading of Eth. '(began to) moan (one to another)' (Knibb) makes poor sense in this context: Eth^{q} 'to live with one another' (Charles, Milik, 'Fragments du livre d'Henoch', 329) is hardly an improvement and is best explained as an inner Ethiopic corruption of the majority text. Has a translator from Aram. confused the verbs נגע (Hoftijzer, 174), Heb. Pi. 'to strike' or even נגח 'to butt' (G 89.43 κερατίζειν) with געי = βοᾶν: ושריו למנגע(ח)/למגעי חד לחד?

(3) **and with them pastured among them** Either 'with them' is a doublet of 'among them' and one or other phrase should be removed (Charles[1906], 165, brackets 'among them' as a ditt.). Could meslēhomu, however, be a corruption of ʾamsālu, 'became bulls like it', the first star?

(4) **elephants, camels and asses** Cf. 87.4, 88.2, 89.6: 89.6 En^{e}4 i 21 (Milik, 240) has the order 'camels and asses/asses and camels and elephants'. Milik finds a correspondence with the three categories of giants at Sync. 7.2 (see above, 125f.).

(5) **bite with their teeth** Cf. 89.11, En^{d}2 i 24 (Milik, 222); Mic. 3.5.

gore with their horns Cf. 89.43 G^{vat} κερατίζειν ... ἐν τοῖς κέρασιν En^{d}2 iii 26, 27 (Milik, 224).

(6) **children of the earth** Cf. 105.1 En^{c}5 i 21 (Milik, 206).

began to tremble Cf. 89.35 En^{c}4 7 (Milik, 205) שריו למרעד.

CHAPTER 87

THE SEVEN ARCHANGELS

Vs. 1-3 have been identified by Milik in G^{2069}; unfortunately only a few words are decipherable so that reconstruction of the original Greek text is, for the greater part, dependent on Eth. Verse 1 develops further the ideas of 7.2-6 interpreted in 'zoomorphic' imagery. While no mention is made of the crimes of the giants against all animate life, birds and beasts, reptiles and fishes, the Judaean author does allude particularly to 7.6 by declaring that 'the whole earth' (so G^{2069}, Eth. 'the earth') raised an outcry against them: the phrase with βοᾶν, βοή recalls the earlier passages, 8.4, 9.2,10 (Eth., G and Sync.).

(2) **I raised my eyes** for the expression, cf. 86.1.

beings who were like white men As humans are represented by animals, the archangels are represented by humans. The reference is to the seven archangels who are named and their functions described at Ch. 20, a list belonging to the hellenistic period. See Milik, 25,174. The four archangels who first emerged from heaven in human likeness were probably Michael, Sariel (or Uriel), Raphael and Gabriel. Uriel is a later form of Sariel, favoured in Christian sources. Their role includes that of executing the divine judgement (e.g. Raphael, 10.4 and all of them at the last judgement 90.21). Of the seven, this leaves Raguel, Saraqael and Remiel as the three subordinate archangels whose duty was to transport and escort Enoch to the heavenly places (v. 3), or, with Elijah, back again to the earth to the scene before (or after) the last judgement (90.31). If the hellenistic list at Ch. 20 is in order of precedence, it is possible that Uriel, Raphael and Michael, who are mentioned first, had now become the four principal archangels; and this would leave Saraqael, Gabriel and Remiel as the three directing angels. Variations in the names and order of the four and seven are frequent in the later tradition. See further Yadin, *Scroll of the War*, 229f., and especially 237f. and above, 199. In the Book of the Parables (Ch. 40) the order and names are: Michael, Raphael, Gabriel and Phanuel. Their functions too differ from those at Ch. 20; differences of this kind clearly point to a later period for the Parables.

Angels are represented as clothed in white garments, in both Jewish and Christian tradition, e.g. 2 Mac. 11.8, Mk. 16.5 par. (cf. Jn. 20.12). 'White' has many connotations in both hellenistic and Jewish writings: in the latter it is not only the priestly and liturgical colour, but also, and especially in apocalyptic writings, the colour associated with heaven and heavenly beings. The Ancient of Days is clad in a 'robe white as snow' (Dan. 7.9). It is, of course, also a symbol of purity and renewal of innocence and is generally so interpreted of the white garments of Rev. 4.4, 6.11, but in these, as in other New Testament uses of the symbol, the idea of angelic or heavenly being or character is also present, e.g. Mk. 9.3 (Transfiguration), Rev. 3.5,18; 7.9,13. See further ThWNT IV, especially 255.12 'it is angels who are to be so recognised (by their white apparel), in a manner not attested in the Old Testament' ... but attested in I En. 87.3.

(3) **grasped me by my hand** cf. 90.31 where Enoch is snatched back from his 'high tower' by the three escorting angels along with Elijah who had been brought up to him in heaven (89.52), to the scene of the last judgement and the creation of the new People of God. See further below, 279.

away from the children of the earth Eth. reads 'generations of the earth', an inner-Eth. corruption (weluda to tewledda) for 'children of earth', as at 86.6: G^{2069} reads υἱῶν [τῆς γῆς]; cf. Aram. בני ארעא at 91.14, 92.2, 105.1.

raised me up to a lofty place, and showed me a tower high above the earth According to Charles we are to regard this place as paradise, although it appears to imply a conception of paradise differing from any that has preceded. Cf. 60.8. Milik takes the view that this 'high tower', from which Enoch will behold the destruction of angels, giants and men, unites into a single place the first celestial abode of Enoch, the heavenly Palace and the mountain-throne of God (Milik, 43 and 15f.) (see above, 148) One need not be so specific, however, about the actual location of this mythical 'tower'. What the Judaean author is concerned to describe is a lofty place, a tower high above the earth, which would supply a vantage point from which Enoch (later to be joined by Elijah) could obtain a panoramic view of the unfolding history of Israel, and first of all of the punishment and destruction of the giants (v. 4). Since 'tower' is used later to symbolise the Temple, we may also think here of a heavenly sanctuary to which Enoch is led (so Dillmann, 257).

all the high places were smaller Eth. wagr, plur. 'awger seems here equivalent to Heb. במה, במות Tg. במתה, designating not only 'mountains' but cultic 'high places'. All such places are smaller than this lofty tower (and so inferior to it)[1]. At the same time the writer is probably emphasising the superior height of this 'lofty tower', since Enoch is to witness from it the waters of the deluge which covered all the 'high hills' (AV), Gen. 7.19,20, cf. 8.5.

CHAPTER 88

PUNISHMENT OF THE FALLEN ANGELS

who had come forth first Cf. 10.4 where it is Raphael who is the avenging angel who punished Asael.

that abyss was narrow and deep, and closed in and dark This place of the temporary imprisonment of the fallen stars is described in greater detail at 18.11-16 and at 21.1-6; 21.7-10 contains an account of their final and eternal place of punishment to which they are to be condemned at 88.3 and 90.24, 'an abyss full of fire and flaming, and full of pilars of fire'. Jude 13 (cf. 2 Pet. 2.17) contains an allusion to this verse: ἀστέρες πλανῆται οἷς ὁ ζόφος τοῦ σκότους εἰς αἰῶνα τετήρηται, ... wandering stars for whom the nether gloom of darkness has been reserved for ever' (RSV). It is only in this verse that the adjective 'dark' is applied to this abyss.

(2) **and one of them drew a sword** refers to 10.9-10. It is Gabriel who arranges for the mutual destruction of the giants' offspring, but nothing is said about how they destroyed each other. Here in this verse a sword is given by the archangel to the animals, so that camels, elephants and asses turn against each other in mutual destruction.

to smite one another Eth. yeguadde'o, the same verb as is used above (85.4) meaning 'to gore'. Did the original here read יתנגחו עמהון 'engaged in thrusting with, fought with'? Cf. Dan. 11.40 and note on 100.1.

the whole earth quaked because of them The Eth. mss have some very curious variants for the simple dibēhomu 'on account of them'. Unless we explain them as

---

[1] Eth. scribes had difficulty with the description of the tower as 'smaller' (ḥeṣuṣ) and alter it to 'strong' (Eth^{tana} ṣenu'). (Eth$^{g\ m\ (q)}$ ḥenuṣ ('built') is probably an inner-Eth. corruption).

the crudest of dittographs, they could conceal a genuine piece of lost text. Did the original have an expression like that at Isa. 24.19 רעה התרעעה הארץ, 'the earth is utterly shattered' (NEB), out of which came the corruption medr diba medr (ארעא על ארעא)? A possible reconstruction of the verse in Aram. would read: ומזע תזוע עליהון ותזדעדע כול ארעא 'And the whole earth quaked and was utterly shattered on account of them'.

(3) **one of those four** refers to 10.11-12. Michael is given the task of binding Semhazah, causing the sons of the fallen angels to turn against each other and slay each other, then binding them all and imprisoning them, like Asael, for seventy generation in the 'valleys of the earth'. In this second version no mention is made of Semhazah, the leader of the fallen stars, in spite of the prominence he is given at 10.11.

stoned (them) from heaven A Biblical punishment for oxen, Exod. 21.28-29.

CHAPTERS 89-90 FROM NOAH TO THE LAST JUDGEMENT

CHAPTER 89

(1) **instructed him in a mystery ... became a man** The words 'in a mystery, trembling as he was: he was born a bull and became a man, and ...' do not appear in En^{e}4 i 14. Similarly at v. 9 there is no space at En^{e}4 ii 5 for the clause 'which became a man' qualifying 'that white bull'. These Eth. 'additions' to the text of En^{e} seem at first to be simply translators' expansions of the shorter original text. But, in fact, Eth. could represent a longer recension of the Aram., in particular in the addition 'he was born a bull and became a man'. An exactly parallel text attested in the Aram. occurs at 89.36: '... I saw in this vision till that sheep (Moses) became a man (En^{c}4 10 'was transformed and became a man' = אתהפך והוא אנוש), and built a tabernacle for the Lord of the sheep ...'.

Cf. Martin, 203 who draws attention to the 'secret' or 'mystery' of the deluge in the Gilgamesh legend, another element in Eth. which could be original.

built for himself a great vessel En^{e}4 i 14 om. 'great': cf. 67.2 where the ark is built by angels.

three bulls Noah's three sons, Shem, Ham and Japheth (Gen. 7.13).

dwelt with him on that vessel En^{e}4 i 15 probably 'entered with him into the vessel'.[1]

it covered over them Cf. En^{e}4 i 15 'and the vessel covered and concealed them'; cf. Gen. 7.16; cf. En. 67.2 'I will place my hand over it'.

(2) **raised my eyes to heaven and saw** En^{e}4 i 16 has a shorter text ('And I was looking and behold ...') both here and in the remaining frgs of vs. 2-6. Milik, 239, notes that this verse is much shorter in Aram. and adds: 'One certainly gets the impression that the original text, such as we can see it in En^{e}, was reworked, following the outline of a more systematic symbolism'. This leaves the question open whether the 'reworking' was done by the translators (into Greek or Ethiopic) or in a longer recension of the Aram. It seems to me unlikely that the original allegory would be expanded by translators to include two quite new features, the 'lofty roof' and the

---

[1] Or could the original have read [יתב]ו עמה לערבא, 'embarked with him on the vessel'?

'enclosure'. Moreover, the longer opening phrase of Eth. is a familiar one (86.1, 87.2) and the symbolism of the 'lofty roof' occurs again at 89.7 reconstructed by Milik as טלילא רמא). It seems probable that we have to do with two original Aram. recensions, a longer one behind Eth. and a shorter recension in En^{e}4i.

a lofty roof Eth. nāḥs le'ul (cf. also 89.7), perhaps = טלילא עליא (cf. 14.11) or even עליתא תלילא 'a lofty chamber'; cf. the use of Heb. עליה, עליות, מעלה = 'roof-chamber(s)' for the 'chambers of the rain', (Ps. 104.3,13, Am. 9.6).

seven sluices Eth. 'asrāb = מרזבין lit. 'channels' = Heb. ארבות (AV 'windows', LXX καταῤῥάκται, Gen. 7.11, 8.2). See above, 175.

into an enclosure Eth. ba'aḥadu 'aṣad = על דירא חד(?) i.e. 'into a pen or fold'; דירא (βουκόλιον) would be an appropriate term for enclosure for cattle (cf. Charles, 190).

(3) **chambers were opened up** En^{e}4i17 חדרין 'chambers'; Eth^M 'anqe'tāt 'springs': Eth^{tana} ne[q]'atāt, 'chasms'(?). Cf. 17.7-8. For Heb. חדר, Job. 9.9 'the chambers of the south (חדרי תימן)' i.e. of the stars, 47.9; cf. the similar use of אוצרות, 'storehouses', 'treasuries' of the winds, stars, etc. See above, 202.

I saw that enclosure ... covered with water En^{e}4i18-19 has a much shorter text than Eth.; I have restored it to read: 'And I looked until the earth was covered with water [(4) and darkness and thick cloud] rose up over it'. The words וקאמין עליה correspond to Eth. waqomu diba medr at the end of v. 4 'and it (the water) rose over the earth'. (For the idiom of the waters 'rising', Jos. 3.16.)

(4) **And water ... increased upon it** Cf. Gen. 7.18, an allusion supporting the theory of a longer recension behind Eth.: (?)ורבו עליה מין וחשך וערפל. Eth. lit. 'I was seeing the height (mal'elt = קומתא?) of that water', or should we read 'I was looking until the water was higher than the enclosure' (see textual note)?

(5) The shorter text of En^{e}4i19-20 reads simply 'And the oxen were submerged and sank [and perished in those waters]'.

(6) En^{e}4i20 has 'And all the oxen and asses and camels and elephants' ירומו̊[ן]. Milik reads ירו מין equating the phrase with a putative Heb. ירו מים 'they sank in the waters' (240): the Aram. writer (he suggests) borrowed the verb from Exod. 15.4: 'Pharaoh's chariots and his host hath he (Jahweh) cast into the sea'. To get the required passive meaning we are to assume that ירו 'is used in an intransitive sense'. Knibb reads ורימ̊י[א] 'wild oxen', noting that Eth. reads 'with all the wild animals'. But since 'asses, camels and elephants' appear as a single group in the allegory, as at 86.4, it seems unlikely that 'wild oxen' would be added at this point, especially since oxen generally are the main subject of the vision. I suggest reading ירומון (pass. Pael = Pual of רמי 'to cast')[1]. Cf. 91.15, En^{g}iiv21. Eth. has 'sank to the ground' tašaṭmu westa medr: was the original וירומון לארעא in the idiomatic sense of 'were laid low'; cf. Tg. 20.21,25.

(7) **sluices ceased** Eth sassalu = ספו, ἀφαιροῦνται(?); Milik התכלאו(?)

chamber ... stopped The original here has been substantially preserved at En^{e}4ii2 ומבועי?] חדריא שכירו] 'and the springs(?) of the (underground) chambers were stopped'; cf. Gen. 8.2 ויסכרו מעינת תהום וארבות השמים. Eth^M 'arayu 'were made level' (if this is the meaning) is an imprecise rendering of שכירו (pass. Pe.). The reading of Eth^{tana} is preferable: '... those sluices ceased from the high roof and the springs of the earth; and other deeps were opened up at the same time'. (wa'arayu ... tafatḥ(u) = ועמקין אחרנין פתיחו כחדא(?). The meaning appears to be that at the same time as the underground springs were stopped, new underground 'deeps' would open up to let the

---

[1] Cf. Bauer-Leander, *Grammatik des Biblisch-Aramäischen*, 93q.

flood-water drain away (v. 8) Cf. Jub. 6.26 'The mouths of the depths of the earth were closed'; Prayer of Manasses 3 ὁ κλείσας τὴν ἄβυσσον.

(8) **till the earth was uncovered** Eth. kašata III.1, patefieri = פתיח. Cf. Milik, 241, 'until the waters vanished (ספו)'.

settled down on the earth Eth. nabara En^e 4 ii 4 תקנת; Cf. Pesh. Prov. 8.25.

(9) **which had become a man** See above on v. 1.

one ... was white ... one of them was red ... and one black The three colours here seem to have a different connotation from the 'red' and 'black' of Abel and Cain at v. 3. In this verse they symbolise the three races, Semites (white), Japhethites (red) and Hamites (black). The identification of the Japhethites is disputed: some think of the Philistines or Phoenicians, but Gen. 10.1f. seems to point rather to tribes in Palestine more closely related to the Semites, but also including Greeks and Persians. The Hamites are the inhabitants of the Nile valley. Cf. Kautzsch, *Die Apokryphen und Pseudepigraphen*, 291 n. a.

(10) The exact identification of all the beasts and birds listed is by no means certain: thus 'leopards' ('anāmert) could also include 'panthers' and 'tigers'. The usual translation of sisit by 'falcons', 'hawks' is doubtful (Dillmann, 259): Tg. ציצא = Heb. תחמס = the male ostrich (so called from its aggressiveness), and I have adopted this meaning. There are variations in the Eth. manuscripts, mostly affecting the order of the names, but there are no significant variants.

The white bull is Abraham.

(11) Fragments of the original of this verse are found in En^d 2 i 24, 25, and En^e 4 ii 10 (Milik, 222, 241).

to bite one another En^d 2 i 24 adds '(to bite) and chase (למדבר) (one another)'.

a wild ass and a white bull with it En^d 2 i 25 ערד ועגל חור עמיה (cf. Milik, 222). The wild ass is Ishmael, progenitor of the Arabs and Midianites, the 'wild asses' of vv. 13 and 16. Cf. Gen. 16.12 where Ishmael is פרא אדם 'a man like a wild ass' (NEB). The white bull born with him is Isaac.

(12) **black wild boar** Esau. Dillmann notes that the fact that Edom and the Edomites are symbolised by the animal most abhorrent to Israel is an indication of the depth of hatred felt for Edom. Vs. 42,43,49,66 and 72 use the 'wild boar' to signify the lesser neighbouring relatives of Israel. Cf. Ps. 80.13.

a white ram En^d 2 i 25 cf. En^e 4 ii 12-13 'a white ram of the flock'. Jacob or Israel symbolised by a ram, the bell-wether of the flock; the twelve symbolise Israel as the flock or 'sheep of Jahweh's fold' (Jer. 23.1, Ezek. 34.31, Ps. 74.1 etc.).

(13) **they gave up one of them** Joseph delivered up by his brethren to the Midianites (vv. 11,16).

those asses in turn gave up that sheep to the wolves The wolves are the Egyptians, so designated throughout the zoomorphic vision.

(14) **And the ram led all the eleven sheep** Eth. 'the Lord' is a theologising substitute; the reference is to Jacob.

many flocks of sheep En^e 4 ii 17 = (?)ענא רבא דאמרין, 'a great flock of sheep', cf. Milik, 241, 'a flock of many sheep'. Cf. Gen. 42-50, Exod. 1.7, the account of the brothers of Joseph in Egypt and their sojourn in the land of Goshen.

(15) Again in this verse Eth. has a longer text: En^e om 'to terrify them' and perhaps also 'they destroyed their little ones' (Milik, 242f.).

began to terrify them, and they oppressed them En^e 4 ii 17 has simply 'began to oppress the flock'.

until they had destroyed ... a river of deep water I read En^e 4 ii 18 (end), 19 = Eth. bawehiza•māy bezuh, 'in a river of deep water', a description of the Nile. For the expression משקע מים Ezek. 34.18 (cf. 32.14) lit. 'a sinking of water', 'deep waters' (AV), 'pure water' (NEB). Cf. also Am. 8.8, 9.5 for שקע of the 'sinking', i.e. the inundations of the Nile.

(16) Again to judge from the fragments preserved at En^{e}4 ii 20, 21 Eth. appears to have a longer text, unless some additions come from a translator.

a sheep ... saved from the wolves Eth. zadeḫna En^{e}4 ii 20 'a sheep snatched(?) (from the wolves)' (cf. Jer. 12.3): Moses.

the Lord of the sheep descended The expression 'Lord of the sheep' is characteristic of these chapters, occurring about 28 times in Eth. The Aram. is fully preserved at 89.33, En^{c}4.4 (Milik, 204), and cf. G^{b} v. 42 ὁ κύριος τῶν προβάτων. For the use of מרא absolutely for 'Lord' 'my Lord' see above, 198f., 238.

at the call of the sheep As Knibb notes (2, 203) the text is deficient. The reading of two mss (Eth$^{i\ ryl}$) ʾemṣerḥ seems a ditt. of the first word in the following ʾemṣerḥ leʿul 'from a lofty abode' (itself an usual expression). We expect a predicate similar to that at v. 20 '(came down) to (ḫabēhomu) the sheep'. Could a phrase similar to that at 9.2 'the sound of their cries' (qāla ṣerāḥātihomu) perhaps lie behind the Eth. corruption, [ἀπὸ φωνῆς βοῆς αὐτῶν] = מן קלא די געתהון? 'The Lord of the sheep came down, because of the sound of their cries, to those sheep'?

from a lofty abode Eth. ṣerḥ is found meaning 'temple' ναός or 'palace' (Dan. 4.26 היכל, Theod. ναός); cf. Martin 'sanctuaire'. Has it arisen in the text from the earlier corruption, and should be omitted?

(17) **harm the sheep** Eth. gasasa = נגע Heb./Aram. 'to touch in order to harm' (e.g. Jer. 12.14); for the verb in Aram. Hoftijzer, 174 (Ah. 165, 166).

(18) **another sheep met it** Exod. 4.27 Aaron meets Moses; Exod. 5.1-5 Moses and Aaron confront Pharaoh.

(20) **to smite those wolves** Exod. 7.14f. the plagues of Egypt.

(21) Verses 21-27 deal with the Exodus.

Cf. Exod. 12.31-39, departure of the Israelites; Exod. 14.5-9 pursuit of the Egyptians.

the eyes of the wolves were blinded Cf. Exod. 14.4 Eth. taṣallalu ʾaʿyentihomu = חשכו עיניהון. For the expression and figure of speech, Ps. 69.24.

with all their forces Eth. bakuellu ḫaylomu, i.e. with all their military might.

(22) **leading them** Exod. 13.21-22, alluding to the pillars of cloud and fire.

his appearance fearful and splendid Cf. 89.30 En^{d}2 ii 29 [וחזוה תקיף ורב וד̊[חיל] 'and his appearance powerful and great and [fearful]'.

(23) **a sea of reeds** Eth. ʿayga māy lit. 'a marsh of waters, a watery marsh', a phrase which corresponds to Heb. אגם מים: Isa. 14.23 = 'marshes', cf. Isa. 35.7 (Tg. אגמין דמיין) Arab. اجمة 'a marshy jungle'. Here the original אגם מיין(?) is the author's equivalent for the ים סוף 'the sea of reeds' of Exod. 13.18 (cf. 10.19).

(24) Exod. 14.19-31: separation of the sea of reeds.

(26) Vs. 26-30, cf. En^{e}iv 3, Milik, 243.

became as it had been created The equivalent of Exod. 14.27 ישב הים ... לאיתנו ... which is variously interpreted (e.g., Gesenius, rediit mare ad perennitatem suam). Does the expression in Eth. perhaps correspond to Heb. כאשר בראשונה, כבראשונה = καθὼς ἀπ' ἀρχῆς, 2 Sam. 7.10. Note Eth^{v} kama qadimu = כבקדמיתא(?), and cf. Isa. 43.9 LXX ἐξ ἀρχῆς, Eth. ʾemfeṭratu. The version of Symm. εἰς τὸ ἀρχαῖον αὐτῆς, probably lies behind the phrase here.

(27) **who pursued ... drowned** En^{e}4 iii 14 '(all the wolves) were pursuing the flock'. En^{e}4 iii 14 adds 'and the waters covered them', as at the end of v. 26 in Eth.

(28) **went forth into a wilderness** En^{e}4 iii 16, 'and they went forth into a wilder-

ness, a place where there was no water and no grass'. The reference is to the wilderness of Shur and Sinai, Exod. 15.22-26.

and they began to open their eyes and to see En^{e} 4 iii 17 'and their eyes were opened'. The receiving and losing of spiritual vision is a recurring motif in these chapters; 89.32,33,44,54; 90.6,9,10.

(29) **And the ram ... sent it to them** Cf. En^{e} 4 iii 19, En^{d} 2 ii 27, Moses' ascent of Sinai and his return at God's command, Exod. 19.

This verse may have been longer in En$^{d\ e}$: traces of a few letters are visible but reconstruction is precarious (cf. Milik, 243f. who supplies a clause from Exod. 20.18,21, 'and they all stood at a distance').

(30) Cf. Exod. 19.11.

(31) **And they cried out to the sheep ... saying** The Eth. text is in a state of disarray (Eth II is untranslatable). The original is only fragmentarily preserved at En^{c} 4.1-2, and the first clause is missing. Milik supplies [וזעקו לאמרא די הוה תנינה], rendering'and they cried to the Sheep who was second (in command) to Him', interpreting the words as referring to Moses (205, Ll. 1-2). The better Eth. text (which I have adopted) is that preserved in Eth^{tana} where the sheep cry out in distress, first to Moses and then to Aaron. En^{c} has a much shorter form of text which I would suggest reading [וזעקו לאמרא די הוה בתריה דכרא דן], lit. 'and they cried out to the sheep which was next (in rank) to that ram', i.e. to Aaron, Moses' second in command. Moreover, it does not seem improbable that it was, in fact, this text which has been misunderstood by translators (and led to the confusion in the mss) and which should have been rendered by the corresponding Greek phrase καὶ ἔκραξαν τῷ δευτερεύοντι τῷ προβάτῳ ὃ ἡγεῖτο αὐτῶν. Cf. 48a which should also probably read ὁ δευτερεύων ἐκείνῳ τῷ κριῷ, 'the second one (in rank) to that ram'; cf. Pesh. 2 Kg. 25.18 ܟܗܢܐ ܕܒܬܪܗ for כהן משנה. The 'second sheep' in Eth. would then be an explanatory gloss (cf. Flemming, 126).

(32) Cf. Exod. 24.12f., 31.18, 32.1f.

without that sheep knowing about them Cf. Exod. 32.7f.

(34) **desired to return to their folds** There is nothing corresponding to this in the Exodus narrative at Exod. 32.25f.

(35) Cf. Exod. 32.26f. This verse is represented by a number of fragments in En^{c} 4 in a longer form of text which, unfortunately, can, in several places, be only conjecturally reconstructed.

took other sheep with it Exod. 32.26f.: the other sheep are the Levites who carry out the death penalty on the guilty Israelites, according to Exodus, three thousand of them.

fell upon ... and began to slay them The fragmentary text of En^{c} 4.7 cannot originally have differed substantially from Eth.: En^{c} 4.7-9 '[... and he fell upon them, and of the flock he slew those who had gone astray. And they (the flock) began to tremble ...']. Behind the supplement is the conjecture that wa'emze = ומן כען, καὶ (ἀπὸ) τότε lies an original ומן ענא, 'and from the flock'. For אתא על cf. Bar Heb. Chron. 21.66, P. Sm. 414 and Heb. בוא על, Gen. 34.27.

that sheep brought back ... fallen away En^{c} 4.8 '... that sheep brought back all the straying flock to their folds'. En^{c} 4.9 adds a line of which only a few words have survived. I have taken the line to be based on Exod. 32.31-32, when Moses returns to the Lord to deplore the people's sin: 'Thereupon that sheep returned to the Lord, to groan and call out and bleat and cry, and the Lord of the sheep ...'. Perhaps we should complete from Exod. 32.35 ... נגף אמרין דטעין '... and the Lord of the sheep smote the sheep which had gone astray'.

(36) **that sheep was transformed and became a man** En^c 4 10. The leading sheep (Moses) becomes a man in order to erect the tabernacle, just as the white bull (Noah) became a man to build the ark. Cf. v. 38. The attestation of these words in the original Aram. is supporting evidence for their presence in the original at 89.1,9. The tabernacle אהל, משכן = LXX σκηνή.

he made all the sheep to stand at that tabernacle Eth. reads lit. 'he made all the sheep to stand in that tabernacle'. This seems a most improbable original. Dillmann, 261, understands the text to mean that they were all under an obligation to worship in that 'house'. The text is alluding, however, to Exod. 33.8: 'Whenever Moses went out to the tent, all the people would *rise and stand* ... until he entered the tent', (NEB, italics mine). Did an original perhaps read כולהון אמרין אקים(קמו) עד דעל במשכנא דן 'he made all the sheep to stand/all the sheep stood until he entered that tabernacle'? They rise and stand in awe of Moses.

(37) **that sheep ... fell asleep** The death of Aaron and the dying out of the older generation in the Forty Years' wanderings in the wilderness. Cf. Num. 20.26 and 32.1-5 (the 'stream' is Jordan).

(38) The death of Moses, Dt. 34.1-8.

which had become a man Cf. En^c 4 10 (above, 262, 264, 89.1,9 and 36).

withdrew from them Cf. Dt. 34.1 'And Moses went up from the plain of Moab unto the mountain of Nebo'. As at Sinai Moses leaves the flock to ascend here 'to the top of Pisgah' to survey the promised land: Dt. 34.5 'So Moses the servant of the Lord died there in the land of Moab'.

all the sheep sought The implication seems to be that they sought but failed to find. Is there a hint of the tradition of the 'assumption' of Moses (cf. Josephus, Ant., iv.8.48 (320-326)?

(39) **they left off their crying** With Eth. 'armama 'em lit. 'to be silent from' i.e. 'to cease from' cf. the use of Heb. חרש מן (e.g. Jer. 38.27).

crossed that stream Jordan.

other sheep i.e. Joshua and the judges, or read 'two sheep', Joshua and Caleb (Dt. 1.36f.). The reading 'all the sheep' is impossible.

(40) Entry into the promised land.

glorious land Cf. Ezek. 20.6, Dan. 11.16,41.

the tabernacle was in the midst of them The tabernacle containing the ark of the covenant went with them and was finally brought among them into the Promised Land. This verse covers the period from the Judges to the rise of Samuel ('another sheep'). For the figures of speech referring to spiritual blindness and its opposite, see below, 290.

(41) **another sheep arose** i.e. Samuel.

(42) Vs. 42-49 have been preserved in a Greek fragment, Vatican MS 1809 (G^{vat}).

dogs ... foxes ... wild boars The Philistines (cf. vs. 46, 47), Ammonites and Amalekites (regarded as a branch of the Edomites). See the explanations of the glossator in G^{vat} (Milik, 45).

a ram ... to lead them Vs. 42-43 describe the rise of Saul and his exploits.

(43) G^{vat} has a longer text, some elements of which seem original: Eth. is a shortened version. Only two clearly decipherable words are visible at En^d 2 iii 2-3.

to butt ... those dogs G^{vat} κερατίζειν ... ἐν τοῖς κέρασιν cf. 86.5 note. G^{vat}, ἐνετίνασσεν εἰς τοὺς ἀλώπεκας = נקש בתעלין(?) 'struck at the foxes hard'. See Milik, 225, who cites 1 Mac. 2.36, 2 Mac. 4.41 and 11.11 ('hurled themselves ... against' NEB).

G^{vat} καὶ μετ' αὐτοὺς εἰς τοὺς ὕας. We expect a second verb. Has תבר Pa. (Heb. שבר

Pi.) 'to break, rend' been confused with בתר = μετά (καὶ συνέτριψεν (εἰς) τοὺς ὕας(? G^{vat} ἀπώλεσε ὕας πολλούς = En^d 2 iii 3. Has Eth. 'destroyed them all' been influenced by the idea of the herem at 1 Sam. 15.9f.

G^{vat} καὶ μετ' αὐτοὺς ἤρξατο τοὺς κύνας. Milik, 225, (also reading ἤρξατο) suggests a mistranslation of שרי = 'let loose', 'free', and suspects a reference to 1 Sam. 14.46: 'and thereafter (it) let the dogs go free'. But this is not a real equivalent of 'Saul broke off the pursuit of the Philistines' (NEB for ויעל שאול מאחרי פלשתים). I suggest rather a mistranslation either of שדי = 'cast out', 'banished' (cf. Dt. 29.28 Pesh.) or of שדד 'to despoil, utterly destroy (the Philistines)' (e.g. Jer. 47.4): ישוד לכלבין read as ישרי, ἤρξατο τοὺς κύνας. The μετ' αὐτούς could be a ditt. or a mistranslation of מן בתרה = thereafter, then'. The verse in G^{vat} would read: 'And that ram began to butt ⌜and pursue,⌝ and it assailed the foxes, attacked (lit. 'broke, shattered') (?) the wild boars, then destroyed the dogs(?)'.

(44) **the sheep ... opened** refers back to Samuel, 'the sheep who arose and led them and brought them all back' at v. 41, and who is specifically mentioned at v. 45. The plural in G^{vat} refers to the Israelites whose 'eyes were opened' at v. 41, but is clearly an inferior text. (Cf. Knibb, 2, 207).

abandoned his lead(?) The phrase, with the same Eth. variant 'renounced its glory', occurs at v. 45. The reading of G^{vat} viz. ἀφῆκεν τὴν ὅδον αὐτοῦ is itself not without difficulty: if 'αὐτοῦ refers to the subject of the verb, i.e. Saul, then 'forsook his way' cannot be right' (Charles, 196, suggesting an original '(forsook) the way of the Lord'). It is also difficult to find a common denominator, in Greek or Aram., to account for the variation of Eth. ὅδον/δόξαν. Did the original read שבק דבריה 'abandoned/forfeited his lead'? Cf. ܫܒܩ ܡܕܒܪܢܘܬܐ abdicare regimen, P.Sm. 4039, and for דבר 'to lead' of a flock, 89.14 En^e 4 ii 16; דברא, Levy, CW I,161, Jg. 5.21, 20.31,32 (Heb. מסילה LXX ὅδος). Did the Eth. v. l. come from reading רב(נ)ותיה (cf. Levy, CW II, 401)?

began to walk ... not straight G^{vat} ἀνοδίᾳ 'in a no-way'. Eth. omits the equivalent of באורח[א] cf. En^d 2 iii 30, implied by G^{vat} ἀνοδίᾳ. For a similar expression, Job 12.24 לא דרך (Symm. ἀνοδία) cf. Isa. 65.2 ההלכים הדרך לא־טוב, Tg. דאזלין באורח לא תקינא. Cf. also Jer. 18.15, Isa. 65.2.

(45) **sent the sheep** 1 Sam. 16.1-13. It seems unlikely that any special significance is to be attached to the use of ἄρην instead of πρόβατον (cf. Charles, 196). The original was probably in both cases אמרא not טליא 'lamb'. (David might be described as טליא in this context before being raised to 'a ram', but not Samuel who has, in any case, already been described as a 'sheep' (אמרא) at v. 4 (bagʿ)). The Greek translator may have felt that ἄρην was more appropriate to David's youthful status, but the fact that G^{vat} uses the same word for Samuel shows that the choice of ἄρην instead of πρόβατον was purely arbitrary.

(46) Cf. 1 Sam. 16.1f.; 17.1f.

G^{vat} σιγῇ κατὰ μόνας 'in secret' may be an addition of the Greek translator and a 'doublet' of κατὰ μόνας. It seems to be an interpretation, however, of 1 Sam. 16.1f. and this would support its originality; cf. Tg. בחשאי Isa. 26.16, Ps. 41.8, Levy, CW I, 287.

but during all these things G^{vat} ἐπὶ πᾶσιν τούτοις, 'in addition to all these things' is hardly suitable in the context but G^{vat} is almost certainly translating על כול אלן 'during, on the occasion of all these things'.

prince and leader G^{vat}εἰς ἄρχοντα καὶ εἰς ἡγούμενον = לראש ולרבן, cf. En 6.7, En^b 1 ii 17 רבני עשרתא (Milik, 166). Cf. also Exod. 2.14 = Ac. 7.28.

(47) Saul pursues David (1 Sam. 21-30) and dies in battle against the Philistines (1 Sam. 31).

brought down the former ram Eth. 'awdaqewwo = אפיל i.e. 'caused to fall (in battle)'; G^{vat} ἔπεσεν. Eth. seems preferable.

(48a) G^{vat} has the first sentence only of this verse (48a): καὶ ὁ κριός ... τῶν προβάτων. The rest of the verse (48b), telling of the death of David and the rise of Solomon, belongs more appropriately after v. 49, which is clearly reciting the exploits of David. The omission of 48b by G^{vat} confirms this order; unfortunately G^{vat} stops short at v. 49, but it almost certainly would have continued with 48b. Charles rearranges the vs. 48a, 49, 48b and I have followed this order.

the second in command (David) to that ram (Saul) Eth I = אמרא די בתריה זכרא דן, ὁ δευτερεύων ἐκείνῳ τῷ κριῷ cf. LXX Est. 4.8 Ἀμαν ὁ δευτερεύων τῷ βασιλεῖ. G^{vat} ὁ κριὸς ὁ δεύτερος. Τεε αβοχε, ξουε οξ χ. 31.

arose and led $Eth^{q\ m\ (g)}$ naš'omu = אקים 'roused up (and led)'. G^{vat} ἀναπηδήσας, lit. 'leaping up', renders קום several times in the LXX (e.g. 1 Kgds. 20.34, 25.10).

(49) **grew and multiplied** The verse describes the prosperity of Israel under David and David's triumphs over his enemies. No one could mistake 'that ram' in this verse for anyone other than David: cf. 2 Sam. 5.17-25, 21.15-22 (the 'dogs', Philistines); 2 Sam. 10.1-12.31 (the foxes, Ammonites); 2 Sam. 8.13 (the wild boars, the Edomites).

feared, and fled G^{vat} has a shorter text ἔφυγον ἀπ' αὐτοῦ καὶ ἐφοβοῦτο αὐτόν Eth. 'were afraid and fled' is a more logical order.

(48b) **a little sheep** Was this an attempt to render טליא 'lamb'? The verse describes the accession of Solomon.

(50) **a house great and broad ... low** 'a house' lit. 'this (the) house', 'the well-known house'. I have read 'was built' (kona taḥanṣa), and 'and a tower [wa]māḫfad (cf. Flemming, 129), = Aram. [... והוא בית רב ורחב מתבנא לענא ומגדל]. Eth^{M} has arisen by the confusion of בית 'house', vid. Jerusalem, with the earlier 'house' (Eth. bēt for 'tabernacle', v. 36): hence the rendering 'And that house became large and broad and for those sheep a high tower was built on that house for the Lord of the sheep' (Knibb). As Dillmann has argued, the most consistent interpretation of the imagery is that of Test. Levi 10.5: ὁ γὰρ οἶκος ὃν ἂν ἐκλέξεται κύριος Ἱερουσαλὴμ κληθήσεται, καθὼς περιέχει ἡ βίβλος Ἐνὼχ τοῦ δικαίου, a probable allusion to this verse. The 'lofty and great tower' is Solomon's Temple. The 'house' is said to be 'low' (tateḥḥeta): Goldschmidt may be right in suggesting the root שפל as the Semitic equivalent verb: Jerusalem does not only lie beneath the Temple on Mount Moriah, but its humble position, in spite of its size, is also thereby indicated, in relation to the lofty tower.

a full table The offerings and sacrifices of the Temple cultus. Cf. Exod. 25.30.

(51) Vs. 51-67 deal with the division of the two kingdoms of Israel and Judah.

forsook their house ... began to slay them 'House' here must mean Jerusalem with the Temple as cultic centre (cf. v. 54 'they forsook the house ... and tower ...); after the division of the kingdoms the ten tribes abandoned the worship of the Temple in Jerusalem. God sent his prophets to bring them back, but they were put to death: the reference is to the massacre of the prophets by Ahab and Jezebel at 1 Kg. 18.4, as v. 52 with its reference to Elijah clearly shows.

many ways Eth. babezuḫ fenāwāt באורחות שגיאת. This could be a mistranslation of 'in the ways of error'; cf. Ps. 19.13 שגיאות (ܛܘܥܝܐ = vagatio, Tg. שגא errare).

(52) **and one of them was saved** The escape and translation of Elijah, 1 Kg. 19.2, 2 Kg. 2.11; cf. also En. 93.8 '... and in it (the sixth week) a man shall ascend'. Other

allusions appear at Mal. 4.5; Mt. 17.3,12; Rev. 11.3-6. See Str.-B. Bd. IV.2. Exkurs 28, 764-98; M. Black, 'The 'Two Witnesses' of Rev. 11.3f.' in *Donum Gentilicium, New Testament Studies in honour of David Daube*, 227f.

it fled Eth. qanaṣa, salire, here = ἀποπηδᾶν rendering נדד, Heb. Aram. 'to flee'; cf. LXX Hos. 7.13, Prov. 9.18.

(53) **many other sheep** The prophets who succeeded Elijah in their equally profitless missions.

(54) **they forsook ... invited that slaughter** A description of the abandonment of the worship in the Temple accompanied by apostasy (cf. 2 Kg. 16.1f.). The straying sheep are also accused of surrendering Jerusalem itself to their enemies; the reference is probably to 2 Kg. 16.7f., where Ahaz calls in Tiglath-Pileser, King of Assyria, to help him against Rezin, king of Syria. Nothing in this Biblical narrative, however, corresponds to the surrender of Jerusalem; Rezin besieged but failed to capture Jerusalem. Eth. qatl = φόνος (cf. 12.6 G). There is no difference in consonants, in Heb. or Aram., between קֶטֶל 'slaughter' (Aram. קטול) and קֹטֵל (Aram. קטול) 'murderer'. Did the original perhaps read here '(until those sheep invited) those murderers (the Babylonians)' (coll. sing.) (perhaps even 'that murderer (Tiglath-Pileser)'? The two forms seem to have been confused in Eth. also: e.g. qatel (qātel nom. ag.?) renders φονευταί at Prov. 22.13 = 26.13 v.l. Cf. Eth[tana].

they fell away entirely Eth. 'emkuellu seḥtu = πανταχόθεν ἐπλανήθησαν = כול תעו(?).

betrayed his place For gab'a II.1 = 'betray' cf. Jdt. 7.26 = מסר(?). Did the original use the term מקום lit. 'place' but also 'sanctuary' (cf. Hoftijzer, 165).

(55) Vs. 55-56 have much in common, suggesting a literary 'doublet', probably expansions by translators.

he gave them over into the hands of Eth. ḫadagomu westa 'eda lit. 'he abandoned them into the hands of ...' = Heb. עזב ביד Ps. 37.33, Neh. 9.28 Tg., שבק ביד.

lions ... foxes For this period lions and leopards and wolves are Assyrians, Babylonians and Egyptians; the other tormentors of Israel are local enemies which have here been variously identified: the 'hyenas' could be the Syrians (Martin), but precise identification is not possible. (Cf. 2 Kg. 17.24).

(56) **devour them** The same figure of speech is used by the prophets to describe the fate of Israel, Jer. 12.9, Isa. 56.9, Ezek. 34.5. An allusion to this verse occurs at Barnabas 16.5: λέγει γὰρ ἡ γραφή· καὶ ἔσται ἐπ' ἐσχάτων τῶν ἡμερῶν, καὶ παραδώσει κύριος τὰ πρόβατα τῆς νομῆς καὶ τὴν μάνδραν καὶ τὸν πύργον αὐτῶν εἰς καταφοράν. In spite of its being introduced as a scriptural quotation, the verse in Barnabas looks more like a free reminiscence of vs. 55, 56 here. It appears to be drawing on a Greek version: νομή = מרעיתא, πύργος = מגדלא(?).

(57) **to call to the Lord of the sheep** While admitting some kind of intercessory role to the Enoch in this verse, Dillmann denies any idea of the heavenly Enoch pleading for the apostate Israel; he does compare, however, 2 Mac. 15.12-16, the vision to Judas of Jeremiah in just such a role. For the theme of abandonment of Israel by God, cf. Jer. 12.9, Ezek. 34.5f.

(58) **remained unmoved** Eth. 'armama lit. 'was silent'; cf. the Biblical use of חרש Hiph., of God permitting evil in silence: Hab. 1.13, Isa. 42.14, Ps. 50.21.

carried off Eth. tahayda = Aram. חטף Ithp.(?).

(59) **seventy shepherds** In an early form of the text of Dt. 32.8 each of the seventy 'nations', i.e. ethnic groups, into which mankind was divided at the creation (Gen. 10), was allocated to a guardian angel (or 'son of God'): Israel alone had the Almighty himself as her protector (Dt. 32.9) (see Milik, 254 and cf. Jub. 15.31f). The author of

the Dream-Visions combines this notion with the idea of the 'seventy periods' of history, here extending from the Assyrian invasion and the fall of Samaria (721 B.C.) to the writer's own time, the 'eschatological' era (c. 200 B.C.). This 'historiography' obviously has been influenced by the 'seventy years' of Jer. 25.11, 29.10, and the 'seventy heptomads' of Dan. 9.23f. as well as probably deriving from even older traditions of 'seventy ages' (Milik, 248 and above, 137).

The 'Lord of the sheep' has abdicated (v. 56), but at the same time has handed over the straying flock to the wild beasts, not only for condign punishment, but also to come under the aegis of the 'guardian angels' of the lions, leopards etc., each one to take responsibility in his time for pasturing the sheep.

The allegory is referring to the 'dispersion' of Israel among the nations of mankind, and, now that God has abandoned Jerusalem and his temple, the people have come under the sway of the Gentile 'nations' and so of their guardian angels. (For the uniquely original theodicy based on his 'seventy shepherds' apologetic, see further below, 272).

drove off Eth. gadafa = Heb. גרש Pi. ܛܪܕ Pa.(?) (Aram. גרש 'to divorce', Levy, NHCW, I 366). The term here suggests the forcible deportation of the Israelites at the time of the Assyrian invasion.

servants Eth. ḍammād = Heb. נער (Gen. 14.24 etc. Isa. 37.6 Tg. עולם)(?).

(60) **exactly numbered** Eth. baḫuelqu = במספר lit. 'by (exact) number', 'genau abgezählt,' Dillmann); viz. 2 Sam. 2.15, Dt. 25.2 etc. (Tg. במנין).

(61) **[a watcher, one of the seven white ones(?)]** All texts read 'And he called another' (kāle'). This kāle' is clearly the recording angel of 90.14 and 22, and one of the seven archangels ('beings who were like white men') introduced into the Dream-Visions at 87.2 (cf. earlier 81.5). Charles may be right in identifying this archangel-scribe with Michael, the protector of Israel (Charles, 201). The Eth. 'another', while construable, really makes little sense and points to a seriously deficient text. I suggest a confusion between חור 'white' and אחרן, ἄλλος as the source of the trouble. Did the original read simply: (וקרא) לחורא '(and he called) a white one' i.e. a white-clad archangel, misread as לאחרנא, (καὶ ἐκάλεσε) τὸν ἄλλον? Or an original עירא חורא, ἄγγελον λευκόν becoming ἄγγελον ἄλλον and ἄγγελον or עירא lost by scribal error? There may have been a longer original text, similar to that of 90.22, which has been shortened by hmt: וקרא ל[עירא חד מן שבע] חורין 'And he called a watcher, one of the seven white ones'.[1]

Observe and mark Eth. labbu ware'i. Did the original read חור = 'observe' (חור־נא with a word-play on חור 'white')?

(62) **all the excess and the destruction** Eth. ṣegāb = Heb. Aram. שבעא: 'all the excessive destruction'.

of their own volition Eth. bare'somu (here and at v. 63 only) sua sponte cf. Heb. כרצנו Dan. 8.4, Est. 9.5 (Tg. כרעותהון) etc.

(63) **reckon them up** The majority reading 'deliver them up' raises difficulties since elsewhere it is always the sheep which are 'delivered up' to destruction (see Dillmann). The repetition of 'deliver them up' in the same sentence suggests corruption and the emendation of Charles[1893] (accepted by Flemming and Beer) offers a satisfactory solution (the corruption occurs again at 61.2; cf. Knibb). The meaning of the verb, maṭana 'reckon up' lit. 'measure' seems more appropriate than 'comprehend'

---

[1] Eth[m t] reads walakuellomu before walakāle': is this a corruption of walamal'ak kāle', ולעירא אחרנא?

(Charles); cf. Flemming-Radermacher 'ihnen nachzurechnen'. (The noun maṭan renders ἀριθμός at Rev. 13.18). The object (understood) is 'all their deeds'; cf. Isa. 65.7 ומדתי פעלתם lit. 'And I will measure their works ...'. The basic idea is the same as at Isa. 65.7, namely, 'to take the measure of' their wicked deeds with a view to condign punishment.

(64) **and bring it all up before me** Cf. 89.70,76. A secret register is to be kept of the misdeeds of the shepherds for which they will be punished at the final reckoning.

(65) **into the hands of the lions** The lions appear here to be the Assyrians and the reference to their conquest of the northern kingdom, unless this has been described at 55, 56, in which case some later events are meant.

(66) **the lions ... devoured** The lions are still the Assyrians (above and cf. 89.10), the leopards the Babylonians and the wild boars the Edomites (for Israel's intense hatred of the latter, cf. Isa. 34.5f., Ezek. 25.12f., Ps. 137.7, Ob. 9-16, Lam. 4.21; according to I Esd. 4.45 it was the Edomites who burned down the Temple.) Dillmann thinks that perhaps the other neighbouring smaller peoples hostile to Israel (see v. 72), which took part in the destruction of Israel at this period, were also included, as at Ezek. 25. The first part of the verse appears still to be referring to the fall of the northern kingdom to Shalmanesar, King of Assyria (2 Kg. 17.1-41) and to the campaigns of Sennacherib or of Nebuchadnezzar against the kings of Judah (2 Kg. 18.13f., 24.8). See the note in Martin for further detail.

demolished Eth. karaya = Heb. הרס(?) cf. Ezek. 36.35 Eth. takarya = נהרס Tg. אתפגר. Ezr. 5.12 uses סתר in this connection.

(67) Does this verse imply that there was a remnant left behind in Jerusalem, but that the author had no knowledge of it or does not want to think about it.

grieved Eth. ḥazana can be used of grief and sorrow or of anger and vexation = LXX λυπεῖσθαι which can render אבל, חרה or Aram. באש ל (e.g. Dan. 6.14 '... the king was sore displeased').

(68) **that white watcher(?) ... destroys** The reference to 'the other' (lakāle'u) recalls v. 61. If the text is rendered literally as it stands, it is necessary to take the verb as passive yeṣṣaḥḥaf and translate 'it was written by the other (lakāle'u)'. (Since la presents difficulties Charles[1906] emends to ba; Dillmann, Beer delete.) Behind a plainly corrupt text lies again (see above v. 61 note) the confusions by the translator (Greek or Ethiopic) of the adj. חור 'white' with אחרן, ἄλλος. I suggest that an original עירא חורא = 'the white watcher' (i.e. the one referred to at v. 61) has been rendered by ὁ ἄλλος (ἄγγελος) and corrupted to ὁ ἄλλος simpliciter, and an Ethiopic mal'ak kāle' to lakāle'u. The majority text yeṣeḥḥef can then be read.

(69) **more than was decreed for them** Eth. 'emšer'ātomu lit. 'their commands' (v. 61) = מצותם, προστάγματα αὐτῶν, פקודיהון(?). The two verses (61 and 69) imply that there was a divine limit set to the number of Israelites to be punished for their transgressions by the angel-guardians of the 'nations'. But this limit had been passed by these 'shepherds' to whom Israel had been handed over for punishment. Here the writer introduces an original theodicy: Israel's sufferings had far exceeded any condign punishment for her sins. The responsibility for these excessive afflictions did not lie at her door, but belonged to those angel-guides who had taken over responsibility when the Lord had abandoned his people to them.

(70) **day by day** Eth. bakuellu 'elat: cf. En. 78.6 Enastr[c] 1 iii 4 (Milik, 292)

laid down Eth. yā'ref: 'arafa II.1 = Heb. הניח 'place, deposit'(?); cf. Ezek. 37.1, 40.2 etc. and cf. below v. 71. (No other instance of 'arafa II.1 is attested in Dillmann, *Lex.*, in this sense.)

(71) The reading and the sealing of the book kept by the archangel scribe, one of the seven white ones (89.10), is intended to mark the end of the first historical period. Chapter 89.72-89.77 is a new period, from Cyrus to Alexander, suitably closed by a similar celestial 'ceremony of the book' (89.77).

laid it ... away Eth. 'anbara = הניח(?). Cf. above v. 70; here = conservandum deposuit (Gesenius).

(72) **twelve periods** Eth. sa'āt lit. 'hours', cf. 90.5 where the term gizē is used: both mean 'periods of time' and probably both render ὥρα = Aram. עידנא (Dan. 4.13,20 etc.). Here the twelve periods refer to the duration of the first era of history under the angel-guides of the nations, at the end of which the three sheep return to restore the fallen house and tower.

three of those sheep Zerubbabel and Joshua (Ezr. 2.2) are the two certain identifications, but who is the third sheep? Dillmann thinks 'three' is a mistake for 'two': but if not, he suggests Haggai or Zechariah for the third sheep. Others thinks of Ezra or Nehemiah, in spite of the interval of time separating them from the first two. A third proposal is that it is three tribes which are here represented, Levi, Judah and Benjamin; and this is supported by the similar allegory at Test. Joseph 19.3. The reference is certainly to the period of the Return and the rebuilding of the Temple; and since the history is in any case being 'telescoped', both periods may have been included; perhaps it was Ezra the Scribe who was in the author's mind as the third sheep. On other references to the Return and rebuilding of the Temple it is sometimes Ezra, sometimes Nehemiah who is singled out for mention (Sir. 49.11-13, 2 Mac. 2.13). There is no problem about the identification here of the wild boars; they are unmistakably the Samaritans and the mixed peoples who supported them of Ezr. 4-5, Neh. 4-6.

(73) **raised up that tower** The rebuilding of the Temple: Ezr. 3-6.

a table Cf. above on v. 50 and cf. Ezr. 3.2-3.

bread on it ... polluted An allusion to Mal. 1.7 (Eth. ḫebest ... rekus = לחם מגאל 'polluted bread'). Cf. also Ezr. 9-10. The offerings were impure because those who brought them were still a subject people, and were mixing freely and intermarrying with the local inhabitants and neighbouring peoples; cf. Mal. 1-2, Ezr. 9-10.

(74) **above all** Eth. diba kuellu, or 'furthermore', 'besides' (Dillmann).

the eyes ... were blinded Nothing is said of the reforms of Ezra; in fact the entire Persian era is regarded as a state of 'blindness' for Israel. (Cf. further Dillmann's comments.)

(75) **remained unmoved** See the note above on v. 58.

the wild sheep According to 90.16 below, these are the Jews of the dispersion corrupted by Gentile ways. The majority text with 'all the sheep' as subject ('... until all the sheep were scattered abroad' Knibb) leaves 'and they had intercourse with them' with a predicate undefined (supplied by the scribes of Eth^{e} and Eth^{b} as 'with all the shepherds' 'with every shepherd and sheep').

they (the shepherds) did not save them. This seems preferable to a possible impersonal plural = 'they were not saved from them', since the shepherds were supposed to watch over the sheep.

(76) **(their) presumptuous deeds** The text of Eth II 'and he read (it) out in the houses (dwellings) of the Lord of the sheep' is the generally accepted version (Dillmann renders 'und jener welcher das Buch schrieb, brachte es hinauf zu den Wohnungen des Herrn der Schafe'). The more difficult text is that of $Eth^{tana(t)}$ where Flemming had earlier suggested reading 'abiyāta 'great things' for 'abyāta 'houses' and rendered (Flemming-Radermacher, 118 n.) 'las die grossen Dinge dem Herrn der

Schafe vor'. The reading of Eth[tana] confirms this conjecture: I suggest interpreting 'abiyāta = רברבן, גדולות, in the sense of things too great and so presumptuous (Jer. 45.5, Ps. 12.4, Dan. 7.8,20; cf. I En. 1.9 En[c] 1 i 17 (Milik, 184). In the present context the 'great things' are the arrogant and presumptuous deeds of the shepherd-guides.

(77) **And he took the book** A shortened form of the account at the close of the first period at v. 71.

CHAPTER 90

(1) **thirty-five shepherds** Eth. reads 37, a figure which is usually explained as a mistake for 35. Dillmann regards it as a scribal slip (Dillmann, 273). It would bring the total number of shepherds to 72, a figure which alternates elsewhere with 70 for the number of the 'nations' of the earth or the rulers of Israel. (See B. M. Metzger, 'Seventy or Seventy Two Disciples' in NTS, Vol. V, 302f. and especially, 304.) It is not surprising that the Eth. should follow the LXX reckoning (72), in spite of the discrepancy when its individual figures for each period are added together and make a total of 72, not 70. The number 35 includes the shepherds of the 12 periods at 89.72 above, and is made up to 70 by the 23 of 90.5 plus the final 12 of 90.17. The schematic historiographical scheme of the author envisages two main periods of post-exilic history each divided into two parts: I. The Persian Period, from Cyrus to Alexander; the first part of 12 periods covers the Return and attempts at reform under Ezra and Nehemiah, followed by 23 shepherd-oppressors, Persian, Egyptian etc. Israel's enemies are represented in this period as wild animals. II. The Greek Period, from Alexander to the Maccabees; this period is again schematically divided into 23 rulers, the oppression under the Ptolemies and the Seleucids, followed by a climactic 12, bringing the scheme down to the writer's own time, probably under the Hasmonaeans, but after the victories of Judas. Chapter 90.28f. envisages the building of a new Temple not just the rededication of the old Temple; and this suggests a date prior to the rededication of the Temple by Judas (165 B.C.). In this period Israel's enemies are symbolised by birds. Identifications of individual shepherds and their peoples and precise dating are scarcely possible in work of this kind, but the broad chronological divisions were probably sufficiently well known to have been recognised by the readers.

completed ... others received them The 'first' are the 12 shepherds of 89.72; the 'others' (23) are the Greeks, Egyptians, Syrians of the second period.

(2) **all the birds of heaven** The world-powers of the Greek period, represented by birds of prey, as the earlier enemies of Israel were symbolised by animals.

the eagles, the vultures The eagles are probably the Macedonian Greeks, the vultures the Egyptians and the 'ravens' the Seleucids. Cf. Dan. 11.6-16.

(3) **flesh was being devoured** Eth. (yetballe'u) šegāhomu = σάρκες αὐτῶν (cf. Dillmann, 62.2 who explains the noun as meaning 'bodies' and therefore taking a plural verb).

that shepherd who pastured the sheep Dillmann takes vs. 3-4 to refer to the oppression and suffering of Israel as the result of the wars of the Diadochi and their struggles for the possession of Palestine (cf. also Martin). The reference to 'that shepherd' suggests that the reader would recognise the identity of the especially severe oppression under this angel-shepherd; Dillmann suggests either Antigonus or

Ptolemaeus Lagi. It may also be possible to explain the phrase as a collective singular: 'over those (the) shepherds'(?).

(4) **by the dogs** If this reading is original the author is then mixing his symbolism, for elsewhere the wild beasts belong to the first period of Israel's post-exilic history, the birds to the second Greek period: 90.18 which mentions both together is no exception since it sums up both periods. In the first period the dogs symbolise the Philistines. Could they refer here to the ἀλλόφυλοι of 1 Mac. 3.41 or to 'the inhabitants of Seir and Philistia' of Sir. 50.26 mentioned along with the Samaritans? But this would mean that small local enemies of Israel were given precedence of the world-powers of the Diadochi, Syria and Egypt. The reading seems to me to be entirely out of place in this context and is possibly best explained as an inner Eth. corruption which has arisen from the previous two words. See textual note, 369

left neither flesh nor skin nor sinew Cf. Mic. 3.2-3; Eth. šegā = בשר/שאר, mā's = עירם, šerw = גיד(?); ḫadagu 'they left over' = Aram. אשאר(?).

(5) **twenty-three shepherds ... fifty-eight periods** The fourth period (beginning at v. 6) differs from earlier periods in that these were marked by a change from one world-power to another: the fourth period is a prolongation of the Greek dominion under the Diadochi, progressively characterised by the worsening behaviour of the Egyptian and Seleucid tyrants towards Israel. The climax is reached in the Maccabaean period, the active persecution by the Seleucids, which is told in the narrative of the final twelve shepherds. The third period closes with the reckoning here of the twenty-three shepherds of the Greek period added to the previous thirty-five of the first Persian period, the total giving fifty-eight shepherds and periods — to be com pleted with the climactic twelve periods of the Maccabaean era, thus giving seventy shepherds and seventy (Eth. seventy two) periods in all. See Dillmann, 282, for a list of names for the Diadochi.

(6) **a few rams(?)** Eth. maḥaseʿ is used for the young, male or female, of the flock, and in this context 'rams' = Heb. איל Aram. דכר (cf. 89.44 En[d] 2 iii 29. Eth II μικροὶ ἄρσενες is usually understood as referring to the size of the rams, but, as a rendering of זעיר could refer to their number: דכרין זעירין = ὀλίγοι ἄρσενες (cf. Schodde, 236).

to those white sheep The 'white sheep' are usually understood to refer to the parents of the Hasidim, 'the faithful adherents of the theocracy' (Charles, 207); and the view that the writer intends us to understand the 'rams' as the Hasidim (or the maskilim of Daniel) is widely agreed.

to open their eyes Cf. 89.28.

to cry to the sheep i.e. to warn them against apostasy by yielding to the policies of Antiochus.

(7) **importuned them** Eth[m] sarḫa II.1 = Heb. הציק Aram. אעיק e.g. Jg. 14.17, 16.16 (LXX παρενοχλεῖν).

exceedingly and terribly blinded Eth. fadfāda waḫāyāla. The variations in the manuscrits show that the expression gave difficulty to scribes (it occurs only here). Charles, 207f., thought it rendered σφόδρα σφόδρα (= מאד מאד(ב), לחדא ולחדא). At 106.19 En[c] 5 ii 25 fadfāda 'emenna, = magis quam renders תקיף מן, and ܬܩܝܦܐܝܬ = σφοδρῶς (P. Sm. 4491); תקיף is rendered by ḫāyāl at 89.30 En[a] 2 ii 29. Did an original read ותקיפית (ית)שגיא? (For ת[קיפ]ית 89.16 En[e] 4 ii 21 Eth. bakuellu ḫaylomu (Milik, 241).)

(8) **swooped down on the rams** Eth. sarara diba = Heb. עוף ב of armies attacking enemies Isa. 11.14.

seized the leader(?) of the rams Eth. 'one of the lambs' (with variations which may point to corruption). The reference has been taken to be to the martyred Eleazar (2 Mac. 6.18f.) or, more probably, to Onias III. The translation proposed assumes a misunderstanding of an original Aram. אחיד in the sense of 'leader, ruler' (ܐܚܝܕܐ P. Sm. 119 and cf. Eth., 'aḫāzi = ὁ κρατῶν Wis. 14.19; cf. also the use in Heb. of אחז (pass. ptc.) = 'taken out (of a number)', 'appointed' Num. 31.30). Did the original perhaps read: (אלין) אחזו לאחידא לדכרין 'they seized the leader of the rams'?

dashed the sheep in pieces Eth. qaṭqaṭa דקק Haph. Dan. 2.34,45, 6.25 (LXX ἔθλασαν (τὰ ὀστᾶ αὐτῶν)).

(9) **broke their horns** Eth^{tana} yādaqqeqewwomu, = שברו קרניהון cf. Dan. 8.7. The imagery in this verse comes from Dan. 8.3f.

horns grew upon those rams waḍ'a = יצא, נפק; cf. Dan. 8.9. The reference is unmistakably to the rise of the Maccabees.

there sprouted a great horn Eth. baquala 'aḥadu qarn 'ābiy = רבת קרנא חדא רברבא(?). It is now generally agreed that the reference is to Judas Maccabaeus; and it would be surprising, indeed, if no mention of the hero of the Maccabaean resistance was ever made. Cf., however, Dillmann who identifies the 'great horn' with John Hyrcanus; for his reasons see further below, 277.

of one of the sheep The reference is no longer to the Asidaeans, 'the rams' but to the Maccabees.

(10) **it had regard to them** Eth. re'ya bomu, חזא להון lit. 'saw to them'; cf. Exod. 18.21 (Heb. and Tg.) and P. Sm. 1233; ראה ב is similarly used, e.g. Ps. 106.44. Charles emends re'ya to re'ya (bomu) and renders 'it pastured with them.'

it cried to the sheep i.e. it appealed to them.

the rams ... ran to it Eth. dabēlāt is a more general word for maḥase' 'rams'. It serves, however, to distinguish the young rams of the whole flock who rally to Judas from the 'rams' of v. 6, the Asidaeans.

(11) **what is more** The phrase seems to be the equivalent of Heb. עם זה or בכול זאת Isa. 5.25, 9.11; cf. Lk. 24.21 σὺν πᾶσι τούτοις. It can also mean 'during all this' or 'in all these circumstances'.

eagles and vultures Not only the ravens (Syrians) but the Greeks and Israel's local enemies ('kites') were harassing the people; the reference could be to the foreign mercenaries from Greece or Egypt employed by the Seleucids: cf. 1 Mac. 6.29, Josephus, *Ant.*, xiii 4.9 (121). The author may also be thinking of similar persecutions in the Diaspora; cf. 1 Mac. 5.38, 6.53f., 2 Mac. 6.8f.

and the sheep suffered(?) Eth. yārammemu = 'and the sheep kept silent'. A confusion of חשא with the root חשש 'to suffer' seems a likely explanation of the text of Eth. which otherwise yields an odd sense in this context; in such circumsyances sheep are unlikely to keep silent.

(12) **strove and fought with it** Eth. gadala III.3 (Eth^{g} nad'a III.1) = נצח Niph., Aram. נצא Ithp.(?) Eth. be'sa mesl = Heb. נלחם ב(?). The manuscripts alternate between a singular and a plural, the former referring to Judas, the latter to the struggle of the followers of Judas.

to lay low its horn Eth. 'atata II.1 = ἀναιρεῖν Heb. הרים Aram. ארים(?). If these are the right equivalents, the expression is ambiguous: it can mean that the 'ravens' (the Syrians) sought to remove the 'horn' (Judas) or 'horns' (his followers), but, with the plural suff., 'their horn' (Eth^{q}), it can also mean 'to exalt their horn' i.e. the 'horn' of the Syrians, 'to triumph'. (Cf. 1 Sam. 2.1, Ps. 89.25.)

(13) Charles took vs. 13 and 16, 14 and 17, 15 and 18, to be 'doublets' and printed them in parallel columns. Vs. 13 and 16 are not, however, exactly parallel, and seem to

represent two phases in the Gentile operations against Israel, first, an appeal by the shepherd-oppressors to the ravens, the Syrians, to conquer the horned ram, then later at v. 16 a concerted campaign of the 'eagles and vultures' (the Egyptians and Macedonians?) with the Syrians (the ravens), and supported by the 'wild sheep' (the apostate Jews of the Diaspora), to destroy the Maccabees. Certainly it would be more logical if v. 16 had followed immediately on v. 13. V. 15 does appear to be a shorter recension of v. 18 and may, therefore, be a 'doublet' of v. 18 but with features not represented in v. 18. V. 18 looks like a climactic verse, so that v. 19 may be out of place and have originally stood after v. 16 or 17.

shepherds and eagles ... kites The presence of the shepherds at this point seems inappropriate: Charles explains it from an original ערבין = 'ravens' misread as רעים (the same would apply to Aram.). The reading could, however, be original: the author may have intended the last three shepherd-oppressors to participate personally in the climactic destruction of Israel. At v. 17 it is said that the destruction wrought by them was greater than that of their predecessors; instead of 'pasturing' the flock for 12 periods, they joined up with all Israel's enemies (cf. v. 16) in their final assault on the forces of Judas. Alternatively, a redactor may have introduced the word in order to incriminate the shepherds themselves in this last attempt to destroy the followers of Judas.

break the horn of that ram Again generally interpreted with reference to Judas.

cried that its help might come The cri de cœur of Judas. See next v.

(14) **that man ... shepherds** the archangel, probably Michael.

helped ... that ram The answer to the prayer of the ram at v. 13. At 1 Mac. 4.30f. Judas prays for divine help against the overwhelming forces of the Syrian Lysias. This is further developed at 2 Mac. 11.6f. where Judas prays that a good angel will be sent to deliver Israel: the response was the appearance of a horseman clad in white with golden weapons, again probably Israel's champion Michael, who fought side by side with the inspirited and thus invincible Israelites. It seems very unlikely that it was to any other occasion this verse is referring: but cf. Dillmann who identifies the ram with John Hyrcanus and cites the latter's vision from Josephus, *Ant.* xiii. 10.3 (282-283) in support. For his reasons see note below on v. 17.

(15) **fell down in darkness**(?) Flemming-Radermacher give the ingenious rendering: '... und verfielen alle in den Zustand des Geblendetseins vor seinem Antlitz'. The longer recension of v. 18 has 'fell ... and were submerged (tasaṭmu) in the (cleft) earth': this recalls not only Num. 16.31-33 but Exod. 15.10 where the verb tasaṭmu renders צללו, ἔδυσαν 'they were submerged'. I suggest that it is from this rare verb in the original that the corruption westa ṣelālot(u)/ṣelmat 'into the (his) shadow/darkness' comes.

(16) **the wild sheep** Eth. 'abāge'a gadām, lit. 'the sheep of the open country', probably to be identified with the hellenised, and so apostate, Jews of the Diaspora. See Dillmann, 282 and above, 273.

came together Does Eth[q] preserve the original text, viz. 'went forth', waḍ'u = יצאו, 'went forth in battle together'? Cf. v. 19.

(17) **those twelve last shepherds** Dillmann, 271., 282f., considered the period from the rise of the Asidaeans (c. 200 B.C.) to the triumphs of Judas (c. 160 B.C.) too short for the 'pasturing' of the twelve last shepherds. Consequently he argued for the identification of the 'great horn' with John Hyrcanus, thus bringing the date of the twelve last shepherds down to the attacks of Antiochus Cyzecenus and his successors (c. 195 B.C.). It seems a mistake, however, to look too closely at the schematic 'historiography' of the apocalyptist. It is also possible that the writer is intentionally

representing this shorter period as a climactic onslaught of all Israel's enemies (cf. vs. 13-16) in which the twelve last shepherds or 'guardian angels' of the Gentiles themselves take an active part, thus incriminating them still more in the excessive afflictions of Israel (cf. above note 274).

(20) **the pleasant land** See note on 89.40.

the white one(?) Eth. kuello does not make sense: read (with Charles, Flemming) kāle'u = ὁ ἄλλος i.e. חורא 'the white one', the archangel in charge of the sealed books (89.61,68). See note above, 271, 272.

(21) **the seven chief white ones** See above on 87.2; 'chief' rather than 'first', which makes poor sense; cf. ܡܕܒܪ P. Sm. 3490. The variant 'six' reported by Dillmann, 283f. may come from Ezek. 9 (the six angels of destruction accompanied by a heavenly scribe clad in white linen). Cf. Dan. 10.5, 12.6, Rev. 15.6, Tob. 12.15.

the first star ... which fell The Eth. text is confused. Eth[g] '(that they should bring before him) the first star' is probably a scribal improvement (but cf. Knibb, 2, 214). The command cannot have referred to the first star to fall only, so that the emendation of 'elku to lakuellu is not only justifiable but necessary (Dillmann, 284). Omit 'beginning with the first star which led the way' as an Eth. expansion. Perhaps the original order of clauses was 'And the Lord ... commanded that they should bring before him the ⌜first⌝ star which fell first, then all the stars ...'.

(24) **flaming fire** Eth.[tana] 'essāt walāheb, אש להב, Lam. 2.3, Ps. 105.32.

over the stars The watchers are the first to be judged and condemned.

an abyss ... fire Here this 'abyss' is not to be confused with the preliminary place of punishment at 18.12-16, 19.1-2, 21.1-6. This is the final place of punishment mentioned at 10.6, 18.11, 21.7-10, 54.6.

(26) **in the midst of the earth** i.e. in Jerusalem, cf. 26.1.

on the south side of that house The Israelite renegades are punished in a different place, clearly Gehenna, on the south side of Jerusalem.

(27) Charles may be right in suspecting corruption through a misunderstanding of the original. Did a translator read כד מתדלקין גמירא 'as they were completely consumed (by fire)', but mistook גמירא for גרמיה = ὀστᾶ αὐτῶν. The verb גמר is used in the Tg. for the total destruction of the herem, e.g. Num. 21.3; cf. also Tg. Exod. 15.7.

(28) **the old house** The 'old house' and the 'new house' are symbols of the old and the new Jerusalem based on scriptural sources such as Ezek. 40-48, Isa. 54.11,12, Hag. 2.7-9, Zech. 2.6-13. No explicit mention is made here of the Temple (the 'tower' of 89.50) but it is no doubt included. The idea of a new Jerusalem is a commonplace of the apocalypses, e.g. 2 Esd. 7.26, 13.36, 2 Bar. 32.3, Rev. 21.2,10. See Martin's note on v. 29 for additional references.

was removed Eth[tana] mēṭewwo (impers. plur.) Eth. mēṭa = שני lit. 'change', 'change place, migrare' (ܥܢܕ loco movit, P. Sm. 4234). '(And I stood up to see) till the old house was removed' ([μετέθηκαν τὸν παλαιὸν οἶκον]?). The variant tasaṭma submergi is meaningless. Eth. taṭawma = אתכרך (Tg. Est. I 8.15 (end), Levy CW I, 386)(?) for the 'wrapping up' of the content, the columns, pillars, ornaments of the house.

pillars Eth. takl ᵉ נציבה(1) 'planting', 93.2, Eth[g] 1 iii 19, Milik, 263; (2) 'pillar' στήλη (Hoftijzer, 184). Translators follow Dillmann's 'Balken'; see his note, Dillmann, 285.

ornaments Eth. senn = Heb. תפארת Zech. 12.7 or הדר Aram. הדרא(?).

in the south of the land Had the writer 1 Mac. 4.46 in mind with its provision for the disposing of the stones of the old Temple? Or does it mean into the wilderness of Sinai (where the Great One dwells, cf. 77.1 and note) or simply to some ritually pure place 'outside the camp' (Num. 19.9). Or is it even conceivable that the writer had the Jewish settlement and Temple at Leontopolis in Egypt in mind?

(29) **erected** Eth. 'amṣe'a = אקים (Zech. 11.16 מקים LXX ἐξεγείρειν)(?) But see textual note.

removed 'demolished'(?) Eth. ṭablala = סחף(?) cf. Heb. סחב Jer. 22.19 and P. Sm. 2591, Eusebius *Hist. Ecc.* x. 4. 57, ܣܚܒ Ethp. = καταβληθῆναι.

the Lord of the sheep was in the midst of it Eth[M] 'all of them (the sheep) were in its midst', i.e. the house was large enough to accommodate all the flock. Eth[b (m) ryl 5 mss] seems preferable, especially in view of the Biblical promises that God would dwell in the midst of Israel, Isa. 8.18, Ezek. 43.9, Jl. 3.17, Zeph. 3.17 etc.

(30) **all the animals on the earth** The conversion of the Gentiles who had not oppressed Israel, and their submission. Cf. Isa. 14.2, 66.12 etc. (An exhaustive list of Biblical references on this theme in Martin and Schodde.)

in everything Eth. bakuellu qāl = בכול דבר/מלה

(31) **those three who were clothed in white** See 87.2,3.

the hand of that ram also seizing hold of me The 'ram' is generally identified with Elijah who had been brought up to Enoch in the 'lofty tower' high above the earth, presumably paradise itself (89.52). It is true, Elijah is there described as one of the sheep and not as a ram, but this need not be an objection, since previous and later leaders are called 'sheep'. A more serious difficulty is the context of v. 31 where, earlier (90.9f.), 'that ram' is almost certainly Judas Maccabaeus. Milik argues for Judas in this verse: '... this ram is almost certainly Judas Maccabaeus who, along with the three angels, accompanies Enoch to the place of the last Judgement (v. 31)'.[1] We have then, however, to explain how this ram, Judas, came to be with Enoch in the high tower to which he had been elevated (87.3), and where he was joined by Elijah. In later traditions it was Elijah, not Judas, who was with Enoch to witness the last judgement. (Cf. Milik, 45, M. Black, 'The 'Two Witnesses' of Rev. 11.3f.' in *Donum Gentilicium, New Testament Studies in honour of David Daube*, 227f.).

led me in ... without condemnation Eth. 'a'raga is hardly suitable to describe Enoch's descent from the high tower to the place where the sheep were gathered. (cf. Dillmann, 285f.) If the author used עלי Pa. = 'a'raga, the second verb may have been אעיל Aph. על 'to lead into', mistranslated as if from עלי 'to cause to ascend'. Eth[M] 'enbala tekun kuenannē is usually rendered 'before the judgement took place'. After the previous 'And thereafter' this is a difficult clause (cf. Knibb), and this has led to suspicious of its authenticity. Did the original have (Eth[m] כול)|די בלא דינא 'who were without (any) condemnation', referring to the innocent sheep of the flock which are then described in the next verse as 'white and pure'?

(32) **white** The Israelites in their new Jerusalem; cf. Isa. 1.26, 4.3, 11.9, 60.18,21.

(33) **destroyed and dispersed** This refers not only to the return of the dispersed Israel but to the bringing back to life of the 'good sheep', the resurrection of the faithful dead (cf. Dan. 12.2. See M. Black, 'The New Creation in I Enoch' in *Creation, Christ and Culture, Studies in honour of T. F. Torrance*, 19).

all the beasts of the field The conversion of the Gentiles and their assembling in Jerusalem.

with great joy Cf. v. 38 and Zeph. 3.17, Isa. 62.3-5, 65.19 etc.

(34) **laid down that sword** Eth. sakaba II.1 = Heb. השכיב Aram. אשכיב 'put it to rest'.

it held them not Cf. v. 36 and Zech. 2.4,8, 10.10.

(37) **a white bull was born** The 'white bull' has generally been understood as the

---

[1] 'Problèmes de la Littérature Hénochique, 359.

Messiah, but, since it is here clearly parallel to the white bull at 85.3 symbolising Adam, the image seems rather to refer to the birth of a new or second Adam, more glorious than the first, for 'his horns are large'. So Milik interprets (45). See also M. Black, 'The New Creation in I Enoch', 19f. This view of the symbolism need not rule out altogether a 'messianic' allusion; the new Adam may have been the apocalyptist's idea of the coming Messiah; the Gentiles fear and revere him as they are to fear the traditional Messiah.

beasts of the field ... birds of the air The Gentiles.

(38) **a buffalo** I adopt Dillmann's conjecture that behind nagar = ῥῆμα lies a transcription of Heb. or Aram. ראמה(א) 'wild ox', 'buffalo', a word which evidently had not been understood by the translator, who may also have been a Christian, familiar with the use of the Johannine 'Word'. Is it possible to see in the buffalo with the large black horns an allusion to the descendants of Ham now, like the Gentiles, brought within the new family of the second Adam? See M. Black, 'The New Creation in I Enoch', 20. An alternative explanation of the difficult epithet 'black' in this context is that the adjective has crept into the text of the Greek version by corruption of μεγάλοις repeated by a scribe as μέλανοις.

(40) Cf. 22.14.

(41) **each according to his destiny** Eth. kefl = Heb. Aram. 'allotted portion'; ܦܠܓܐ = μοῖρα, εἱμαρμένη P. Sm. 1294. A free paraphrase of this verse oocurs in a Greek Manichean writing (the writer introduces it as an actual quotation from the 'Apocalypse of Enoch'); see M. Philonenko, 'Une Citation Manichéenne du Livre d'Hénoch' in Revue d'histoire et de Philosophie religieuse, 52, 1972, 337-340.

(42) **the first dream** i.e. the vision of the Deluge, Chh. 83-84.

CHAPTER 91.1-10, 18-19

ENOCH ADDRESSES METHUSELAH AND HIS FAMILY

(91.1-10, (11-17 see Ch. 93), 18-19: Eth., En[g] 1 ii 17-21 (18-19)

91.1f. is generally regarded as introducing a new section of the book, usually referred to as 'the Epistle of Enoch' (Milik, 47f.), where Enoch again addresses his son Methuselah (see note above on 81.5), but now together with his whole assembled family. (For the title 'Epistle of Enoch' see below on 93.1f.)

(1) **all your brother ... the children of your mother** According to 2 En. 1.10, 57.2, their names were Methuselah, Regim, Riman, Uchan, Chermion and Gaidad.

a voice calls me So Flemming-Radermacher, Martin, Knibb; Dillmann, Schodde, Charles 'the word'. Dillmann argues that the parallelismus membrorum supports 'word': as the 'word of the Lord' came to the ancient prophets, so 'the word' came to Enoch, and 'the word' is parallel to 'the spirit', i.e. the holy spirit of prophecy; for the 'outpouring' of the spirit cf. Jl. 2.28-32 (LXX 3.1-5), Ezek. 39.29, Zech. 12.10. But 'voice' seems more natural in this context and still preserves the parallelism with 'a/the spirit'.

(3) **children of righteousness** I have adopted the reading of Eth I, which, if original, would give an expression parallel to IQS 3.20, 22 בני צדק. The alternative reading of Eth II, however, has much in its favour, especially in the use of nagara III.3 = διελαλεῖτο? (cf. Eth. Lk. 1.65).

I admonish you Eth. samʿa II.1 = העיד ב, אסהיד ב (Dt. 32.46, 2 Kg. 17.15).

love rectitude ... walk therein Cf. 94.1. Eth. faqara II.1 = אחיב, ἀγαπᾶν, giving a word-play with '(my) beloved ones' (חביבי(ן, ἀγαπητοί.

(4) **with a double heart** (בלב(א ולב(א: Ps. 12.2 and cf. Sir. 1.28 μὴ προσέλθῃς αὐτῷ ἐν καρδίᾳ δισσῇ, Jas. 4.8 δίψυχοι.

associate not ... companion Eth. ḫabara I.1 = חבר Hithp./Ithp. Eth. sutāfē, חבר (Prov. 28.24 LXX κοινωνός).

(5) **a great chastisement** i.e. the deluge; Eth. maqšaft = מכה a divine 'affliction' (like the 'plagues' of Egypt).

accomplished ... come to an end Eth. = Heb. כלה 'to be accomplished, fulfilled' (of divine purposes), 'to be consummated' (συντελεῖσθαι), 'to come to an end, cease'; cf. Aram. כלא, 107.1, En[c] 5 ii 39, 'oppression will come to an end'.

It shall be up-rooted Note the repetition of this phrase at vs. 8 and 11. V. 11 comes from the Apocalypse of Weeks (see below 291f.), where, in the Seventh Week, the elect are to 'root out (עקר) oppression' (v. l., En[g] 1 iv 14, 'unrighteousness'). This repeated idea and the phrase 'to up-root from its foundations' oppression or unrighteousness seems to stem from v. 11 in the Seventh Week of the Apocalypse of Weeks. Our author appears to be utilising its phrases and ideas as material for his paraenesis in these verses. See further M. Black, 'The Apocalypse of Weeks in the Light of 4QEn[g],' in VT, Vol. XXVIII, 465f. and below, 291f.

its whole structure ... disappear Cf. 94.6, 99.13, Eth. ḫalafa = עבר (Job 30.15).

(6) **shall possess it all(?)** Eth. 'aḫaza probably renders κρατεῖν, κατακυρεύειν, ירש, ירית 'to seize and occupy by force', used especially of 'possessing the earth' (38.4, 48.8, 62.3). Do we have here a mistranslation of עבדי חמסה 'workers of oppression' as 'works of oppression', as at 91.14, En[g] 1 iv 20 (see below 294).

for a second time Eth. kā'ebata 'for a second time' referring to the growth of wickedness after the Deluge, seems preferable to 'in a twofold degree' (Charles).

(7) The paraphrastic character of vs. 7 and 9 make it very difficult to reconstruct even a tentative Aramaic basis. Some may even doubt if there ever was one to these verses. It is clear that v. 8, like v. 5, draws on v. 11 from the Apocalypse of Weeks for its content and phraseology. It was probably originally composed in Aramaic. Similarly the basis of vs. 7 and 9 was probably also an original Aramaic composition, again drawing its principal motifs, so far as content is concerned, from the Apocalypse of Weeks, e.g., the disappearance of wickedness and of the ungodly and the final judgement of mankind, from 91.14f. But some of the Eth. variants noted show something of the free expansion and contraction which has taken place in this tertiary version. For a parallel verse in the later 'Noah apocalypse', cf. 106.19.

(8) **shall be up-rooted ... foundations of oppression** Like v. 5 above this verse is drawing on ideas and phraseology from v. 11 describing the Seventh Week in the Apocalypse of Weeks; for foundations of oppression' cf. אשי חמסא En[g] 1 iv 14.

(9) **And everything ... blazing fire** Charles, following Dillmann, takes 'tower' (māḫfad) to mean 'temple', and this would go well with 'idol', but if the original was πύργος = מגדל, the meaning is that of a fortified tower, not of a 'temple' or 'shrine', for which we would require מקדש. It is true that māḫfad 'tower' is used to designate the Temple in Jerusalem in the Dream-Vision at 89.50,54,56, but 'tower' there is used symbolically (Jerusalem is the house, its tower the Temple).

The text could be at fault: the majority text is awkward, lit. 'And everything shall be given over, the idols of the nations, towers, to the blazing fire'. Knibb takes māḫfad 'tower(s)' as the subject of the following tenadded, rendering 'And all the idols of the nations will be given up; (their) towers will be burnt in fire'. But the first verb 'will be given up', yetwahhab, requires a predicate, and the expression 'essāt tenadded =

πῦρ φλεγόμενον Dan. 7.9 (cf. Dillmann, *Gramm.*, §201, 528). Eth. maḫfad 'tower' looks like a gloss which has been incorporated into the text. Or is it a mistranslation of צירא, ܨܘܪܬܐ 'effigy (cf. Isa. 45.16 צירים = 'idols'), a duplicate of mesl = דמו εἰκών(?), misread as from צור, מצור πύργος, e.g. Eth. Isa. 29.3? Perhaps the original had 'the images and effigies of the heathen' (דמותא וצירתא די עממא).

They shall be cast into the judgement of fire The absolute rejection of the heathen seems to be taught here. This reprobation of the heathen does not appear to agree with the teaching of v. 14, where after the judgement of the wicked, 'all men' are 'to look to the true eternal path'. That verse, however, belongs to the Apocalypse of Weeks which has all the appearance of being an earlier source incorporated in his work by the author-redactor of 91-104. There is no difficulty if the subject is the idols of the heathen, and not the heathen themselves. A 'judgement of fire' means a divine verdict of destruction by fire.

(10) For the relationship of this verse to 92.3 and 91.17-18, see below, note on 91.17.

the righteous shall arise i.e., in the resurrection. In Eth. 'the righteous' is a collective singular as at 92.3.

wisdom shall ... be given unto them A motif from 91.10, the gift of wisdom to the elect in the Seventh Week.

[For 91.11-17 see below 291f.]

(18) **paths of righteousness** The theme of the two paths/ways of righteousness and of wrong-doing is further elaborated at 94.1f. Charles finds here evidence for the earliest 'non-canonical' reference to the 'Two Ways', doctrine. It is doubtful, however, if there is any specific reference here to this idea; it is a familiar enough contrast in the Old Testament (e.g., Dt. 30.15-16, Jer. 21.8, Ps. 1.6).

(19) **Walk in the paths of righteousness ... paths of wrongdoing** En[g] 1 ii 19-20 has a longer text than Eth.: the fragment preserved קושטא למהך בהון, requires the kind of supplement supplied by Milik, 'choose (בחרו)' or perhaps 'seek (בעו) the paths of righteousness to walk in them'. Cf. 94.3,4. Similarly for the negative parallel, En[g] had a longer text: 'and keep away from the paths of violence, from walking in them'(?) (so Milik, 261). Cf. 94.2.

shall perish everlastingly אבד is a prophetic perfect. En[g] 1 ii 21 לסוף אבדנא means literally 'to a finality of destruction', i.e., to utter and complete annihilation (אבדנא = ἀπώλεια). The idea of everlasting perdition as the fate of the ungodly — the opposite of 'life everlasting' — is central in Jewish apocalyptic and the New Testament. It is much more than simply the extinguishing of life, the meaning of the related Heb. verb and noun (אבדון, אבד, Job 26.6, Ps. 88.12). The apocalyptic term means absolute perdition, total destruction in Gehenna: cf. Mt. 7.13, Phil. 1.28, 1 Tim. 6.9, Heb. 10.39, 2 Pet. 2.3, 3.7. See the discussion in ThWNT Bd. I, 395 (Oepke). (Oepke remarks that the word does not occur in this eschatological sense in the Apocrypha and Pseudepigrapha, but he has failed to notice this verse.) Cf. further 5.5 (En[a] 1 ii 15), 84.5 (of the destruction by the Flood), 91.14 (En[g] 1 iv 21) בור עלם 'the eternal Pit', 16.1 G ἀπὸ ἡμέρας ἀπωλείας; and Tg. Isa. 53.9 'the death (leading to) destruction' parallel to 'Gehenna'. The Heb. noun has given rise to the proper name Abaddon, Ἀβαδδών Rev. 9.11. The Aram. expression has so far not been reported elsewhere from Qumran sources.

CHAPTER 92, 93.1-2

ENOCH'S EPISTLE

(92.1-5, 93.1-2, Eth. En^{g} 1 ii (92.1-2), En^{g} 1 iii (92.5), Milik, 260f.)

(92.1) In its Eth. version this verse reads like a book title, and may well have been intended to be so understood by the translator; Charles, in fact, considers it to be the formal title of the whole of Chh. 91-104, and rearranges the order of the chapters so that 92 appears as the opening chapter. Chh. 91-108 are identified by Milik as the so-called 'Epistle of Enoch' (Milik, 47f.). All translators give what is basically the same version of v. 1: thus Knibb now offers a literal rendering of the traditional text: 'Written by Enoch the scribe—this complete wisdom teaching, praised by all men and a judge of the whole earth—for all my sons who dwell upon the earth and for the last generations who will practise uprightness and peace'.

Fragments of the original Aram. of the verse have been preserved at En^{g} 1 ii 22 (Milik, 260). The identification is quite certain, and shows that 92.1 followed immediately on 91.19 (the last verse of the chapter) exactly as in the Eth. version: there can be no question of 92.1f. having originally preceded 91 as the opening chapter of Chh. 91-104. The restoration of the first fragment by Milik as [די כת]ב ויהב למ̊ת[ושלח] corresponds roughly to Eth II 'that which was written by Enoch'. Eth^{g}, however, reads, lit. 'that which was written by Enoch, the writing (book, epistle)' ('Das von Henoch verfasste Buch', Flemming-Radermacher), perhaps better ἡ γεγραμμένη ὑπὸ τοῦ Ἑνὼχ ἐπιστολή. A sentence beginning with די is odd (but cf. Milik, 262), and, in my opinion, requires a subject such as that in Eth^{g}; I suggest an original ספר חנוך די כתב 'Epistle of Enoch which he wrote...' perhaps rendered as ἡ γεγραμμένη ὑπὸ τοῦ Ἑνὼχ ἐπιστολή (= Eth^{g}). The restored text 'and gave to Methuselah' seems secure, even if it has been lost without trace in Eth.

The second Aram. fragment חכים אנשא is obviously the source of the noun 'wisdom' in Eth. Milik rightly sees that it is an epithet of Enoch, lit. 'the wise one of men', i.e. 'the wisest of men'. Milik, 260, restores [ספר פרשא], 'distinguished scribe', as at EnGiants^{a} 8.4, but the epithet which would account for the Eth. rendering 'scribe of all skill' (ṣaḥāfi zakuellu temhert, [γραμματεὺς ὀξυγράφος] v. l. te'mert crpt.?) would be the conventional phrase ספר מהיר found in the Old Aramaic (Cowley, Aramaic Papyri, 212) as well as at Ezr. 7.6, Ps. 45 (44).2 (LXX γραμματεὺς ὀξυγράφος 'an expert scribe' NEB); P. Sm. 2027, Galen 33.v. ὄγε θαυμασιώτατος (ܦܗܝܪܐ) Πάμφιλος[1].

the chosen of the sons of men ... and judge of all the earth The reading בחי̇[ר] seems quite certain, but does the text of Eth. represent the original? Cf. Milik, 263. At 22.14, En^{d} 1 xi 2 דין 'Judge' is used of deity (Milik wrongly 'judgement'). Could the original here have read 'the chosen of the Judge (בחיר די דין) of all the earth'? The Judge of the earth is God (Ps. 94.2 cf. 96.13 = I Chr. 16.33). The reconstructed Aram. text (following Milik, 260) is necessarily conjectural. The epithet 'judge' as applied to Enoch would anticipate the role of the Son of Man at 69.27.

who observe uprightness and peace 'Uprightness and peace' is an unusual combination, and the sentence 'Enoch ... to all my children ... to all ... who observe

---

[1] Milik may be right in placing the frg. noted by C-B, 11, here: did it read 'Enoch, the scribe skilful in his craft and the wisest of men' where the phrase, lit. 'quick in works' is modelled on Prov. 22.29 מהיר במלאכתו (LXX ἄνδρα) ὀξὺν ἐν τοῖς ἔργοις αὐτοῦ? See below, Appendix B, 421.

uprightness and peace' lacks a predicate. Was this supplied by שלום 'peace', i.e. 'greeting', taken into the predicate of the last clause by a translator who failed to recognise its original force? (Eth[tana] reads salām nominative instead of salāma accusative, perhaps a relic of the original reading). If this is correct, we have then in שלם the formal greeting concluding what may originally have been the epistolary prescript (see further below on 93.1f.). An alternative possibility would be to explain the reading 'who observe uprightness and peace' as a scribal expansion of the greeting טב ושלם, 'prosperity and peace'. Cf. 2 Mac. 1.1 εἰρήνην ἀγαθήν and the formula בטב שלם in inscriptions (CIS ii 811, 1305; Hoftijzer, 304) and the similar formula שלם ושרות, Driver, *Aramaic Documents*, III 1, XIII 1.

(2) **Be not grieved** Read בבאשתא: Milik בבהשתא = בבהתא lit. ('do not be) in shame'; but 'grief' not 'shame' is required by the context; cf. 102.5,7 (μὴ λυπεῖσθε). For the verb באש in this sense cf. Dan. 6.14(15) מלכא באש עלוהי, LXX λυπούμενος ὁ βασιλεύς (Theod. ἐλυπήθη); Hoftijzer, 32, Cooke, *North-Semitic Inscriptions*, 190, 191; cf. ܒܐܫܬܐ = infortunium, P. Sm. 440.

times ... things Cf. Ec. 3.1, Job 24.1 (the 'times' and 'days' of the Almighty): 'The day of reckoning is no secret to the Almighty, though those who know him have no hint of its date' (NEB). The idea here is that, though the 'times' are evil, there will be other 'days' (including the 'day of reckoning').

(3) **the righteous shall awake from sleep** See above, 282 for the connection of this verse with 91.17. 'The righteous' is here a collective singular. For the idea of the resurrection of the righteous, cf. 91.10, 100.5 and especially Dan. 12.2; in the Parables 51.1, 61.5. See also 91.17 note, 295.

He shall arise and proceed Eth. ḫalafa I.1 is not the usual equivalent of Aram. הלך. Has the translator read יחלפון for יהלכון? Or could יחלפון be original? In Heb. חלף is used intransitively = 'to change, be revived, renewed' (Job 29.20, Ps. 90,5,6). The verb חלף is well attested in Aram. as in Heb. e.g., 32.2, Eth[e] 1 xxvi 19,21 = διέβην, ἦλθον, and with the meaning of 'change' Hoftijzer, 89; the 'change of death' (חלף מותא) is a phrase from the inscriptions (Cooke, *North-Semitic Inscriptions*, 236).

The more difficult reading is Eth[g] 'wisdom shall awake from sleep, righteousness shall arise and proceed' (yaḫālef). Should we read ṭabib for ṭebab and ṣādeq for ṣedq and translate: 'The wise (חכימין) shall awake from sleep, the righteous (קשיטין) shall arise and be changed (יחלפון) and walk (יהלכון) in the ways of righteousness'? (Cf. 1 C. 15.51).

(4) **he will ... give him ... uprightness** 'uprightness' is a divine gift; cf. 48,1, 58.4.

he shall execute judgement The majority text has lit. 'he will be in goodness and righteousness', an unusual construction: Charles 'he will be (endowed) with ...', Flemming-Radermacher, Knibb, 'he will live in ...'. The reading of Eth[tana t] (noted but not adopted by Martin, 242) makes superior sense. For the idea of the 'righteous' acting as 'judges', cf. Wis. 3.8 and 1 C. 6.2 (οἱ ἅγιοι).

walk in eternal light See note on 1.8.

(5) **sin ... no more ... appear** See above on 91.17; if 92.5 is drawing on this stanza (91.17) from the Apocalypse of Weeks, then this use of this Apocalypse is to be attributed to an original Aram. 'recension'; the Aram. fragments are just sufficient to identify 92.5, which is immediately followed by 93.1. See Milik, 264.

CHAPTER 92, 93.1-2

(93.1) **And after he had given over** Eth[g] 'gave' or 'had given' requires an object, and Milik proposes to read 'his Epistle' (אגרתה or כתב ידה; or read ספריה; cf. above on 92.1). For 'after this' ʾemdeḫraze, he suggests כדי 'And when (he was transmitting his Epistle)'. But, as Dexinger points out (107), the regular equivalent of ʾemdeḫraze is מן בתרה, and he suggests בתר די in this context (so Eth[q]). If we read a pluperfect, we can then render ובתר די יהב הוה ספריה 'And after he had given over his Epistle (i.e., to Methuselah), Enoch took up his discourse and said'.

On such an interpretation of the text, the 'Epistle of Enoch', which the patriarch handed over to his son Methuselah, would appear to have consisted only of 92.2-5, and this conflicts with the traditional understanding of the ʾΕπιστολὴ ʿΕνώχ as comprising Chh. 97-108, the section of Enoch so designated by G[b]. In view of the evidence at 92.1 that we may have an epistolary prescript of the type familiar at 2 Mac. 1.1, it is possible that it was just such a short letter which appeared in the original Aramaic Enoch; 92.1 states that the letter which is to follow was intended not only for Enoch's family but for all generations of mankind on the earth: what follows in vs. 2-5 is just such a universal message. What the extent and full contents of such an 'original epistle' were it is impossible to say, except that they probably were not confined to 92.2-5; they may even have included the Apocalypse of Weeks. If this is the correct explanation, we would then have to conclude that it was from some such an original shorter 'epistle' of Enoch that the Greek ʾΕπιστολὴ ʿΕνώχ took its title for the closing chapters of the book (91-108?).[1]

The translator of Eth. has produced a garbled version, in which he makes Enoch address his family 'from the books', no doubt understood as the 'books' which Enoch wrote, or the 'heavenly tablets' which he read (cf. 33.3f., 40.8, 74.2, 81.1f.); v. 3a is a repetition of the same phrase.

(2) **eternal elect ... uprightness** Eth. ḫeruyāna ʿālam (Eth[ull] laʿālam) is usually rendered 'the elect of the world', an expression otherwise unparalleled in Enoch (so Charles). 'The elect of eternity' could mean either 'the elect from (all) eternity' or 'the elect to (all) eternity', or both, 'the elect from eternity to eternity, from everlasting to everlasting'. Milik conjectures 'the elect ... who have grown up from the plant (... מן נצבת ... [די סלק]ו) of truth and justice': if we read [די על]ו, we can account for the Eth. 'and concerning (baʾenta = על) the plant of righteousness'; for Heb. עלה 'spring up, burgeon' of trees and plants, Isa. 55.13 etc.; cf. Aram. עלה 5.1 En[a] 1 ii 9 τὰ φύλλα; Enastr[d] 1 i 4, 6, idem. Cf. 10.16 G τὸ φυτὸν τῆς δικαιοσύνης καὶ τῆς ἀληθείας. For יצבתא 93.2 and Dan. 7.16 יציבא LXX Theod. ἀκρίβεια, 'truth'.

the watchers and holy ones Perhaps restore קדישי קדישין 'the most holy ones' rather than עירין וקדישין (Milik). Cf. 14.23 (note).

I ... know everything For this idea of the omniscience of the wise man or prophet, and of Enoch in this omniscient role, one given by heaven, see W. C. van Unnik, 'A Greek Characteristic of Prophecy in the Fourth Gospel', in *Text and Interpretation Studies in the New Testament* presented to Matthew Black, ed. by E. Best and R. M. Wilson (C.U.P., 1978), 211f.

tablets(?) of heaven See on 47.3 for a note on this and related expressions.

---

[1] For Aramaic epistolography, see J. A. Fitzmyer, *A Wandering Aramaean*, 183-196, P. S. Alexander, 'Remarks on Aramaic Epistolography in the Persian Period', in JSS, Vol. 23.2 (1978), 155-170.

CHAPTER 93.11-14

A NATURE POEM
(Eth., En^{g} 1 v, Milik, 269f.)

These verses, on natural theology, preserved in a few quite substantial fragments in Aramaic, occur at the end cf. Ch. 93. In Eth. they follow 93.10, the description of the Seventh Week, but since this is followed by the Eighth Week in the original Aramaic (91.10f.), 93.11-14 clearly belong elsewhere in the original, most probably after the Apocalypse of Weeks at the end of Ch. 93. Dillmann thought that they illustrated 91.10, the promise in the Seventh Week to the elect of sevenfold instruction in 'all his Creation'. This last phrase, however, is probably a gloss linking 91.10 with 93.11-14, inserted at this point by the redactor or scribe responsible for the reversing of the two parts of the Apocalypse of Weeks (see note below, 291).

Most commentators regard the verses as more of the nature of an 'interpolated pericope', perhaps excerpted from an independent source, imitating, if not actually alluding to, similar passages in the Old Testament or extra-canonical sapiential literature (cf. especially Isa. 40.12-13, Job. 38.33f., Prov. 30.4, Wis. 7.15-21). Were the verses an original part of the 'Epistle of Enoch' (91-108), composed by the same author as the rest of the 'Epistle', or have they come from an independent mind and source? As with the Apocalypse of Weeks the evidence is not such as can provide a definite answer either way, but once again the unity of subject matter and literary form — the repeated rhetorical questions — create a certain presumption in favour of the 'interpolation' theory. The fact that the verses are found in the manuscripts of the Aramaic does not itself prove that they are the work of the author of the 'Epistle'.

Milik contends that 'the passage constitutes a eulogy of Enoch (delivered by himself!) who had just accomplished things inaccessible to simple mortals, in particular the journeys described in En. 1-36' (270). This understanding of the passage depends on his interpretation of it as referring to the 'ultra-terrestrial journeys of Enoch, in particular to that in the abode of souls' (271. L. 19); the specific reference is to Enoch's ascending into heaven and his vision of a 'disembodied soul or spirit' (v. 12). But the rhetorical question about 'who has ascended into heaven' is imitating Prov. 30.4 and the 'soul or spirit' can be interpreted (along the lines of Dillmann, reading 'his breath or spirit') as referring to purely natural phenomena, in keeping with all the other natural phenomena in the passage. The whole pericope can, in fact, be explained as a poem on natural theology which could have come from a source quite independent of Enoch. At the same time, the very nature of the subject matter, the wonders of God in creation, would render it entirely suitable for inclusion in the Book of Enoch. Whether composed by the author of the 'Epistle' or some earlier unknown poet may be impossible to determine; the verses themselves, in the original now recovered, are not unworthy, as literature, of the Hebrew sapiential tradition to which they belong.

(11) **who ... is able to know what are the ordinances of heaven** The first question is preserved in En^{g} only, and the bracketed words are, for the most part, a conjectural supplement based on a possible allusion to Job 38.33-34 'Knowest thou the ordinances (חקות) of heaven ... Canst thou lift up thy voice to the clouds?'; cf. also Ezr. 7.23 (perhaps the original had 'the ordinances of the God of heaven' par. to 'the voice of the Holy One'?); cf. Milik, 269f. 'who ... can understand the command of God?'.

the voice of the Holy One En^{g} 1 v 16 קלי קדשא, with hirek compaginis (contrast Milik, 269).

(12) **or [show forth] the glory which [they are declaring]** The reading 'brightness,

splendour, glory', is quite certain; the text clearly contained some reference to the 'glory' of the heavenly bodies ('the works of heaven'). The conjectured reconstruction assumes an allusion to Ps. 19.2.

see his breath or spirit The usual rendering '(who can see) a soul or spirit', is taken to refer to Enoch's vision of disembodied 'souls' or 'spirits' during his heavenly journeys (cf. Ch. 22; so Charles, Martin, Milik). In a poem, however, dealing otherwise exclusively with natural phenomena — and reminiscent of such passages as Isa. 40.13, Prov. 30.4 — we would expect some physical manifestation of divine power. Dillmann (followed by Schodde) may be right in equating nafs with נשמה 'breath' and manfas with רוח (Schodde 'his breath and his spirit'). But are these expressions employed in their 'full spiritual sense' (Dillmann) or simply of natural phenomena as at Isa. 30.33, 2 Sam. 22.16 (cf. Ps. 18.15), describing a destroying wind, or as a 'blast of Jahweh' which produces ice (Job 37.10)? Did the original perhaps read נשמת רוחו, 'the blast of his spirit'?

be able to return and tell Cf. Prov. 30.4. There is no need to explain these words as referring to Enoch's extra-terrestrial journeyings.

(13) **all its extent and its shape** Cf. Job 38.5.

(14) **the length of the heavens** Cf. Jer. 31.37, Isa. 40.12, Job 11.8,9.

number of the stars Cf. Ps. 147.4.

come to rest Eth. ʿarafa II.1 = Heb./Aram. נוח, 'settle down', 'have their abode'.

CHAPTERS 93.3-10, 91.11-17

THE APOCALYPSE OF WEEKS

Eth. En^{g} 1 iii 22 (93.3), En^{g} 1 iv 11-13 (93.9-10), En^{g} 1 iv 14-26 (91.11-17a), En^{g} 1 ii 13-17 (91.17b) Milik, 259f.

The 'discourse' of Enoch known as 'The Apocalypse of Weeks' has been the subject of much detailed study and interpretation[1] since it first became known with the publication of the Ethiopic Enoch, in which alone it had survived until the discovery of (substantial) fragments of the original Aramaic at Qumran.[2] The most remarkable feature of the text, as it has been reproduced in the Eth. version, is that, in the course of its transmission, it has suffered a 'severe dislocation'[3] in which the last Three Weeks of the Apocalypse appear first, at 91.11-17, before the first Seven Weeks at 93.1-10. The Aramaic fragments, on the other hand, present a text which gives the Ten Weeks in their correct numerical (and chronological) order.[4] Editors generally transpose the two sections in order to recover the original sequence of verses. This problem is discussed below, 288f.

---

[1] Usually, but not exclusively, by editors and commentators of 1 Enoch. The most recent monograph (which includes rich bibliographical material) is that of F. Dexinger, *Henochs Zehnwochenapokalypse und offene Probleme der Apokalyptikforschung*, in *Studia Post-Biblica*, Vol. 29 (Leiden, 1977). (Dexinger claims to have provided the first literary-critical — including redaction-historical — treatment of this section of 1 Enoch (102). See also M. Black, op. cit. (above, 91.5 note)).

[2] Milik, 263f. En^{g} 1 ii and 1 iii, iv.

[3] Charles, 218. In Charles's view this was not the only 'dislocation' in this part of Enoch.

[4] Milik, 48, 247. See further below, 291.

As it stands in all texts, the Apocalypse of Weeks appears as a literary unit within the last major section of Enoch, and is expressly attributed to Enoch himself (93.3). Earlier editors were all inclined to view it as an original source document used by the author of the book, not only earlier than the work in which it had been incorporated, but possibly even pre-dating the Book of Daniel.[1] A different view is argued by J. T. Milik: 'No serious evidence exists to disprove that the author of this Apocalypse of Weeks is the same author as composed the rest of the Epistle, towards the end of the second century or at the beginning of the first century B.C.' (255f.). There are cogent arguments against this view and in favour of the older theory. (1) The first Seven Weeks are historical, looking back on the history of the world and of Israel; the last Three Weeks are visionary, looking forward to the still unknown future of mankind; and the last historical period, in the Seventh and the beginning of the Eighth Week, like the last of the Dream-Visions (90.19f.), is best interpreted as referring to the Maccabaean victories, but before the reconsecration of the Temple (see below, note on v. 12). Dream-visions and Apocalypse of Weeks belong, in that case, within the period to which the composition of the Book of Daniel is usually assigned. The 'Epistle of Enoch' (as identified by Milik), on the other hand, within which the Apocalypse of Weeks occurs, is a much later composition (Milik himself suggests a date towards the end of the second century or at the beginning of the first century B.C. (256)). (2) Milik has convincingly argued (254f.) that the Apocalypse of Weeks has compressed a traditional cycle of seventy weeks of years (10 Jubilees) which lies behind the chornology of the Dream-Visions. Daniel and other apocalyptic sources, into Ten Weeks of Years, thus giving the history of the world, starting from Adam, in Ten Weeks, or one complete Jubilee, and then sub-dividing the eschatological era into Three Weeks. These calculations may be earlier than Daniel and the period of the Maccabees, but it was at this time that they appear to have become popular and were most widely disseminated.[2] (3) The Enoch paraenesis to his family within which the Apocalypse occurs, draws freely on imagery and phraseology from it (see the notes on 91.11,17; 92.3,5); and this suggests the use by the author of the 'Epistle of Enoch' of the Apocalypse as a basic source document, one not only incorporated in his work, but commented on and expanded in his paraenesis. Incidentally, this use of the original Apocalypse as source material for the paraenesis enables us to fill out several lacunae in the Aramaic text (see especially note on 91.17).[3]

In the Ethiopic version (or recension) the Apocalypse of Weeks has, as it were, been torn in two parts in the middle of the Seventh Week and the two pieces reversed (with intervening text). There are two possible explanations for such a drastic 'rearrangement' of the text: (1) it may have been accidental; a scribe may simply have got his pages out of order. In that case the 'disarrangement' may have taken place either in the Ethiopic version or in its Greek *Vorlage*. (2) Alternatively, the Eth. order may be explained as a recensional rearrangement of the text, now attested in its correct order by the Aram. fragments.[4] There certainly appear to have been 'recensional' expansions, most probably from the hand of the Ethiopic translator (see the note below on v. 11). The most likely explanation is that the rearranging of the verses and the paraphrastic expansions are the work of the translator(s) of the Greek

---

[1] Cf. Hofmann, *Henoch*, 5f.; Volz, *Eschatologie*, 75; Charles, 218, 224; Milik, 248; Dexinger, 102.

[2] Hengel, *Judaism and Hellenism* (London, 1974), 187f.

[3] See also my article (above, 91.5, 281), 465f.

[4] Dexinger, 103f.

or Ethiopic versions. So far as the Aram. is concerned, we can, no doubt, speak of two recensions: (1) an independent Apocalypse of Weeks, followed by (2) the Aram. text as now attested, incorporating the Apocalypse of Weeks into the Book of Enoch.[1]

(93.3) **took up his discourse** See on 1.2.

Enoch is not only the seer and speaker; he dates the First Week as the Week in which he was born. If the Apocalypse of Weeks began its existence as an independent work, it must still be regarded as part of the Enoch saga.

justice was delayed Eth. taʿaggaša. Milik reconstructs קשטא כבר הוה מתקים 'justice was still enduring', a meaning for ʿagaša III.2 rejected by Dillmann who renders 'while judgement and righteousness were still delayed (verzogen)' (Flemming-Radermacher 'still held back' (noch zurückhielten)). The rendering 'was still enduring' ('still lasted' Knibb) is open to the further objection that Enoch was born, in that case, before the demoralisation of mankind, whereas, according to En. 6, mankind's degeneration began with the watchers in the days of Jared, Enoch's father. The Coptic fragment (Milik, 81) reads (in Garitte's Latin version) justum opus (= ?קושטא) permansit compositum, where permansit (ⲁϥϭⲱ) corresponds to Eth. taʿaggaša (ἐπέχειν?). An equivalent Aram. is כתר, Pa., Syr. (ܟܬܪ), mansit, tardavit (P. Sm. 1859). 'Justice' seems preferable in this context to 'righteousness' for קושטא, since the reference is to the coming 'judgement' of the Deluge; hence Eth.'s 'judgement and justice'. Does ⲉϥⲥⲙⲟⲛ̄ⲧ, compositum = συνθέμενον and ἐπέσχεν συνθέμενον, כתר מתקימה 'waited to be established', referring to the coming 'covenant (קימה) of Noah'? Cf. v. 4.

(4) **And thereafter** All Eth. mss read 'after me', i.e. after Enoch, a reading which has given rise to difficulties, since 'falsehood and violence' presumably had 'sprung up' since the time of Jared and the fall of the angels (see Dexinger, 122). It seems likely, however, that the reading 'after me' is a corruption: each of the Ten Weeks begins with the same temporal conjunction, e.g., 91.12, En^{g} 1 iv 15 ומן בתרה 'and thereafter'; its eleventh occurrence is to introduce the 'Weeks without end' of eternity (91.17, En^{g} 1 v 25-26).

the former End This translation is to be preferred to 'the first End', since both 'the former End' (the deluge) and 'the latter End' (the Last Judgement) are clearly in the author's mind. 'Former' ((קדמי) has also the sense of 'ancient', i.e. that which happened in the days of the 'ancient ones' (Aram. קדמיא = οἱ ἀρχαῖοι). For the use of End (feṣṣāmē = סוף) in an eschatological sense, see the note on 10.3. Cf. Gen. 8; Mt. 24.37; 2 Pet. 3.6-7.

And after it is ended ... shall increase Charles notes that 'the time order in the close of this sentence is not observed'. In other words, mention of a growth of wickedness seems out of place immediately after the deluge and before the covenant and commandments of Noah. Dillmann felt that the last line, which he read as 'he (Noah) will make a law for sinners', should have preceded 'after it is ended', but he accepts the order as it stands as not unsuitable since this law was given in anticipation of the sinfulness of mankind (Gen. 8.21); for the same reason it is named 'a law for sinners'. For an alternative explanation, see next note.

he (Noah) shall make a law for sinners Dexinger (112, n. 11) prefers the passive 'a law shall be made' on the grounds that the tendency in apocalyptic is to favour the impersonal construction (cf. v. 6). But the attestation of the passive here is weak.

It is usually assumed that the reference is to the covenant with Noah (Gen. 8.20f.)

---

[1] Cf. Dexinger, 108f.

and the Noachic commandments (Gen. 9.4), and this would be supported by the possible reference to a 'covenant' at v. 3. The author may have in mind, however, the same scheme as at Jub. 6.8, 7.20, according to which, in the Twenty-Eighth Jubilee, Noah began 'to enjoin upon his sons' sons the ordinances and the commandments', since they had sinned in the post-deluge period. This would account for the specific mention of a 'law for sinners' after the deluge. See Dexinger, 124.

(5) **A man shall be chosen** Abraham and his seed are to be chosen as the race through which God would bring 'truth and righteousness' into the world; cf. 10.16 (note).

a plant of righteous judgement Cf. 93.2 and En^{g} 1 ii 19-20 'plant of truth and righteousness'.

his posterity(?) Eth. 'and after him' = ומן בתרה is stylistically clumsy following the opening formula 'And thereafter' ומן בתרה. Charles suggested a misreading of Heb. אחריו as אחריתו. The same mistake is possible in Aram. (e.g. אחרתה = 'posterity', Cooke, *North-Semitic Inscriptions*, 65, 190). For אחר prep. 'after' and אחרה 'posterity', Hoftijzer, 10.

shall come forth as a plant of eternal righteousness Eth I 'shall come forth', yewaḍḍeʾ = יצא/יפוק (Eth II 'shall come' yemaṣṣeʾ). For 'plant of eternal righteousness', 93.10, En^{g} 1 iv 13-14 [נצבת] קשט עלמא.

(6) **a vision of holy ones and of righteousness** Dexinger (following Knibb) prefers 'visions of the holy and righteous ones' as the more difficult reading (113). The reading of Eth^{g}, however, is more likely to be original: the 'holy ones' are not the elect, the 'holy and righteous ones', but the angels in the Sinai theophany (Dt. 33.2) and the 'righteousness' refers to the giving of the law to Moses on Mount Sinai. The idea that the law was given by angels on Sinai is a later rabbinical development; see Str.-B., Vol. III, 554f. and cf. Gal. 3.19, Heb. 2.2, Ac. 7.53(38), Josephus, Ant., xv. 5.3 (136).

a law for generations upon generations The law given on Sinai which is for ever; cf. 99.2.

the court (of the Tabernacle) Eth. ʿaṣad lit. 'enclosure' = דרתה, Heb. חצר, referring to Exod. 27.9 חצר המשכן Tg. דרת משכנא, LXX (καὶ ποιήσεις) αὐλὴν τῇ σκηνῇ. Has the text lost [τῆς σκηνῆς] in transmission? The Coptic fragment (Milik, 82) interprets of the Tabernacle (ἡ σκηνή); Dillmann refers to the 'enclosure of the national life by the law'.

(7) **the House of glory and dominion** (מלכותה) The reference is usually taken to be to Solomon's Temple, but possibly including Mount Sion, city of David (Dillmann). There could also be an allusion to the House of David and his perpetual sovereignty —2 Sam. 7.16, 'Thy house and thy kingdom (ממלכתך) shall be established forever'. Cf. Dexinger, 129.

(8) **will become blind** The figurative use of חשך, σκοτίζεσθαι and its cognates, the 'darkening' of the mind, to describe any state of mental, moral or spiritual insensitivity, obtuseness or just plain ignorance, is common to most languages: it occurs, with the emphasis on spiritual blindness, at Test. Reub. 3.8, Levi 14.4, Gad 6.2 (+ τὸν νοῦν) and in the New Testament at Rom. 1.21, Eph. 4.18. (For Aram. Test. Levi^{a} 8 iii (14.4) see Milik, 23).

godlessly forsake wisdom Eth. lit. 'the hearts of all of them will become impious (yetrassāʿ) away from wisdom'. Eth. rasʿa is here given a Heb./Aram. meaning; normally it means 'to be forgotten' (see Dillmann, *Lex.*, 281 and Flemming-Radermacher, 125); cf., however, Charles^{1906}, 195 n. 25, citing Ps. 18.22 as a parallel. The original perhaps reads: ולבהון כולהון ירשע מן חכמה 'and the hearts of all of them will become impious and forsake wisdom'. The reference is to the period of the divided kingdom and of its sequel, growing degeneracy and darkness. Cf. 99.8.

a man shall ascend As at 89.52 Elijah, like Enoch himself, is to ascend miraculously to heaven; cf. 2 Kg. 2.11 and note on 89.52.

House ... burnt with fire The destruction of Jerusalem and the Temple in the Babylonian invasion of 586 B.C., with the subsequent dispersion or exile of the people (2 Kg. 25.9-11). The Coptic fragment reads 'Temple' for 'House of dominion', a reading preferred by Milik (81.n.2).

the whole people (and) the captains of the host Eth II offers a simple and straightforward text, 'the whole race of the chosen root' (Knibb). Martin and Flemming-Radermacher choose the better attested and more difficult text of Eth I 'the whole race of the powerful root', which is, to say the least, an unparalleled description of Israel; it is not surprising that translators prefer the simpler text. Eth I is a corruption of sarwē/sarāwita ḫayl 'captains of the host', and the reference is to the dispersion into Egypt of the 'whole people' of Israel (for zamad in this sense Jos. 4.14) along with their captains after the fall of Jerusalem to the Babylonians, described at 2 Kg. 25.26, Jer. 41.16f. (sarāwita ḫayl = שרי החילים Tg. רבני חילותא ἡγεμόνες τῆς δυνάμεως.)

(9) **the Seventh Week** This Week appears to extend from the Captivity to the time of the author. It is an apostate period on which the same judgement is passed as at 89.73-75. The apostasy has been variously identified: with the influence on Israel of foreign cults during the Exile (Hofmann), or with the 'mixed marriages' condemned by Ezra (Martin). Those who think of apostasy in the writer's own time identify it with the general state of irreligion during the period of the Second Temple. See Dexinger, 132f.; J. P. Thorndike compares CD 1.3-8 in 'The Apocalypse of Weeks and the Qumran Sect', Revue de Qumran, Vol. 3 (1961/2), 171.

a perverse generation For the expression cf. Dt. 32.20 דור תהפכת (LXX γενεὰ ἐξεστραμμένη), Tg. (P-J) דר דהפכנין, Syr^{vg} ܕܪܐ ܗܦܟܬܐ; cf. Mt. 17.17, Lk. 9.41, etc.

(10) **the elect shall be chosen** As at 10.16, 93.5 the 'eternal plant of righteousness' is the national (or ethnic) Israel from whom 'the elect', a 'remnant' or corps d'élite, is to be selected. According to En^{g} 1 iv 12 its role is to be that attributed to the 'servant of the Lord' in Second Isaiah, viz., to be 'true witnesses' (i.e., of the Lord); cf. Isa. 43.10,12 and 1 QS 8.6 (עדי אמת); see Black, *Scrolls and Christian Origins*, 128f. The group has been identified with the Hasidim of the Maccabaean period (Schubert), even with the Qumran sect itself. (Geoltrain); see Dexinger, 133, n. 69.

eternal plant of righteousness See above at 10.16.

concerning all his creation There is no room for these words at En^{g} 1 v 13. They are probably a gloss, preparing the way for Enoch's reflections on natural theology at vs. 11-14. See above, 286.

(91.11) Ch. 93.9-10 (En^{g} 1 iv 11-13) describing the Seventh Week (in substantial fragments for v. 10) is followed immediately by 91.11-17 (En^{g} 1 iv 14-26) concluding the Seventh Week (91.11) and going on the describe Weeks Eight to Ten (91.12-17); 91.17 is completed at En^{g} 1 ii 13-17 (cf. Milik, 260 and below, 294f.), which is then followed directly by 91.18,19 (further paraenesis by Enoch) and 92.1 (Enoch's 'epistle' handed over to Methuselah).

This analysis[1] makes it quite certain that the Apocalypse of Weeks in En^{g}

---

[1] Cf. Milik, 48, 265.

proceeded in numerical (or chronological) sequence, Week Seven (Eth. 93.10) followed by Weeks Eight to Ten (Eth. 91.11-17). Unfortunately Weeks One to Six are not represented in the fragments, but there can be little doubt that they followed each other in the same numerical sequence.

Whatever explanation for the major textual dislocation in Eth. is given, how are we to account for its longer text in v. 11? Has it been 'elaborated within the Ethiopic version because of the dislocation of 91.11-1,7 in order to smooth over the harsh juxtaposition of 91.1-10 and 91.11-17' (Knibb)? If so, then it is no different from other verses containing translators' expansion, except that, in this case, the versional additions seem more numerous, although some of them are also shared with vs. 7 and 12. The translator into Greek or Eth. appears to have understood the first verb, 'they will uproot', lit. 'cut off', as an impersonal plural and rendered by a passive. Did he read עבד שקרא (עולה) (see Milik, 266) as עבדי שקרא (עולא), lit. 'the doers of falsehood (iniquity)' and interpret '(they (impersonal) are to cut off the blasphemers', rendered as a passive, 'from blasphemers they will be cut off'? In other respects the verse seems to be nothing more than a free expansion, based on the verse as it has been preserved in its original form in En^{g} and drawing largely on motifs and phraseology from other verses. Thus the 'cutting off' of the 'roots of unrighteousness' is to consist of the destruction of sinners by the sword (an idea borrowed from v. 12), and the destruction of blasphemers (note the clumsy repetition 'shall be cut off') may be no more than an explication of the 'cutting off' of the 'structure of falsehood' (or of the 'works of deceit'). Whether any part of this expanded version could represent a longer Aram. recension than the one preserved must be doubtful. It is perhaps worth noting, however, that two Biblical expressions occur in this material, namely, 'to plan wrong-doing' (Mic. 2.3, Nah. 1.11) and 'to commit blasphemy' (Neh. 9.18,26) both possible in Aramaic.

foundations of oppression En^{g} 1 iv 13 אשי חמסא, cf. Ezr. 4.12 אשיא (LXX θεμελίους).

the structure of falsehood So Milik (עבד שקרא), continuing the figure of a building; Knibb takes as a coll. sing. 'the work(s) of deceit'; cf. 93.4 and 1QpHab. 10.12 מעשי שקר. The reference may be to the heresies of the enemies of the elect.

to destroy it utterly(?) Milik's reconstruction 'to execute judgement' is stylistically clumsy since the same phrase occurs again in the next verse. The expression עשה כלה 'to make a complete end of', 'to destroy utterly', is familiar in the prophetic literature (e.g., Ezek. 20.17). The noun is not attested in Aram. but the verb כלא = Heb. כלה is found (107.1, En^{c} 5 ii 28); the noun is frequently used in Qumran Heb. (Kuhn, *Konkordanz*, 103) and later Heb. has כילוי 'destruction' (Levy, NHCW, II, 334). This verse is alluded to at 91.5, '... it (oppression) shall be uprooted from its foundations and its whole structure will disappear'.

(12) **a sword ... to all the righteous** The Eighth Week marks the transition from history to 'transcendental eschatology'. Older commentators, while not denying further references to historical events, tended to describe this Week in general terms (and virtually identical language) as 'the beginning of the messianic age' and the first act of the last judgement; Charles can speak of 'the setting up of the messianic kingdom'.[1] This idea receives even stronger support from the next verse, especially if the promised 'royal House' 'for all generations forever' includes the notion of the Davidic Promise of 2 Sam. 7. Such theological considerations, however, do not

---

[1] *Apocrypha and Pseudepigrapha*, II, 264.

prevent commentators from finding a historical reference in 'the sword' to be 'given' to all the righteous; and there is a large measure of agreement that it refers to events in the Maccabaean period, more specifically to the struggle for freedom under the Maccabees. As Dexinger has noted (138),[2] the same terminology is employed as at 1 Mac. 3.3 (cf. 3.12), where Judas 'protects the camp with a sword (ἐν ῥομφαίᾳ: cf. also 90.19 in the Dream-Visions where the same events are almost certainly being described. The 'wicked' in this verse no doubt include the 'renegades' in Israel (cf. 1 Mac. 2.44 ἄνδρες ἄνομοι). Cf. also Isa. 34.5 and for similar passages, 94.7, 95.7, 96.1, 98.12; and in the Parables 38.5.

(13) **possessions righteously** Eth. 'abyāt = οἰκίας(?) 'household goods'. See Liddell and Scott, s.v. οἶκος II. Attic law distinguished between οἶκος 'property' and οἰκία 'house', but both meanings are found in semitic בית. (At. Jg. 2.6 LXX οἶκος is used to translate נחלת 'heritable property'). Cf. Mk. 12.40 κατεσθίοντες τὰς οἰκίας τῶν χηρῶν, and especially Mk. 10.29-31, Mt. 19.29 the rewards of 'houses', i.e. 'possessions' and 'eternal life'. Our verse is the closest known parallel to this idea of a reward in property for the righteous. Could this verse lie behind the Gospel passage? For בקשוט, 'righteously' cf. 10.18, ἐν δικαιοσύνῃ

royal(?) House of the Great One For 'the Great One' see above note on 14.2. The vision of the writer goes far beyond any mere purification and rededication of the Second Temple such as was undertaken by Judas in 165 B.C. He is clearly thinking in terms of Ezek. 40-48, of the rise of a new Temple in a new Jerusalem; it is not only to be a Temple or Palace of surpassing splendour, but also one that will last 'forever and ever'. Was the situation of the writer similar to that at 90.28 where we are told that the old Temple is to be removed and replaced by an entirely new edifice (the old Temple in this case also being the Second Temple of Ezra-Nehemiah and not the Temple of Solomon)? For the expression, but applied to a heavenly Temple, cf. IQ**28** Col. IV., lines 25-26, DJD, Vol. I, 126f., 134f. היכל מלכות; cf. also Test. Levi 18.6. See further M. Baillet, 'Fragments araméens de Qumran 2: Description de la Jérusalem Nouvelle', in RB, Tom. 62 (1965), 222-245, R. J. McKelvey, *The New Temple: the Church in the New Testament*, (Oxford, 1969), 30f.

Did the writers of both apocalypses, the Dream-Visions and the Apocalypse of Weeks, compose their visions during the Maccabaean struggles, but before Judas recaptured Jerusalem and rededicated the Temple? If so, their intention may well have been to encourage and inspire their contemporaries to resist their Seleucid oppressors by reviving the vision of Ezekiel of a New Temple in a New Jerusalem: they were no doubt convinced that the desecrated Second Temple had to be removed and replaced in such a visionary New Jerusalem. Could the thought of a new House of God to last forever perhaps be intended also to recall the divine Promise made to David at 2 Sam. 7?

(14) **a righteous judgement ... of the whole earth** Eth. freely 'for the whole world'. 'Righteous judgement', in the Sixth and Seventh Weeks, has been executed on the enemies of Israel, both external and internal ('the sinners'); now justice, in the Ninth Week will be available for Gentiles as well as for Jews, all over the world. The 'righteous judgement' exacted by God is universal in its range. Note how this universalism is emphasized: the righteous judgement is to be for 'all the children of the whole earth'; the total destruction of the wicked is similarly stressed; they are to

[2] Cf. also H. H. Rowley, *The Relevance of Apocalyptic*, (London, 1963), 98f.

vanish 'from the entire earth', lit., 'from all of the whole earth'. The idea that this 'righteous judgement' is simply an extension of the judgement of sinners in v. 12 (Dillmann) is to be rejected. The judgement proceeds in stages: (1) a judgement is given for Israel (or the elect Israel) against her foes, external and internal; (2) this is followed by a universal judgement of mankind; and (3) the last stage is the judgement of the angels and the condemnation of the watchers or fallen angels. (See below on v. 15). There is a similar three-stage judgement depicted at 90.24f.: first the 'stars', i.e., the fallen watchers are condemned (in this scheme first not last); secondly, the Seventy Shepherds are condemned; the third stage is the condemnation of the 'blinded sheep'. (90.33f. implies the conversion of the Gentiles, but there is no suggestion of this in the Apocalypse of Weeks.)

work(er)s of iniquity ... eternal pit A personal subject is required by the Peil ירמון[1] 'will be cast'; עבדי רשעה 'evil-doers', lit. 'workers of iniquity', rather than 'works of iniquity (impiety)' = ἔργα τῆς ἀσεβείας, עובדי רשעה (so Milik). The erudite explanation offered by Milik of Eth's 'will be written down' (269) is unconvincing. A misreading of ῥιφήσεται as γραφήσεται (with the previously mistranslated ἔργα as subject) seems a more likely source of the impossible Eth. text. Heb. בור is a familiar term for the pit of the grave (LXX ᾅδης); here it is the Pit of Gehenna, the place of the final punishment and destruction of the wicked; cf. Tg. Isa. 14.15,19, 38.18) גוב בית אבדנא.

all men shall look to the true, eternal path A further extension of the writer's universalism.

(15) **the seventh part** The author is following the traditional schema of the 'Seventy Weeks' (of years)' (7 × 70 = 490) or Ten Jubilees (10 × 49 = 490) (Milik, 254). The 'seventh part' of the 'Tenth Jubilee' is, therefore, the final 'Week (of years)'.

an everlasting judgement and the (decreed) time of the great judgement On the Enoch doctrine of the last judgement, see above on 22.4, and for the eschatological concept here of the '(decreed) time' (קץ), see F. Nötscher, *Zur theologischen Terminologie der Qumran-texte*, 167f., IQS 4.20, IQH 6.29.

Will be exacted from all the watchers of heaven One reading of Eth^{M} seems closer to the original (ποιηθήσεται = יתעבד cf. 22.4 En^{e} 1 xxii 3). Eth^{g} provides a shorter version of the clause and one closer to the Aramaic; Eth^{M} looks like a prosaic expansion.

(16) **the first heaven** Cf. Isa. 65.17, 66.22, Ps. 102.25f., 2 Pet. 3.10-13, Rev. 21.1. Is Rev. 21.1 alluding to this verse? The idea of a 'new creation' (and the actual expression) occurs at 72.2 in the astronomical section, and at 1 QS 4.25 עשות חדשה lit. 'the making of the new'; cf. also 45.4, Jub. 1.29. See further G. Dalman, *The Words of Jesus*, (Edinburgh, 1902), 178 and M. Black, 'The New Creation in I Enoch' in *Creation, Christ and Culture: Studies in Honour of T. F. Torrance*, 17.

seven-fold Cf. Isa. 30.26: '... the light of the sun shall be sevenfold ...'; Isa. 60.19,20.

(17a) **shall be no more seen**(?) Eth. 'will no more be mentioned' makes poor sense. The correct reading has been preserved in the verse based on this stanza of the Apocalypse of Weeks at 92.5 'sin ... shall no more be seen'. The suggestion of a misreading of φανήσεται as φωνήσεται would account for the corruption: Aram. חטיתה לא עד תתחזי.

(17b(?) **And the righteous ... for ever** The text of the first half (17a) of this concluding stanza at En^{g} 1 iv 25-26 can be fairly confidently reconstructed with the help of Eth. from the extant fragments. The second half at En^{g} 1 ii 13-17 (see the analysis at

[1] Or Pael pass. Above, 263.

91.11, above 291f.) is more difficult to decipher, since not more than a small fraction of the five lines of text has survived, and, apparently, without anything corresponding in Eth. Milik notes, however, that 'Ll. 13-17 (beginning) seem to correspond to En. 91.10' (261), and proposes to insert 91.10 into these lines (260). 91.10 (beginning) is in fact practically identical with 92.3 (beginning):

91.10a And the righteous shall arise from their sleep
92.3a And the righteous (coll. sing.) shall arise from their sleep.

The second line of 91.10 'And wisdom shall arise and be given to them', which Milik proposes also to read in En[g] 1 ii 13-14, appears to echo 93.10 (the gift of wisdom to the elect in the Seventh Week); 92.3 has, as its second line 'he (the righteous) shall rise and walk in the paths of righteousness'. Instead of assuming that 91.10 originally stood at this point in the text (and presumably was subsequently displaced to its present position in Eth.), an alternative possibility is that 91.10/92.3—a doublet at least for the first line—are drawing their principal motifs (and their expressions) from this concluding stanza of the Apocalypse of Weeks, i.e. from 91.17, En[g] 1 ii 13-17. I suggest that these verses supply the clues for the deciphering of the fragmented text at En[g] 1 ii 13-14 which would then reads:

[וקשיטי]א̊ [יתעירון מן שנתהון יקומו]ן ויהלכ̊[ון באורחות קושטא]
'And the righteous shall awake from their sleep,
They shall arise and walk in the paths of righteousness'.

There remains the problem of En[g] 1 ii 15-17 where there are two parallel verbs of which ותנוח אר[עא] is certain and תשב[ת] virtually so, both with the meaning of 'rest' or 'cessation'. Again the clue comes from the paraenesis at 91.5 'and all unrighteousness shall cease' which could supply the missing portion of our text [ועו]לה תשב[ת כולה] 'And unrighteousness shall altogether cease': as a parallel perhaps we should read ותנוח אר[עא מן חמסה 'And the earth will be at rest from oppression' (Milik 'from the sword').

the righteous shall awake Commentators since Dillmann have understood the reference here to be to the resurrection of the righteous; and this would certainly make a fitting conclusion to the vision of the new heaven and the perfected state of mankind ('they shall practise goodness and righteousness'); all that is needed to complete this idealistic picture of mankind's perfection is the resurrection of the righteous who will then continue 'to walk in the paths of righteousness'. The closest parallel is En. 100.5: '... though the righteous sleep a long sleep, they have nought to fear', and 103.4, 'the spirits ... who have died in righteousness shall live and rejoice ...' (51.1,2 in the Parables may represent much later traditions). The closest canonical parallel is Dan. 12.2, where we find the same figure of speech as in our verse, '... many of those who sleep in the dust shall awake ...'.

CHAPTER 94

(1) **worthy of acceptance** Cf. ܡܩܒܠܐ = εὐπρόσδεκτος (P. Sm. 3469), 1 Tim. 1.15. For the two paths, cf. 91.18 note.

shall ... be destroyed and fail Eth. ḥaṣaṣa = Heb./Aram. כלה(א), cf. Job 4.9(?). (Milik, 270, חסר). Cf. 107.1, En[c] 5 ii 28, Milik, 210.

(2) **to illustrious men** Eth. ʾemurān = ידיעין (P. Sm. 1555). The reference is to the future leaders of Israel, Moses, Aaron and the prophets.

a generation to come For Eth. tewled = דר meaning 'a future generation', cf.

Ps. 22.30, LXX 21.31, γενεὰ ἡ ἐρχομένη. Was the original here perhaps לדר אחרן 'for a later generation': cf. Ps. 102.18, LXX 101.19 εἰς γενεὰν ἑτέραν: Eth. fenāwāt 'ways' = Aram. אורחתא follows in this verse, so that אחרן could have fallen out by scribal error. (For אחרן in Aram., Hoftijzer, 11).

paths of ... death Dt. 30.15, Prov. 14.12, 16.25, Jer. 21.8.

(3) **draw not nigh to them** Cf. 91.4, 104.6.

(4) **a life of goodness** Eth^tana^ = ἀγαθωσύνη = טבותא(?) seems more appropriate in the context.

(5) **hold fast ... hearts** Eth^M^ reads lit. 'Hold fast (te'ḫezu) in the thoughts of your heart and let not my words be effaced from your heart'. The text is usually construed as if 'my words' was the object of the first verb, e.g., 'And hold my words firmly in the thoughts of your heart, and let (them) not be erased from your heart' (Knibb, cf. Charles). This is, however, to force the syntax in the two clauses. Eth. ḫellinnā lebb = ἡ μελέτη τῆς καρδίας (LXX Ps. 19.14), but 'Hold fast to the meditation of your heart' cannot be right. Eth^g^ supplies the full predicate to the main verb (omitted accidentally by the scribe of Eth^g^ by hmt), but should be emended to read 'Hold fast to my precepts (par. to 'my words') in the meditation of your hearts'.

to entreat wisdom evilly Eth. lit. 'to do wisdom evil' probably = κακοποιεῖν (Charles), Aram. אביש 'to harm'. Could this be abstract for concrete, i.e., 'do harm to the wise'? Dillmann, 305f., thinks the reference could be to the dangers from hellenism; Martin identifies 'the sinners' here with the Sadducees and the hellenisers.

no place will be found Cf. 42.1.

(6) **unrighteousness and wrong-doing** See above, note on 91.11. Here 'wrong-doing' for gefʿ = חמסה seems more appropriate than 'violence' (above at v. 2).

have no peace a recurring formula; cf. 5.4, note, 113, and below, 306.

(7) **build their houses with sin** Cf. Jer. 22.13.

those who acquire gold and silver Is this line an interpolation from 97.8 (Charles)?

(8) **you will be parted** Eth. waḍ'a cedere possessione (Dillmann, *Lex.*, 944 = Lk. 14.33), but cf. ܢܦܩ ܡܢ P. Sm. 2421 = 'to lose'.

Most High Above, 9.3 note.

(9) **commited blasphemy** See note on 91.11 for this expression.

day of slaughter Eth. keʿwata dam lit. 'the shedding of blood' = Aram. שפיכת דם (Tg. (P-J) Gen. 28.20).

(10) **He who hath created you** For similar sentiments cf. Ps. 2.4, 37.13, and En. 89,58, 97.2: but contrast Ezek. 18.23, 33.11. (The idea is scarcely 'anti-biblical' (widerbiblisch, Dillmann, 306). Cf. also Lk. 15.7,10, 2 Pet. 3.9.

will overthrow you Eth. gafte'a = הפך (Gen. 19.25, Heb. and Tg. Am. 4.11, LXX καταστρέφειν, of personal subjects).

(11) Charles considers this a defective stanza.

your righteous ones i.e., the descendants of Methuselah and his family. The variant 'your (sing.) righteous ones' would refer to Methuselah alone.

CHAPTER 95

(1) **a cloud of waters**(?) An odd expression, even if it is not altogether impossible (cf. Charles, 236 note). The verse is clearly reminiscent of Jer. 9.1 (LXX 8.23), 'Would that my head were all water, my eyes a fountain of tears ...' (NEB). LXX τίς δώσει

κεφαλῇ μου ὕδωρ καὶ ὀφθαλμοῖς μου πηγὴν δακρύων. Another close parallel is 2 Bar. 35.2: 'Become ye springs, O mine eyes, and ye, mine eyelids, a fount of tears' (Charles). Although the text is corrupt at a number of points, the allusion to the verse in Jeremiah is unmistakable; and it seems likely that it was Jeremiah's imagery which was original and adopted, perhaps also adapted, by the author of Enoch.

Charles suggested a mistranslation of πηγή (עינא)[1] by νεφέλη (עננא). Did the original perhaps read (cf. Jeremiah) עין דמעין 'a fountain of tears', misread as νεφέλη (ענן) ὕδατος (דמין)?: מי יתין עיני להון לעין דמעין 'Oh that my eyes were a fountain of tears'. Eth[g] appears to have read (cf. again Jeremiah) lit. 'who will give water to my eyes': τίς δώσει μοι ὕδωρ τοῖς ὀφθαλμὸῖς μου. Is this the remains of a longer text τίς δώσει κεφαλῇ μου ὕδωρ καὶ τοῖς ὀφθαλμοῖς μου νεφέλην ὕδατος (πηγὴν δακρύων)?

trouble in my heart Eth. ḥazana lebbeya = עצבת לב Prov. 15.13 (Tg. כיבא דלבא) ('sorrow of heart' RSV).

(2) **hatred and wickedness** Eth. gabra ṣel'a wa'ekaya: 'to practise wickedness' = Heb. עשה רע, but 'to do hatred' seems an unusual expression. I have, accordingly, adopted the reading of En[g1] lit. 'Who will give you hatred and wickedness ...'.

(3) **again ... will deliver them** Is the writer thinking of the deliverance of the Exodus (Martin) or of some recent Maccabaean victory (Dillmann, Charles), or in more general terms: the Lord has always delivered the righteous.

that you may execute judgement upon them Cf. 91.12.

(4a) **pronounce anathemas ... loosed** The reference seems to be to the practice of 'anathematising' or 'cursing', probably to the accompaniment of magical formulae, incantations and spells. The 'woe' is on 'sinners' who issue such anathemas with the intention of 'loosing' them, i.e., 'undoing' the spell. Because of their sins, however, there will be no 'remedies' (φάρμακον = כשף?) available to them to undo their curse or spell (Eth. reḥuq = Heb./Aram. רחיק/רחוק 'to be lacking' (Ec. 7.23)). The text of Eth II reads 'that you do not loose': but this makes poor sense, since the purpose of such 'banns' or 'spells' was that they should be binding and so effective. For the expression 'to bind (by imprecations)', אחרם En. 6.5, En[a] 1 iii 3, Milik, 150. Was Heb./Aram. פתח (Pi. Pa), like Eth. fatḥa, used of 'loosing' spells, and does εφφαθα Mk. 7.34 mean 'be loosed' as well as 'be opened'?

(4b) **because of your sin** Eth[tana] adds lit. 'his word where (ẖaba) it has been rejected' which makes little sense as it stands. I suggest reading ẖabr(a) = 'the words of (your) spell-binding have been rejected(?)'. For ẖabr = חברו, En. 65.6 and 8,3 En[a] 1 iv 1, Milik, 157. We could then read as a separate line: 'Because of your sins the words of (your) spell-binding ((your) incantations) have been (will be, proph. perf.?) rejected'.

(5) Cf. Prov. 17.13, 20.22, 24.29, Rom. 12.17.

required according to your works Cf. 100.7.

(6) **weigh out injustice** This seems an odd expression. Perhaps we should read with Eth[m] = lit. 'those who follow, favour injustice'; cf. Dillmann, *Lex.*, 1079 and Sir. 19.26, zayādallu la'ekuy = πονηρευόμενος.

(7) **delivered up ⌜and persecuted⌝ unjustly** The bracketed words look like a ditt. and should probably be omitted. Charles rendered ba- or 'em'ammaḍā by 'because of injustice', but the phrase means lit. 'with or by injustice', i.e., 'unjustly' ('vom Unrecht', Flemming-Radermacher). According to the usual understanding of this text the sinners are to be 'delivered up' 'unjustly', hitherto the fate of the righteous. This

---

[1] Cf. EnGiants[b] 1 ii 9, Milik, 304.

seems most unlikely to have been original; and Eth II, 'you will be delivered up (you) who are (men) of injustice', looks like an attempt to get rid of the offensive idea of 'injustice' being meted out even to sinners. This may also be why En[g] changed the verbs from passive to active, 'you will deliver them up and persecute (them, i.e., the righteous)' the more usual role of the wicked, but we are left then with the contradiction of the next line that 'their yoke' (i.e., the yoke of the righteous or 'its yoke' the yoke of injustice) 'will lie heavily on you (the sinners)'. The interpretation suggested in the translation assumes a rapid change of person: the subject of the last clause 'antemu is emphatic, 'as for you', and refers to the 'righteous' as antecedent. Alternatively, the text may be considered to be seriously at fault, and a stanza referring to the 'righteous' misplaced along with a line referring to the 'sinners'. I would suggest that the 'Woes' on sinners ends at 7a, 'Woe to you, sinners, for you persecute the righteous', giving a climactic single line instead of a full stanza. The words beginning 'For (Eth[v] 'And) as for you ...' mark the transition to the next section on the righteous (cf. the opening of 96.3 'And as for you, indeed ...').

(95.7b) For (And) as for you (the righteous) you will be delivered up and persecuted unjustly;

And their yoke will lie heavily upon you.

CHAPTER 96

(2) Cf. Isa. 2.19,21, 40.3, Jer. 49.16, from which the text is drawing its expressions: here, however, the 'crevices of the earth' and the 'clefts of the rock' are places of refuge and safety. Could the reference be to the early stages of the Maccabaean struggles when the 'lawless' (οἱ ἄνομοι = ʿammāḍeyān, רשיעין) hunted down the bands of loyal Jews under Judas who found safety in the caves and valleys of the Judaean wilderness?

you shall ... enter ... the clefts if the rock, all times The adv. laʿālam, εἰς τὸν αἰῶνα, לעלמה presents difficulties in this context. Could it be a corruption of עולמין νεανίσκοι (cf. Isa. 40.30 Tg.) parallel to 'eguāl (coll.) παιδία, ילודין? Eth[g] has two variants hitherto unaccounted for: yetnassatu, [καθαιρεθήσονται](?) which makes no sense, and yetbawwe'u '(they) shall be brought in' a variant of 'you shall ... enter'. The first could be a mistranslation of יתנצחון 'shall triumph' misread as יתנסחון καθαιρεθήσονται. If the subject was an original עולמין, we obtain a parallel line to 'Your young ones shall mount and rise as eagles', viz. 'And (your) youths shall triumph, and make their way into (ויתנצחון עולמין ויעלון) the clefts of the rock, like rock-badgers ...'.

desert-owls Eth. ṣēdanāt = LXX σειρῆνες = בנות יענה(?) Job 30.29, really the (wild) desert ostrich noted for its plaintive cry (NEB 'desert-owl'). Cf. Knibb, 2, 229, and above, 161.

(3) **healing shall be yours** Cf. 95.4.

a bright light Cf. 38.2,4, 92.4.

a voice of rest Eth. qāla ʿeraft = קלא ניחא vox benevolentiae (dei) (?). The phrase does not seem to have any Biblical counterpart, unless it could be קול דממה, lit. 'voice of silence', 'a still small voice' of 1 Kg. 19.12.

(4) **make you appear righteous** The belief that prosperity was the reward of righ-

teousness, and, therefore, evidence of prosperity proof of righteousness: nevertheless, their consciences convict them of their sinfulness.

for a memorial Enoch's words in this verse are 'an inscribed memorial', as it were, to their wicked deeds. Cf. Job 13.12.

(5) **the finest of the wheat** Cf. Ps. 81.17, חלב חטה, 147.14, Dt. 32.13-14 '(he satisfied him) with the finest flour of wheat; and he drank wine from the blood of the grape' (NEB) (cf. next verse), Neh. 8.10. Could there be an original word-play with חיטין, 'wheat', חטאין 'sinners' and תחתיא = Eth. teḥḥutān, 'the poor'? (cf. P. Sm. 4425) Note also the verb kēda = דוש 'to tread underfoot', but also 'to thrash' of wheat, etc.

drink new wine, the choicest of the wine(?) Eth. 'drink the strength of the roots of the spring', a practically unintelligible text. (Dillmann, followed by Knibb, interprets 'the best of the water'). Eth. ḫayl could represent an original Aram. חוליא (Syr. ܚܘܠܝܐ) 'sweet wine'.[1] The author repeats the consonants later in the verse, and could be playing on the word: 'And tread underfoot the poor in your might (בחילכון)'. Charles, 239, thinks of an allusion to Am. 6:6, 'Woe to you who ... drink wine in large bowls' where יין, οἶνος has been confused with עין, πηγή. Did a translator perhaps misrender καὶ πίνετε τὸ γλεῦκος τὴν ῥίζην τῆς πηγῆς (שרש עין) for τὸν πρῶτον οἶνον (ראשי יין,[2] the choicest of the wine)?

(6) **who drink waters ... fail**(?) Eth. lit 'who have water at all times to drink', which seems to imply that water was such a rare commodity that its regular use for drinking was available only to the rich. Or is Charles right in suspecting corruption? Perhaps a less difficult alternative to his proposal to read 'from every fountain' for 'at all times', is a text 'Woe to you who drink waters that at all times fail', מיין דבכול זבן מכזבין = ὕδατα πάντοτε ἐκλίποντα, in which the verb has fallen out by hmt. Cf. Isa. 58.11 to which the verse may be alluding: 'thou shalt be ... like a spring of water whose waters fail not' (לא־יכזבו מימיו).

exhausted ... wither away The verbs continue the figure of the water that at all times fails; its consequence is the exhaustion and death of those who satisfy their thirst at such wells.

the fountain of life Cf. Ps. 36.9, Jer. 2.13, 17.13.

(7) **unrighteousness and deceit and blasphemy** Cf. 94.6,9.

(8) **oppress the righteous** Eth. kuarʿa = Heb./Aram. עשק(?) (Mal. 3.5).

CHAPTER 97

Chapters 97.6b-107.3 are preserved in G^b. A detailed study of the text of 97-104 has been made by G. W. E. Nickelsburg, 'Enoch 97-104: A Study of the Greek and Ethiopic Texts', in *Armenian and Biblical Studies*, edited by M. E. Stone, Jerusalem, 1976, 90-156 (hereafter Nickelsburg.)

(1) **the day of violence** Eth. ʿammaḍā = (in this context) (cf. 91.11, En^g 1 iv 14, Milik, 265). The day of judgement is to be one of violent destruction of the wicked. Cf. Charles's note on 45.2.

---

[1] Tg. חוליא = 'sweetness'; in the old Aram. חליא is 'vinegar' (Hoftijzer, 88), but Syr. ܚܘܠܝܐ = mustum, γλεῦκος, 'sweet wine' = עסיס (Am. 9.13, Jl. 3.18).

[2] Cf. Ezek. 27.22 (Tg) Levy CW II 397.

(2) **Be it known unto you (sinners)** Note the abrupt change of subject and cf. above 95.7. Cf. 94.10 for the same harsh sentiments and contrast Lk. 15.10.

(3) **prayers of the righteous** Cf. 9.2f., and 97.3, 5; also 47.1-4, 99.3f., 104.3 and Rev. 6.10.

(4) This verse is addressed to a group of the author's fellow-countrymen who have been in allegiance with 'the sinners', presumably the Gentiles; they are 'the fellow-travellers'. Eth II, by introducing the negative, makes the words apply to the sinners themselves of whom it could hardly then be said that they had been 'associated with the sinners'.

(6) **he will do away with ... iniquity** G^b ἀναφελεῖ from ἀναφαιρεῖν = ἀφαιρεῖν (cf. C-B, 32). Could the verb be used here as at Lev. 10.17, Heb. 10.4 'to expiate, require propitiation for'? Cf. Ezek. 16.63: 'You will remember, and will be ashamed ... once I have accepted expiation for all you have done.' (NEB) The clause 'deeds which partook of iniquity' seems unsemitic: did the original perhaps read 'he will require expiation for all the deeds of those who participated in iniquity (oppression)', the reference still being to the 'collaborators' of v. 4?

(7) **in the midst of the sea or on land** A polar expression intended to include all sea-farers and no doubt also the inhabitants of the 'islands' of the sea as well as all who dwell on the continents of the earth.

Their memory of you ... ill for you G^bμνημόσυνον εἰς ὑμᾶς κακόν. Cf. 100.10, 104.8. Both sea and land will retain the memory of their sinfulness and will testify against them. Eth. yabs = Aram. יבשה, G^b ξηρά, and Eth. 'ekuy = Aram. בישא, with the meaning not only of 'evil' but also 'harmful'.

(8) **acquired possessions ... everything we have wished** Cf. G^b. This longer text appears to have arisen 'from a double rendering of ὃ ἐὰν θελήσωμεν with the preceding as well as with the following verb' (C-B, 33). On the other hand, the longer text of Eth. may be original, since two different verbs are used, '(everything) we wished, faqadna = צבינא(?) and '(everything) we planned ḫallayna = חשבנו(?). G^b may have omitted the first clause by hmt.

(10) **like water they will be poured out** The same phrase occurs at 98.2 in a similar context; for the expression cf. Ps. 22.14.

quickly be taken away Eth. ya'āreg ^{e} ἀναβήσεται, Aram. שקל Ithp. read as סלק Ithp., cf. Levy, CW, II, 168, 512, Tg. 1 Chr. 5.1(?).

great curse G^b κατάρα = Eth. margam; cf. Prov. 3.33 'The curse of the Lord is on the house of the wicked'. LXX κατάρα also renders שממה = 'desolation' (Isa. 64.10(9)).

CHAPTER 98

(1) **I swear unto you** A formula occurring here for the first time, but repeated in this part of the book, e.g., vs. 4,6, 99.6. Does it serve to mark off this section as a separate piece (Dillmann, 309)?

to the wise and to the foolish The variant 'not to the foolish' may be a scribal 'improvement'.

iniquities G^b ἀνομίας = רישעין (ܪܘܫܥܝܢ) scelera(?).

(2) **you men ... maidens** Eth. ḫebr = χρῶμα, perhaps translating Heb. רקמה, Aram. רקם (Ezek. 16.18, ἱματισμὸς ποικίλος)(?).

In royalty ... power The reference could be to the gorgeous finery and splendid raiment worn by the kings and rulers of the period, most probably the Seleucids.

and priceless things Eth. alone, kebr = τίμια. Along with silver, gold and purple, is this coll. sing. יקרא (cf. ܐܝܩܪܐ) in the sense of ξένια (cf. 2 Sam. 8.2, Ezr. 1.6 מגדנות 'precious things' (AV), 'valuable gifts' (NEB))(?).

are destined to pass away(?) G^b ἔσονται εἰς βρώματα, Eth. 'and food (mabāle´t)' (sic), both impossible readings: G^b, however, points to an original להון למכלא 'are destined to pass away', mistranslated as 'will be for food'; for the construction see note on 1.1 and for the verb כלא(ה) 'be spent, used up, come to an end' (e.g., of water, Gen. 21.15, meal, 1 Kg. 17.14,16), cf. En. 107.1, En^e 5 ii 28, Milik, 210, and 292, 323.

and your household goods(?) will be poured away like water G^b 'and in your houses they will be poured out ...' seems less satisfactory than to assume that ἀγαθά has dropped out before ἐν ταῖς οἰκίαις ὑμῶν parallel with the previous subjects, 'gold and silver etc.', and corresponding to the same phrase in 97.9.

(3) **they ... have neither knowledge ... shall perish** Eth. has the 3rd person throughout.
C-B takes the first clause with v. 2. Either we should read 2nd or 3rd persons consistently, or the rapid change from 3rd to 2nd person is the idiom noted above, 104. If the change seems too abrupt here, perhaps we should read 'Because you have neither knowledge etc.'.

desolation and great slaughter G^b ἐρήμωσις Eth. nedēt lit. 'poverty'; G^b seems more original: Eth. underlines the contrast from the sinners' previous wealth. For the 'great slaughter' at the judgement, see Charles on 45.1, Dillmann, 127.

Your spirits ... into the furnace of fire The wicked are cast into Gehenna apparently as incorporeal spirits; cf. 103.8, 108.3; Dillmann, 310. The 'fiery furnace' recalls Dan. 3.6,11, but is here used of Gehenna; cf. above on 10.6,13, Mt. 13.42f.

(4) Eth. has the extraordinary text, 'I swear to you, sinners, that a mountain dabr (ὄρος) does not become a slave, and a hill does not become the hand-maid of a woman, even so sin was not sent on earth ...'. At v. 5 G^b has the clause οὐχ ὡρίσθη δο[ῦλον] εἶναι ἢ δούλην '... it has not been decreed that there be slave, male or female'. Has the Eth. clause arisen from a variant for ὡρίσθη (ἐγένετο ὅρος?), where ὅρος 'rule, regulation' was read as ὄρος, 'mountain'? The remainder of the resultant bizarre simile may be Eth. expansion. I suggest as a possible original: לא הוה גזיר די להוא עבדא או אמתא, 'it was not decreed (i.e. from above) that there should be slave, male or female'. The translation omits the Eth. version of this clause at v. 4 and follows G^b by taking it as belonging to v. 5.

(5) G^b has a much fuller version here; Eth. confines the subject of the verse to ἀτεκνία, giving the first sentence of G^b (without στείρᾳ) and the last two words of the verse only, viz., [καὶ γυναικὶ οὐκ ἐδόθη ἀτεκνία ἀλλὰ διὰ τὰ ἔργα τῶν χειρῶν ἄτεκνος ἀποθανεῖται]. The stanza on ἀτεκνία, however, clearly begins with the words ὁμοίως οὐδὲ στεῖρα γυνή, so that this first sentence on ἀτεκνία has obviously been displaced, and should come within the verse (5a post 5d) and not at its beginning.
This observation has an important consequence for the restoration of the lacuna in G^b (C-B, 35, 2, lines 1-2): since the clause on slavery 5b (οὐχ ὡρίσθη δοῦλον εἶναι ἢ δούλην) then follows v. 4, I suggest that the subject is not ἁμαρτία (as in Eth.) but δουλεία, and I have restored the papyrus accordingly. Several considerations support this proposal. The opening stanza on ἁμαρτία in Eth. is redundant, since v. 5 has a stanza on ἀνομία, virtually the same subject. The expression οἱ ποιοῦντες αὐτήν (דעבדין לה) gives a word-play with עבדא, δοῦλος. (The original may well have read די מעבדין, 'those making slaves, enslaving'; cf. Kraeling, *Brooklyn Papyri*, 8.6). Those

who ‘practise slavery’ are then said ‘to come to’ i.e. ‘become subject to, fall under’ a great curse, which seems an appropriate punishment for this crime. (For ἀφικνεῖσθαι εἰς, אתמטי ל cf. P. Sm. 2074).

(6) **are revealed in heaven** Cf. 97.6, 100.10, 104.7,8, and for Old Testament parallels Job 22.13,14, Ps. 73.11, 94.7.

(7) **do not imagine to yourselves or think** Eth. lit. ‘do not think in your spirits/souls’ = G^{b} μὴ ὑπολάβητε τῇ ψυχῇ ὑμῶν, and ‘do not say in your heart’ = Aram. לא תאמרון בלבכון; cf. G^{b} μηδὲ ὑπολάβητε (ditt.) τῇ καρδίᾳ ὑμῶν.

that no sins are known ... Most High The majority text of Eth. reads a 2nd person ‘that you do not know nor see ...’. This is a less satisfactory text, the point being that sinners imagine that their iniquities go unseen and unheeded. The impersonal plural of G^{b} = Eth^{ull} conveys this sense.

(9) **to the wise you do not look** I have adopted the reading of Eth^{m} lit. ‘you do not know’ but rendered as equivalent of לא תסתכלון ‘you do not know’, i.e., ‘recognise, look to, pay heed to’. The original Aram. would then play on the word סכלין, ἄφρονες. (Was מסכילין used for ‘the wise’?) Among the Eth. variants Eth$^{tana\ g^{1}\ g}$ is adopted by Flemming-Radermacher and is also close to the original, ‘not concern oneself with’ (Charles’s rendering of this reading gives the verb the Heb. sense of רשע, ἀσεβεῖν found at 100.9, but less appropriate in this context; cf. Dillmann, 280). The version of Eth^{M} = G^{b} is a free interpretation (cf. Nickelsburg, 98, who prefers this reading).

will not come to you Eth. rakaba = ἀπαντᾶν = Heb. מצא Aram. אשכח.

evil ... encompass you C-B cj. περιέξει = Ps. 40 (LXX 39).13, περιέσχον με κακά, Heb. אפפי עלי רעות.

(10) **hope to be saved** Eth. ḥaywa renders σωθῆναι, but the original Aram. would use the same verb חיא: the verb in this context appears to mean to escape the judgement and enjoy eternal life.

for you are unaware, although you are prepared I have adopted the reading of Eth$^{tana\ q}$: sinners are unaware that they are ripe for condemnation. If the reading of Eth^{M} is preferred, we may compare Ps. 49.7-9.

(11) **obstinate of heart** Eth. gezufāna lebb renders ἀναιδεῖς at Isa. 56.11; here it corresponds to G^{b} σκληροτράχηλοι τῇ καρδίᾳ. This construction is unknown in Heb. or Aram., but σκληροτράχηλος alone = קשי ערף (Aram. קשיי קדלא); cf. Ac. 7.51 Pesh. ܩܫܝ ܩܕܠܐ and 2 Kg. 17.14 Tg. The addition of τῇ καρδίᾳ seems a translator’s expansion (Eth^{q} om.), but cf. 100.8 where Eth II reads geftuʿāna lebb = ‘perverse (διεστραμμένοι) of heart’. The original probably read קשיי קדלא only (see note on v. 12).

eat blood Cf. 7.5. This breach of the Levitical ritual code may be the result of hellenising influences (cf. Dillmann, 311).

have good things to eat Cf. Ps. 103.5: ἐν ἀγαθοῖς goes strictly with ἐμπίπλασθε.

you shall have no peace The formula at 5.4 (see note).

(12) **delivered into the hands of the righteous** Cf. 95.3,7.

behead ... you Eth. matara kešāda lit. ‘cut the throat’ = גזר קדלא ‘to behead, execute’, Cowley, *Aramaic Payri*, Ah. 134 (cf. τραχηλοκοπεῖν, τενοντοῦν, Heb. ערף, of animals. The punishment fits the crime of being ‘stiff-necked’ (Aram. קשיי קדלא).

(13) **tribulation** G^{b} κακά = Eth. mendābē, Heb. צרה, Aram. עקתה. See note on 1.1.

no grave shall be dug for you The Hebrew horror of an unburied corpse; cf. Isa. 14.19,29, Jer. 8.2, 22.19, etc.

(14) **seek to set at nought** G^b βούλεσθαι = Aram. צבא, בעא Heb. בקש(?) petivit, voluit.

(15 **lying words ... of heresy** Eth. nagara ḥassat, cf. Jer. 7.4, דברי השקר Aram. מלי שוקרא(?) G^b πλάνησις = טעו 'apostasy, idolatry': טעו is also used for the object of idolatrous worship, the false deity itself, 'idol', (Levy, CW, I, 312), like כזב 'lie', res mentita (Am. 2.4). Did the author perhaps write מלי כדבין ומלי טעון '(Woe to you who write down) the words of false deities and the words of idols'? Cf. the note on 99.1.

they write G^b οἱ γράφοντες ... γράφουσιν can hardly be right. Was the second verb יכדבון 'they deceive' read as יכתבון (yeṣeḥḥefu, γράφουσιν)? (Could the prominent αὐτοί (Eth. we'etomu) refer to the 'idols' rather than to the sinners themselves? 'They (the idols) deceive ...').

men apostatize ... fall away The reading of Eth^{tana} is closest to G^b: yārasse'ewwo labā'ed 'they will cause others to fall away' corresponds to G^b πολλοὺς ἀποπλανήσουσιν. (The first verb 'they (men) apostatize' (yerse'u) in Eth^{tana} is perhaps redundant.) The variations on the verb ras'a, 'to apostatize' (see C-B 41), make poor sense, e.g., Eth^{M} '(that men may hear) and not forget ('iyerse'ewwo) folly (la'ebad)' (Knibb).

you shall not have peace ... die For χαίρειν as a rendering of שלום Isa. 48.22, 57.21; cf. En. 99.13, 102.3, 103.8 and see G. Zuntz, JThSt, 45 (1944), 163 n. 1, Knibb 1, 370. For the expression 'of a sudden they shall die', Job 34.20 (רגע ימותו).

CHAPTER 99

(1) Cf. 99.7f., 104.10.

Eth^{M} 'Woe to those who ...' note again the confusion of second and third persons. Is this peculiarity perhaps traceable to the misreading of Aram. suffixes (e.g., להון for לכון) rather than to an idiosyncrasy of the author's style? Cf. C. C. Torrey, 'Notes on the Greek Text of Enoch', JAOS 62 (1942), 59.

The meaning of the Greek text of this verse is not in doubt: the text of Eth., on the other hand, contains some unfamiliar, if not strange, expressions, which suggest a different original from G^b.

who causes apostasies G^b πλανήματα seems to be used in the same sense as πλάνησις at 98.15, where a form of ras'a, res'at is again employed to render this term. But the expression 'to cause (lit., make) apostasies' is strange. Does πλανήματα here = Aram. טעון 'false deities' and was the original וי לכון פלחי טעון, 'Woe to you worshippers of false deities ...'?

by your deceitful works receive honour and glory The expression in G^b 'lying works' seems to correspond to Heb. פעלת שקר at Prov. 11.18. Could ἔργον here be equivalent to Heb. מעשה, Aram. עובד 'idol' (or perhaps פעל?)? Eth. may then have preserved a better text, if we read ἔργα (nagar may have arisen as an inner Eth. corruption or through the influence of the previous λόγους ψευδεῖς of 98.15): '(Woe to you) who ... glorify and honour false works (idols)', די ... משבחין ומיקרין עובדי שקרא). The verse then resumes the subject of 98.15 viz. idolatrous apostasy.

you shall surely perish G^b ἀπολώλατε, perfectum confidentiae or perfectum propheticum.

no good life G^b lit. 'salvation unto good', Eth. ḥeywat šannāyt 'good life', both unusual renderings of an otherwise unparalleled expression. Have the versions arisen from the corruption of σωτηρίαν (ζωὴν) αἰώνιον, where αἰώνιον has been corrupted to ἀγαθόν?

(2) **words of truth ... everlasting law** i.e., the Scriptures of the Old Testament. G^{b} τὴν αἰώνιον διαθήκην probably refers to the Mosaic law.

count themselves ... without sin There is no foundation for Charles's rather fanciful interpretation that this verse refers to the adoption of hellenistic customs, in particular to the removal of the 'sign of circumcision'. For mistranslations of this line, Nickelsburg, 94, n. 23.

swallowed up in the earth G^{b} = Aram. יתבלעון a Biblical phrase (cf. Num. 16.32, Exod. 15.4 LXX κατεπόθησαν), more likely to be original than Eth. 'they will be trampled underfoot'. Cf. Nickelsburg, 108.

(3) On the mediatorial function of the angels, see Charles on 9.10, and on Test. Levi 3.5., Test. Dan. 6.2. The idea has its roots in the Old Testament at Job. 5.1, 33.23, Zech. 1.12, and cf. Rev. 8.3,4.

At that time G^{b} τότε, Eth. 'in those days': so also at v. 4, but at 99.5 G^{b} has ἐν αὐτῷ τῷ καιρῷ ἐκείνῳ = בה בעת ההיא Aram. בה בעידנא ההוא which seems original. Cf. Knibb, 2, 233, S. Aalen, 'St. Luke's Gospel and the Last Chapters of I Enoch', NTS 13, (1966-67), 3.

Most High See note on 9.3.

to remind him lit. 'as a memorandum, reminder', i.e., the petitions of the righteous, mediated to the Most High by the angels, are to come before God as a reminder of the sins of the wicked.. The second occurrence of the expression seems redundant.

(4) The shorter text of G^{b} is probably the result of an omission by hmt ('peoples' ... 'peoples'; cf. Eth^{d}). The omission thereby of any subject makes the verb συνταραχθήσονται appear to be an impersonal plural, 'men will be thrown into confusion', but this is purely accidental due to the omission of the subject (λαοί). For the expression 'kindreds of the peoples', cf. Ps. 22.28 (LXX 21.28). משפחות גוים, αἱ πατριαὶ τῶν ἐθνῶν.

destruction of wickedness So G^{b} but the ἀδικίας could be a translator's addition (cf. Eth^{M} $Eth^{tana\ q}$, 'sinners').

(5) G^{b} has a longer and superior text with the original parallelismus verborum better preserved.

At that time See above on 99.3.

women bearing children ... infant babes Eth. ʾella yeṣṣēnasu (= yeḏḏēnasu) = αἱ συλλαβοῦσαι / τίκτουσαι (ḏansa = συλλαμβάνειν cf. Nickelsburg, 95). G^{b} ἐκβάλλειν 'to expose children' = Heb. שלך, Aram. טלק, Gen. 21.15; cf. Euripides, Ion 964. An alternative original is suggested by Eth. if it read αἱ συλλαβοῦσαι ἐκβαλοῦσιν, הריות תפקון (for הרי = הרה 'a pregnant woman', Hoftijzer, 69), where נפק = יצא, means 'to miscarry': 'pregnant women miscarry' or (Aphel) 'cause to miscarry' (cf. Exod. 21.22). Eth. yemaššeṭu, is a verb which regularly renders ἁρπάζειν and its compounds: G^{b} should be read as ἐξαρπάσουσιν (cf. Nickelsburg, 95, C-B, 43; Knibb, 2, 233 for other proposals.) The verb renders שלך 'to cast forth' in the LXX (Job 29.17, 1 Mac. 7.29), but with the special force in the present context of 'to expose'.. This would support the proposal to take ἐκβάλλειν earlier as 'to miscarry.'

pregnant women will abort ... cast away their children There is no need to emend Eth. yedeḫḫedu to yādeḫḫeḍu (Nickelsburg, 95) since deḫḏa I.1 as well as II.1 = Heb. שכל Pi. Aram. אתכיל, abortum pati (see Dillmann, *Lex.*, 1114). G^{b} ῥίψουσιν 'will cast away', i.e. as corpses, or 'expose' (cf. LXX Gen. 21.15)?

(6) Tertullian (*de idol.* iv) cites verses 6 and 7 in an 'edited' version: et rursus iuro vobis, peccatores, quod in diem sanguinis perditionis poenitentia parata est. The idea

that 'repentance' is prepared for the wicked is foreign to the author of the original Enoch and looks like a Christian importation. V. 6 is lost in G^b.

sin is destined Eth. dalawa III.2 = Aram. עתד Ithp. or עתיד ל.

the day of unceasing bloodshed Eth. dam za'iyahadde'. The old Aram. is familiar with the Biblical term דמי 'cessation, pause, rest' (Hoftijzer, 58). Did an original read יום דם בלא דמי? $Eth^{t^2, e\ 3\ mss}$ ke'wata dam = שפוך דם).

(7) Tertullian (ibid.) qui servitis lapidibus, et qui imagines facitis aureas at argenteas et ligneas et lapideas et fictiles, et servitis phantasmatibus et daemoniis et spiritibus infamis (infamibus) et omnibus erroribus non secundum scientiam, nullum ab eis invenietis auxilium. Cf. also Jub. 1.11, 11.4.

phantoms ... no help from them G^b φαντάσματα cf. Wis. 17.14, Mt. 14.26, etc. Eth. om. Eth. has 'unclean spirits' = πνεύματα ἀκάθαρτα: cf. Mk. 1.26, Zech. 13.2 (LXX). This could be original or a scribal alteration of 'evil spirits' to conform to the Gospel phrase? Although [βδελύγμασιν] is a restored reading, and is omitted by Eth., it is singularly appropriate = Heb. שקוצים, Aram. שקוצין lit., 'detested things, abominations' (2 Kg. 23.13, Ezek. 5.11, 8.10, 20.7,8) or תועיבן (Ezek. 8.6). Eth. ṭā'ot, εἴδωλον טעו, is ambiguous and can mean πλάνη or εἴδωλον: G^b has wrongly rendered by the former—it is 'idols' which are worshipped not 'errors' or 'apostasies'. The words 'not according to knowledge' seem odd,[1] although making sense and are an old reading found in Tertullian. Has there, however, been a primitive mistranslation of the Aram. idiom כל מנדעם (Hoftijzer, 119, 158, Cowley, *Aramaic Papyri* 21.7, 49.3) ומנהון מנדעם עדרא לא ישתכח לכון כולא, '... and no help at all will be obtained for you from them'(?), where מנדעם has given rise to κατ' ἐπιστήμην.

(8) V. 8 implies an 'idolatrous apostasy'. Cf. Wis. 14.12,27, Rom. 1.25. The line 'And their eyes ... hearts' has fallen out of G^b by hmt: note the reading καὶ καρδίας αὐτῶν (correctly emended by C-B to τῆς καρδίας αὐτῶν); the καὶ points to the omission of the lost line, perhaps καὶ [τυφλωθήσονται οἱ ὀφθαλμοὶ αὐτῶν ἐν φόβῳ] τῆς καρδίας αὐτῶν.

(9) **all their works (idols?) ... fashioned (served?)** G^b has a much shorter, probably abridged text. Are 'works' here עובדין again = פעלים 'idols', and gabru = עבדו, פעלו 'served'? See note above on 99.1. Eth. baḥassat = לשוא LXX ἐπὶ ματαίῳ Exod. 20.7.

worshipped G^b ἐλατρε[ύσατε], cf. v. 7 λατρεύοντες (C-B wrongly ἐλαερ[γήσατε]).

In a trice ἐπὶ μιᾶς Eth. bame'r = Aram. כחדא. See Aalen, NTS 13,3, Dan 2.35. The same phrase is differently (but also correctly) rendered at 98.3 by κοινῶς.

(10) **receive the words of the wise** Eth. maṭawa III.2 = Heb. לקח, Aram. קבל (Prov. 24.32); G^b οἱ ἀκούσαντες seems a free rendering: Eth. 'words of wisdom' could be original.

learn them ... Most High Cf. Ps. 119.71. Eth. 'paths of the Most High' has arisen from the following 'walk in the paths of ...'. For G^b ποιεῖν τὰς ἐντολάς cf. Dt. 6.25, 15.5.

they will be saved Eth. here renders G^b literally; cf. 98.14, 99.1.

(11) Verses 11 and 12 are omitted in G^b.

plot evil against your neighbour Eth. safḥa la'ekit = Heb. נטה רעה על, Aram. נגד ביש על(?) (Ps. 21.12 LXX ἔκλιναν εἰς σὲ κακά and cf. Gen. 39.21).

---

[1] The inner Eth. corruptions or corrections make sense in the context but this is coincidental: Dillmann, 72, 'in the idol temples', bameḥrāmāt; the same word can also mean 'idols' (Dillmann, *Lex*., 84).

you shall be slain in Sheol Cf. 22.13. Sheol is now Gehenna as in later usage; cf. 63.10 and Charles, 127f.; see ThWNT, Bd. I, s.v. ᾅδης 146f. (J. Jeremias).

(12) **use false and deceitful weights** Eth$^{tana\ g\ g^1\ (q)\ i}$ lit., 'make balances (masfart[1]) of sin and deceit'. Cf. Am. 8.5 לעות מאזני מרמה, lit. 'to bend balances of deceit', 'to tilt the scales fraudulently' (NEB). Cf. Am. 8.5 in Syr^{vg} and Prov. 11.1, Tg. מסאתא דניכלא.

who prosper(?) The majority text yāmarreru understood since Dillmann as 'cause bitterness' is more than doubtful. The readings of Eth$^{tana\ g\ q}$ yā'ammeru and Eth$^{g^1}$ yāmakkeru could both go back to ישכלון: סכל/שכל has also the special sense in Heb. (and Aram.?) of 'to succeed, prosper', (always as the result of wise (or wily) dealings). (See S. R. Driver on 1 Sam. 18.5). A conjecture like this would certainly suit the present context.

an end will be made of ... (you) Eth. yetwēde'u from תכלון, כלא = Heb. כלה 'come to an end, be destroyed'(?), perhaps [ἐξαναλωθήσονται]. See note below on 107.1.

(13) **who construct ... peace** 'by the toil of others' seems a free translation. The text proposed follows Eth$^{tana\ t,\ x}$ [ποιεῖτε]?) and Eth$^{g^1}$, with Nickelsburg, 95f.

no peace G^b οὐκ ἔστιν ὑμῖν χαίρειν. See note above, 113.

(14) **foundations** Eth. mašarrat = Aram. אשה (cf. 91.11, En^g 1 iv 14, Milik, 265). The 'foundations' of the Torah and Covenants.

no rest Eth. 'eraft G^b ἀναπαῦσαι = Heb. מנוחה Aram. ניחא (cf. Levy, CW, I, 108, Gen. rabba 87 (beginning) 'he has no rest').

(15) **assist wrong-doing** Abstract for concrete, 'assist the wrong-doers', i.e., the oppressors. The reference is probably to the 'collaborators' with the (Gentile) oppressors; cf. 100.4, note.

great judgement See note on 10.12.

(16) **overthrow your glory** G^b ἐκτρέψει (ἐκστρέψει) (C-B wrongly ἐκτρίψει) = Eth. yāwaddeq = Heb./Aram. הפך or השליך (cf. Lam. 2.1, Tg. טלק).

put evil into your hearts At. Ec. 7.4 'ekaya lebb renders πένθος, אבל. Cf. Prov. 15.13 עצבת לב, Tg. כיבא דליבא, LXX λύπαι. Has this line, not found in G^b arisen by a dislocation of the text which has been substantially preserved in Eth$^{g^1}$, 'and arouse anger against your hearts', but reading 'the wickedness in your hearts', the full expression found in the expanded majority text?

arouse his fierce anger For the expression cf. Ps. 78.38. Eth. manfasa ma''ātu = Heb. רוח אפו, Aram. רוח אפוהי (Exod. 15.8 and Tg. Neoph., Job 4.9).

will remember your wrong-doing At 99.3 the righteous bring their intercessions for sinners before the angels to be placed before the Most High: here they are to recall the wickedness of the sinners, no doubt to convey this to the mediating angels.

CHAPTER 100

(1) **fathers shall attack their children** Eth. guade'a III.3 mesla = Heb./Aram. נגח Hithp./Ithp. lit., 'to engage in thrusting with' and so 'to fight with, attack' (Dan. 11.40), LXX Theod. συγκερατίζεσθαι.

shall fall in (mutual) destruction Eth. lit. 'shall fall in death', but mot can render

---

[1] The reading of Eth II 'foundations (mašarrata)' is followed by Knibb, 'two lay foundations of sin and deceit' (En^m maqšafta is meaningless).

ἀπώλεια (Eth. Job 27.7, Dillmann, *Lex.*, 203). If the original had read 'shall fall and die' it would have been so rendered (cf. 2 Sam. 2.23). LXX Lev. 26.7 has πεσοῦνται … φόνῳ (MT ונפלו … לחרב), and this would also be an appropriate expression here.

flows like a river The Biblical expression is 'like water' (Dt. 12.16,24, Ps. 79.3); the original may have been כמיא 'like water'.

(2) **from his son(s) … beloved one** Eth. 'nor his sons' sons' could be original and omitted in G^b by hmt. For ἀγαπητός = חביב cf. 14.6, En^c 1 vi 16, Milik, 193.

the sinner from his dear one G^b ἔντιμος = Heb./Aram. יקיר Jer. 31.20, Lk. 7.2; G^b 'nor from his brother' could be a translator's expansion. Eth. 'dear brother': kebur = ἔντιμος.

together G^b ἐπὶ τὸ αὐτό = כחדא. See above, note at 99.9.

(3) This type of imagery appears again at Rev. 14.20 and in Talmudic sources; see Charles, 248, Schürer, *History* I, 552, n. 171.

shall be submerged to its axles The translation combines the text of Eth. 'shall be submerged' with G^b μεχρὶ ἀξόνων. Instead of καταβήσεται Eth. read [καταποντίσεται] = tessaṭṭam (cf. Exod. 15.4). (G^b καταβήσεται may have arisen through the influence of καταβήσονται in the next verse.) For μεχρὶ ἀξόνων Eth^M has 'up to its height' ('eska mal'eltā), or 'up to above its height' ('eska lā'la (mss 'elata) mal'eltā, $Eth^{tana\ g\ g^1\ q}$). LXX ἄξων = Heb. אופן Aram. גלגלא 'wheel'.[1] Did the original read עד עלא מן גלגלוהי μεχρὶ ἐπάνω ἀξόνων 'to above its wheels (axles)'? For עלא מן, ἐπάνω, mal'elta 32.2, En^e 1 xxvi 20, 21, Milik, 232, cf. 86.1 En^f 1 1, Milik, 244. Has Eth. lost or failed to reproduce the key words ἀξόντων αὐτοῦ and been left with a text 'eska mal'elta (μεχρὶ ἐπάνω) read as 'eska mal'eltā 'up to its height'?

(4) **angels shall descend into (their) hiding-places** G^b καταδύνοντες seems to be a doublet of the main verb. The idea is of the angels of punishment searching out all those who have assisted the oppressors, the 'sinners', οἱ ἁμαρτωλοί or 'renegades' of Israel, in their hiding-places (τὰ ἀπόκρυφα = Heb. סתר, מסתר Aram. סיתרא (e.g., Tg. Dt. 27.15). The text of Eth I 'all who brought down sin' (for Eth II 'all who were assisting sin') is an inner Eth. corruption helped by its implied reference to the watchers; so rightly Nickelsburg, 97, cf. Charles, 248. The same expression 'those who assist sin' for the 'renegades' occurs above at 99.15.

(5) **as the apple of an eye** Cf. Dt. 32.10, Ps. 17.8.

thereafter G^b ἀπ' ἐκείνου (sc. καιροῦ). The variations of Eth. stem from an original 'amēhu or 'amēhā. Cf. Nickelsburg, 97f., especially 98, n. 43.

sleep a pleasant (lit., sweet) sleep Cf. Jer. 31(38).26, Ec. 5.11 (ἡδύς = מתוק misread by Eth. as מתית?). Charles interprets of the pious dead guarded by angels in their promptuaria, as at 2 Esd. 7.85: videntes aliorum habiticula ab angelis conservari cum silentio magno, and 7.95 requiescent cum silentio multo ab angelis conservati; cf. also 2 Esd. 4.35, 7.32, 2 Bar. 30.2. Dillmann thought of the righteous who die a violent death whose sleep is prolonged, but who are watched over safely by angels until their resurrection. In this context, where 'thereafter' has an eschatological significance, there is certainly more intended than simply the quiet rest of the righteous watched by their guardian angels in this life.

(6) **Then** G^b τότε. Read Eth. 'amēhu(ā) and cf. v. 5.

---

[1] Aram. גלגלא is used for any round object, e.g., $Enastr^b$ 6.9 Milik, 284, for the 'disc' of the moon, Tg. Ps. 17.8 for the 'pupil of the eye', and most probably also for the axle of the wheel; cf. Heb. גליל 'cylinder, rod' and Sir. 33(36).5 ὡς ἄξων (גליל?) στρεφόμενος par. to τροχὸς (אופן) ἁμάξης.

children of earth ... words of this book For the expression בני ארעא 92.1, En[g] 1 ii 23, Milik, 260. G[b] κατανοήσουσιν ἐπί, cf. the use of ידע ב. G[b] ἐπιστολή = Aram. ספרא, כתבא 'book, writing'. The use of the term by G[b] helped to create the title of this part of Enoch; see above note on 92.1.

(8) **obstinate of heart** See above note on 98.11.

fear shall lay hold on you Eth. appears to have read [λήμψεται ὑμᾶς φόβος], the usual phrase (cf. Isa. 33.14 LXX λήμψεται τρόμος τοὺς ἀσεβεῖς), and I suggest λήμψεται rather than περιέχει (C-B)[1] to fill the lacuna in G[b]. LXX περιέχειν renders אפף, סבב (Aram. נקף Aph.), 'to encompass', usually of 'evils', = 'the pangs of Sheol', but nowhere with 'fear' as subject.

(9) **works of your hands** Probably not only evil deeds, following on wicked words, but idols going together with blasphemous words(?).

the service (works) of the Holy One you have forsaken C-B emended G[b] to ὅτι ἀπὸ τῶν ἁγίων ἔργων ἀπεπλανήθητε 'because ye have gone astray from the deeds of holiness'. As Nickelsburg commented, 'Bonner's reconstruction is essentially correct' (99): the reading of Eth II has gebra rasāʿkemu = [ἔργον ἀπεπλανήθητε], and Eth[tana t] has ʾemgebra = ἀπὸ τῶν ἔργων: αγιων has fallen out by its similarity to απο (cf. 1.2 απο λογων read by editors as αγιολογων). An expression 'holy works' seems unlikely, and a rendering of the C-B reconstruction into Aram. suggests a more probable original: די טעיתון מן עבד(י)/עבודא קדישא, οἱ/ὅτι ἀπὸ τῶν ἔργων/τῆς λειτουργίας(?) τοῦ ἁγίου ἀπεπλανήθητε 'because you have forsaken the work/service of the Holy One'. (Aram. עבודא is especially used in the sense of λειτουργία, and this could be the meaning here).

in blazing fire For the phrase Exod. 3.2, Ps. 29.7 לבת אש = LXX φλὸξ πυρός, Dan. 7.9, En. 14.12. The fires of Gehenna are meant.

(10) **he will enquire** The reading with God as subject supports the conjectural reconstruction of v. 9 with 'the Holy One' as predicate.

the sun ... and ... the stars Cf. 97.7, 98.6-8, 104.8. Since these heavenly bodies circle the earth where the wicked execute an unjust judgement on the righteous, they see and can bear testimony against them; cf. Heb. 2.11, Lk. 19.40.

executed judgement Cf. Ezr. 7.26, En. 91.12, En[g] 1 iv 16 Milik, 266 for the expression עבד דינא.

(11) **be watchful over your sins** I have adopted the reading supported by Eth[t,b] yeḫēllewu = ישקדון, [ἐγρηγορηθήσονται], 'be watchful over' (ἐπὶ) rather than yeḫēlleyu = 'think about' (Knibb); the negative makes no sense. For this use of שקד, cf. Jer. 31 (LXX 38).28 and LXX Lam. 1.14 ἐγρηγορήθη ἐπὶ τὰ ἀσηβήματά μου, reading נשקד על פשעי, and understanding the text as 'watch was kept over my transgressions'.

(12) **and to the dew ... come down (upon you)** 'Spoken ironically' (Charles). The first half of the verse is virtually identical in G[b] and Eth. The fuller text of G[b] in the second half-verse gives a perfect parallel to the first half-verse and is certainly original. The Eth. text here is seriously defective, but preserves a clause, 'if it (the dew) will accept gold and silver from you', which may contain an original element, reinforcing the irony—these gifts may prove to be unacceptable to the dew and the mist. For LXX διαγράφειν = Heb. שקל Aram. תקל Est. 3.9: perhaps לטל ... תקלו דהבא וכספא די יחותון עליכון אין יקבלון מנכון

---

[1] C-B reads π[ε]ρ[ιεχε]ι: traces of letters are visible, but none can be identified any longer.

(13) Cf. Ps. 147.17.

snow and hoar-frost ... and all their afflictions Eth. 'asḥatyā = χίων; cf. 77.4 Enastr^b 23.10, Milik, 289. Eth. ḥamadā = πάχνη pruina, 'hoar-frost', Heb. כפור. Does παγετός in G^b, where Eth. has a second ḥamadā = πάχνη (the two Greek nouns mean 'rime' or 'frost') render an original Aram. קרחא = turbo, tempestas (P. Sm. 3739) as if = Heb. קרח 'frost'? We should then render '(all?) the winds and their storms'. The repetition of 'their afflictions' in G^b suggests that its first occurrence is a scribal expansion. G^b δύνασθε is translation Greek for δυνήσεσθε.

drive down upon you G^b, ἐπιρρίψη is here intransitive; cf. C-B who cites LXX Job 27.22: it could, however, be a mistake for ἐπερρίφη or have arisen as a mistranslation of רמי the passive participle.

CHAPTER 101

(2) **windows of heaven** Gen. 7.11 and cf. above, on 89.2. See also 2 Chr. 7.13.

(3) Cf. 5.4, En^a 1 ii 13. The author of this section seems to be drawing on the motifs and phraseology of chapter 5, and especially here, 5.4. Did Eth. read δικαιοσύνη for μεγαλωσύνη (see Nickelsburg, 120)? Note the close agreement of Eth II with G^b, καὶ ἐπὶ (πάντα) τὰ ἔργα ὑμῶν. Did Eth$^{m\,(g^1)}$ read ἐν (πᾶσι) τοῖς ἔργοις ὑμῶν = בעובדיכון 'on account of (all) your works'.

(4) For the language and imagery, cf. Jer. 5.22f., 46.7, Ps. 107.23-27, Sir. 43.24, Wis. 14.1, IQH III.6.

Behold the sea-captains G^b ὁρᾶτε = Aram. חזו, 'observe, consider', e.g. En. 2.2, En^a 1 ii 1, Milik, 145, G ἴδετε, 5.3. Eth. nagašta 'aḥmār, lit. 'ships' captains'. The conjecture that an original Aram. מלחי 'sailors' has been mistranslated as מלכי 'kings', thus accounting for the phrase in Eth., lit. (apparently) 'kings of ships', has been accepted by Charles, Flemming, Martin, Knibb (who cites this as evidence for an Aram. original, 2, 236). The original was probably מלחן (cf. Hoftijzer, 152) or רישי מלחא (P. Sm. 2134), but is there any need to assume a mistranslation? The Eth. phrase means 'chiefs of ships', a translation of ναυκλήρους, which can mean both 'ships' captains' and 'shipowners', and the context suggests that both meanings were intended.

tossed ... by the waves Eth. tahawka = Heb. געש Hithp., Aram. טרף Ithp.(?) (cf. Zech. 10.11, Tg. Levy CW 1321).

They are in distress Eth. yetmanadabu = G^b χειμαζόμενοι(?). Eth. tamandaba = LXX κινδυνεύειν, but the Eth. verb also translates LXX θλίβεσθαι. Did the original here perhaps read יצטערון, 'they are in distress', referring to the sea-captains (and ships' owners) as subject; v. 5 continues 'and ... they are afraid', but, as the verse explains, their distress and fear is not only occasioned by the dangers of the sea, but lest they lose their precious cargoes as well as their lives. Cf. Ps. 73.4 Tg. where 'to be in distress' (מצטערין) follows 'to be afraid' (תווהין), Levy, CW I, 334. (Did the Greek translator connect צער with Heb. סער, סערה, 'tempest'?)

(5) **and on this account ... shipped abroad** G^b [οὕ]τως Eth. ba'enta zentu, Aram. לקבל דנה (Ezr. 4.16), כל קבל דנה (Dan. 3.22, 6.10, LXX καὶ οὕτως) 'All their goods and possessions'; for טב/טוב in this sense, 1 Sam. 15.9. G^b ἔξω ... ἐκβάλλουσιν cf. נפק לבר lit. 'to go forth abroad' and so 'to be shipped, exported abroad'; for this use of נפק = יצא, 1 Kg. 10.28,29, 2 Chr. 1.16,17. Eth. is closer to the original:

according to G^b 'all their goods and possessions they cast into sea', but G^b = אפקו (Aph.).

they are anxious in mind Eth^{g^1} ḥallaya I.2 =μεριμνᾶν and for the expression cf. Prov. 12.25, דאגה בלב־איש 'an anxious heart' (NEB), Isa. 57.11 Tg. יצף. Eth^{g^1} has two versions of this clause, the first practically = Eth^M (where there is clearly a corrupt text, 'and they think nothing good in their hearts' (Knibb)), and the true reading, given above, inserted before 'will swallow them up', where it is a meaningless intrusion.

(6) Cf. Jer. 5.22, Job 38.8, Ps. 65(64).7.

turbulence Eth. ḥusat cf. Heb. געש (Jer. 5.22) Aram. רגוש(?).

he has determined their bounds Eth. 'sealed' is 'a little strange' (Knibb, 2, 236), but cf. Job 9.7, 'God 'seals (חתם)' the stars'. The reading 'he has determined (Eth^e ʿaqama = Aram. תחם) all their works' seems preferable, except that 'works' (gebrat) seems a mistake for 'boundaries, limits', the reading of G^b πέρατα which best suits the context. C-B, 57, suspects a corruption from πέρατα to πράγματα, πράξεις as the source of Eth. 'works'. Alternatively, an original עבריהון 'their boundaries' may have been misread as עבדיהון 'their works'. G^b συνεστήσατο could be explained from חתם = 'sealed', 'closed' (cf. Const. Apost. 8.12, ὁ συστησάμενος ἄβυσσον), or from תחם 'determined'. Eth^M [ἐσφαγίσατο?] has read חתם instead of תחם. An original, 'he has determined their boundaries and waters' (cf. Eth I) is no doubt also possible.

(7) Cf. Isa. 50.2.

(8) **who travel upon the seas** G^b τοῖς κινουμένοις ἐν τῇ θαλάσσῃ. Eth. lit. 'who move on the land and in the sea'. This is usually interpreted to refer to all living creatures. 'He too has given instinct to animals and reason to man' (Charles, 252). But to understand 'intelligence' (ἐπιστήμη) as including 'instinct' is to force a meaning on the word. G^b supplies the clue to the original: κινεῖσθαι = Heb./Aram. נוע, occasionally = κινεῖν in LXX, which can = 'be tossed about' (Am. 9.9) or = vagari, 'to travel', and this seems the more appropriate sense in this context; cf. Am. 8.12 ('they shall wander (נעו) from sea to sea'), Ps. 59.15, '... they wander to and fro in search of food'. The following words, moreover, 'the sea-captains fear the seas' point to the same context as v. 4; it is the sea-captains who possess skill, since they fear the sea.

CHAPTER 102

Chh. 102-104 bear a close resemblance to Wis. 2-3. Literary dependence of Wis. on En. (or on a common source) is argued by C. Larcher, *Études sur le Livre de la Sagesse*, (1969), 106-112; cf. G. W. E. Nickelsburg, *Resurrection, Immortality and Eternal Life in Intertestamental Judaism*, 128f., 128, n. 69.

(1) **a wave (tempest?) of blazing fire** LXX κλύδων = Heb. סער(ה) Aram. עלעולא 'tempest' (cf. Heb. גלגל 'whirl-wind'), and for the wrath of God, Ps. 83.16, Am. 1.14, Jon. 1.4. And we should perhaps so render. The second part of the phrase τοῦ πυρὸς τῆς καύσεως = Aram. יקדת אשא Dan. 7.11, LXX Theod. καῦσις πυρός. Eth. 'a raging fire' is either a short, free paraphrase or, more probably, the remains of the longer expression 'the raging tempest (עלעולא קשה) of blazing fire'; (for קשה of 'winds' 'tempest', Peshitta Ac. 27.18 P. Sm. 3768). The repetition of the personal pronoun ἐφ' ὑμᾶς and (τῆς καύσεως) ὑμῶν suggests corruption: Eth^{tana} reads dibēhomu =

עליהון which could be a corruption of עליון = ὁ ὕψιστος giving a characteristic word-play with עלעולא(?); behind the second ὑμῶν may lie an original עליכון. Alternatively, the original may have read: 'And then, when the Most High(?) hurls against you a raging tempest of blazing fire—you, where will you flee and escape?'.

(1-2) **with a mighty sound ... afraid** The phrase ἤχῳ μεγάλῳ is found in G^b only. For the expression, cf. Ps. 68.34 יתן בקולו קול עז 'he utters with his voice, a mighty sound'. In a convincingly argued discussion of the problems of these verses (and there have evidently been several dislocated clauses), G. Zuntz suggested reading ὅταν δῷ ἐφ' ὑμᾶς φωνὴν αὐτοῦ ἐν ἤχῳ μεγάλῳ.[1] No less difficult than the position of this last phrase, are the accusatives in the clause which follows viz. τὴν γῆν σύμπασαν ... συνταρασσομένην. Cf. C-B, 58, 'The accusatives in (lines) 9-10 (τὴν γῆν ... συνταρασσομένην) can stand only if some verb such as ὄψεσθε is inserted after καί (9), as James suggested'. A solution would be to posit the omission of ὅταν ἴδητε due to hmt with the previous ὅταν δῷ, and to render 'when you see the whole earth shaking and trembling ...'. A more satisfactory alternative, however, is offered by the text behind Eth. which takes the clause, in the nominative case, with the following verse.

(3) Here there is evident confusion in the text. In the order of clauses in Eth. we meet with the curious idea that the angels, after executing their commands (as angels of punishment as at 53.3, 100.4 etc.?), then seek to hide themselves.[2] This latter clause, although omitted in G^b, would come most appropriately after 'heaven and all the heavenly luminaries shall shake and tremble ... with a great fear': the heavens themselves seek to hide before the 'glory of the Great One'. Similarly, the clause which speaks of the earth's shaking and trembling follows this clause most appropriately, and goes closely with the last clause, 'And the children of men shall tremble and quail ...'.

I suggest the following rearrangement on the basis of the combined testimony of G^b and Eth.:

(a) (G^b Eth.) καὶ 'πάντες' οἱ ἄγγελοι συντελοῦντες τὸ συνταχθὲν αὐτοῖς
(b) (G^b Eth.) καὶ ὁ οὐρανὸς καὶ οἱ φωστῆρες σειόμενοι καὶ τρέμοντες φόβῳ μεγάλῳ
(c) (Eth. only) καὶ ζητοῦντες ἀποκρυβῆναι ἀπὸ προσώπου τῆς δόξης τῆς μεγάλης
(d) (Eth. cf. G^b) καὶ πᾶσα ἡ γῆ σειομένη καὶ τρέμουσα καὶ συνταρασσομένη
(e) (G^b Eth.) καὶ τρέμοντες ἅπαντες οἱ υἱοὶ τῆς γῆς ...

the whole earth ... disturbed Eth. hoka, III.1, Heb./Aram. זוע Ithp. 'be shaken', and so 'be terrified', G^b σείεσθαι, Eth. re'da = G^b τρέμειν, Heb./Aram. רעד and Eth. guague'a I.2 συνταράσσεσθαι, Heb./Aram. בהל (Niph. Ithp.), cf. Jer. 51.32, Dan. 5.9.

the glory of the Great One Cf. 14.20.

(4) **Fear not ... died** G^b θαρσεῖτε is one LXX equivalent of Heb. לא תיראו, Aram. לא תדחלון. If this Aram. was the original here, then it has been reproduced verbatim by Eth., but not necessarily from an original mediated by a Greek μὴ φοβεῖσθε, as Nickelsburg, 122, suggests, since an Eth. translator may well have been familiar with the LXX usage. Moreover, at 104.2 tasaffawu, lit. 'be hopeful' also renders θαρσεῖτε, so that it seems probable that the longer Eth. text here has arisen out of the two renderings of θαρσεῖτε, 'fear not' and 'be hopeful'. A similar 'doublet' is τῶν

---

[1] JThSt 45 (1944), 161-70, especially 165f.; also in JBL 63 (1944), 53f.

[2] Zuntz, *ibid.*, 167 'If they go into hiding, how can they possibly be said at the same time to carry out their task?'.

δικαίων καὶ τῶν εὐσεβῶν of G^{b} both = Aram. קשיטין. At 103.3 where G^{b} reads ταῖς ψυχαῖς τῶν ἀποθανόντων εὐσεβῶν, Eth. has simply 'the souls of those who died in righteousness', which is evidently the translator's equivalent of the reading of G^{b}. An unmistakable half of the same Eth. phrase occurs here at v. 4, viz. 'those who have died in righteousness', possibly also here the end of the complete phrase [ψυχαὶ] τῶν ἀποθανόντων εὐσεβῶν. In Aram. the phrase is נפשת מיתין קשיטין, the souls of the righteous dead'. Both G^{b} and Eth. have arisen by elaboration of this original expression on the basis of these 'doublets' in the versional tradition, θαρσεῖτε twice rendered in Eth. and קשיטין giving rise to δίκαιοι and εὐσεβεῖς. In that case it is the shorter text of G^{b} which is closest to the original: לא תדחלון נפשת מיתין קשיטין, 'Fear not, souls of the righteous dead'. Cf. Nickelsburg, 121f.

(5) **grieve not ... in tribulation** Eth. 'iteḥzanu = μὴ λυπεῖσθε Heb. עצב Niph. Aram.(?) ביש Ithp. (Dan. 6.15).

in your lives your body of flesh ... piety G^{b} lit. 'it did not happen to the body of your flesh according to your piety'. Cf. Prov. 5.11 בשרך ושרך, LXX σαρκὸς σώματός σου: Eth. rakaba, G ἀπαντᾶν = Heb. קרה Aram. קרא Tg. ארע. Eth. ḫirutekemu = G^{b} τὴν ὁσιότητα ὑμῶν, cf. Prov. 14.32 LXX ὁσιότης = (ה)תמ Tg. תמימות.

days of your lives G^{b} only αἱ ἡμέραι ἃς ἦτε crpt. ex יומי חייכון(?). The Greek here has preserved a line lost in Eth.

But wait patiently(?) for the day on which is the judgement of sinners Here there has been extensive corruption: Eth II gives an unfinished sentence, and none of the variants cited in the apparatus can be regarded as improvements. Charles, 212, emended the difficult 'enka ('therefore, now') to ṣeneḥu = 'expect, hope for', the phrase found in a similar context at 108.2,3. A similar emendation is to read አኑኍ 'anuḫu for እንከ 'enka = Aram (ל) אוריכו Tg. for (ל) יחל Hiph. 'to wait patiently for, hope for'. Charles emends konkemu ḫāṭe'ān 'you became sinners' to kuennanē ḫāṭe'ān 'the judgement (of sinners)' and this gives excellent sense; I suggest reading yekawwen kuennanēhomu laḫāṭe'ān, ἔσται ἡ κρίσις τῶν ἁμαρτωλῶν.

(6) **The righteous die like us** The original text lies behind 7a: ὁμοίως ἡμῖν = Heb. כמונו Aram. כמתנא read as from מית by Eth. (כמתנא 'as we die') and apparently by G^{b} as כמני, κατὰ τὴν εἱμαρμένην (Isa. 65.11 מני LXX τύχη, parallel גד 'fortune, god of fortune').

what did they gain G^{b} περιγίνεσθαι = Heb. יעל Hiph., Aram. הני Ithp.

(7) This verse could be an expanded doublet of v. 6 (7a, as noted above, preserves the original opening phrase ὁμοίως ἡμῖν). Against this is its fuller form which could be original, and the parallel, but not identical expressions, 'what did they gain', and 'what advantage did they have'.

in grief and darkness For σκοτία = Heb. חשך, Aram. חשוכא, = 'mourning' (Isa. 47.4), 'evil, sin' (Isa. 5.20), 'perplexity' (Job 5.14) etc.

what advantage do they have over us Eth. fadfād, Heb. Aram. יתר, LXX περισσεία, περισσός.

(7 (end) 8) The Eth. text is confused, and G^{b} is a shortened form of the whole verse which makes some sense, but is clearly, on comparison with Eth., a truncated version of the text behind Eth. In the first half v. Eth. appears to have read a Greek text where σωθήσονται (misread as ἰσώθημεν and connected wrongly with the previous verse) preceded ἀναστήσονται, restoring the natural order: the two words go closely together, but, instead of the G^{b} edited version 'let them rise from the dead and be

saved', Eth. asks an ironical rhetorical question: 'Shall they (τι, מה) be saved (i.e. from death: ἀπὸ τοῦ νῦν crpt. ex ἀπὸ τοῦ θανάτου?) and rise (from the dead)?.

(8) **Truly** Eth. 'esma = כי (Hoftijzer 118)(?)

(9) In G^b's re-editing by shortening the text, the word order has been changed: the καλῶς ... πεῖν rightly belongs, as the Eth. order confirms, with the sentence after τοιγαροῦν (= כל קבל דנא 'in view of this, accordingly'). Eth. 'I say to you, sinners' is an expansion: G^b rightly continues the direct speech of the sinners, '... it is well for us' (read ἡμῖν for ὑμῖν). There is no need to account for καλῶς as due to a corruption of ἱκανόν (Nickelsburg, 123): Eth. 'akalakemu may well be a rendering of καλῶς (sc. ἐστιν) ὑμῖν (read ὑμῖν): ... טב לנו למאכל.

to ... rob ... see good days Cf. LXX 1 Esd. 4.24. Does ἁμαρτάνειν here render רשע Heb. Hiph./Aram. Aph. = ἀδικεῖν. Above, 126.

(10) **Consider then** G^b ἴδετε οὖν, see above, note on 101.4.

those who are truly righteous Eth. simply 'the righteous'. I take G^b nom. pendens οἱ δικαιοῦντες ἑαυτούς as either a mistake for the accus. τοὺς δικαιοῦντας ἑαυτούς, or, as = Heb. צדק Hithp., Aram. קשט Ithp., e.g. 10.21, En^a I vi 3 Milik, 162, יתקשטון = Eth. '(all men) shall become (completely) righteous'. Milik, 163, gives the Ithp. here 'a declaratory nuance', '... all men shall be declared, proclaimed righteous.' This could also be the force of the Ithp. in this verse; I have opted for the simple intensive use of the Ithp. in this context: 'those who are truly, perfectly righteous' parallel to 'for no unrighteousness has been found in them till they die (i.e. all their lives)'. For this use of Hithp./Ithp. cf. G-K § 54d (3) 149).

their end G^b καταστροφή, which renders סוף in the LXX: Eth. tafṣāmēt(a) = סוף 91.13, En^g iv 17, Milik, 266; cf. also 91.19 En^g 1 ii 21, Milik, 260.

no unrighteousness Eth. read πᾶσα ἀδικία which has been changed to πᾶσα δικαιοσύνη in G^b by a scribe who understood the words to refer to the 'sinners' not the 'righteous'.

CHAPTER 103

(1) For v. 1 we are dependent on a confused Eth. textual tradition, which seems, at first, almost impossible to sort out. The translation follows mainly Eth^{tana}, which offers a shorter text, which can not only best explain the variations, but which also employs attested Enoch and Biblical terminology. The first expression, 'by the glory of the Great One' (the reading of Eth I and Eth II 'the great glory') corresponds to G^b 104.1, ἡ δόξα τοῦ μεγάλου (see above, note on 14.10); the second, 'by the glory/honour of his kingdom' recalls the phrase at Dan 4.33 (36) ליקר מלכותי, Theod. εἰς τὴν τιμὴν τῆς βασιλείας μου. The final term 'magnificence' = μεγαλωσύνη, cf. 9.3 Sync. A different form of Eth. text is favoured by Flemming-Radermacher (followed by C-B): 'by the glory of the Great and Honoured and Mighty One in dominion, and by his greatness I swear to you'. ('The expression is suspiciously full' C-B).

(2) **the tablets of heaven ... books of the holy ones** Cf. 18.1,2, 93.2, and Charles, 91 (note to 47.3). I have adopted the reading 'the books of the holy ones' as giving a closer parallel to 'the tablets of heaven' than 'the holy books'. Cf. En^g 108.3 'the books of the Holy One' i.e. God, which could be original, perhaps also here.

inscribed Eth. leku'a = G ἐγκεκολαμμένα = Heb./Aram. פתיח/פתוח.

(3) **the righteous dead** G^{b} ταῖς ψυχαῖς τῶν ἀποθανόντων εὐσεβῶν. See above, note on 102.4.

(4) Cf. especially Wis. 3.1-6, 5.15-16.

nor their memorial Cf. Ps. 9.7, Job 18.17.

the Great One See above, note on 1.3.

(5) **deceased sinners** Eth I mewetān ḫāte'ān cf. G^{b} οἱ νεκροὶ τῶν ἁμαρτωλῶν = חטיין מיתין. Eth II 'sinners who have died'.

your ill-gotten wealth Eth. be'la ḫāṭi'atekemu = הון רשעה; cf. CD 8.12 (Charles, 256, *Apocrypha and Pseudepigrapha* II, 809). Cf. Lk. 16.9, μαμωνᾶ τῆς ἀδικίας.

your associates Eth. 'those who are like you' = οἱ ὅμοιοι ὑμῶν 'your (social) equals', Heb. חבורים, Aram. שותפין(?).

enjoyed properity Eth^{g1} re'yu šannāya lit. 'see good' = ראה טוב Ps. 34.12, Job. 7.7, Ec. 2.24 'to enjoy good, prosperity' (cf. Nickelsburg, 125f., who thinks Eth^{g1} is a gloss). G^{b} ἐν τῇ ζωῇ αὐτῶν occurs also in the next verse, where it is repeated. Here, in view of 'all their days' it seems redundant.

(6) **in splendour** Eth. basebḥat = G^{b} ἐνδόξως, LXX for Heb. גאון, גאה Tg. גיאותא (e.g. Prov. 8.13), and perhaps here 'in arrogance'? Is G^{b} an abridged form of text?

(7) **Sheol** Sheol is here identified with Gehenna as the place of final punishment.

(8) **in darkness** See note above, on 102.7.

in toils Eth. bamarbabt = G ἐν παγίδι. The meaning 'chains' is usually accepted here for marbabt (e.g. Charles, Knibb, following Dillmann, *Lex.*, 287). It is clearly rendering G^{b} παγίς, 'snare', as elsewhere it translates σαγήνη (Ec. 7.26) and ἀμφίβληστρον (Hab. 1.15). The use of παγίς in this context suggests Heb. מוקשי מות (Ps. 18.6 = 2 Sam. 22.6, Prov. 13.14, Prov. 14.27 (ἐκ παγίδος θανάτου, Tg. מן פחא דמותא). Did an original read בפחא דמותא = ἐν παγίδι θανάτου? An alternative would be to explain ἐν πάγιδι as a mistranslation of a noun from √פחח; cf. P. Sm. 3080 ܦܚܝܚܘܬܐ = debilitas, infirmitas.

the great judgement See above, note on 1.7.

Eth. has an expanded text: the original was probably simply 'will come to the great judgement which is for ever and ever'. For the expression לכול דרי עלמין 91.13 En^{g} 1 iv 18, Milik, 266.

you shall have no peace See above, note on 5.4.

(9) In the opening sentence G^{b} is closer to the original. To read a prohibition μὴ γὰρ εἴπητε, followed by the long speech of the righteous describing their afflictions (vs. 9-15) does not make sense. An introductory question μὴ γὰρ οὐκ εἴπετε 'Did you not say etc.?' is required.[1] (Perhaps οὕτως has fallen out of the text by hmt after δίκαιοι: 'Did you not thus say etc.'; cf. the use of אמר כן at Dan. 2.24,25, 4.11, introducing direct speech.)

have met with much evil and been consumed Eth. 'have found many evils' = Heb. מצא Aram. אשכח = 'experienced many evils (misfortunes)'? An alternative Heb./Aram. 'and much evil befell us' (Heb. קרה, Aram. ארע, see above on 102.5) (cf. Eth^{tana}) is also a possible original. Eth. tawaddā'na = G^{b} ἀνηλώμεθα; LXX (ἐξ-)αναλίσκειν = Heb./Aram. כלה (Ezek. 35.15 Pu.; cf. 107.1, En^{c} 5 ii 28, Milik, 210).

have become few and lost heart Cf. Dt. 28.62, Ps. 107.39. The text could mean

---

[1] Perhaps μὴ γὰρ εἴπετε by itself לו גיר תאמרון / אמרתון? can be given this force? 'Were you not saying'?/'Do you not say'? Cf. Blass-Debrunner, § 427.2.

'become insignificant' (cf. below v. 15). Eth. ne'sat manfas = Heb. (קצר נפש(רוח; cf. Num. 21.4, Jg. 16.16 LXX ὀλιγοψυχεῖν.

(10) **we perish ... have we found** Behind Eth II se'nna, could lie ἠσθενήσαμεν, a misrendering of אתחללנא 'we are slain' (cf. Ezek. 32.26, Hoftijzer, 89, Cowley, *Aramaic Papyri*, Ah., 168), but read as from חלה, ἀσθενεῖν. Or should we read taṣ'erna (from ṣer'a III.2 = עצר Ithp.) ἐβασανίσθημεν (cf. Wis. 16.1)? We obtain then a suitable parallel to 'we perish'. G^b has again lost a substantial part of the text by hmt. LXX ἀντιλήμπτωρ = Heb. משגב (Pi.) 'protector', attested also in the old Aram. (Hoftijzer, 291). It is usually assumed that ἀντιλήμπτωρ here lies behind Eth. zayeradde'ana, but rad'a = βοηθεῖν, a different, if parallel verb to שגב. I have adopted the longer text of Eth II and combined the reading with G^b: καὶ οὐδένα ἀντιλήμπτορα οὐχ εὑρήκαμεν = Aram. ומשגב מנדעם לא אשכחנא 'and no protector of any kind we have found'. (For the indef. pron. מ(נ)דעם in the old Aramaic, Hoftijzer, 158, and above, 305).

(11) This verse has been influenced by Dt. 28.13,33,44,48.

fruits of our toil Eth. ṣāmā κόπος = Heb. יגיע (Tg. ליאות); cf. Dt. 28.33 'product, fruits of labour'. Cf. Nickelsburg, 127.

we have become the food for sinners It is possible to read kona as 3rd. sing. masc. with ṣāmā as subject: 'it (the produce of our labour) has become food for sinners'; cf. Dt. 28.33: 'A nation whom you do not know shall eat the fruit of your land and all your toil ...' (NEB). G^b ἐγενήθη(μεν) could have arisen through the influence of the earlier ἐγενήθημεν.

(12) **to those that goaded us** I have accepted the reconstruction of Charles, 258, and C-B, 67: 'to those who hated us' is a redundant ditt. displacing 'to those that goaded us'.

beheaded us Eth^M om. G^b καὶ περικ[υκ]λοῦσιν ἡμᾶς (= Eth^{tana} waya'awwe-duna) misrenders Aram. נקף I by περικυκλοῦν instead of נקף II Heb., Aram. 'to strike off', used in the Tg. to render Heb. ערף, Exod. 34.20 Tg; גזר קדל, Cowley, *Aramaic Papyri*, Ah. 134; cf. Job. 19.26 'to strike off' the skin; Eth. naqafa 'to peel, flay'. The latter is a possible meaning in this verse, continuing the metaphor of the yoke and goad, but the previous phrase about 'bowing their necks' supports the proposed conjecture. They bowed their heads under the yoke only to have them severed. The translation further assumes that G^b, with this addition, supplies an extra line before Eth. 'and shown us no mercy'.

(14) **we made complaint to the rulers** Eth. sakaya = G ἐντυγχάνειν Aram. קבל En. 22.5, En^c 1 xxii 4, Milik, 229. Are the 'rulers' the Seleucids?

slandered ... us G^b τοὺς καταβάλλοντας: Eth. 'ella yeballe'una 'those who devoured us' (cf. v. 15?) or does Eth. bal'a have the force here of Aram. אכל קרצא, 'to slander'; cf. Dillmann, *Lex.*, 489. G^b βιάζεσθαι = Heb. חמס Aram. אחמס Aph.(?). At v. 15 Eth. hēda seems to represent a word like גזל 'to rob' rather than חמס 'to do violence to, to wrong'.

they did not receive our petitions For this clause Eth. has 'they did not pay attention to our cries'. This gives a good parallel to the last clause 'And would not hearken to our voice', and could be original: ṣerāḥa 'iyerē'eyu = לא חזו לזעקתנא, [οὐκ ἐπεῖδον τὰς βοάς ἡμῶν] (cf. Ps. 31.7, Ac. 4.29). Is this an additional line omitted by G^b rather than a free rendering of καὶ τὰς ἐντεύξεις ἡμῶν οὐκ ἀπεδέξαντο?

(15) There are substantial differences between G^b and the underlying Greek version of Eth. in this verse. The sentence in Eth. 'And they helped those oppressing and devouring and making us few ...' reads like an expanded form of the more succinct Greek στερεοῦσιν αὐτοὺς ἐφ' ἡμᾶς. On the other hand, the three lines which follow

and are omitted by G[b] seem original 'And they concealed their wrong-doing ... scattered us'. While Eth. has omitted the line (quite certainly original) 'And they were not informed ...', it may supplement G[b] from its Vorlage in the clause 'and they concealed our murder'. The last full line may have read in the original: 'Nor were they reminded of the sins of the renegades, that they had lifted up their hands against us'.

oppressed and devoured us Cf. v. 14 above. For this use of κατεσθίειν = Heb. Aram. אכל lit. 'to devour' and so 'to destroy, exterminate'; cf. Jer. 10.25 (parallel to and with a word-play on כלה 'to finish off, destroy'). The original could have been Aram. כלא (see 292, 323) or some form of Heb./Aram. נכל 'to deceive', 'to defraud' (Aram. Pa.?) and this would be even more appropriate in the context, especially if the original of the first verse was גזל 'rob'.

supported G[b] only στερεοῦσιν C-B 'they harden against us them who slew us'. But στερεοῦν LXX = Heb. חזק Pi., Aram. תקף Aph. 'to support, encourage'.

informed ... murderers G[b] πεφονευμένων = 'our murdered ones; Eth. qatlana 'our murders' (but cf. Prov. 22.13, 26.13, where qatl renders LXX φονευταί (above, 270)). Has an original קטלין = 'murderers' been read as pass. ptc. קטילין 'murdered ones'?

lifted up their hands against us Cf. 2 Sam. 20.21, Ps. 10.12.

CHAPTER 104

(1) **the angels remember you for good** Cf. Neh. 5.19. For the intercession of the angels, see note on 15.2.

and your names ... Great One G[b] om. This could be an omission due to hmt, but it may be explained as an expansion of Eth.; it seems unlikely that the phrase 'the glory of the Great One' was repeated in the original in two parallel lines, but cf. Eth[g1] 'before the angels of the Great One'. (C-B, 70, 'an alternative version of the preceding sentence'(?)).

the glory of the Great One See above on 14.20.

(2) **Be of good courage** See above on 102.4.

you were worn down by evils and afflictions G[b] ἐπαλαιώθητε translates a pass. of Aram. בלי, 'to become worn out, old', but which, in the Pa., can have the more general meaning 'to wear down, afflict' (Dan. 7.25 LXX κατατρίψει Theod. παλαιώσει). Heb. בלה Pi. is rendered by ταπεινοῦν at 1 Chr. 17.9, and it is this meaning of the semitic root which Eth. ḫasra has given here.

you shall shine and appear G[b] reflects an original תזהרון ותתחזון/ותיפעון: cf. Dan. 12.3, and En. 2.2, En[a] 1 ii 2, Milik, 146, En. 1.4, En[a] 1 i 6, Milik, 142. For the relationship to Dan. 12.3, see G. W. E. Nickelsburg, *Resurrection, Immortality and Eternal Life in Intertestamental Judaism*, 120f. Cf. also 2 Esd. 7.97, 125.

(3) **your cry will be heard** i.e. the cry of the righteous against their persecutors.

your judgement i.e. judgement for you against the sinners.

vengeance shall be required Eth. ḫašaša III,1,3 = תבע (ܐܬܒܥ) Ithp. ἐκζητεῖσθαι, ἐκδικεῖσθαι. Cf. below, v. 7.

from all who have helped ... plundered you Cf. Nickelsburg, 131: συλλαβήσεται ('an unrecorded vulgarism' C-B) and μετέσχεν are virtually synonymous (Nickelsburg, 131 n. 181). Certainly their meanings complement one another, but we cannot account for 'the double reading' as 'a stylistic variant on the Greek level', since the clauses in which the verbs occur have different meanings. Eth. om. ἐφ' ὅσα ... ὑμῶν and reads [οἵτινες συνελάβοντο ...] for ὅστις μετέσχεν. Is the impossible '... your judgement ... will also appear inasmuch as (ἐφ' ὅσα) it will assist (συλλαβήσεται) you concerning your affliction' a rewriting of this last clause? Did the original read: 'and from all who have helped and collaborated with (ὅστις μετέσχεν) those who plundered and devoured you (κατεσθόντων ὑμᾶς).

(4) This verse is omitted in G^b; it seems unlikely that it is an Eth. interpolation, especially in view of the number of omissions in the Greek version.

angels of heaven Cf. 104.6, Mt. 22.30, Mk. 12.25.

(5) **Are you about to commit iniquity ...?** Eth. I menta Eth^m ment) = τι, מה asks a rhetorical question. There is nothing apparently in Eth. corresponding to τὰ κακά (Knibb). It can hardly be coincidence, however, that the second sentence in Eth. begins 'neither ('akko አኮ) will you be going to hide ...' when 'ekuy እኩይ) = τὰ κακά. I suggest reading 'Are you about to commit iniquity'? (τὰ κακά = 'ekuya), where G^b has preserved the object of the question in Eth.: G^b [τί μέλλετε ποιεῖν] τὰ κακά. The 'righteous' is the subject throughout, the irony being maintained till the end of the rhetorical questions.

to be found out ... like sinners G^b εὑρ[εθ]ῆτε = Heb. מצא, Niph., Aram. שכח Ithp., deprehendi. G^b σκυλ[λ]ήσεσθε may preserve an original verb: 'Are you about to be found out to be in trouble, like sinners', Aram. משתכחון תצטערון(?); or read σκυλεύσετε (תבזון) 'to be robbers'(?)

(6) **prospering** Eth I 'prospering in their ways' is a rendering of εὐοδουμένους (Eth II 'in their desires' is an inner Eth. corruption).

from all their evil-doings G^b ἀπὸ πάντων τῶν ἀδικημάτων αὐτῶν: G^b omits 'for you shall become companions of the angels of heaven'.

companions of the angels of heaven I have chosen the more difficult reading of Eth I and adopted the conjecture of Flemming. Eth II seems more likely to be a scribal correction to the more familiar Biblical expression, צבא השמים. For the idea, 2 Bar. 51.5,10,12, and cf. Black, *Scrolls and Christian Origins*, 139f.

(7) **And if you sinners say ... day by day** It is still the righteous who are the subject (cf. Nickelsburg, 134, and in *Resurrection, Immortality and Eternal Life in Intertestamental Judaism*, 114f., 120, n. 31: C-B, 72 emended G^b to μὴ γὰρ εἴπητε οἱ ἁμαρτωλοὶ ὅτι [1] ...). For ἐκζητεῖσθαι = תבע Ithp. see above on v. 3.

(9) **Be not godless in your hearts** G^b μὴ πλανᾶσθε ἐν τῇ καρδίᾳ ὑμῶν see above, 303. The impiety is apostasy, especially idolatry.

charge with lying Eth. ḥasawa I.2 = Heb. כזב Hiph., Aram. כדב Pa. (Job 24.25).

great Holy One See above on 1.3.

count on your idols The more difficult and probably the true reading here is preserved in Eth$^{g\ g^1}$ ḥasaba = Heb. Aram. חשב 'to hold in esteem, value' and so 'count on'. Eth II sabbeḥa = G^b μὴ δότε ἔπαινον has read the original as שבח 'praise'.

all your false gods and all your godlessness Eth. ḥassat = ψεύδη = Heb. כזבין (Aram. כדבין) res mentitae 'idols' (Am. 2.4, Ps. 40.5 etc.) Eth. res'ān = πλάνη. See above, 303.

issue not in ... great sin G^b οὐ γὰρ εἰς δικαίωμα εἰσά[γουσιν] Eth. lit. 'are not for' = ... לא הוו ל(?), i.e. 'will not turn to ...'.

---

[1] The ms reads μὴ γὰρ εἴπητε ὅτι κ.τ.λ. Is this another μὴ γὰρ εἴπετε, לו גיר תאמרון / אמרתון 'Do you not say?' as at 103.9? 'Were you not saying?'

(10) **And now I know this secret ... make apostates of many** Cf. 103.1,2. Did the read אודעת = 'I make known'? G^{b} ἐξαλλοιοῦσιν = Eth. mēṭa, Aram. שנא Pa.; ἀναστρέφειν = Heb. הפך, Jer. 23.26 (Syr[vg] ܗܦܟ), Eth. om.; ἀλλάσσειν = חלף, שחלף(?); cf. ܣܥܠܐ pervertere a fide (P. Sm. 1287).

speak wicked words Cf. Ps. 109.20.

fashion great graven images Eth. faṭara = πλάσσειν feṭrat = πλάσμα (LXX Ps. 102(103).13 = יצר). The original was almost certainly יצרין used in the sense of 'graven images, idols' (cf. Hab. 2.18).

write books Eth[tana] write my books.

according to their words and in their names Eth. has only 'according to their words': G^{b} only ἐπὶ τοῖς ὀνόμασιν αὐτῶν: both clauses seem original, but could also be doublets.

(11) **Would that they would write down** Eth. sobasa = ὄφελον, Heb. אחלי, Aram. לוי (ܠܘܝ) (Exod. 16.3 (מי יתין) Ps. 119.5, 2 Kg. 5.3).

in their languages Eth. = ἐν ταῖς φωναῖς αὐτῶν(?) G^{b} ἐπὶ τὰ ὀνόματα αὐτῶν, a ditt. from verse 10 replacing the original(?).

(12) **And again another secret I know** 'I make known'(?) (cf. above, v. 10).

(13) **shall be recompensed** En^{c}5i (Milik, 206f.) [ו]ישמחון כ[ול...] : the identification of the Aram. fragment is quite certain, as is the following fragment at 105.1. If it seems unlikely that the original read 'shall rejoice and be glad' = G^{b} Eth II, Eth I may preserve the original verb '(shall rejoice) and be recompensed' (wayetʿaššayu = ישתלמון(?)).

CHAPTER 105

(1) The attestation of this verse by the Aram. disposes of Charles's theory that Ch. 105 did not belong to Ch. 91-104, a theory which seemed to be confirmed by the absence of any trace of this chapter in G^{b}; the Eth. version has, on the contrary, reproduced the original sequence of chapters and verses. Cf. Milik, 208, Knibb, 2, 243.

they (the righteous) shall summon ... wisdom If the original was קרא 'proclaim' or 'read', the latter meaning may have been intended: 'that they should read (them,' i.e. 'the books', continuing the theme of 104.13); the following verb could then be rendered 'and make them heard'. For the expression 'sons of earth' (בני ארעא), 12.4 G^{b} οἱ υἱοὶ τῆς γῆς), 15.3, 100.6, 102.3.

Explain to them Eth. ʾarʾeyu = אחוי cf. 33.4 En^{e}1 xxvii 19, Milik, 235, 106.19 En^{c}5 ii 26, Milik, 209.

their teachers and leaders(?) ... earth Eth. marāḫi, in this context = παιδαγωγός, מלפנא, or ἡγούμενος, Heb. נגיד, Aram. רב: Eth. ʿesēyāt lit. 'recompenses' is clearly wrong: what we require is a masc. noun, parallel to 'their teachers'. Eth. ʿasaya = Aram. פרע, and Tg. פורען can mean either 'recompense' or 'official, officer' (Dt. 16.18 = שטר, 2 Chr. 34.13 = מנצחים). Heb. פרע, פרעות means 'leader(s)' (fem. plur. for the office), e.g. Jg. 5.2, Dt. 32.42 (LXX ἀρχόντων.) In Syr. ܦܘܪܥܢܐ means 'avenger' (at Ezek. 9.1 it renders פקודות, qui de urbe vindictam exigunt, P. Sm. 3286). Was there an equivalent use of פורענין/פורענות in Aram. for 'leaders, overseers, avengers'? The fem. plur. would account for Eth. ʿesēyāt.

(2) This verse is usually taken to be part of the oratio recta of v. 1, and the latter attributed to 'the Lord': it is not surprising that it has been almost universally treated

as a Christian interpolation. From a close study of the Aram. fragments, it seems clear that there must have been at least a full line at this point (cf. En[c] 5 i 23-24, Milik, 207). Behind Eth., at the beginning of v. 1, lies the familiar Old Testament LXX expression [λέγει κύριος] (in the older recension of Eth I without the following [ὅτι] of Eth II). Without [λέγει κύριος] v. 1 would naturally be construed as a continuation of the oratio recta of Enoch from the preceding chapter, and the following v. 2 'for I and my son etc.' would refer to Enoch himself and his son Methuselah (106.1 which follows immediately after 105.2 in the Aramaic text at En[c] 5 i 26, similarly refers in the oratio recta of Enoch to 'my son Methuselah'.) The first main verb of v. 2 Eth. damara I.2 usually renders μιγνύναι which itself translates Ithp. חבר at 10.11 (En[b] 1 iv 9), referring to the (illicit) union of the Watchers with the daughters of men. In the present context it would mean 'associate with', 'become associates (חברים) with' the 'righteous' i.e. the Enoch *societas* (חבר), 'in the paths of righteousness in their lives.' Did part of the lost original line read די אנא ובני נתחבר עמהון, 'for I and my son will be associated with them etc.'? An alternative original for μιγνύναι is semitic ערב , used meaning *inter alia* 'to sponsor, stand surety for'. Did the Aramaic read 'for I and my son shall stand surety for them (נערב בהון) etc.?[1] This would also appropriately describe the relationship between Enoch and his son Methuselah with future generations of their followers: they are to be the patrons and sponsors of the righteous for ever, just as the latter are to be the teachers and leaders of mankind. The first meaning, however, 'to be associated with' seems, on the whole, the more likely alternative. Did it refer to the association of the righteous with Enoch and Methuselah through the 'books' of Enoch and their transmission and dissemination by Methuselah? Or is it to be understood as a 'mystic' union? The [λέγει κύριος] can be explained as a parenthesis introduced by Enoch to give divine authority to his commission of the righteous. Or was the familiar formula a Christian gloss or interpolation to produce the singularly appropriate divine promise for generations of Christian believers?

CHAPTER 106

(1) Although barely two or three words are preserved in Aram. (cf. Milik, 352), there can be no doubt that 106.1 followed immediately on 105.2, (after a vacat) and thus belongs to I Enoch. The chapter has also been preserved in G[b] and in a Latin version (see Charles, 264f., described as a fragment of 'the Book of Noah'). Some fragments related to v. 10 have been preserved in **IQ19** (see below, 320). Eth. offers an abridged text which omits a whole generation in this first verse (the generation of Enoch and his son Methuselah), and is also, elsewhere, a somewhat free rendering (cf. v. 2); the Latin version is even more apocopated. The translation follows the longer text of G[b] but (with Milik, 207) supplying 'Methuselah' as subject of ἔλαβεν αὐτῷ γυναῖκα: it is his father Methuselah who takes a wife for his son Lamech, as earlier Enoch had for his son Methuselah.

(saying) 'Brought low has righteousness been' See Milik, 207. We require a λέγουσα = למאמר to introduce the etymological explanation of the name, as in the Biblical

---

[1] A Greek original of Eth. ἐγὼ καὶ ὁ υἱός μου συμμείξομεν αὐτοῖς could be a rendering of either Aram. text. Cf. Prov. 11.13 LXX.

parallels (e.g., Gen. 5.29): Milik conjectures לו מך 'brought low indeed'; also possible is לם מך 'truly brought low'. (For this etymology, Milik, 215).

(2) Cf. verse 10 below.

the tresses ... glorious Knibb om. καὶ ἡ κόμη αὐτοῦ (demdemāhu) as a gloss. It is more probably part of an attempt to render the Biblical 'the tresses of the hair of the head', e.g. Num 6.5 פרע שער ראשו, LXX κόμην τρίχα κεφαλῆς, Tg. פרוע שער רישיה. Cf. also Dan. 7.9 '... the hair of his head like pure wool (כעמר נקא)'. G[b] lit. 'thick (C-B 'curly') and glorious' (עבי והדיר Milik, 207).

(3) **taken up** G[b] καὶ ἀνέστη Eth. naš'a = Heb./Aram. נשא 'lift up' (Eth. III.1 = Niph. Ithp.). G[b] is almost a misrendering—or is it a heightening of the miraculous ('he arose (by himself) from the hands of the midwife')?

blessed the Lord of righteousness G[b] εὐλόγησεν τῷ κυρίῳ. The 'Lord of righteousness' is original: it occurs in the parallel passages at 22.14, 90.40. V. 11 repeats the description here with variants: there Eth. has bāraka = εὐλόγησεν; here Eth. 'conversed with' tanāgara could be a corruption, or perhaps come from tagānaya = ἐξομολογήσατο (Charles). The Latin fragment has both adoravit and laudavit (oravit at v. 11 for adoravit) and dominum viventem in secula.

(5) **children of the angels** Cf **IQ19** (DJD I, 85). See above on 69.4.

his type Eth. feṭrat = G[b] τύπος Heb. תבנית, Aram. דמות (Dt. 4.16)(?).

like the rays of the sun IQ 19 (DJD I, 85), כחדודי השמש.

(6) **it seems to me** Eth. yemasselani = δοκεῖ μοι, cf. Heb. דמה Pi. Aram. דמי Pa. (Tg. Est. II 4.13) G[b] ὑπολαμβάνω.

from the angels The abridged Latin version reads et timuit Lamech ne non ex eo natus esset nisi nuntius dei (sic).

lest ... a wonder will be wrought Eth. manker = Heb./Aram פלא (Ps. 26(25).7, LXX θαυμάσια, Tg. Jer. II Exod. 8.15,18); cf. G[b] μήποτέ τι ἔσται.

(7) **amongst the angels** According to v. 8 and 65.2f. at the ends of the earth.

(9) **a great distress** All mss of Eth. I read ṣāheq 'concern', but ṣā'q = G[b] ἀνάγκη 'distress' seems more appropriate in the context. (Eth II 'a great matter (nagar)').

dreadful vision Eth. rā'y 'eḍub, perhaps Aram. חזות תמיהא(?). Eth. 'eḍub = θαυμαστός (Dan. 8.24, Theod. נפלאות = θαυμαστά). This meaning seems preferable to 'eṣub = σκληρός, קשה, lit. 'hard, difficult' (Knibb = Charles, 'disturbing').

(11) **blessed the Lord of heaven** Cf. v. 3. G[b] τὸν κύριον τοῦ αἰῶνος as at 9.4 (cf. 81.10) (see note on 9.4). 'Lord of heaven', if original, occurs here only in Enoch, but cf. Dan. 5.23 מרא שמיא and ap.Gn. 7.7, 12.17, 22.16 (Fitzmyer *Genesis Apocryphon*, 99, 101, 177): 22.16 'Lord of heaven and earth' (מרא שמיא וארעא) = Gen. 14.19(?); cf. also En. 84.2 'Lord of the whole creation of heaven', Jub. 32.18 'the Lord who created heaven and earth' (= Gen. 14.19), Test. Benj. 3.1 'the Lord of heaven and earth'. (For Phoenician and Nabataean parallels, see Fitzmyer *Genesis Apocryphon*, 99.) Lat. frg. oravit dominum viventem in secula.

(12) **and behold ... the truth** G[b] om. 'and behold, ... make known to me' and in its place has the unintelligible (τὴν ἀκρίβειαν) εχιει which C-B suggested might be a corruption of ἣν ἔχεις, but this is an unconvincing guess. Eth. assumes a verb such as ὑποδεικνύειν, Aram. אחוי. We could have a corruption of (ἵνα) ὑποδείξης (μοι) '(that) you might tell (me) the exact truth'. Or could εχιει be an attempt to transcribe Aram. אחוי by a translator who failed to understand the original? For אחוי cf. 106.19, En[c] v ii 26, and for יציבתא, ἀκρίβεια, 93.2 En[g] 1 iii 20.

(13) **will make a Promise** I read ἀνακαινίσει with C-B for ms αναξκαινισει. Eth. presupposes אנא חנוך, from which αναξ could be a corruption. Eth. is probably

influenced by Isa. 42.9, 43.19, 48.6; for a similar phrase to that in G^b cf. 2 Mac. 4.11. Milik's version (210) 'Truly [The Lord] will restore [His Law on the earth ...]' is based on a reconstruction of En^c 5 ii 17 (frg. c): [בׄל יחׄ]דת מריא לדתא על ארעא; יחדת for ἀνακαινίσει seems right, and דתא 'Law' 'is not impossible' (Barr, 'Aramaic-Greek Notes, II, 181), but Barr prefers מלה for πρόσταγμα and renders 'the Lord will do a new thing (יחדת מריא מלה) upon the earth.' In that case, however, the familiar Biblical expression 'to do a new thing' (עשות חדשה Tg. עביד חדתא Isa. 43.19) would have been used. The conjectural reconstructed text proposed 'Truly the Lord will make a Promise ([... יכׄ]רת מריא מילה) requires us to assume that יכרת (Hoftijzer, 127) has been misread as יחדת (Did Aram. read יחרת, Hoftijzer, 97, Milik, 282?). The reference would then be to Gen. 9.9 and for the expression מילה for the covenant with Noah cf. Hag. 2.5 (cf. RSV). The Greek mistranslation may have been influenced by the idea of the Noachic commandments (G^b πρόσταγμα). The omission in this context of all mention of the Covenant with Noah would have been surprising.

and according as I was shown Fragment c of En^c 5 ii (Pl. XV) seems to me to belong here and not at the beginning of the verse (Milik, 209): ... [כלק]בל די = G^b τὸν αὐτὸν τρόπον (cf. Dan. 2.40 Theod.) For τεθέαμαι 'I was shown' see En. 32.1, En^c 1 xii 30, Milik, 202.

in the generation of my father Jared G^b om. Jared. Cf. En^c 5 ii 17, fragment d, Milik, 209 (d is missing on Plate XV).

exalted ones(?) of heaven ... covenant of heaven We may take the verb here 'aḫlafu as an impersonal plural and render: '(in the generation of my father Jared) the word of the Lord was transgressed from heaven above'. It seems more likely, however, that the subject is contained in Eth. 'emmal'elta samāy, lit. 'some/ones from the height of heaven' (ἐκ ὕψους οὐρανοῦ?) = Aram. ממעלי שמיא, 'some exalted ones of heaven'; or perhaps we should read simply מעלי שמיא 'exalted ones of heaven'(?) (cf. Eth^{ehk} mal'elta samāy). For the expression, Levy, CW II, 57 מעלי = ܡܥܠܝ P. Sm. 2883; cf. עליא = coelestes, P. Sm. 2889. The original מעלי has been confused by translators with some form of the root עלי (cf. Aram. מלעיל, Heb. מעלה). The reference, as the context shows, is to the fallen 'sons of God', the 'watchers' (as Eth^n spells out by reading 'angels of heaven'). C-B explains G^b παρέβησαν τὸν λόγον κυρίου ἀπὸ τῆς διαθήκης as 'a double construction, once with a direct object, once with a preposition and genitive'(81f.). This is an unlikely construction, however, in Heb. or Aram.: I suggest a second παρέβησαν on the grounds that παραβαίνουσιν τὸ ἔθος in v. 14 is a 'doublet', a more general statement of the same idea, (cf. Eth^{tana} which uses šer'āt for both διαθήκη and ἔθος). Moreover, G^b requires a subject, which is supplied from Eth. from this clause, and ἐκ ὕψους ... παρέβησαν could have been lost by scribal error. I suggest παρέβησαν τὸν λόγον κυρίου [ἐκ ὕψους τοῦ οὐρανοῦ καὶ παρέβησαν] ἀπὸ τῆς διαθήκης τοῦ οὐρανοῦ. The same verb need not have appeared in the original Aram.: παραβαίνειν τὸν λόγον = עבר מילה; παραβαίνειν ἀπό = סטא מן, Heb. סור מן, LXX Dt. 17.20 ἵνα μὴ παραβῇ ἀπὸ τῶν ἐντολῶν. For earlier conjectures and discussion of this verse, see Torrey, *Notes*, 60, Knibb, 2, 245f. Cf. also above, 196, on 39.1.

(14) **transgressed the law** G^b τὸ ἔθος = Eth. šer'āt. The word is not used in the LXX. Cf. Ac. 6.14 and 15.1 where it is a comprehensive and general term embracing the 'customs' of Moses, i.e. the covenants, law and commandments. The clause seems to be a doublet of the previous clause in G^b παρέβησαν ἀπὸ τῆς διαθήκης. En^c 5 ii 18 fragment e ועברין = wayaḫallefu = καὶ παραβαίνουσιν.

they had intercourse with women Eth. tadammaru = G^b μετὰ γυναικῶν συγγίνονται. The reading of Milik En^c 5 ii 18 'perverted their nature to go in (to women)'

(שנין למעל) does not appear on Pl. XV; cf., however, 7.1, En[b] 1 ii 18 'and they began to go in (to them)' (ושריו למעל). Aram. על ל = Heb. בוא אל coire cum femina. Above, 257.

they bore-children ... creatures of flesh G[b] καὶ τίκτουσι οὐχ ὁμοίους πνεύμασιν ἀλλὰ σαρκίνους. V. 17a certainly belongs at the end of this v. 14 (cf. Charles, C-B, Milik, Knibb); Milik follows C-B and the shorter text of G[b], the latter explaining Eth's 'and they begot giants on the earth' as an interpolation from 7.2. But, in view of the apocopated text of G[b] elsewhere it seems more likely that, after τίκτουσι, the first part of the accusative has dropped out by scribal error, viz. τέκνα αὐτοῖς ἐπὶ τῆς γῆς γίγαντας (οὐκ ὁμοίους κ.τ.λ.). Did the original read גברין די לא דמין לרוחין להן לבשרין (Milik, 209); בשר is used for 'man' as a creature of flesh and blood, over against God, or, here, in contrast to 'spirit'.

(15) For possible traces of Aram. of this verse at En[c] 5 ii 20, see Milik, 209.

And there will be great wrath ... earth Eth. has this clause at 17b and a ditt. of 15c 'and there will be great destruction' replaces it here. Something has clearly gone wrong in the Eth. version (cf. Knibb, 2, 246) and we should follow G[b] here.

for one year The duration of the Flood has been calculated on the basis of Gen. 7.11 and 8.14.

(16) **shall be left ... all mankind ... shall die** Milik's reconstruction of En[c] 5 ii 21 is short by 13 letters from the average 55 letter line of this part of En[c]. He omits Eth. 'shall be left on the earth', G[b] καταλειφθήσεται an allusion to Gen. 7.23 (LXX κατελείφθη μόνος Νωε), a clause which must have been an integral part of the original text, and for which there is space in the line. I suggest a line incorporating this clause:

[ודן עלימא] Line 20

די יליד[לכו]ן]ישתאר על ארעא ו]יפ[לט הוא ובנוהי כד ימותון כול בני] ארעא Line 21

('And this child) which is born to you will be left on the earth, and[1] he and his sons will escape[1] when all mankind on earth[2] shall die ...'. (For the connection of G's καταλειφθήσεται with the etymology of Noah, see below on v. 18. Milik's attempt (214 top) to relate פלט to Gen. 5.29 is unsuccessful.)

his three sons See Gen. 6.10.

the earth ... of great corruption See Milik, 213, note Ll. 21-2; G is a foreshortened text and πραΰνει = ותנוח ארעא, wrongly given a transitive sense. For this word-play on the name 'Noah', see below on v. 18.

(17) This verse is now incorporated in verses 14 (17a), 15 (17b) and 16 (17c).

(18) **call his name Noah** G[b] reads δικαιωσαιωσιων which C-B reads as δικαίως καὶ ὁσίως omitting ιων in his text: the second αιως, however, is a ditt. and ιων the relics of [Νῶ]εον, (cf. Josephus Νῶεος, *Ant.* xx.2(25) etc., H-R III, 121) (CB's καὶ ὁσίως is the source of Milik's 'pious call this boy who is born' (213 note Ll. 22-23).

your remnant That the original here was ניח = ἀνάπαυσις (as Milik proposes) is borne out by the explanation of the etymology in G[b] ἐφ' οὗ ἂν καταπαύσητε καὶ υἱοὶ αὐτοῦ (= בדיל די תנוחון הוא ובנוהי, cf. Milik, En[c] 5 ii 23). For this etymology of Noah, see Milik, 215, C-B, 80f. G[b] = Eth. ὑμῶν (ὑμῖν) κατάλειμμα introduces a second etymology of the name 'Noah', based on another sense of the word נוח, (Hiph. 'to leave behind'; cf. Martin, 282, thus, Sir. 44.17, Νωε εὑρήθη τέλειος δίκαιος ... ἐγενήθη, κατάλειμμα τῇ γῇ ὅτε ἐγένετο κατακλυσμός. Cf. also v. 16 above.

1-1 The alternative Eth. form of the clause at the end of the verse.

2 Cf. 91.14, En[g] 1 iv 20.

from all the sins ... in his days Cf. C-B ἀπὸ πάντων τῶν ἁμαρτωλῶν καὶ ἀπὸ πασῶν τῶν συντελειῶν ἐπὶ τῆς γῆς. Eth. is a better guide to the original text.

(19) **the secrets of heaven(?) ... informed me** The reading of En[c] 5 ii 26 'the angels (holy ones) have shown me' is quite certain. Milik, 216 note L. 26, suggests that Eth. has transposed the subjects of the two clauses, reading 'the secrets of the holy ones, for the Lord has shown (them) to me'. There is a lacuna in the Aram. where the object of the first clause stood; ('the secrets of the heavens' (ברזיהון דשמיא)(?) cf. 41.4); cf. Knibb, 2, 248.

CHAPTER 107

(1) **I beheld ... shall wrong them (the descendants of Noah)** Milik reads και ειδων τοδε in G[b] and explains it as crpt. for κατ' εἶδος τόδε = בכדין, where I propose reading בהון. These word seem more likely, however, to be a duplicate of τότε τεθέαμαι (καὶ εἶδον τότε) at the beginning of the verse. The reading יבאש בהון 'shall wrong them', referring back to the descendants of Noah (the Remnant) of 106.18, is supported by the reading ἐπ' αὐτούς = עליהון (En[c] 5 ii 29) at the end of the verse.

injustice cease En[c] 5 ii 28 חמסא יכלא par. to רשעה יסוף where Aram. כלא has the sense of Heb. כלה 'to be brought to an end'. See note on 91.11, 292.

(3) **he will rejoice the earth** Eth. faššeḥa IV.2 = G[b] εὐφραίνειν, another interpretation of נוח but in this case again along the lines of the Biblical exegesis of Gen. 5.29, perhaps from ינחנו, (LXX διαναπαύσει ἡμᾶς); cf. C-B 86f.

CHAPTER 108

This last section of Enoch (no trace of which is found in Greek or Aramaic) is clearly an independent addition of later date and composition. It is still written in a language which points to 'translation-Greek' (and, therefore, in this part of the book, an Aram. original). The writer is familiar with the earlier parts of Enoch, e.g. at v. 3 he is developing ideas about Gehenna on the basis of the description of the fiery place of punishment of the disobedient stars at Chh. 18, 21. See Charles, 269.

(1) **keep the law** i.e. the law of Moses; cf. 99.2, unless šerʿat = διαθήκη (so Dillmann, 328, comparing 93.4). The expression, referring directly to the Mosaic ordinances, is of later date than corresponding phrases in the earlier chapters, e.g. 'to practise righteousness' (Dillmann, ibid.).

(2) **You who did good ... evil-doers** Biblical expressions (Ps. 14.1,3, 53.3,4, 2 Kg. 21.9) and in this v. polar expressions, and this makes the reading of Eth II lit. 'you who have performed it, (the law?) 'most probably a scribal slip or alteration.

(3) **blotted out of the Book of Life** Cf. Ps. 69.28. Eth. 'Book of Life' may be an alternative for 'books of the holy ones', the similar expression used at 103.2. On the other hand, it was in the Book of Life that the names of the faithful were recorded (Ps. 69.28). Note the reading of Eth[g] 'books of the Holy One'.

their spirits shall be slain Cf. 22.13, 98.3, 99.11 and Mt. 10.28.

a place deserted and void This is clearly intended by the author to be the 'chaotic flaming hell beyond the limits of the earth' (Charles), described at 21.1-2 and 18.12:

21.2 G τόπος ἀκατασκεύαστος καὶ φοβερός recalls Gen. 1.2 LXX ἡ δὲ γῆ ἀόρατος καὶ ἀκατασκεύαστος, and, in recalling za'iyāstare''i this verse reproduces the ἀόρατος of the LXX, not in its familiar Greek meaning of 'invisible', but in the sense of Heb. תהו, 'wilderness' (cf. 60.8 and Dillmann, 184). The original may have read צדיא ותהיא (באתרא) 'deserted and turbulent' (cf. the use of תהא in the Tg., Levy, CW, II, 530). The Greek translator appears to have interpreted the words of the original in terms of the LXX of Gen. 1.2. Eth$^{g'}$ has the variant bamakāna dayn za'iyāstare''i 'in a place of judgement without form ...'.

in the fire shall they burn Eth I naddu, proph. perfect (Flemming).

(4) **something like a cloud that was not discernible** Eth. re'ya III.1 (cf. III.3) = Aram. שכל Ithp. (cf. Heb. שכל)(?), 'to recognise', = ἀθεώρητος? Does this mean that the object in question, resembling a cloud-mass, became lost to sight because of its immense distance ('depth'), the latter preventing recognition of what it truly was?

because of its depth The seer is contemplating, not the heights of heaven, but the depths of hell.

I was unable to observe (it) Eth. = [οὐκ ἐδυνάμην ἐπισκοπεῖν(?)] = lā'la naṣṣero) The seer was unable to take proper sightings of it (ἐπισκοπεῖν is a technical astronomical term).

flames of fire ... shining ... shining mountains ... thither For sebuḥ, ἔνδοξος, in this sense, cf. P. Sm. 4025, B. O., iii.1.227 columnae igneae ... [Syriac] splendentes. The writer is drawing on 18.15 and 21.3: 18.15 refers to the stars κυκλούμενοι ἐν τῷ πυρί (En^{c} 1 xx, Milik, 228) and 21.3 to seven stars 'like great mountains and burning with fire (G ἐν πυρὶ καιομένους)'. The 'burning stars' like great mountains would support the conjecture that the vision in this verse was simply of the blazing fires of this hell beyond earth and of 'burning' mountains. Cf. Isa. 33.14.

(6) **everything that is to happen** It seems unlikely that the Greek read (τῶν προφητῶν) τῶν μελλόντων γενέσθαι. See Dillmann, 329.

(7) **some of them are to be written** This seems an odd expression, whether it refers to the spirits of the sinners, the words of the prophets, or the deeds to come. Eth^{g} has 'some of you' 'emennēkemu. Was the original perhaps מנך, 'For they (the deeds of the wicked) are to be by you written down and inscribed above in heaven ...'? Enoch is the celestial recording scribe; cf. 12.3f., 15.1, Jub. 4.23, 2 En 40.13, 53.2, 64.5, Test. Abrah. (ed. James), 115.

the humble i.e. the ענוים of the Psalms and the Qumran Essenes (e.g. IQH 5. 21). The reading of Eth^{tana} 'apostates' does not fit the context which, from this point on, is chiefly concerned with the rewards of the righteous.

afflict their bodies Does this refer to some form of self-castigation? Cf. Lev. 16.29. Comparison of Eth^{tana} with the traditional text suggests that a scribe at some time has taken considerable liberties with the transmitted text. (Charles suspected Essene influence, but this seems unlikely.)

(9) **a passing breath** Cf. Job. 7.7f. and Jas. 4.14. The righteous hold this life of no account: they loved heaven more than their life (v. 10 lit. 'breath').

the Lord tried them much Cf. Wis. 3.5.

(10) **I have recounted in the books** Enoch is the speaker: the speech of the angel ends at verse 9.

loved heaven ... their life in the world Cf. 48.7. Is 'heaven' samāy a mistake for semeya 'my Name'?

yet they blessed me (the Lord) From the next verse till at least v. 12 all editors assume that the speaker is God (or the Lord); cf. especially v. 12 'I will bring forth in shining light those who have loved my holy Name'. Thus Dillmann, 330 on v. 11,

'Nun springt die Rede in eine Rede Gottes über die fortlaüft bis v. 12'. While such a change of person is found in poetic discourse (see G-K § 144 p, 462) the 'leap' here may seem too abrupt. Has there been textual corruption?[1] Alternatively, an author/editor of this last 'writing of Enoch' may simply have tacked on vv. 10-12 (15), where the Lord is the speaker from an independent source, leaving the sudden change of person, from third to first person, as in his source. (Note that Eth[g1] omits vv. 11-15, ending the book at v. 10 with the words 'yet they blessed' (without the suffix 'they blessed me').)

one by one Eth. la'ar'estihomu = Heb. לגלגלותם, Num. 1.2 LXX κατὰ κεφαλὴν αὐτῶν Aram. לראשיהון capitatim, Syr.[vg] Num. 1.18,20.

(11) **I will summon ... the generation of light** For 'generations of light' cf. 61.12 ('every spirit of light') and the בני אור of Qumran; for similar expressions in the New Testament, Lk. 16.8, Jn. 12.36, 1 Th. 5.5, Eph. 5.8f.

(12) Cf. Dan. 12.3.

(13) **times without number** i.e. degrees of shining not periods.

(15) **days and times** i.e. presumably for their punishment.

---

[1] Perhaps λέγει κύριος has been accidentally dropped from the text after 'yet they blessed me'?

TEXTUAL NOTES

These notes consist of (1) Eth. variant readings, where the Ethiopic version has been collated against the Greek translational traditions, and (2) Eth. variations within the Eth. textual tradition itself. The Greek versional base for the collation has been the Brill edition of the Greek Enoch, *Apocalypsis Henochi Graece*, edidit M. Black (Leiden, 1970), supplemented by *Appendix* B *Apocalypsis Henochi Graece: Addenda et Corrigenda*, the latter largely the fruit of the collation of the Eth. version. For the Eth. version itself I have used the editions of Flemming and Charles, supplemented by that of Knibb (Introduction, 2f.).

As is explained fully in the Introduction, instead of an Eth. or English apparatus, I have rendered these Eth. *variae lectiones* back into their equivalent in Biblical (or Septuagint) Greek. In many cases these 'Greek' variants are bound to be identical with the actual Greek text read by the Eth. translator — no alternative equivalents are possible — but it must be emphasised that these are putative Greek variants only; to prevent confusion with genuine Greek variants (e.g. those of Sync.) I have cited these equivalents in the commentary in square brackets; in the Textual Notes they appear in unaccented Greek, with ms attestation where the tradition divides (Eth I, Eth II, Eth^{M}, Eth^{mss}). Care has been taken to employ only vocabulary, grammatical forms or syntax fully attested in Biblical Greek, chiefly from the LXX, but including other surviving 'translation-Greek' literature from the intertestamental period.[1] In addition, the extant Aram. fragments have been collated from the editio princeps of J. T. Milik (Introduction, 1 n. 4).

CHAPTER 1

(1) En^{a} 1 i 1 [מלי ברכתא כדי ברך] חנוך לבח̇[ירין קשיטין] εκλεκτους ⌜και⌝ δικαιους | $Eth^{tana\ u}$ τους ασεβεις (rasiʿān Eth^{M} ʾekuyān warasiʿān) | om. και σωθησονται δικαιοι

(2) En^{a} 1 i 2 ונסב מתליה [ואמר אימר חנוך] και απεκριθη Ενωχ και ειπεν | Eth I ανθρωπος δικαιος ου παρα του θεου οφθαλμοι αυτου ανεωγμενοι και ορων ορασιν του αγιου του εν τοις ουρανοις ην εδειξεν μοι οι αγγελοι | En^{a} 1 i 3 ומן מלי̊ [עירין] וקדישין כלה [שמעת] και ηκουσα παρ αυτων παντα και εγνων εγω ο εθεωρουν | En^{a} 1 i 4 [ולא להדי]ן דרה להן לד[ר ר]חיק אנה אמ̊[לל] Eth II αλλα επι γενεαν (Eth. I om.)εσομενην πορρω ουσαν om. διενοουμην et εγω λαλω

[1] The main 'lexicon' has been the Hatch-Redpath *Concordance*. Thackeray's *Grammar of the Old Testament in Greek* (Cambridge, 1909) is also an indispensable tool, but should be supplemented by later studies, such as M. Zerwick's *Biblical Greek* (Rome, 1963). See further M. Black, The Biblical Languages, in *The Cambridge History of the Bible*, *From the Beginnings to Jerome*, edited by P. R. Ackroyd and C. F. Evans, 7-11. For extra-LXX literature, consult Albert-Marie Denis, *Introduction aux Pseudépigraphes Grecs d'Ancien Testament* (Leiden, 1970), *Apocalypsis Henochi Graece* edidit M. Black, *Fragmenta Pseudepigraphorum quae supersunt Graeca* ... collegit et ordinavit Albert-Marie Denis (Leiden, 1970).

(3) Eth^tana cf. G περι των εκλεκτων λεγω ⸢και περι αυτων⸣ απεκριθην την παραβολην (mesālē pro mesla mesālē) | ελευσεται ο αγιος ⸢και⸣ μεγας | + και ο θεος του αιωνος post αυτου

(4) και εκειθεν πατησει επι το Σεινα ορος | εν τη παρεμβολη αυτου | En^a 1 i 6 [תיה]גבור[בכוח] ויופׄע Eth I c. G εν τη δυναμει (Eth II ισχυι) της ισχυος (Eth II δυναμεως) αυτου | εκ του ουρανου

(5) (επι)σεισθησονται G πιστευσουσιν | om. και 3° ... τα ακρα της [γης] (G ασωσιν leg. ζητησουσιν?) | En^a 1 i 7 [בכל קצות] ארעא ויזׄו[עון כ]ל קצ[ות ארעא] | φοβος και τρομος

(6) Eth^M και (Eth^tana + πεσουνται cf. G.) και σεισθησονται ορη υψηλα om. ⸢και πεσουνται και διαλυθησονται⸣ | om. ⸢του διαρυηναι ορη⸣ | En^a i 8 [וישפלון] רמן | απο φλογος

(7) Eth I c. G διασχισθησεται (teššaṭṭat Eth II tessaṭṭam) | om. σχισμα ραγαδι | + (post κατα παντων) και κατα παντων των δικαιων (ex v. 8 cf. Flemming, 2)

(8) τοις δικαιοις | και συντηρησει τους εκλεκτους | om. και ειρηνη | και γενησεται ελεος επ αυτους | ευοδωθησονται και ευλογηθησονται | om. και παντων ... ημιν | φως ⸢θεου⸣ | om. και ποιησει επ αυτους ειρηνην

(9) Eth^tana ιδου c. Jude 14 Eth^M και ιδου G οτει = οτι (Milik οτε) | Eth^tana e c. G ερχεται Eth^M ηλθεν c. Jude 14 | En^c 1 i 15 [עם רבו]את קדישׁי[ן] εν ταις μυριασιν αγιων cf. Jude 14 א (1846) P 72 pc. εν μυριασιν αγιων αγγελων (Nestle-Aland[26] εν αγιαις μυριασιν αυτου) | Eth^tana c. G κατα παντων (Eth^M κατ αυτων) | απολεσαι τους ασεβεις και ελεγξαι (Eth I yezallef/yezlef / yezzālaf Eth II yetwāqaš) πασαν σαρκα En^c 1 i 16 [לכל ב]שׂרא | Eth^tana περι παντων των εργων ασεβειας zagabru leg. zagebra [zarasʿu]) ων (Eth^M και) ησεβησαν κατ αυτου αμαρτωλοι και ασεβεις | En^c 1 i 17 [ועל כול מלין] רברבין וקשין om. και σκληρων ... κατ αυτου

CHAPTER 2

(1) En^c 1 i 17-18 [אתבוננא] בכל עׄוׄ[בדוהי?] וחז]וא לכון לעובד ש[מיא] Eth^tana c. G κατανοησατε (ṭayyequ Eth^M ṭayyaqqu) | πως ουκ ηλλοιωσαν τας οδους αυτων οι φωστηρες οι εν τω ουρανω En^c 1 i 18-19 [די לא חלפו? אורחתהון נהוריא] במסרת [דב]ריהון | Eth II οτι τα παντα ανατελλει και δυνει (Eth I yaʿaqqeb φυλασσει? | om. τεταγμενω et και ταις εορταις αυτων φαινονται | En^a 1 ii 1 ולא מעׄבׄ[רין] בׄסׄרכן και ου πăραβαινουσιν την ταξιν αυτων

(2) En^c 1 i 20 (cf. En^a 1 ii 1-2) חזוא לכון ל[א]רעא ואתבוננא[ו]/וׄאׄ[תבו]ננו בעובד[ה]/ בעבדה Eth I c. G και διανοηθητε περι (Eth II εκ) των εργων | om. ως ουκ εισιν φθαρτα | ως ουκ αλλοιουνται παντα τα εργα του θεου φαινομενα cf. En^c 1 i 21, En^a 1 ii 7 וכׄול מתחזא ל[כ]וׄן

(3) En^a 1 ii 3 עליה ....ׄ........ [קיטא] חזו לדגלי ως πασα η γη πληρης υδατος και νεφελη και δροσος και υετος αναπαυουσιν επ αυτην[1] En^a 1 ii 3-4 [ובדגלי שתוא ד[כל ארעא [תתמלא מין ו]עננה...

---

[1] ועננה וטל ומטרא מחתין עליה Cf. Enastr^d i, below, Appendix B, The 'Astronomical' Chapters of the Ethiopic Book of Enoch (72 to 82), 419.

CHAPTER 3

Ena 1 ii 4 cf. Enc 1 i 23 חזו דכל אילנ[יה] כ̊להן מיבשין Ethtana t c. G καταμαθετε (ṭayyequ EthM ṭayyaqqu) και ιδετε (re'eyu EthM re'iku) παντα (Eth II pr. οτι) τα δενδρα πως φαινεται ως (Ethq om.) ξηρα | και παντα τα φυλλα εκπιπτοντα/αποῤῥεοντα εκτος διατεσσαρων δενδρων Ena 1 ii 5 cf. Enc 1 i 24 [ונתירין כול עלי]הון ברא מן ארבעת עסר אי̊ל̊נ̊י̊[ן] | α ουκ εκπιπτει/αποῤῥει αλλα μενει απο του παλαιου ('embeluy (leg. za'enbala ⌜'eska⌝ yemaṣṣe' ḥaddis?) εως γενηται το νεον ειτε δυσιν ειτε τρισι ετεσιν Ena 1 ii 5-6 = Enc 1 i 25 [עד מתחדתין מבלי] מתקימין ד̊ע̊ליהון ד̊תרתין ודתלת ש̊נ̊ין

CHAPTER 4

Ena 1 ii 6 [קיטא] הזו לכון לדגלי Ethu και παλιν καταμαθετε (ṭayyequ EthM ṭayyaqqu) τας ημερας | τον καιρον του θερους | Ena 1 ii 7 + Enc 1 i 26 [די הות חממה ב]הון כוייה ושלק̊ה ως γινεται ο ηλιος επ αυτην (sc. γην) | Ena 1 ii 7 cf. Enc 1 i 26-27 ואנתון מ̊ט̊ל̊ל̊י̊ן̊ ו̊ט̊ל̊ל̊י̊ן בעין מן קדמיה [על חממתיה וארעא חרת מן כויתה] Ethtana και (EthM om.) εναντιον αυτου (sc. του ηλιου) υμεις ζητειτε σκεπην και σκιαν κατα το καυμα του ηλιου και η γη καιεται εκ θερμης (Ethm + και) καυσωνος | Ena 1 ii 8 cf. Enc 1 i 27 [ולמ]דרך על ע̊פ̊ר̊ה [ועל כיפיה] לא תשכחון מן [חמתה] και υμεις ου δυνασθε πατειν επι την γην η επι την πετραν κατα το καυμα αυτου

CHAPTER 5

(1) Enc 1 i 28 + Ena 1 ii 9 [אתבוננוא] בכול אילניא כולהון [די יעלון] בהן ירוקין וחפין [להון] Ethg u καταμαθετε (ṭayyequ EthM ṭayyaqqu) ως τα δενδρα εν φυλλοις χλωροις σκεπεται | Ena 1 ii 10 [וכול פיריהון להוא לה]דר תשבחה και καρποι/καρπωσι (yefarreyu) om. και πας ο καρπος αυτων εις τιμην και δοξαν | Ena 1 ii 10 + Enc 1 i 29 חק[רו ו]אתבוננו בכול עבדיה א̊ל̊י̊ן και διανοηθητε περι παντων και γνωτε | Ena 1 ii 11 + Enc 1 i 30 [ואשכילו די אלהא די]ן חיא לעלם/לכול עלם עבד כל עבדיה אלין ως εποιησεν αυτα παντα ουτως ? ο ζων εις τον αιωνα cf. Flemming, 3 leg. gabra kamaze (Eth I crpt gabarkemu Eth II gabra lakemu) Knibb 2, 65.

(2) Ethm και τα εργα αυτου ενωπιον αυτου απ ενιαυτου εις ενιαυτον γινονται (EthM zayekawwen Ethtana wayekawwen) | και παντα τα εργα αυτου δουλευει (yetqannayu) αυτω και ουκ αλλοιου(ν)ται | αλλα ωσπερει επεταξεν ο θεος ουτως τα παντα γινεται Ena 1 ii 12 וכלהון עבדין ממרה

(3) Ena 1 ii 12 om. v. per hmt? Ethtana t1 u ιδετε (EthM pr. και) πως αι θαλασσαι και οι ποταμοι ομου (ḫebura) αποτελουσιν τα εργα αυτων

(4) Ena 1 ii 12-13 ואנתון שניתון עבד כן [. . . ותע]ב̊רון עלוהי [ותמללון עלוהי מלין] רברבן וקשין בפום (ביום ms) טמתכן [על רבותה קשי לב]ב̊ן לת שלם לכן | ουδε εποιησατε τας τας εντολας του κυριου | om. οτι κατελαλησατε εν τοις ψευσμασιν υμων

(5) Ena 1 ii 14 [. . חייך יבדן] ושני תלותון יומיכון אדין απολειτε | Ena 1 ii 15 ושני

אבדנכן יסגון בל]וט ע]ל[מין ורח]מ̊ין Eth I c. G και [ετη απωλειας υμων] πληθυνθησεται εν (Eth II om.) καταρα αιωνων cf. Flemming, 4 | om. και ειρηνη

(6) Eth I τοτε δωσετε τα ονοματα (sema Eth II salāma) υμων εις καταραν αιωνιον πασιν τοις δικαιοις En[a] 1 ii 16 [ללוט עלם ל[כל קשיטין ... | και υμας (leg. εν υμιν) καταρασονται οι αμαρτωλοι παντοτε και εν υμιν (leg. bekemu) ομου μετα των αμαρτωλων leg. ομουνται οι αμαρτωλοι cf. Charles[1906], 10, Flemming, 4. | om. G και παντες οι αναμαρτητοι ... καταρα

(7) En[a] 1 ii 17 [ו]לכל [בחירין]? | χαρα

(8) και τοτε δοθησεται τοις εκλεκτοις σοφια (G τοτε δοθησεται ... γην ditt.) | ου κατα ληθην | om. G και εσται ... πλημμελησουσιν et leg. αλλ οι την σοφιαν εχοντες εντραπησονται (yeganneyu) cf. Charles[1906], 11.

(9) ουδε μη αμαρτωσιν ουδε κριθησωνται ετι | εν οργη ουδε εν θυμω | cf. En[b] 1 i 1 כול יומי חייהון

CHAPTER 6

(1) En[b] 1 ii 2 והווא כד[י][1] Eth[g] c. G ου αν Eth[M] c. Sync. οτε | αυτοις c. Sync. | En[b] 1 ii 3 שפירן ו[טבן]

(2) En[b] 1 i 4 ואתמ̊[ללו] | απο των υιων των ανθρωπων

(3) ποιηθηναι | Eth I c. G οφειλετης αμαρτιας (Eth II pr. ταυτης της) μεγαλης

(4) En[a] 1 iii 1-2 + En[b] 1 ii 6 [ענו] ואמרו כלהן נמ[א ... די לא] נסיר כלנה מן מלכא ד[ן] και απεκριθησαν αυτω παντες λεγοντες | om. παντες 3° | Eth[tana] cf. G μη αποστρεψαι την γνωμην ταυτην και ποιησαι το πραγμα τουτο (Eth[M] την γνωμην ταυτην)

(5) En[a] 1 iii 2-3 ... [אדין ימו] כלהן כחדא ואחרמ[ו ανεθεματισαν παντες αλληλους εν αυτω c. G om. Sync. v 6 per hmt ανεθεματισαν (5) ... ανεθεματισαν ... εν αυτω (6)

(6) En[a] 1 iii 4 ביומי ירד על [ראש חרמון]̊ן Eth II και ησαν παντες διακοσιοι και κατεβησαν εις Αρδης (sic) ο εστιν η κορυφη του Ερμων ορους (crpt ex εν ημεραις Ιαρεδ εις την κορυφην) του Ερμων ορους (Eth I om.) και εκαλεσαν το ορος Ερμων καθοτι ωμοσαν και ανεθεματισαν αλληλους εν αυτω (Eth I om. εν αυτω)

(7) En[c] 1 ii 24 + En[b] 1 ii + En[a] 1 iii 5 ואלין שמהת [רבניהו]ן variae lectiones nominum angelorum in *Table of Dekadarchs*, 118.

(8) En[a] 1 iii 13 + En[b] 1 ii 17a אלין אנון רבנין ורב[ני] עס[רתהן] Eth I ουτοι εισιν δεκαδαρχαι αυτων Eth II ουτοι εισιν αρχοντες διακοσιων αγγελων και παντων των λοιπων μετ αυτων (cf. 7.1)

---

[1] = והוא + waw mater lectionis. Cf. Fitzmyer, *Genesis Apocryphon*, 200, A. Sperber, *A Hebrew Grammar: a New Approach* (New York, 1943), 177, P. Kahle, *Die hebräischen Handschriften aus der Höhle* (Stuttgart, 1951), 42 (Isa. 5.3 שפטו = שפוטו)

CHAPTER 7

(1) En^{a} 1 iii 13 כלהן [אנון וא]חרנין/וי]תיריה[ן] cf. Sync. En I c. Sync. [ουτοι] και οι λοιποι παντες μετ αυτων cf. Eth^{M} 6.8 Flemming, 6. | En^{a} 1 iii 13-14 = En^{b} 1 ii 18 ונסבו להון נשין מן כל די בחרו ושריו cf. Gen. 6.2 | ανα μιαν pro γυναικας 2° | και συνεκοιμηθησαν αυταις G και μιαινεσθαι εν αυταις cf. G 9.8. | En^{a} 1 iii 15 (cf. En^{b} 1 ii 19) ולאלפא אנון חרשה

(2) En^{a} 1 iii 16 = En^{b} 1 ii 21 והויה בטנן מנהן ויל]דה גברין .. [ולא] הוו מתילדין על ארעא ילדי]ן די רביו כגבורתהן] και το υψος εκαστου αυτων πηχων τρισχιλιων

(3) ουτοι G οιτινες | παντας τους κοπους En^{a} 1 iii 17 + En^{b} 1 ii 21, 21a ואלין ה]ווא] [אכלין עמל כל בני אנשא ולא י]כילו לכלכלא אנון] εως G ωστε

(4) En^{a} 1 iii 19 ושריו לקטלה לאנשא | Eth^{q} cf. G και επεστρεψαν (tamayṭu) οι γιγαντες επ αυτους και (Eth^{M} om.) κατησθιοσαν τους ανθρωπους

(5) Eth^{q} και επεστρεψαν (tamayṭu) ημαρτον | En^{a} 1 iii 19-20 ולמקם מן ק]ובל] ושריו למכל | En^{a} 1 iii 21 ונני ימה | En^{a} 1 iii 21a רחש[יא] En^{b} 1 ii 24 כל כנף ו]חיו]ת ארעא En^{b} 1 ii 25a והוו שתין דמא [בשר[הן] και το αιμα επινον εξ αυτης (sc. σαρκος) Eth^{tana} cf. G om. εξ αυτης

(6) En^{b} 1 ii 25 מתעבד?

CHAPTER 8

(1) En^{b} 1 ii 26 [למ]עבד חרבן די פרזל ושר]ינין די] נח[ש] ποιειν ρομφαιας και μαχαιρας και ασπιδας και θωρακας om. και οπλα et διδαγματα αγγελων Sync. + και παν σκευος πολεμικον | En^{a} 1 iii 23 (Plate III) + En^{b} 1 ii 26-27 [ואחזיא] להון [מטל]ון אר[עא] מא יעבדון ד[הבה] למעבדה 27 [תיקו]נא ועל כספא למעבדה לצמידין ל[נשא] και υπεδειξεν αυτοις τα μετ αυτα (za'emdeḥrēhomu) leg. τα μεταλλα cf. Dillmann, SAB, 1892, 1047, Charles^{1906}, 16, n. 3. | En^{b} 1 iii 28 [ואחזי]א על כוחלא ועל צדיד[א] | το στιλβειν και το καλλωπιζειν c. Sync. | και παντοιους λιθους πολυτελεις και εκλεκτους (cf. I Chr. 29.2) και παντα βαφικα ποικιλιας (cf. Jg. 5.30). και αλλαγματα κοσμου (tawlāṭa ʿālam) = חליפת תיקונין Eth$^{g\ (tana\ q)}$ tawallaṭa ʿālam

(2) Eth$^{tana\ (g)}$ και εγενετο μεγαλη ασεβεια και επορνευσαν (Eth^{M} πολλη πορνεια) En^{b} 1 iii 1 p. ? [והוו]א פח[זין] | Eth I c. G και ηφανισθησαν εν πασαις ταις οδοις αυτων (Eth II πασαι αι οδοι αυτων) Flemming, 7. Sync. ηφανισαν τους οδους αυτων.

(3) En^{a} 1 iv 1 + En^{b} 1 iii 1 שמיחזה אלף חבר[ו] Eth^{M} Αμηζαρακ (sic) εδιδαξε παντας επαοιδους και ριζοτομους | En^{a} 1 iv 1 + En^{b} 1 iii 2 [חרמוני] אלף חרש למשרא וכשפו [וחרטמו ותוש]יה ברקאל] אלף [נחשי ברקין] Eth^{tana} Βαρακηλ αστροσκοπιας cf. Sync. Eth^{M} αστροσκοπους En^{a} 1 iv 2-3 + En^{b} 1 iii 3 [כוכבאל א]לף נחשי כוכבין Κωχαβηλ τα σημειωτικα | En^{a} 1 iv 3 [זיק]אל אלף נחשי זיקין Ταμιηλ (G Σαθιηλ) εδιδαξεν αστεροσκοπιαν | En^{b} 1 iii 4 [ארע]תקף אלף נחשי ארעא Eth G om. sed cf. Sync. | En^{a} 1 iv 4 שמשיאל אלף נחשי שמש Eth G om. sed cf. Sync. | En^{a} 1 iv 4 [שהריאל אלף נחשי] שה[ר] Ασραδηλ (G Σεριηλ) σεληναγωγιας | En^{a} 1 iv 4-5 + En^{b} 1 iii 5 cf. Sync. וכולהון שריו לגליה רזין לנשיהון

(7) και εν τη απωλεια των ανθρωπων εβοησαν και αφικνειτο/ανεβη η κραυγη αυτων εις τους ουρανους En^{a} 1 iv 5 + En^{b} 1 iii 5-6 [ובמבד אנשא] מן ארעא [יזעקון] וקלה[ון] סלק ק[דם שמיא]

CHAPTER 9

(1) En^{a} 1 iv 6 + En^{b} 1 iii 6 אדיק מׄיכאל ושריא]ל ו[רפאל [וג]בׄרׄי[אל] מן קדשׁ[שמיא וחז]ו דם סגי שפ]יך על ארעא [וכל ארעא אתמלית [רשעה ו]חמסה די הווא חטי[ן] ל[ה] και τοτε παρεκυψαν Μιχαηλ και Γαβριηλ και Συριαλ και Ουριαν εκ του ουρανου και εθεασαντο | Eth$^{g\ (tana)}$ Μιχαηλ και Συριαλ και Γαβριηλ (om. και Ουριαν) Eth II c. G επι της γης Eth I + κατω | + και πασαν ανομιαν γινομενην επι της γης cf. Sync. G om. per hmt

(2) En^{a} 1 iv 8 + En^{b} 1 iii 9, 10a, 10 [ועללו . . .] ואמרי קדם [קדשיא . . ד[קׄלה וז[עקו]תׄ בנ[י ארעא סלק[י]ן עד תרעי שמיא φωνη/κραυγη βοων αυτων ανεβη (ʿerāqā leg. ʿarga)· ανεβοησε η γη μεχρι πυλων του ουρανου

(3) Eth I και νυν υμιν αγιοις (Eth II ω αγιοι) του ουρανου εντυγχανουσιν En^{a} 1 iv 10 cf. En^{b} 1 iii 10-11 [וכען לכן קדי]שי ש[מיא די קבלן נׄפׄשׁ[ת אנשא ואניחן ואמרן] | εισαγαγετε ημιν την κρισιν

(4) En^{b} 1 iii 13 ומיכׄ[אל] | [και προσελθοντες ... Sync.] ειπον τω κυριω αυτων τω βασιλει· συ (ʾesma leg. ʾanta c. Flemming, 8) ει κυριος των κυριων | βασιλευς των βασιλεων | Eth$^{b\ 3\ mss}$ c. G και ο θρονος της δοξης σου Eth^{M} αυτου En^{b} 1 iii 14-15 [אנתה] מרנא רבאׄ [הו]א מרא עלמיׄן . . . [וכורס]אׄ יקרך לכל דר דריא די מן עלמ[א] | Eth I cf. G αγιον και ενδοξον εις παντας τους αιωνας Eth II εις πασας τας γενεας του αιωνος | Eth^{M} + ʿκαι ευλογητον και ενδοξονʾ Eth^{tana} om.

(5) συ εποιησας τα παντα | παντων εξουσια μετα σου | εξουσιαν εχων cf. Sync. | και συ ορας παντα και ουκ εστιν ο δυναται κρυβηναι απο σου cf. Sync. G om. per hmt (v. 6 και παντα ορας)

(6) Eth I συ ορας (reʾika Eth II ορα reʾikē) α εποιησεν Αζαζηλ | Eth$^{n,\ a}$ cf. G ος/οσα Eth^{M} ως | τα εν τω ουρανω πεπραγμενα | En^{a} 1 iv 20 [ול]הן יעׄתׄ[דון] עותדוהי ידועי די בני אנשא]? | om. εγνωρισεν ανθρωποις cf. v. 7 et Sync.

(7) Eth$^{t^2}$ και εγνωρισεν τας επαοιδας (Eth I om. τας επ.) τοις ανθρωποις (Eth II om. τοις ανθ.) (ʾammara sebʾatāta lasabʾ om. per hmt) Σεμιαζα | En^{a} 1 iv 21 [ואנתה יהבת ליה שלטנ]אׄ לכלה[ון חבורין] G Eth ω (Sync τω Σεμιαζα) την εξουσιαν εδωκας αρχειν (Sync. εχειν) των συν αυτω αμα οντων (חבורין)

(8) Eth$^{tana\ q}$ cf. G των ανθρωπων επι (G om.) της γης Eth^{M} om. επι της γης Eth^{gmt} crpt (badiba sabʾ) | Eth^{M} ḫebura ex v. 7? | cf. Sync. μετ αυτων [και] εν ταις θηλειαις «και» εμιανθησαν cf. Flemming, 9 | Eth$^{tana\ m}$ c. G Sync. πασας τας αμαρτιας Eth$^{g\ q}$ + ταυτας Eth^{M} ταυτας τας αμαρτιας

(9) c. Sync. Appendix B, 419 | Eth I c. G και 1^{0} (Eth II om.) | Eth$^{tana\ g}$ baza cf. G υφ ων Eth^{M} wabaze

(10) Eth I c. G αι ψυχαι (nafsa Eth II nafsāt) των τετελευτηκοτων | Eth II δυνανται Eth I c. G δυναται

(11) και συ οιδας (G ορας) ταυτα και [εας αυτους] καθ εαυτους | Eth^{n} c. G τι (Eth^{M} και τι)

CHAPTER 10

(1) Eth I c. G τοτε (Eth II pr. και) υψιστος ο μεγας ⸢και⸣ αγιος ελαλησεν και επεμψεν | Eth^{tana} Ασυριαλ (ex Σαριηλ?) | Eth II και ειπεν αυτω post Λαμεχ Eth I om. (cf. G)

(2) μελλει ερχεσθαι (G Sync. γινεσθαι) επι πασαν την γην | Eth^{m} c. G απολεσει (Eth^{M} απολειται) | Eth^{tana} c. G παντα (Eth^{M} om.) οσα εστιν εν αυτη

(3) και νυν διδαξον αυτον | μενη | Eth I cf. G εις πασας τας γενεας Eth II crpt εις πασαν την γην (Eth^{n} + του αιωνος)

(4) και παλιν ειπεν ο Κυριος τω Ραφαηλ | Αζαζηλ | χερσιν και ποσιν c. Sync. | om. και 3° | Δουδαηλ | Eth^{tana} c. G και (Eth^{M} om.) εκει βαλε αυτον

(5) επιθες αυτω G και υποθες αυτω

(6) εν τη ημερα τη μεγαλη της κρισεως | απαχθητω?

(7) ιασαι την γην c. Sync. | Eth$^{b\ 7\ mss}$ οτι ιασομαι Eth$^{q\ t,\ 9\ mss}$ ινα ιασωμαι (Eth$^{tana\ g\ m\ u}$ ιασηται) την γην (G πληγην) μηδε απολουνται ... | Eth II c. G crpt εν μυστηριω (Eth I lameštira) ολω ω επαταξαν οι εγρηγοροι leg εν μυστ. ολως ο υπεδειξαν οι εγρηγ.?

(8) om. αφανισθεισα | εν τη διδασκαλια των εργων Αζαζηλ

(9) και επι τους υιους της πορνειας | Eth$^{g\ 2\ mss}$ c. G και απολεσον (Eth II + ⸢τους υιους της πορνειας⸣ ditt.) τους υιους των εγρηγορων | και εξαγαγε και πεμψον αυτους καθ αλληλους εαυτοι ⸢και⸣ (Eth^{tana} om.) καθ εαυτους· εν πολεμω απολουνται En^{b} 1 iv 6 בקרב אבדן |

(10) μακροτης pr. οτι | και παντες αυτων ερωτησουσιν σε και ουκ εσται [πασα ερωτησις] τοις πατρασιν αυτων περι αυτων | Eth^{q} οτι ουκ (Eth^{M} G Sync. om.) ελπιζουσιν | Eth^{tana} c. G ζησαι (Eth^{M} om.) ζωην αιωνιον

(11) και τω Μιχαηλ ειπεν ο κυριος | En^{b} 1 iv 8-9 ואודע לשמ̊[יחז]ה ולכ[ול חברוה]י די אתחברו [לנשיא]? εν παση ακαθαρσια αυτων

(12) Eth I c. G και (Eth II om.) οταν Eth^{a} 1 iv 9-10 [וכדי] יבדון בניהון ויח̇[זון לאבדן . .] | παντες οι υιοι αυτων | Eth^{u} c. G και (Eth^{M} + οταν) ιδωσιν | και 3° om. c. Sync. | υποκατω των ναπων της γης | συντελεσμον αυτων | En^{b} 1 iv 10 [על] שבעין ד̊[רין בהלות] ארעא עד יומא רבא [דיסוף דין עלמא]

(13) και τοτε | om. και 2° | εν τη βασανω | Eth I και εν δεσμωτηριω συγκλεισθησονται εις τον αιωνα En^{c} 1 v 1 [בע]י̊קא ול[בית עגון די] עלמא

(14) και οταν κατακαυθη En^{c} 1 v 1 ? וכול די אי[קד] | En^{c} 1 v 2 ובק̊[צא די דינא די] א̊דין יאבדון לכול דריא μεχρι τελειωσεως γενεων των γενεων

(15) En^{c} 1 v 2 [אכרת] (cf. v. 16) και απολεσον cf. G Eth^{m} + τα πανουργηματα

(ṭebabihomu) Eth^{m} παντα τα πνευματα των μυσικων (tawnēt זמרא crpt. ex ממזרין κιβδηλων?)

(16) (En^{c} 1 v 3 ואכרת עולה מן [אנפי ארעא] Eth^{tana} c. G Aram. και (Eth^{M} om.) απολεσον | απο προσωπου της γης | Eth$^{t, 10 mss}$ c. G εκλειπετω (Eth^{M} εκλειψει) | En^{c} 1 v 4 [ותתחזא נ]צבת קושטא ותהוׄ[א לברכה] Eth^{q} αναφανησεται | Eth$^{g t u (q)}$ και εσται εις ευλογιαν· τα εργα της δικαιοσυνης και αληθειας εις τους αιωνας μετα χαρας φυτευθησεται G om. και εσται . . . αληθειας per hmt

(17) En^{c} 1 v 5 . . . [כול קש]יטין יפלטון ולהון [חיין] Eth I c. G εκφευξονται (yeguayyeyu Eth II crpt yeganneyu) | En^{c} 1 v 5 [אל]פׄיׄן וכול יומי [ילדותכון ו]שיבתכון בשלם תׄ[שלימון]

(18) και τοτε | En^{c} 1 v 6 בקשוט וכולה תתנצׄ[ב אילנין ותתמלא] בׄרכה και ολη (c. En^{c} cf. G) φυτευθησεται δενδρα | om. εν αυτη

(19) En^{c} 1 v 6-7 וכול אילנין [רענין ?] και παντα τα δενδρα της αγαλλιασεως φυτευθησεται επ αυτη | επ αυτη post φυτευοντες | η αμπελος η αν φυτευθη επ αυτη | Eth I c. G οινον (Eth II καρπον) εις πλησμονην (cf. LXX Exod. 16.3) | και πας σπορος ος σπαρησεται επ αυτην εκαστον μετρον ποιησει χιλιαδα και εκαστον μετρον ελαιας ποιησει ανα βατους δεκα ελαιου En^{c} 1 v 7, 9-10 [וגופנא די תתנצב בה] . . . [א]לף [סאין . . .]

(20) Eth^{tana} cf. G απο πασης αδικιας και απο πασης ακαθαρσιας (Eth^{M} ανομιας) | και απο πασης ασεβειας | Eth$^{g q u (m)}$ c. G και πασας τας ακαθαρσιας τας γινομενας (Eth^{M} απο πασων των ακαθαρσιων των γινομενων) επι της γης εξαλειψον (’aḥleqomu Eth^{tana} ’arẖeqomu) απο της γης

(21) και εσονται παντες υιοι ανθρωπων δικαιοι cf. En^{a} 1 vi 3 יתקׄשׄטון (Milik) G. om. | ευλογουντες με και παντες με προσκυνουντες

(22) om. πασα | αμαρτιας G ακαθαρσιας | Eth^{M} και απο πασης (Eth$^{u, n}$ om. απο πασ. c. G) οργης και απο πασης μαστιγος | Eth II και ουκετι πεμψω επ αυτην κατακλυσμον (Eth I om. κατακλ. c. G) εις γενεας των γενεων και μεχρι αιωνος

CHAPTER 11

(1) Eth^{q} της ευλογιας μου | om. και 2° | κατενεγκειν αυτα επι την γην επι τα εργα και επι τον κοπον των υιων των ανθρωπων

(2) Eth I και (Eth II om.) ειρηνη και αληθεια | Eth^{tana} c. G κοινωνησουσιν ομου (Eth^{M} om. ομου) | εις πασας τας γενεας των αιωνων G crpt των ανπων

CHAPTER 12

(1) Eth^{tana} c. G και προ τουτων (Eth^{M} kuellu pro ’ellu) των λογων | om. αυτω Eth^{tana} και τον τοπον αυτου (wamakano)

(2) παντα τα εργα | Eth I c. G μετα των εγρηγορων (Eth II αγιων) και μετα των αγιων (Eth II εγρηγορων) εν ταις ημεραις αυτου

(3) και εγω Ενωχ | om. εστως | Eth I c. G τω κυριω της μεγαλωσυνης (Eth II τω κυρ. τω μεγαλω ʿebay pro ʿabiyāt) | om. του αγιου του μεγαλου | Eth^{M} εφωνουν με Ενωχ τον γραμματεα (Eth^{tana} om. τον γραμ. c. G) και ελεγον μοι

(4) Eth^{tana} και ειπε (Eth^{M} om. και) c. G | Eth$^{tana\ (g)}$ την στασιν (leg. c. Eth^{g} meqwāma) του αγιου (Eth^{g} των αγιων Eth^{M} την αγιαν στασιν) του αιωνος | και μετα των γυναικων ηφανισθησαν (māsanu) | και εποιησαν ωσπερ οι υιοι των ανθρωπων ποιουσιν | και ηφανισαν αφανισμον μεγαν επι της γης

(5) Eth I και ουκ εσται αυτοις (Eth II επι της γης) ειρηνη ουτε αφεσις αμαρτιων | Eth$^{g\ (tana\ q\ u)}$ και (Eth^{M} om.) οτι χαιρουσιν (Eth^{M} pr. ου) των υιων αυτων

(6) και ουκ εσται αυτοις ελεος ουτε ειρηνη

CHAPTER 13

(1) πορευθεις ειπεν τω Αζα(ζ)ηλ cf. G

(2) Eth I c. G και ανοχη (Eth II + και ελεος) και ερωτησις | περι ων εδιδαξας αδικηματων | βλασφημιας pro G ασεβειων | Eth II τοις υιοις των ανθρωπων (Eth I c. G τοις ανθρωποις)

(3) αυτοις ομου | Eth$^{tana\ g\ u}$ c. G και 2° (Eth^{M} om.) | φοβος και τρομος

(4) ηρωτησαν με | ινα εγω αναγω G αναγνω | της ερωτησεως αυτων | Eth I c. G ενωπιον κυριου του ουρανου Eth II προς κυριον του ουρ.

(5) απ αισχυνης των αμαρτηματων αυτων ων κατεκριθησαν

(6) Eth$^{t\ u\ n}$ c. G τοτε (Eth^{M} pr. και) | τας δεησεις αυτων | En^{c} 1 vi 1 [עם תחנו]נ̊יהון על כול ר̊[וחותי]הון לכול חד וחד [עובדיהון] περι των πνευματων αυτων και εκαστου των εργων αυτων και περι ων δεονται c. En^{c} G om. και εκαστου των εργων αυτων

(7) εν τη? Δαν | εκ δεξιων δυσεως Αρμων | και ανεγιγνωσκον |

(8) εως εκοιμηθην | Eth$^{u\ n}$ μετα τουτο pro G και ιδου | En^{c} 1 vi 3-5 עד [די נטלת] לשכני עיני לתרעי ש̊[מיא] וחזית חזיון דרגוז או[כחה] om. και ηλθεν φωνη λεγουσα

(9) En^{c} 1 vi 5-6 [.. ואתית] עליהון וכולהון כנישין כחד̊ה ויתבין וא[בלין באבל מיא] Ουβελσειαλ sic | Σενισηρ

(10) En^{c} 1 vi 7 ומללת קודמיהון כול[הון חזיונין] και ελαλησα ενωπιον αυτων πασας τας ορασεις | Eth I c. G κατα τους υπνους (Eth II + μου) En^{c} 1 vi 7 מלי [ושרית ממלל] קושטא וחזיה ומוכח לעירי שמ̊[יא]

CHAPTER 14

(1) αυτη η βιβλος En^{c} 1 vi 9-10 ספר מלי קושט[א] ... בחלמא די אנה [חזית] του αγιου ʼκαιʼ μεγαλου

(2) ο εγω νυν λεγω | Eth$^{tana\ q.\ 3mss}$ εν γλωσση σαρκινη μου | Eth^{tana} c. G και εν τω πνευματι (Eth^{M} + μου) του στοματος (Eth^{M} om. του στομ.) ο εδωκεν ο μεγας (Eth^{M} + το στομα) τοις ανθρωποις cf. 84.1 En^{c} 1 vi 11 די יה[ב] רׄבא לבני [אנשא] | νοησαι καρδια G νοησει καρδιας

(3) ως εκτισεν και εδωκεν τοις ανθρωποις νοησαι λογους εννοιας και εμε εκτισεν και εδωκεν μοι ελεγξασθαι τους εγρηγορους τους υιους του ουρανου G om. τοις ανθρωποις ... εδωκεν μοι per hmt. En^{c} 1 vi 12 [מלין] מנדע [ואנ]הׄ חלק ועבד וברא לי .

(4) εγω την ερωτησιν υμων εγραψα om. των αγγελων | και εν τη ορασει μου ουτως εδειχθη En^{c} 1 vi 13 ובׄחזוה לי אתחזׄית ... | διοτι η ερωτησις υμων ουκ εσται υμιν εις πασας τας ημερας του αιωνος En^{c} 1 vi 13 כלקובל די בעׄו[תכון לא אתקבלת ...] Eth^{tana} και δικαιωμα ετελεσθη (Eth^{M} τελειον) εφ υμας En^{c} 1 vi 13-14 [ודינא משלם] בגזיר[תא עלי]כון Eth^{M} και ουκ εσται ditt.? (Eth$^{n\ ryl^2}$ + ειρηνη)

(5) και [μηκετι] απο του νυν αναβησεσθε εις τον ουρανον En^{c} 1 vi 14 די עוד מן כׄ[ען לא תסקון לשמיא] | και εν τη γη ερρεθη δησαι υμας εις πασας τας ημερας του αιωνος En^{c} 1 vi 15 [אתאמר למאסרכון] על כול יומי עלמא

(6) και προ τουτων G και ινα περι τουτων | En^{b} 1 vi 8-9 + En^{c} 1 vi 15-16 (cf. Milik) [חזיתו]ן לאבדׄנא די חביבי[כון ודי כול] בניהון ובק[ריהון ...] | και ουκ εσονται υμιν κληρονομοι αυτων (ṭer(rā)yānihomu) | En^{c} 1 vi 17 [על חרב לסוף] אבדן (cf. 91.19 En^{g} 1 ii 21)

(7) En^{c} 1 vi 17 כלקובל די ב[עותכון עליהון ..] περι υμων ... ουδε περι αυτων En^{c} 1 vi 18 ותהוון אנתון בעין ומתחנׄ[ין]

(8) και εμοι ορασις ουτως εδειχθη | En^{c} 1 vi 20 [ערפילין] לי זעקין וזיקין וב[רקין] | Eth II εθλιβον/εθορυβαζον (c. G?) με yāṣeʿʿequni Eth I yāṣehhequni cf. 89.46 G εθλιβον | εξεπετασαν με ʿκαι κατεσπουδαζον μεʾ ditt. | En^{c} 1 vi 20-21 [ונטלוני] לעלא ואובלוני ואעׄ[לו]ני לׄ[שמיא] om. και εισηνεγκαν με

(9) και γλωσσαι πυρος εκυκλουν αυτην En^{c} 1 iv 22 [ולשני] נור סחור סח[ור] | ηρξατο

(10) En^{c} 1 vi 23 [עד די] אדבקת לביא ר[ב] | Eth^{m} εν λιθοις μαργαριτου (G χαλαζης cf. 18.7) | ως λιθοπλακες ʿεν λιθοις εξ χαλαζηςʾ ditt. | om. και πασαι ησαν εκ χιονος | και εδαφος αυτου χιονος En^{c} 1 vi 24 ותלגא [ארעיתיה]

(11) αι στεγαι αυτου (om. και)

(12) En^{c} 1 vi 25 [ונור] דלק סחר ל[כול כתליהו]ן Eth II κυκλω των τειχων αυτου Eth I om. αυτου | και θυραι αυτου

(13) και εισηλθον | και θερμον | En^{c} 1 vi 26 כתלגא וכול [תענוג] Eth I c. G τρυφη ζωης Eth II τρυφη και ζωη

(14) om. και 3° | επεσον επι το προσωπον μου En^{c} 1 vi 27 ונפלת | και εθεωρουν εν τη ορασει

(15) Eth^{M} και ιδου αλλος οικος μειζων τουτου και ολη η θυρα αυτου ανεωγμενη κατεναντι μου και οικοδομημενος εν γλωσσαις πυρος Eth^{tana} cf. G και ιδου θυρα κατεναντι μου αλλος οικος μειζων τουτου και ολος οικοδομημενος En^{c} 1 vi 28 מן דן רב וכולה ...

(16) εν δοξη και τιμη και μεγαλωσυνη | εως leg. ωστε? c. G | En^{c} 1 vi 29 לא אכל למׄחׄויׄאׄ לׄכׄוׄ[ן] ..[על די]

(17) και το εδαφος αυτου

(18) ειδον εν αυτω | τροχος αυτου | και φωνη (G ορος) Χερουβιν leg. και [εγρηγ]ορους/ορους Χερουβιν (ועירין וכרובין)?

(19) Eth II του θρονου του υψιστου Eth I c. G om. του υψ. | πυρος φλεγομενου | Eth II ιδειν αυτο (Eth I c. G om. αυτο)

(20) Eth. I c. G η δοξη η μεγαλη (Eth II ο μεγας της δοξης) En^{c} 1 vii 2 [יק]רא די [רבא] | και το περιβολαιον | om. ως ειδος

(21) om. εις τον οικον τουτον | Eth I c. G ιδειν 1° (re'ya Eth II rā'ya) | το προσωπον του εντιμου και του ενδοξου | ιδειν αυτον

(22) Eth I cf. G το πυρ φλεγομενον κυκλω αυτου Eth II φλοξ πυρος φλεγομενου κυκλω αυτου | προειστηκει αυτω | ουδεις εγγιζει αυτω εκ των κυκλω αυτου | om. ειστηκασιν | και αυτος ου προσεδεηθη συμβουλιας (Eth^{M} + αγιας ditt. ex v. 23?) | Eth. om. και πας λογος αυτου εργον

(23) Eth I (wa)qeddesāta qeddusān leg. waqeddusāna qeddusān? οι αγιοι των αγιων Eth II οι αγιοι | νυκτος και ημερας

(24) περιβλημα (gelbābē leg. gelbub) περιβεβλημενος vel περικεκαλυμμενος (cf. 13.9) | και τον λογον μου αγιον leg. ακουσον c. G

(25) om. και προσελθων μοι εις των αγιων | και ηγειρεν με και προσηγαγεν με . . .

CHAPTER 15

(1) ειπεν μοι τη φωνη αυτου | om. ο ανθρωπος . . . της φωνης αυτου | ακουσον pro G ηκουσα

(2) και πορευθητι ειπε (Eth^{tana} c. G pr. και) τοις εγρηγοροις του ουρανου τοις πεμψασιν σε ερωτησαι περι αυτων | Eth$^{tana\ q\ t\ u,\ 8\ mss}$ c. G και (Eth^{M} om.) μη

(3) Eth^{M} τον ουρανον τον υψηλον και (Eth^{tana} om. c. G) τον αγιον του αιωνος | Eth^{M} και (Eth$^{q,\ 2\ mss}$ om. c. G) ωσπερ | και εγεννησατε υιους (Eth^{q} om.) γιγαντας

(4) (Eth I cf. G και υμεις ητε αγιοι πνευματικοι (Eth II πνευμ. αγιοι) ζωντες ζωην αιωνιον | badiba leg. badama εν τω αιματι των γυναικων cf. Flemming, 19, Knibb, 2,100) | εν αιματι 3° c. G crpt leg. τας θυγατερας | και εποιησατε καθως και αυτοι ποιουσιν

(5) Eth I c. G δια (Eth II pr. και) τουτο | σπερματιζωσιν | Eth$^{g\ q\ u}$ τεκνωσωσιν τεκνα (weluda) cf. Gen. 30.3, Flemming-Radermacher 43, n. 1. | Eth$^{tana\ g\ t\ u}$ ινα ουτως μη εκλειπη (Eth$^{g\ (t)\ u}$ 'iyenteg Eth^{tana} 'iyetḫadag Eth^{M} crpt. yetgabar) | Eth$^{tana\ g\ t\ u}$ αυτοις·παν εργον (balāʿlēhon(kemu)) crpt ex kuellu lomu Flemming, 20) εργον = עבדא crpt ex עיבורא συλληψις?

(6) υμεις δε προτερον υπηρχετε πνευματικοι ζωντες ζωην αιωνιον | Eth$^{tana\ (m)}$ c. G και (Eth^{M} om.) ουκ αποθνησκοντα

(7) Eth I c. G και (Eth II om.) δια τουτο | om. εν 1° | Eth I cf. G οτι πνευματικοι του ουρανου Eth II om. του ουρανου

(8) πονηρα c. Sync. G ισχυρα | καλεσουσιν [αυτους] επι της γης c. Sync. | Eth^{tana} cf. Sync. οτι επι της γης και εν [τη γη]

(9) πνευματα pr. και | απο της σαρκος αυτων cf. Sync. | cf. G διοτι απο των ανωτερων (Sync. ανθρωπων) εκτισθησαν | εκ των αγιων εγρηγορων εγενοντο η αρχη αυτων και η αρχη θεμελιου | πνευματα πονηρα εσται επι της γης (cf. Sync.) και πνευματα πονηρα καλεσουσιν [αυτους] cf. Sync. et v. 8

(10) και πνευματα | και τα'πνευματα της γης | εν τη γη η κατοικησις αυτων

(11) νεφελας(ι) leg. Ναφειλιμ c. Flemming, 20 | Eth$^{g\ q}$ c. G αφανιζοντα (yāmāsenu Eth^{M} yemāsenu) | om. πνευματα σκληρα γιγαντων cf. Sync. | τραυματα (ḥazana)? ποιουντα | om. αλλ ασιτουντα | Eth$^{m,\ 3\ mss}$ c. G Sync. διψωντα (yeṣamme'u Eth^{M} 'iyeṣamme'u) | Eth^{tana} προσκοπτοντα ('iyet'aqqafu leg. yet'aqqafu Eth^{M} 'iyet'āwaqu) | om. πνευματα

(12) Eth$^{m\ t^{1}\ u}$ c. G εξαναστησει (yetnašše'u Eth^{M} ⌜'i⌝yetnašše'u) | ταυτα τα πνευματα | επι τους υιους των ανθρωπων και επι τας γυναικας | om. απ αυτων per hmt cf. 16.1

CHAPTER 16

(1) Eth$^{tana\ (q)}$ cf. G αφ ημερων σφαγης (Eth^{M} εν καιρω σφαγης cf. Sync.) και απωλειας και θανατου των γιγαντων εν αις ('enta ẖaba G αφ ων) τα πνευματα εξεπορευθη | Eth^{g} εκ του σωματος της σαρκος αυτων ('emnafesta šegāhomu) Eth$^{tana\ (q)}$ εκ της ψυχης ('emmanfas) και της σαρκος αυτων cf. G | εσται αφανιζοντα ... αφανισουσιν leg. c. Flemming, 21 zayāmāsen ... yāmāsenu | Eth^{M} μεχρι ημερας τελειωσεως (Eth$^{b\ 2\ mss}$ κρισεως) της μεγαλης εν η ο αιων (leg. 'ama 'ālam Flemming, ibid.) ο μεγας τελεσθησεται ⌜απο των εγρηγορων και ασεβων⌝ Eth$^{g\ (q\ t\ u\ tana)}$ + πας τελεσθησεται απο των εγρηγορων και ασεβων ⌜πας απο των εγρηγορων⌝ leg. 'em kuellomu teguhān wara'ayt(?) απο παντων εγρηγορων και γιγαντων cf. Sync. εφ απαξ ομου τελεσθησεται

(2) και νυν [ειπε] τοις εγρηγοροις | οιτινες προτερον εν τω ουρανω ησαν

(3) και νυν υμεις | και τα κρυπτα ουκ ετι (leg. εστιν) [α] ουκ ανεκαλυφθη υμιν cf. Lods, 150 | και εξουθενημενον μυστηριον (mennuna meštira) cf. Dillmann, SAB, 1892, 1049, Charles^{1906}, 47.

(4) ουκ εστιν υμιν ειρηνη

CHAPTER 17

(1) και απηγαγον με (naš'uni cf. v. 4) εις τινα τοπον | Eth$^{g\ m\ q\ (t^{2})}$ c. G. εν ω οι ('ella Eth^{M} om.) οντες εκει (om. γινονται) ως πυρ φλεγον ...

(2) Eth^{g} και απηγαγον με (wawasaduni Eth^{M} wawasadani) | εις γνοφωδη? τοπον | η κορυφη της κεφαλης?

(3) τοπους | om. τους θησαυρους των αστερων και | om. και 4° | Eth^{tana} εις τα ακρα βαθους οπου (westa ʾaṣnāfa ʿemaqa ḫaba Eth^{M} westa ʾaṣnāfa ḫaba ʿemaqu) | + και ρομφαιαν πυρος post τας θηκας αυτων

(4) υδατων ζωης (= מיא חיא crpt ex מיא מחתיא?) ⸢των λεγομενων⸣ | ο εστιν κατεχον (yeʾeḫḫez) cf. Dillmann, SAB, 1892, 1045 (cj. παρεχομενον G παρεχον)

(5) ηλθον | καταρρει το πυρ αυτου | και εκκεχυται (G ρεει) εις θαλασσαν μεγαλην την προς δυσιν

(6) Eth^{M} και ιδον παντας (Eth$^{m\ q\ t}$ c. G om.) τους μεγαλους ποταμους | om. μεχρι του μεγαλου ποταμου και | om. ου

(7) και ιδον ορια (ʾadbāra) των γνοφων χειμερινα | του χειμωνος | την εκχυσιν υδατων πασης της αβυσσου

(8) και ιδον το στομα παντων των ποταμων της γης

CHAPTER 18

(1) και ιδον | ιδον 2° pr. και | τα θεμελια

(2) ιδον 2° pr. και

(3) και ιδον οτι οι ανεμοι εκτεινουσιν (yerabbebewwā) το υψος του ουρανου H om. per hmt | μεταξυ ουρανου και γης | Eth^{M} + αυτοι (Eth^{tana} pr. και) εισιν οι στυλοι του ουρανου G om. per hmt

(4) Eth$^{m,\ 2\ mss}$ και ιδον ανεμους τον ουρανον στρεφοντας δυνοντας (Eth$^{b\ 6\ mss}$ pr και Eth^{m} ʾella yāʿarrebu Eth^{M} yaʿarrebu) τον τροχον του ηλιου

(5) και ιδον | Eth I c. G βασταζοντας εν νεφελαις Eth II τας νεφελας | + (post εν νεφελαις) και ιδον τους οδους των αγγελων G om. | ιδον εις περατα της γης το στηριγμα του ουρανου επανω

(6) Eth^{M} και παρηλθον προς νοτον και [ιδον τοπον] καιομενον (Eth^{g} zayenadded Eth^{M} wayenadded) ημερας και νυκτος | om. βαλλοντα

(7) Eth^{M} om. τα μεν Eth^{g} τα cf. G | leg. το εν post ανατολας | το δε εν απο λιθου μαργαριτου και το εν απο λιθου στιβι (fawwes) | τα δε κατα νοτον

(8) αφικνειτο εις ουρανον | En^{c}1 viii 27 כרסא אל[הא] | pēka φουκα

(9) Eth^{tana} cf. G. και α ην επ εκεινα (ʾellu Eth^{M} kuellu) cf. Knibb, 2,105, Flemming-Radermacher, 47.

(10) κακει ιδον τοπον περαν της μεγαλης γης crpt ex לעבר בריתא דארעא? En^{c}1 viii 28 ותמן אשת̊[יצין שמיא] Eth II εκει συναχθησεται (yetgābeʾu) τα υδατα Eth I οι ουρανοι (Eth II cf. Gen. 1.9.)

(11) Eth II χασμα της γης (Eth I c. G om. της γης) βαθυ εν στυλοις του πυρος του

ουρανου και ιδον εν αυτω (leg. westētu?) στυλους ουρανου (Eth$^{tana\ g\ t^2}$ om.) του πυρος καταβαινοντας En^c 1 viii 29 [וחזית] בה עמוד[י נורא] | ουτε εις υψος ουτε εις βαθος

(12) En^c 1 viii 30 [ולהלא] מן דן [נקרא ?] και επι το χασμα (cf. supra, v. 9) | om. οπου | ουτε θεμελιον γης υποκατω αυτου | επ αυτω G υπο αυτο

(13) και ως πνευμα πυνθανομενον μου crpt ex περιπλανωμενα? cf. Flemming, 24.

(14) Eth I c. G τοις αστροις (Eth II + του ουρανου)

(15) Eth^M και (Eth$^{b^1\ 2\ mss}$ om. c. G) ουτοι εξ αρχης | om. οτι τοπος εξω του ουρανου κενος εστιν

(16) ενιαυτου μυστηριου sic. crpt ex ενιαυτων μυριων

CHAPTER 19

(1) Eth^m c. G και 2° Eth^M om. | Eth^{tana} c. G πολυμορφα Eth^M pr. και | μιαινεται G λυμαινεται | Eth II τους ανθρωπους Eth I αυτους | και πλανησει τους ανθρωπους G αυτους | τοις δαιμονιοις ως θεοις | Eth$^{tana\ (m\ t),\ b}$ μεχρι ('eska Eth^M 'esma) της ημερας της μεγαλης κρισεως

(2) Eth$^{tana\ q\ (g,\ a^1)}$ c. G αι γυναικες αυτων των παραβαντων αγγελων ('asḥitomu malā'ekt Eth^M 'asḥiton malā'ekta samāy 'their wives having led astray the angels of heaven'|(Knibb)) | ως ειρηναιαι crpt ex εις σειρηνας

CHAPTER 20

(1) και ταυτα τα ονοματα των αγιων αγγελων των γρηγορουντων

(2) Eth^{tana} Συραηλ | Eth^M ο της βροντης και του τρομου Eth$^{tana\ g\ q}$ ο του κοσμου και του τρομου cf. G Charles^{1906}, 52 n. 4.

(3) ο των πνευματων των ανθρωπων

(4) Eth^M Ραγουηλ Eth$^{g\ q}$ Ραβουηλ Eth^u Ραμουηλ | ο εκδικων τον κοσμον και τους φωστηρας

(5) ο επι των αγαθων των ανθρωπων | τεταγμενος επι λαω

(6) Σαρακαηλ | ο επι των πνευματων των υιων των ανθρωπων α πνευματα εξαμαρτανει

(7) Eth I c. G ο επι του παραδεισου και των δρακοντων (Eth II ο επι των δρακ. και του παρ.) και Χερουβειν | G^2 + Ρεμειηλ ... ανισταμενων Eth c. G^1 om. | om αρχαγγελων ονοματα επτα

CHAPTER 21

(1) Eth$^{tana\ u}$ c. G$^{1,\ 2}$ εως της ακατασκευαστου (Eth^{M} εως τοπου ακατ.)

(2) Eth^{M} om. εωρακα (Eth$^{g\ t\ (q)}$ hab. c. G) et τεθεαμαι | Eth^{tana} c. G$^{1,\ 2}$ τοπον ακατασκευαστον (za'ikona delewa = Eth Gen. 1.2) (Eth^{M} crpt. zadelew)

(3) om. και ερριμενους | επ αυτω | ομου ως (ḫebura kama) crpt G$^{1,\ 2}$ ομοιους | Eth^{tana} c. G$^{1,\ 2}$ εν πυρι (ba'essāt) (Eth^{M} wakama za'essāt)

(4) αμαρτιαν pro G$^{1,\ 2}$ αιτιαν?

(5) και ειπεν | αυτος ηγειτο μου | om. μοι 2° | Eth I εξακριβαζει (Eth II + και ερωτας) και φιλοσπευδεις

(6) Eth^{tana} c. G$^{1,\ 2}$ των αστερων του ουρανου Eth^{M} om. του ουρ. | Eth I c. G$^{1,\ 2}$ κυριου Eth II + υψιστου | Eth^{tana} μυρια ετη (ʿāmat Eth^{M} crpt ʿālam) | αριθμον του χρονου (Eth^{tana} om. του χρον.) των αμαρτηματων αυτων

(7) Eth I c. G$^{1,\ 2}$ και διακοπην ειχεν ο τοπος (makān Eth II crpt wassanu) | Eth$^{tana\ m\ u\ (t^1)}$ ουτε μετρον ουτε μεγεθος ηδυνηθην ιδειν ουδε ⸢ηδυνηθην⸣ εικασαι (ʿāyyeno) (Eth^{M} crpt. cf. Flemming, 27.) | Eth^{g} + τοις οφθαλμοις μου

(8) ο τοπος ουτος | om. ως 2° |οδυνηρος (ḥemām) G δεινος ?

(9) Eth$^{q,\ 3\ mss}$ c. G τοτε απεκριθη μοι Ουριηλ ... και ειπεν μοι (Eth^{M} τοτε απ. μοι Ουρ. .... ⸢και απεκριθη μοι⸣ και ειπεν μοι) | Eth^{tana} δια τι εφοβηθης και ουτως [εποηθης (dangaḍku leg. dangaḍka); και απεκριθην.] εποηθην (dangaḍku) επι τουτω τω φοβερω τοπω και εμπροσθεν (qedma) της οψεως ταυτης της οδυνης (ḥemām)

CHAPTER 22

(1) επι δυσμων | om. αλλο | και πετρα στερεα

(2) Eth^{M} και τεσσαρες τοποι καλοι ⸢και⸣ εν αυτω | Eth$^{tana\ g\ m\ t^1\ u}$ και τεσσαρους τοπους καλους ⸢και⸣ εν αυτω | Eth^{M} βαθος εχοντες (Eth$^{tana\ g\ m\ t^1\ u}$ εχοντας) και πλατος | om. G τρεις αυτων ... ειπον per hmt. (λειοι ... λεια) | πως λεια τα κοιλωματα (om. ταυτα) (zayānkuarrakuer) | σκοτεινα (ṣelmat leg. ṣelmut)

(3) ουτοι οι τοποι οι καλοι | Eth I c. G εις αυτο τουτο (Eth II lomu 'ellāntu) | En^{e} 1 xxii 1 נפש]ת כל בני אנשא] ωδε επισυναχθησονται πασαι αι ψυχαι υιων των ανθρωπων (G των ανθρωπων)

(4) En^{e} 1 xxii 1-3 [1]והא אלן אנון פחתיא לבית עגון [2]לכדן עב[ידי]ן עד יום די יתד̊י̊נון ועד זמן יום קצא ד[י] [3]דינא רבא די מנהון יתעבד Eth$^{m,\ ull}$ εποιηθησαν Eth^{M} c. G εποιησαν | μεχρι του χρονου διορισμου αυτων ⸢και διορισμος εσται πολυς⸣ μεχρι του χρονου της κρισεως μεγαλης επ αυτους

(5) En^{e} 1 xxii 3-4 תמן חזית רוח אנש מת קבלה [ו]אנינה ע[ד] שמיא סלק ומזעק וקבל ... Eth$^{tana\ m\ q\ t^1\ u}$ c. G τεθεαμαι Eth^{M} pr. και | τα πνευματα των υιων ανθρωπων νεκρων οντων | om. εντυγχανοντας | Eth II και (Eth I om.) | η φωνη αυτων

(6) En^{e} 1 xxii 5 [... ד֯[י וקדישא לעירא ל[רפא]ל שאלת] τοτε ηρωτησα | om. το εντυγχανον | En^{e} 1 xxii 6 ... [רוחא קבל]ה֯ דמן הוא דכד[ן אנ֯נא ου ουτως η φωνη αυτου | Eth^{ull} c. G εως του ουρανου Eth^{M} om.

(7) Eth$^{tana\ q,\ 2mss}$ και απεκριθη μοι λεγων En^{e} 1 xxii 7 ...[וענה לי א[מר c. G Eth^{M} και απεκ. και ειπεν μοι λεγων | Eth II om. Αβελ 2° Eth I om. και Αβελ | Eth I μεχρι του απολεσθαι το σπερμα αυτου παν (Eth II om.)

(8) τοτε ηρωτησα περι αυτου και περι των κριματων (crpt ex κυκλωματων) παντων και ειπον | Eth$^{tana\ g\ m\ q}$ c. G εχωρισθησαν (Eth^{M} εχωρισθη)

(9) Eth^{q} c. G χωριζεσθαι (yetfalaṭ Eth^{M} yefleṭu Eth$^{tana\ g\ m}$ yefleṭ) τα πνευματα των νεκρων | και ουτως εχωρισθη (Eth^{ull} c. G + εις) τα πνευματα των δικαιων | Eth$^{tana\ g\ q}$ c. G ου η πηγη (zawe'etu Eth^{M} zewe'etu) | του υδατος (Eth$^{g\ q}$ + της ζωης) εν αυτω φωτεινη (berhān leg. berhet c. Flemming, 29.) ex ברא דמיא..נהיריא ?

(10) Eth^{tana} wakamāhu και ομοιως sed crpt ex wakamaze και ουτως (Eth^{M} bakama kamāhu) vel leg. και ομοιως ουτως (wakamāhu kamaze)? | εκτισθη τοις αμαρτωλοις leg. [τοις πνευμασιν] των αμαρτωλων (cf. G et vv. 12, 13) Eth^{tana} crpt tafalṭu ḫāte'ān

(11) Eth I c. G ωδε (Eth II pr και) | μεχρι χρονου της μεγαλης ημερας της κρισεως | και των μαστιγων | καταρωμενων cf. G | και της ανταποδοσεως των πνευματων | και εκει | μεχρις αιωνος ⸢ητοι το απο του αιωνος⸣

(12) και ουτως | Eth I c. G οιτινες εμφανιζουσιν Eth II και των εμφανιζοντων

(13) Eth$^{d\ y}$ c. G και ουτως (Eth^{M} om. και) | Eth^{tana} οσοι ουκ ησαν ⸢δικαιοι αλλα αμαρτωλοι οι⸣ ολοι ανομοι (feṣṣumāna 'abbāseyāna Eth^{M} 'abbasā) | μετα των ανομων εσονται ομοιοι αυτων (kamāhomu) חבריהון? | om οτι οι ενθαδε θλιβεντες ελαττον κολαζονται αυτων | και τα πνευματα αυτων ουκ θανατωθησονται εν ημερα της κρισεως En^{d} 1 xi 1 לא י֯תנזקון ביום דינא ק֯[שיטא ולא יתערון] מן תנ֯ה֯

(14) En^{d} 1 xi 2 ואמרת להוה בריך דין קושט[א] ευλογητος εσται ο κυριος μου ο κυριος της δοξης και της δικαιοσυνης En^{d} 1 xi 2 [מרא] רבותא֯ | ο παντων κυριευων εως του αιωνος

CHAPTER 23

(1) En^{d} 1 xi 3 [ומן תמן הובלת (אובלת ms) לאתר א֯[חרן μεχρι των περατων της γης (G om. μεχρι)

(2) πυρ καιομενον και διατρεχον En^{d} 1 xi 4 ? ולא י֯ש֯ר֯י֯ ב֯ר֯ה֯טיה] | En^{d} 1 xi 4 [כח]ד֯א֯ לב[ך] Eth^{g} wa'ella leg weʿul kamāhuma διαμενον αμα? Cf. Dillmann, *Lex.*, 924.

(3) En^{d} 1 xi 5 [די לא אי]ת֯י֯ ל֯ה֯ כ֯ל שליא֯ו֯

(4) και ειπεν μοι post μετ εμου ην | ουτος ον εωρακας ο δρομος [πυρος] το προς δεσμας πυρ το εκδιωκον εστιν (zayenadded leg. zayesadded c. Flemming, 30) En^{d} 1 xi 5 דגלא ונורא הוא?

CHAPTER 24

(1) κακειθεν εφωδευσα εις αλλον τοπον της γης cf. 23.1 G om. | ορη πυρος φλεγομενου ημερας και νυκτος En^{d} 1 xi 6 [טורין די נור די ד]לק הוה ביומ[א ובליליא]

(2) και επορευθην προς αυτο (mangalēhu cf. 18.9) | εντιμα pro G ενδοξα | om. παντα | και παν εκατερον του εκατερου διαλλασσον | και οι λιθοι εντιμοι και καλοι | ενδοξα τη ορασει αυτων και ευειδη | και εστηριγμενα εν επι τω ενι και τρια επι νοτον εν επι ενι

(3) om. τω ορει | Eth$^{h\ o\ b^{1}\ (g)}$ και υπερειχεν αυτων (wanoḫomusa Eth^{M} nuḫomusa τω υψει) leg. υπερειχεν αυτων τω υψει (noḫomusa nuḫomu)? | ομοιωθη (yetmāsalu leg. yetmāsal) ⸢παντα αυτων⸣ καθεδρα θρονου | δενδρα ευωδη

(4) Eth$^{q\ u,\ ull}$ c. G ο (Eth^{M} om.) ουδεποτε |και ουδεν εξ αυτων ουδε ετερον ομοιως αυτω [ηυφρανθη] ο (Eth$^{g\ m\ u}$ om. c. G) οσμην ειχεν . . . | και ο καρπος αυτου καλος και ο καρπος αυτου ωσει βοτρυες φοινικων

(5) και τοτε | ιδου (Eth I om.) τουτο το δενδρον καλον και ευειδες | τα φυλλα αυτου | Eth^{tana} c. G τα ανθη (Eth^{M} ο καρπος) αυτου ωραια(ος) λιαν τη ορασει

(6) και τοτε | και εντιμων post αγιων | Eth^{M} αυτος (Eth^{tana} pr. και c. G) | ο επ αυτων (zadibēhomu)

CHAPTER 25

(1) τι ερωτας με περι της οσμης του δενδρου τουτου | om. και τι εθαυμασας | και εξακριβαζει [την αληθειαν] μαθειν

(2) Eth^{M} και τοτε απεκριθην αυτω (Eth^{g} om.) εγω (Eth$^{tana\ q\ m\ t^{1}\ u}$ om.) | Eth^{M} Ενωχ (Eth$^{tana\ q}$ om.) λεγων (Eth^{q} και ελεγον αυτω) cf. G τοτε απεκριθην αυτω | om. σφοδρα

(3) Eth^{M} και (Eth^{tana} om.) απεκριθη μοι (Eth^{g} om. c. G) λεγων | + (post υψηλον) ο εωρακας | Eth^{M} θρονου κυριου (Eth^{q} θεου c. G) | ο αγιος ⸢και⸣ μεγας | ο κυριος της δοξης | οταν (Eth$^{tana\ q}$ pr. και) καταβη

(4) Eth I cf. G τουτο το δενδρον ευωδιας (Eth II pr οσμης) | κρισεως Eth^{g} pr. ημερας | Eth$^{g\ t\ (tana)}$ εκδικησει παντας και τελειωσει (Eth^{M} τελειωθησεται) | τοδε G τοτε

(5) Eth^{M} εκ του καρπου αυτου ⸢δοθησεται (Eth$^{tana\ q\ m}$ pr. και)⸣ τοις εκλεκτοις ζωη· (Eth^{tana} c. G εις ζωην·) εις βορραν (G βοραν) μεταφυτευθησεται | του κυριου G του θεου

(6) Eth^{M} εις (Eth^{tana} c. G pr. και) το αγιον ελευσονται (leg. yebawwe'u c. Charles^{1906}, 64 n. 5) Eth$^{g\ q,\ 3\ mss}$ c. G και οισουσιν (yābawwe'u) αυτω leg. εις το αγιον ελευσονται και οισουσιν αυτω (οσμας εν τοις οσεοις αυτων) | καθως εζησαν | om. και 5° | Eth II βασανοι και οδυναι και κοπος (Eth II πληγαι) και μαστιγες

(7) Eth II τον κυριον (Eth I θεον c. G) της δοξης | οτι ητοιμασεν τοιαυτα ανθρωποις δικαιοις En^{d} 1 xii 1 [מלכא עלמין כ]ל[קבי]ל די מכ֯[ין הוא . . . ?

CHAPTER 26

(1) τοπον ηυλογημενον πιονα Eth$^{b\ 8\ mss}$ pr. και | εν ω [δενδρα εχοντα] παραφυαδας cf. Flemming, 32 | εκ δενδρου εκκοπεντος En^{d} 1 xii 2 [הובלת למ]צ֯י֯ע֯ א֯[רעא . . . די ביה] א֯י֯ל֯[נין די ינקיהן מתקימין ופרחין מן קיסא גדידא]

(2) και υποκατω En^{d} 1 xii 3 ונפקין מן תחתוהי | Eth I c. G εξ ανατολων ('emmangala ṣebāḥ Eth^{M} mangala ṣebāḥ) | και η ρυσις αυτου προς νοτον

(3) Eth$^{q\ (tana)}$ c. G τουτου ('emze Eth^{M} kamaze) | ανα μεσον αυτων En^{d} 1 xii 5 [רם] מנ֯ה ובניהון חלה ע[מיקה] και ουκ | δι αυτης c. G leg. bātini c. Flemming, 33 | παρα το ορος

(4) και φαραγγα υποκατω αυτου ανα μεσον αυτων En^{d} 1 xii 7 [ו]ב֯י֯נ֯ת֯הון ח֯[לות] | και αλλας φαραγγας βαθειας και ξηρας | επ ακρων των τριων αυτων

(5) βαθειαι και ουκ εχουσαι πλατος (cf. v. 3) | [ουκ] εφυτευετο leg. 'iyettakkal c. Flemming, 33.

(6) και εθαυμασα περι της πετρας και εθαυμασα περι της φαραγγος cf. En^{d} 1 xii 8 [ותמ]הת על ט֯נ֯ר֯[א] (Milik ט֯ו֯ר֯[יא])

CHAPTER 27

(1) τοτε ειπον | Eth$^{g\ t^{1}\ u}$ πασα ('ella leg. kuellā?) ηυλογημενη | εν μεσω αυτων post εστιν

(2) Eth^{M} (G om.) τοτε απεκριθη μοι Ουριηλ (Eth$^{13\ mss}$ Ραφαηλ cf. Charles^{1906}, 67 n. 27, Knibb 2,115) εις των αγιων αγγελων ος μετ εμου ην και ειπεν μοι | αυτη η φαραγξ | om. γη καταρατος Eth^{tana} c. G οι κεκατηραμενοι Eth^{M} om. | το κριτηριον αυτων G το οικητηριον

(3) εν ταις εσχαταις ημεραις εσται επ αυτων (αυτους?) ορασις της κρισεως της εν δικαιοσυνη | εις τον αιωνα τον απαντα χρονον conflatio ex εις τον απαντα χρονον c. G (cf. LXX Est. 8.12) et εις τον αιωνα χρονον (cf. LXX Isa. 34.10) | ευσεβεις

(4) και εν ταις ημεραις | ευλογησουσιν αυτον

(5) Eth I ηυλογησα c. G Eth II pr. εγω | και αυτω ελαλησα (nagarku) και εμνησα (zakarku) G και την δοξαν αυτου εδηλωσα και υμνησα

CHAPTER 28

(1) κακειθεν επορευθην προς ανατολας εις το μεσον ορους Μαδβαρα | Eth I και ιδον ερημον και [τοπον] μονον (wabāḥtito(tu) leg. wa[makāna] bāḥtito?) Eth II c. G.

(2) απο (Eth^{tana} pr. και c. G) των σπερματων και υδωρ ανωθεν ανομβρουν (Eth^{tana} pr. και) επ αυτον ($Eth^{q,\ 3mss}$ + και) φαινομενον (crpt ex G φερομενον)

(3) Eth I c. G ως υδραγωγος δαψιλης (Eth II υδραγ. ως δαψ.) ος κατερρει (yesarreb) | ⌒και παντοθεν αναγει και εκειθεν υδωρ και δροσον

CHAPTER 29

(1) και επορευθην om. ετι εκειθεν En^{e} 1 xxvi 4 אזלת בא[תרא . . .] | απο του Μαδ-βαρα | Eth^{tana} c. G και (Eth^{M} om.) προς ανατολας του ορους τουτου En^{e} 1 xxvi 5 ל[מדנ]ח ט[ורא דן]? | ηγγισα (qarabku) G ωχομην

(2) και εκει ιδον δενδρα κρισεως (leg. κτισεως) πλειονα κοστου (quasquas) αρωματων λιβανων και ζμυρνας (G πνεοντα αρωμ. λιβ. και ζμυρ.) | Eth I και τα δενδρα αυτων ομοια [καρυαις] (yetmāsalu [lakarkā'] Eth II 'iyetmāsalu cf. Flemming, 35)

CHAPTER 30

(1) En^{c} 1 xii 23 . . . [ולהל]א מנהון ארחקת Eth^{tana} και επεκεινα (cf. 18.9) τουτων των ορεων [ωχομην προς] ανατολας | Eth II και (Eth I om.) ου μακραν (leg. εκ μακρων ?) ιδον τοπον αλλον φαραγγας υδατος ως (Eth I ουτως) του αεναου (za'iyetwēdā')

(2) En^{c} 1 xii 24 [. . . דב]ה קניא טביא די בשמא [די דמין לחילפא דימא . . .] Eth II και ιδον δενδρον (Eth^{tana} δενδρα) ωραιον και τα αρωματα αυτου ομοια τω σχοινω Eth I ομοιον δενδρω αρωματος ως του σχοινου

(3) En^{e} 1 xii 25 [חזית] קונם בשמא . . . ולהלא מן נחליא [אלין . . .] και επεκεινα (cf. v. 1, 18.9)

CHAPTER 31

(1) En^{c} 1 xii 26 [ואחזיאת טורין] אחרנין ואף בהון חזית אילנין די נפק . . . και ιδον αλλα ορη εν οις δενδρα | Eth II και εκπορευομενη (Eth I c. G εξ αυτων) στακτη (leg. māya [lebn]?) (Eth I om.) ⌜και εκπ. εξ αυτ.⌝ ως νεκταρ το καλουμενον σαρραυ

(2) En^{c} 1 xii 27-28 ולהלא מן טורין אלן אחזיאת טור אוחרן . . . και επ εκεινο το ορος cf.

30.1,3, 18.9. | om. προς ανατολας των περατων της γης | En^{c} 1 xii 28 [וביה אהליא] ו̊כ̊ו̊ל[הון מלין אהלא] ו̊הוא דמא לקלפי [לוז] + post ορος και εν αυτω δενδρα αλοης (’alwā) και παντα (’elku leg. kuellu) τα δενδρα πληρη [στακτης] (leg. melu’āna lebn) ως αμυγδαλον ⸢και σκληρον⸣

(3) En^{c} 1 xii 29-30 + En^{e} 1 xxvi 15-16 [וכדי מקטעין/מקלפין] באי[לניא אלן נפק מנהון בש]ם רי̊ח̊ וכדי מדקין / מדקק קלפיא/קלפוהי אלן [יתיר מן כול בשמא . . .] και οταν λαβωσιν (crpt ex τριβωσιν?) τον καρπον αμεινον παντων αρωματων

CHAPTER 32

(1) En^{e} 1 xxvi 16-17 + En^{c} 1 xii 30 [ולהלא מן טורין] אלן כלצפ(ו)ן מדנחהון אחזיאת טורין אחרנין [מלאין נ]רד טב וצפר וקרדמ̊ן [ופ]לפלין και μετ εκεινα (cf. 18.9) αρωματα εις βορραν βλεπων επανω των ορεων | om. προς ανατολας | δενδρων αρωματικων G σχοινου

(2) En^{e} 1 xxvi 18-21 ומן תמן הובלת [למד]נח כל טוריא אלן רחיק מנהון למדנח ארעא ואחלפ̇[ת עלא] מן ימא שמוקא וארחקת̇ שגיא מנה ואעברת על[א] מן חשוכא רח[י]ק מנה και εκειθεν εφωδευσα επανω κορυφης παντων των ορεων τουτων (leg. lakuellu ’elku pro la’elku?) | Eth^{tana} c. G μακραν [απεχ]ων προς ανατολας της γης (Eth^{M} om. της γης) | και απ αυτης μακραν εγενομην και διεβην επανω του αγγελου Ζωτιηλ

(3) En^{e} 1 xxvi 21 ואחלפת ליד פרדס קשטא και ιδον κρειττονα (kaḥaktihomu) παντων (leg. lakuellu) δενδρων δενδρα πλειονα ⸢και μεγαλα⸣ φυομενα εκει και ευωδουντα μεγαλα και μεγαλοπρεπη και ενδοξα και το δενδρον της φρονησεως ου οι εσθιοντες επιστανται φρονησιν μεγαλην

(4) και ομοιον κερατεα om. το δενδρον . . . τα δε φυλλα αυτου | om. ομοια | η δε οσμη του δενδρου διηρχετο (G διετρεχεν) και προεβαινε (yebaṣṣeḥ) πορρω om. απο του δενδρου

(5) και ειπον | om. ως 1° | ως καλον και επιχαρι τη ορασει αυτου

(6) και απεκριθη μοι Ραφαηλ | και ειπον μοι post ο μετ εμου ων | ο πατηρ σου ο πρεσβυτερος (cf. LXX Gen. 43.27) και η μητηρ σου η πρεσβυτερα οι προ σου και εγνωσαν την φρονησιν και διηνοιχθησαν οι οφθαλμοι αυτων και εγνωσαν οτι γυμνοι ησαν και εξεβαλλοντο εκ του παραδεισου En^{e} 1 xxvii 10-11 ואמך רבתא ויד̊[עו] ד̊י̊ ע̊רטליין

CHAPTER 33

(1) En^{e} 1 xxvii e [ואחזית תמן] . . . שנין

(3) ως εδειξεν μοι ο αγγελος Ουριηλ ος μετ εμου ην En^{e} 1 xxvii 19 [אוריאל חד מן] ע̊ירין?

(4) Eth II και παντα εδειξεν μοι και εγραψεν και ετι τα ονοματα αυτων εγραψεν

μοι και τας εντολας αυτων και τα εργα αυτων (megbārātihomu) Eth I τας συναγωγας αυτων (maḫbarātihomu) En^{e} 1 xxvii 20 frg. g ... יה למ̊ ... ואחזית עבדין 21 רבר̊[בין]

CHAPTER 34

(1) En^{e} 1 xxvii 21 ואחזית עבדין רבר̊[בין] Eth^{M} και εκει ειδον μεγαλα και ενδοξα θαυμασια (mankera Eth$^{g\ q\ t^{1}\ u}$ mekra crpt)

CHAPTER 35

(1) En^{c} 1 xiii 22-23 .... [וחזית] תרעין פתי̊[חין] Eth II και ειδον τρεις θυρας Eth I + του ουρανου | τοσαυτας θυρας οσους εξοδους (Eth^{u} οσους εισοδους και εξοδους | En^{c} 1 xiii 24 חשבוניהון

CHAPTER 36

(1) En^{c} 1 xiii 25 מן תמן הובלת לדרום ס[יאפי ארעא . . .] לרוח דרומא לטל ומ[טר ולרוח קידומא] και δροσον και ομβρον και ανεμον [καυσωνα?]

(2) Eth^{v} εις τα περατα της γης = 34.1, 35.1 (Eth^{M} του ουρανου) | En^{c} 1 xiii 27 אחזית תרעין ת̊ל[תהון]

(4) En^{c} 1 xiii 30 באד̊י̊ן̊ [ברכית בכל עידן] Eth I ηυλογησα (Eth II + και) δια παντος και (Eth II om.) ευλογησω | Eth I τοις πνευμασι και τοις ανθρωποις (Eth II τοις πνευ./ταις ψυχαις των ανθρωπων) | το εργον αυτου και (Eth$^{tana\ q\ t\ u}$ om.) παν κτισμα αυτου ινα ιδη τα εργα της δυναμεως αυτου

CHAPTER 37

(2) Eth^{M} ακουσον ... τους λογους του αγιου ους λαλησω (’enagger Eth^{tana} γνωρισω ’a’ammer) ενωπιον του κυριου των πνευματων

CHAPTER 38

(2) Eth II και οταν φανισθη ο δικαιος (ṣādeq (Eth I η δικαιοσυνη (ṣedq)) κατα

προσωπον των δικαιων και εκλεκτων (’ella ḫeruyān leg. walaḫeruyān Eth^{y} laḫeruyān) [ων] (Eth^{q} η ελπις αυτων και) τα εργα αυτων επικρεμαμενα τω κυριω των πνευματων | Eth^{tana} το ονομα του κυριου των πνευματων (Eth^{M} τον κυριον των πνευμ.)

(3) Eth^{tana} και οταν αποκαλυφθη τα κρυπτα αυτων τω δικαιω (laṣādeq Eth^{M} τοις δικαιοις laṣādeqān) κρινει (yekuēnen Eth^{M} κριθησονται yetkuēnanu) τους αμαρτωλους (ḫāṭe’āna Eth^{M} ḫāṭe’ān οι αμαρτωλοι)

(4) Eth^{m} την γην και τον ουρανον | Eth^{q} τα προσωπα των δικαιων και αγιων

(5) Eth II βασιλεις (Eth I + και) κραταιοι

CHAPTER 39

(1) Eth^{m} και εσται εν ταυταις ταις ημεραις καταβησονται [ινα] γενωνται ως ([kama] yekun kama) τα τεκνα (leg. daqiqa c. Eth$^{g\ q\ u\ t^{2}}$) των εκλεκτων και αγιων εκ των υψιστων των ουρανων (leg. ’emle‘ulāna samāyāt) και το σπερμα αυτων εν γενησεται μετα των υιων των ανθρωπων

(3) Eth II νεφελαι και καταιγις Eth I καταιγις

(4) Eth II κατοικητηρια των δικαιων (Eth I των αγιων) και κοιμητηρια των αγιων (Eth I των δικαιων)

(5) Eth II μετα των αγγελων Eth^{tana} + αγιων Eth I + της δικαιοσυνης αυτου | Eth^{m} δικαιοσυνη ως υδωρ ερρει εν ταις ημεραις αυτων

(6) Eth I και εν εκεινω τοπω (Eth II εν εκειναις ημεραις) ειδον οι οφθαλμοι μου τον εκλεκτον της δικαιοσυνης και πιστεως (Eth II τον τοπον των εκλεκτων της δικ. και της πιστ.) και η δικαιοσυνη εσται εν ταις ημεραις αυτου (Eth II αυτων) και οι δικαιοι και αγιοι αναριθμητοι εσονται ενωπιον αυτου εις τον αιωνα του αιωνος

(7) Eth^{M} κατοικητηριον αυτων (Eth$^{g\ m\ (tana)}$ αυτου) | Eth^{M} φανουσιν / εκλαμπουσιν (yetlaḫḫayu Eth$^{tana\ g\ q\ t\ u}$ crpt yetḫēyalu) | Eth^{M} και δικαιοσυνη ⸢ουκ αφανισθησεται εμπροσθεν αυτου⸣ και η αληθεια ουκ αφαν. εμπ. αυτου (Eth$^{b\ 4\ mss}$ om. και η αληθ .... αυτου)

(9) Eth^{tana} + (post υψωσα) και εδεηθην | Eth^{ull} τον κυριον των δυναμεων

(11) Eth I τι γινεται εις τον αιωνα (Eth II τι γινεται ο αιων)

(12) Eth^{M}πληροι την γην τοις πνευμασιν Eth^{tana} πληρης η γη των πνευμ.

(13) Eth^{q} ειδον οι οφθαλμοι μου παντας τους εγρηγορους τους μη κοιμωμενους | Eth^{tana} το ονομα του κυριου των πνευματων Eth^{M} om. των πνευμ. Eth^{v} το ον. του κυριου του αιωνος / εις τον αιωνα

CHAPTER 40

(1) Eth II ενωπιον της δοξης (Eth[g m q, 7 mss] om. της δοξ.) του κυριου των πνευματων

(2) Eth[M] ⸢ειδον⸣ και επι τεσσαρων πτερυγων του κυριου των πνευματων ειδον Eth[c v] om. ειδον 1° | Eth I τεσσαρα προσωπα εξαλλα (kāle'a) των μη κοιμωμενων ('ella yenawwemu Eth II crpt? 'ella yeqawwemu των εστηκοτων)

(3) Eth[d y] ενωπιον του κυριου των πνευματων

(4) Eth[tana] το ονομα του κυριου των πνευματων

(5) Eth[q] crpt τον εκλεκτον των εκλεκτων

(6) Eth[d] και τριτην φωνην ηκουσα δεομενην ('enza 'ese''el leg. yese''el) και προσευχομενην (yeṣēli) Eth[M] δεομενων ('enza yese''elu) και προσευχομενων (yeṣēleyu) cf. Flemming, 44 | παραιτουμενην leg. yāstabaqquā'.

(8) Eth[b 6 mss] + (post παντα κρυπτα) λεγων αυτω

(9) Eth I ο (Eth II + αγιος) Μιχαηλ | Eth I ο (Eth II + αγιος) Γαβριηλ | Eth[M] ο επι την μετανοιαν εις (Eth[b 4 mss] και) ελπιδα των κληρονομουντων την αιωνιον ζωην Eth[q] την ελπιδα της αιωνιου ζωης

(10) Eth I του κυριου των πνευματων Eth II του κυρ. του υψιστου

CHAPTER 41

(2) Eth[M] κατοικητηρια των εκλεκτων (ḫeruyān) Eth[tana] κατ. των αμαρτωλων (ḫāte'ān) | Eth[tana] και τας συναγωγας των αγιων (māḫbarātihomu laqeddusān) Eth[M] τα κοιμητηρια των αγιων ex 39.4? | Eth[a] το ονομα του κυριου της δοξης

(3) Eth[M] και εκει ειδον ποθεν εξερχονται (Eth[g m] ερχονται) | Eth[M] και εκειθεν (Eth[q] εκει Eth[v] + ειδον) εμπιμπλαται ο κονιορτος της γης (ṣabala medr Eth[g t¹ u] ṣebula medr)

(4) Eth I τους θησαυρους της χαλαζης και των ανεμων (Eth II om. και των αν.) | Eth[tana g u²] om. per hmt και τους θησαυρους των νεφελων | και αι νεφελαι αυτων μενουσιν (yaḫāder) επανω της γης προ του αιωνος

(5) Eth[tana] και ως ετερος ενδοξοτερος εστιν του ετερου και απο των συντελειων αυτων τας εορτας (wa'emmeḥwāromu ba'āla? Eth[M] wameḥwāromu be'ul 'and their magnificent course' (Knibb) / και πιστιν τηρουσιν ετερος μεθ ετερου εν τη διαθηκη / των ορκω η / ω εμμενουσιν (Eth I zaḫādaru Eth II zanabaru Eth[k] ḫabru)

(7) Eth[u (g q, 2 mss)] και μετα ταυτα ειδον (Eth[M] om.) τας οδους cf. Charles[1906] 83, n. 8 | Eth[m t¹ u (tana)] ενωπιον της δοξης του κυριου των πνευματων | Eth[tana] ανατολαι αυτων (kuenatomu) pro αινεσεις αυτων ('akwatētomu)

(8) Eth I ο (Eth II + λαμπρος) ηλιος | Eth[tana] εν ονοματι του κυριου των

πνευματων Eth^{M} om. των πνευμ. | εκτισεν faṭara leg. falaṭa? | Eth^{M} εν ονοματι δικαιοσυνης αυτου En^{g} om. αυτου Eth$^{t^1}$ εν ονοματι πνευματος δικαιοσυνης Eth$^{a^1}$ (Garrett) κατα (bakama) την δικαιοσυνην αυτου

(9) Eth^{M} και ου δυναμις (Eth^{tana} Σατανας) δυνησεται κωλυειν | Eth^{g} οτι δικαστην (leg. makuannena cf. Eth$^{a^1}$ (Garrett)) πασιν αυτων καταστησει (yerēsi) Eth^{M} crpt yerē'i 'the Judge sees them all' Knibb cf. Charles^{1906} 83, n. 32.

CHAPTER 42

(2) Eth I εστερεωθη (taṣan'āt) Eth II εκαθισε? (taṣe'nat)

CHAPTER 43

(1) Eth^{M} εκαλεσε παντας Eth$^{g\ m\ q\ tana}$ om. παντ.

(2) Eth^{M} την ημεραν της γενεσεως αυτων (Eth$^{g\ q\ u}$ 'eluta konatomu leg. helluta = hellāwē conflat. ex το ειναι αυτων et το γενεσθαι αυτων? cf. Flemming, 47.) | Eth$^{g\ q\ t}$ [ως] κινησις αυτων αστραπην (mabraqa Eth^{M} mabraq mabraqa vel leg. [ως] κιν. αυτ. αστραπει αστραπην (yebarreq mabraqa)) ⌜γεννα⌝ cf. Ps. 144(143).6 LXX αστραψον αστραπην

(4) Eth I ταυτα τα ονοματα των αγιων (Eth II των δικαιων) των κατοικουντων επι της γης και πιστευοντων (Eth^{tana} pr. μη) εν τω ονοματι του κυριου των πνευματων (Eth^{q} om. των πνευμ.)

CHAPTER 44

(1) Eth^{tana} μενειν (ḫadira) μετ αυτων (Eth^{M} crpt ḫādiga?)

CHAPTER 45

(1) Eth$^{tana\ (n)}$ και αυτη η παραβολη δευτερα εις τους αρνουμενους το ονομα του κυριου των πνευματων (Eth^{M} το ον. των οικιων των αγιων (lamāḫdara qeddusān)) και τας συναγωγας των αγιων (māḫbara qeddusān) (Eth^{M} του κυριου των πνευματων (la'egzi'a manāfest))

(3) Eth I ο εκλεκτος μου (Eth II om. μου) | Eth I οταν ιδωσι τους εκλεκτους μου Eth II τον εκλεκτον μου | και αναπαυσεις ⌜αυτων⌝ ουκ εισιν αυτοις (leg. me'rāfomu ⌜ḫualqua⌝ 'albomu?) | βαρυνθησεται | Eth I το ονομα (Eth II + αγιον και) ενδοξον μου

(4) Eth^{M} και κατοικισω εν μεσω αυτων τον εκλεκτον μου (Eth$^{tana\ g\ q\ u,\ 2\ mss}$ τους ἐκλεκτους)

(6) Eth I η κρισις (Eth$^{ryl\ d\ e\ y}$ + μου)

CHAPTER 46

(1) Eth I ωσπερ εις των (Eth II + αγιων) αγγελων (Eth^{ull} ωσπερ αγγελος ειρηνης)

(2) Eth I και ηρωτησα ενα των (Eth II + αγιων) αγγελων Eth^{ull} τον αγγελον ειρηνης ος επορευετο μετ εμου frt leg. c. Flemming, 49 και ηρωτ. τον αγγελον ος επορ. μετ εμου | περι εκεινου / του (zeku) υιου του ανθρωπου | Eth^{M} παντα τα κρυπτα Eth^{q} πασας τας ορασεις | Eth^{M} συν τω κεφαλαιω των ημερων Eth$^{ryl^{1}}$ συν τω παλαιω (beluya) των ημερων cf. Eth^{ull} 47.3 Eth^{t} 50.1 (ex Dan. 7.9)

(3) Eth I ενικησεν (mo'a) Eth II + παντας

(4) Eth^{tana} ο υιος του ανθρωπου ον εωρακας Eth^{M} wazentu (Eth$^{t,\ e}$ + we'etu) walda sab' Eth^{tana} om. wazentu

(6) Eth^{tana} και στραφησεται (yetgafattā') Eth^{M} καταστρεψει (yegafatte') προσωπα των δυνατων | και η αισχυνη καλυψει αυτους (yemalle'omu) | Eth^{tana} εκ των οικιων αυτων και των κοιτων

(7) και ουτοι εισιν οι κυριευοντες (yekuēnenu) των αστερων του ουρανου | και οι καταπατουντες επι την γην και κατοικουντες εν αυτη | om. tagbāromu ʿammaḍā 20 ditt. | Eth$^{b\ t^{2},\ 4\ mss}$ om. wakuellu

(8) Eth$^{g\ t^{1}}$ και εκδιωκουσιν (Eth^{tana} και επισκεπτονται (yefaqqedu)) τας οικιας των συναγωγων αυτου / τας συναγωγας αυτου Eth^{M} και εκδιωκονται απο ⌜των οικιων⌝ των συναγωγων αυτου

CHAPTER 47

(1) Eth^{M} το αιμα του δικαιου (dama ṣādeq) Eth^{m} cf. v. 2 το αιμα των δικαιων

(2) Eth$^{tana\ g\ m,\ 4mss}$ το ονομα Eth$^{t,\ 10mss}$ εν τω ονοματι

(3) Eth^{ull} ειδον τον παλαιον (labeluya) των ημερων cf. 46.2, Eth^{t} 50.1

(4) Eth$^{tana\ m}$ ο αριθμος των δικαιων ηγγισεν (Eth^{M} ηλθεν) Eth^{M} ο αριθ. της δικαιοσυνης

CHAPTER 48

(1) Eth$^{m,\ 3\ mss}$ ειδον την πηγην δικαιοσυνης την μη εκλειπουσαν / ανεκλιπη (za'iyetẖua(uē)llaqu Eth^{M} wa'iyetẖuallaqu c. Dillmann, 160, Charles, 112, cf.

58.6) | συν τοις δικαιοις ⸢και αγιοις⸣ (Eth^{u} om. και αγ. Eth^{tana} om. και Eth^{g} crpt και δικαιοις) και τοις εκλεκτοις

(2) Eth^{tana} baqedma εμπροσθεν (Eth^{M} maqdema)

(4) Eth I τοις δικαιοις Eth II + και τοις αγιοις

(5) Eth I δοξασουσιν και ευλογησουσιν (Eth II ευλογ. και δοξ.) και ψαλουσιν το ονομα του κυριου των πνευματων (Eth$^{g\ u}$ ψαλ. τον κυριον των πνευματων)

(6) Eth^{q} προ/απ αρχης του αιωνος εμπροσθεν αυτου

(7) Eth$^{t\ u,\ 2\ mss}$ τοις δικαιοις και αγιοις | Eth^{M} και εκδικητης (faqādē Eth$^{g\ (q)}$ bafaqādu Eth^{tana} wafaqādu) εγενετο cf. Charles 94, Flemming-Radermacher 70.

(9) Eth I εμπροσθεν των αγιων (Eth II δικαιων) | Eth I εμπροσθεν των δικαιων (Eth II αγιων)

(10) Eth^{M} αναπαυσις (ʿeraft Eth^{tana} ʿeqeft = σκανδαλον) Cf. E. Isaac, JAOS 183.2 (1983), 403. Eth I εμπροσθεν αυτων (Eth II αυτου)

CHAPTER 49

(1) Eth$^{b\ 2\ mss}$ το πνευμα της σοφιας (ex v. 3)

(2) Eth^{m} εν πασαις ταις οδοις της δικαιοσυνης

(3) Eth$^{g\ q\ u}$ πνευμα σοφιας και ⸢πνευμα⸣ συνεσεως (manfas zayālēbu Eth^{M} manfasa zayālēbu ‘the spirit of one who gives insight’ cf. Dillmann, 25, Lib. Nat., Tom. 41,57 om. | manfasa ’ella nomu baṣedq ‘the spirit of those who sleep in righteousness’ crpt ex manfasa lebbunnā waṣedq πνευμα γνωσεως και ευσεβειας Isa. 11.1 Eth LXX?

(4) Eth^{m} + (post εμπροσθεν του κυριου των πνευματων) και η δοξα αυτου εις τον αιωνα του αιωνος

CHAPTER 50

(1) Eth^{m} om. τοις αγιοις και | φως (berhāna) των ημερων leg. των εθνων? Eth^{t} ο παλαιος (beluya) των ημερων cf. 46.2, 47.3

(2) Eth^{g} εν (Eth^{M} pr και) τη ημερα της θλιψεως η θησαυρισεται κακα (zatazagba ’ekuya leg. ’ekuy vel ’ekit Eth^{M} tezzaggab ’ekit) | Eth^{m} εν ⸢τη τιμη και⸣ τω ονοματι του κυριου των πνευματων

(3) Eth II εμπροσθεν του κυριου των πνευματων (Eth I crpt. (ditt.) εν ονοματι του κυρ. των πνευμ͂

CHAPTER 51

(1) Eth^{M} την παραθηκην αυτης Eth^{g} τα τεθησαυρισμενα (ʾella tazagbu) εν αυτη | Eth^{g} και Αδης αποδωσει (Eth^{M} + ditt. την παραθηκην και) τα παραδοθεντα αυτω cf. Flemming, 54

(3) Eth$^{q\ t^{1}}$ ο εκλεκτος μου | Eth II επι του θρονου αυτου (Eth I μου) | εκ της μελετης του στοματος αυτου

CHAPTER 52

(1) Eth$^{g\ m\ t^{1}\ u}$ πασας τας ορασεις των εν κρυπτω (Eth^{M} των κρυπτων)| Eth^{M} εν καταιγιδι (banakuarkuāra nafāsāt) (Eth^{n} πυρος) Eth^{tana} εν αρματι ανεμων (bamankuarākuera nafāsāt) ex 2 Kg, 2.11?

(2) Eth^{M} παντα τα κρυπτα του ουρανου τα (Eth^{tana} pr. και) μελλοντα γενεσθαι επι της γης (Eth$^{tana\ g\ m\ t^{1}\ u}$ om. επι της γης)

(3) Eth I τι εστιν ταυτα (Eth II om.) α εν κρυπτω εωρακα

(4) και ειπεν μοι . . . ταυτα παντα . . . τη (Eth$^{tana\ q\ u}$ pr εν) εξουσια του Χριστου αυτου εσονται

(5) Eth I αναμεινον χρονον επι μικρον (Eth II + και οψει) και αποκαλυφθησεται σοι παντα τα κρυπτα | Eth II α εφυτευσεν takala Eth I kallala (= עטר crpt ex עתד) ο κυριος των πνευματων

(6) Eth I α εωρακασιν οι οφθαλμοι σου (Eth II α εωρακας) | Eth^{M} ως υδωρ καταβαινον ανωθεν επι (Eth^{tana} om.) ταυτα τα ορη ⸢και⸣ γενησεται (wayekawwenu leg yekawwenu) μαλακα | Eth II υπο / υποκατω (Eth I εμπροσθεν) των ποδων αυτου

(7) Eth I ουδε δυνησονται (Eth II + σωζεσθαι και) εκφυγειν

CHAPTER 53

(1) Eth II και εκει ειδον οι οφθαλμοι μου φαραγγα βαθειαν και ανεωγμενον το στομα αυτου (Eth I αυτων)

(2) Eth^{m} και την ανομιαν (און) (gēgāya Eth^{tana} gēgāyāta) ποιουσιν αι χειρες αυτων και παντες οι ποιουντες την ανομιαν (און) (zayeṣāmewu lagēgāy) ⸢αμαρτωλοι⸣ βρωθησονται (yeballeʿu leg. yetballeʿu?) | [ου] μενουσιν (leg. [ʾi]yetqawwamu Eth^{tana} ⸢ʾiʾyeqawwemu) και αφανισθησονται (wayaḫallequ c. Eth$^{tana\ t^{1}}$ Eth^{M} pr. ουκ) εις τον αιωνα του αιωνος

(3) Eth II επορευοντο (yaḫawweru Eth I ẹμεινον yaḫadderu crpt?) και επετηδευον

(4) Eth$^{ryl^{2}\ n}$ + (post μετ εμου) λεγων αυτω

(5) Eth^{tana} και απεκριθη μοι λεγων Eth^{M} και ελεγε μοι

(6) Eth^{tana} φανερωθησεται ο δικαιος (sedq leg. sādeq) και εκλεκτος και η οικια της συναγωγης αυτου απο του νυν ου κωλυθησεται Eth^{M} φανερωσει ο δικαιος και εκλεκτος την οικιαν της συναγωγης αυτου (Eth^{q} και φανερωσει την δικαιοσυνην και τον εκλεκτον της συναγωγης αυτων)

(7) Eth$^{tana,\ b\ 3\,mss}$ και (Eth^{M} + ου) ταυτα τα ορη γενησεται | Eth^{m} ως η δικαιοσυνη αυτου ως η γη (sic) Eth^{M} προ προσωπου αυτου ως η γη Eth$^{tana\ g\ t^1\ u}$ προ της δικαιοσυνης αυτου ως η γη crpt ex כארץ המישור ως η γη η πεδινη(?)

CHAPTER 54

(1) Eth^{q} και στραφεις εβλεψα (Eth^{m} + και ειδον) αλλο μερος της γης |

(2) Eth I εις ταυτην την φαραγγα βαθειαν (Eth II om. βαθ.)

(4) Eth^{g} ουτοι οι δεσμοι (ʾellu ʾesrat ⌜mabāleʿt⌝ cf. Charles^{1906} 97, n. 44, 98 n. 1)

(5) εως Αδου κατω (matḥetta kuellu leg. siʾola? ⌜dayn⌝) | Eth$^{a^1\ Garrett}$ και λιθοις τραχεσιν καλυψουσιν επανω αυτων (malʿeltihomu Eth^{M} crpt malāteḥihomu cf. Charles^{1906} 98, n. 13.)

(7) Eth^{M} ανοιχησονται (yetraḫḫāwu) Eth^{tana} ανοιξει (yāreḫhu) Eth$^{g\ t^1\ u}$ ανοιξεις (tāreḫḫu) | Eth$^{g\ m}$ και πασαι (ʾella leg. kuellu) αι πηγαι υποκατω (Eth^{M} + των ουρανων και) της γης om. Eth^{M} diba ... samāyāt c. Eth$^{b^1}$ (crpt. ditt.) cf. Flemming, 58.

(8) Eth^{tana} συν τοις υδασιν (mesla (Eth^{M} zamesla) māyāt) | Eth^{tana} τα επανω των ουρανων (Eth^{M} + υδατα et ditt. τα επ. των ουρ.)

(10) Eth^{M} διοτι γινωσκουσιν (Eth^{tana} pr. ου)

CHAPTER 55

(2) Eth II ου ποιησω ουτως (Eth I om. ουτως)

(3) Eth$^{m\ (q)}$ και η εντολη μου αυτη εστιν [τη παρεμβολη του Αζαζηλ] (wateʾzāzeya ze weʾetu [lateʿyenta ʾazāzēl] Eth^{M} crpt waʾemze bateʾzāzeya weʾetu) οτε ηυδοκησα κρατησαι αυτων εν τη χειρι των αγγελων Eth$^{tana\,g\,m}$ επαφησω (ʾaḫāder Eth^{M} yaḫāder) | Eth I om. ο κυριος 1°

(4) Eth^{tana} τους εκλεκτους μου | Eth$^{tana\ g\ m\ t^1}$ επι του θρονου της δοξης (Eth^{M} + μου) Eth$^{4\ mss}$ εκ δεξιων του θρονου της δοξης μου

CHAPTER 56

(1) Eth^{M} εχομενοι (ʾeḫuzān Eth$^{g\ (q)}$ yeʾeḥḫezu Eth$^{tana\ t^1\ m\ u}$ om.) δεσμους

σιδηρους και χαλκους (mašāgera ḥasin wabert?) Eth$^{g\ (m\ u)}$ μαστιγας και δεσμ. σιδ. και χαλκ. cf. Flemming, 60

(2) Eth^{g} οι εχομενοι μαστιγας (Eth^{M} om. μαστιγας)

(3) Eth I εις (Eth II pr. εκαστος) τους εκλεκτους ...

(4) Eth^{M} η ημερα της πλανης (Eth^{tana} της δοξης) αυτων)

(5) Eth^{M} συναχθησονται (yetgābe'u) Eth$^{g\ q}$ επιστρεψουσιν (yegabbe'u) | Eth^{M} εξαποστελουσιν (yewaddeyu Eth^{g} yewaddequ) τους αρχοντας αυτων | εν μεσω leg. εκ μεσου (mā'kala leg. 'emmā'kala?) των νομων αυτων crpt ex מערותיהם των σπηλαιων αυτων?

(6) Eth^{tana} και αναβησονται και καταπατησουσιν την γην των εκλεκτων μου (Eth$^{g\ q\ t}$ αυτου Eth^{M} αυτων) και γενησεται η γη των εκλεκτων μου (Eth$^{t^2,\ 2\ mss}$ αυτων Eth^{M} αυτου) εμπροσθεν αυτων καταπατημα και τριβος

(7) Eth^{q} [ουκ] ενισχυσεν (teṣanne' leg. ['i]taṣan'a) η πιστις αυτων εν αυτοις | Eth^{n} ου πιστευσει ('iya'ammeno Eth^{M} crpt 'iya'ammero) ανηρ ετερω και (Eth I om. ετ. και) αδελφω αυτου | Eth$^{t^2}$ εως ουκ (Eth^{M} om.) εσται αριθμος των νεκρων εκ των θανατων αυτων ('emmotomu leg. 'emmetomu (Eth^{tana} 'emwe'e'etomu (sic) εκ των ανδρων αυτων)

(8) Eth^{q} και τὴν απωλειαν αυτων ουκ αφησει ('iteḫaddeg) Αδης και καταποθησονται οι αμαρτωλοι (Eth^{M} crpt και απωλεια αυτων Αδης καταπιει τους αμαρ.)

CHAPTER 57

(2) Eth I εκ των περατων του ουρανου (Eth II της γης) εως των περατων της γης (Eth II του ουρανου) | Eth^{M} εν ημερα μια (Eth^{tana} εν ωρα μια)

CHAPTER 58

(1) Eth^{tana} παραβολην δευτεραν (ex. 57.3)

(3) Eth^{q} και εσται τοις εκλεκτοις εν τω φωτι της ζωης αιωνιου

(4) Eth I εν τω ονοματι του κυριου του αιωνος Eth II παρα (baḫaba) τω κυριω του αιωνος (Eth^{b} των πνευματων)

(5) Eth II και μετα ταυτα ερρεθησεται τοις αγιοις (Eth I + εν τω ουρανω) οτι ζητησουσιν εν τω ονοματι αυτου (bašamāy leg. basemu?) τα κρυπτα της δικαιοσυνης

(6) φως μη εκλειπον / ανεκλιπη (za'iyetḫuallaqu leg. za'iyetḫallaq cf. 48.1 | Eth^{q} οτι το προτερον σκοτος απολειται Eth^{M} οτι πρωτον το σκοτ. απολ.

CHAPTER 59

(2) Eth^{m} οταν ηχηση (yedanagged̩ Eth^{M} yedaqqeq) | η φωνη αυτων ακουεται εν ταις κατοικιαις της γης (wamāḫdarāta yabs leg. bamāhdarāta yabs) | εδειχθησαν μοι αι φωναι της βροντης (waqāl leg. qāl zanaguadguād)

CHAPTER 60

(1) Eth I ειδον οτι εσεισε τον ουρανον των ουρανων σεισμος μεγας Eth II ειδ. οτι εσεισθη ο ουρ. των ουρ. σεισμον μεγαν

(2) Eth I και κεφαλαιον των ημερων εκαθισεν Eth II και τοτε ειδον κεφ. των ημερ. καθιζον

(3) Eth I και οι οσφυες μου ετρομησαν / διερραγησαν (tafatḫa Eth II pr. taqaṣʿa wa ωδινησαν και (cf. Sir. 31(34).5)) και οι νεφροι μου (kuelyāteya Eth II kuellantāya) ετακησαν

(4) Eth I και επεμψε Μιχαηλ (Eth II pr. ο αγιος) αλλον αγγελον (Eth II pr. αγιον) εκ των αγιων (Eth II +αγγελων)

(5) Eth I Μιχαηλ (Eth II pr. ο αγιος) | Eth$^{tana\ t}$ τι εωρακας (Eth^{M} δια τινα ορασιν) οτι ουτως εταραχθης (taḥawka leg. c. Flemming, 64 taḥawkaka)

(6) Eth^{u} τοις μη (Eth^{M} om.) προσκυνουσιν / λατρευουσιν τον δικαστην της δικαιοσυνης (lakuennanē leg. makuannena ṣedq) και τοις αρνουμενοις τον δικαστην της δικαιοσυνης (leg. makuannena sedq)

(7) Eth^{g} δυο κητη μεγαλα

(8) Eth^{u} Duidain (Eth^{M} Dendayn) | Eth^{M} ου παρεδοθη (tamaṭṭawa Eth^{tana} tamayṭa μετεστραφη) ο προπατωρ μου 'emḥēweya Eth^{b} σου Eth^{ull} + Ενωχ Eth^{tana} υιος ανθρωπου ('eguāla 'emma ḫeyāw)

(11) Eth$^{tana\ g\ m}$ υποκατω της γης (Eth^{M} επι της γης) | Eth^{M} εν τοις θεμελιοις του ουρανου (Eth$^{e\ x}$ της γης)

(12) Eth I yetkaffalu 2° ditt. om. | Eth I πηγαι ('anqeʿt leg. 'anāqeḍa θυραι c. Flemming, 65) των ανεμων (Eth II crpt 'anqeʿt wanafāsāt) | Eth I εκαστη κατα την δυναμιν του ανεμου (babaḫayla Eth II baḫayla manfas) | και κατα (wakama leg. wabakama) την δυναμιν της δικαιοσυνης | και παντα τον διαμερισμον τον διαμεριζομενον (leg. yetkaffal vel c. Eth^{tana} takafla cf. Eth^{M} v. 13)

(13) Eth^{M} και παντας τους διαμερισμους τους διαμεριζομενους (Eth^{q} om. τους διαμεριζ.)

(14) Eth^{M} οτι τη βροντη εισιν καταπαυσεις εν τη ανοχη τη φωνη αυτης δοθειση (Eth^{u} om. δοθ.) | Eth^{tana} και εκ ενος πνευματος (wa'im'aḥadu manfasa leg. wa'em'aḥadu manfas) αι δυο πορευονται και ου διαχωριζονται

(15) Eth^{q} baḫoḍā εν αμμω Eth^{M} zaḫoḍā leg. (ba)kama ḫoḍā ως αμμος = כחלא

(16) διασπειρεται εν πασιν τοις οριοις ('adbāra) της γης

(17) αγγελος πονηρος (mal'aka zi'ahu leg. mal'ak za'ekuy?)

(18) Eth^{tana} και το πνευμα του χιονος ουκ εκλειπει / ανεκλιπες (ẖadga leg. ẖadgat 'albo? Eth^{M} ẖādaga)

(19) Eth I οτι τη πορεια αυτου εστιν η δοξα Eth^{M} οτι η πορ. εν τη δοξα | Eth$^{t^1 u}$ εν τη αποθηκη αυτου εστιν αγγελος Eth I και η αποθηκη αυτου (Eth II + φως και) αγγελος

(21) και οτε διασπαρησεται επι πασαν γην συζευχθησεται πασιν τοις υδασιν τοις επι της γης | Eth^{M} και ει συζευχθησεται εν παντι καιρω τοις υδασιν τοις επι της γης Eth$^{q, 6mss}$ om. και ει ... επι της γης Eth$^{tana\ g\ (q),\ 4mss}$ και ει συζευχ. εν παντι καιρ. τοις υδ. τοις επι της γ. 22 ικανουσθη αν τα υδατα τοις κατοικουσιν την γην ('emma kona leg. 'emma konu māyāt yaẖadderu diba yabs) cf. Flemming, 67.

(22) Eth II οτι τροφη τη γη εστιν απο του υψιστου ⌜του⌝ εκ του ουρανου (Eth I εν τω ουρανω) | και οι αγγελοι μετρουσιν αυτο (yetmēṭawewwo 'receive it' leg, yetmēṭanewwo)

(24) Eth^{tana} εσται ταυτα τα δυο κητη εις την ημεραν μεγαλην (la'elat (Eth^{M} crpt lā'la) 'abāy) του κυριου ετοιμα του χορηγηθηναι (delewān yessēsayu) | Eth^{tana} ινα η οργη του κυριου των πνευματων (Eth II om. των πνευμ.) επ αυτους επαναπαυσηται (tā'āref) [και] ⌜ινα⌝ μη εξελθη η οργη του κυριου των πνευματων [εικη] (cf. Eth^{M} v. 25)

(25) Eth^{tana} οταν επαναπαυσηται / καταπαυση η οργη του κυριου πνευματων επι παντας εσχατον (deẖra) κρισις εσται εν ελεει και εν ανοχη Eth^{M} + ditt. (post πνευματων) επ αυτους επαναπαυσεται ινα μη ελθη η οργη του κυριου των πνευματων εικη επ αυτους

CHAPTER 61

(1) Eth^{q} και ειδον εν ταυταις ταις ημεραις [οτι] εδοθη σχοινια μακρα δυσι (la'elku leg. lakel'ē Eth^{M} la'elketu) αγγελοις

(2) ηρωτησα τον αγγελον [τον μετ εμου οντα] (cf. Eth^{m} et v. 3) | Eth I ταυτα (Eth II + μακρα) σχοινια | Eth$^{q, 2\ mss}$ om. και ειπεν μοι· επορευοντο μετρησαι

(3) Eth^{tana} ουτοι εισιν οι ('ella Eth^{M} om.) τα μετρα των δικαιων (Eth^{M} + πατερων) ... οισουσιν | Eth I τοις δικαιοις (Eth II om.)

(4) Eth I ταυτα τα μετρα ⌜α⌝ δοθησεται τη πιστει/αληθεια και τοις leg. (wala'ella?) προσεχουσιν τη δικαιοσυνη (Eth II τω λογω της δικαιοσυνης)

(5) Eth^{M} τους καταφαγεντας εκ των ιχθυων της θαλασσης και εκ των θηριων Eth$^{t\ u}$ τους καταφαγ. εκ των θηριων Eth$^{q\ m}$ κητων? ('emmazāgebt leg. 'emmanābert c. Flemming, 69) και εκ των ιχθυων της θαλασσης | οτι επισρεφουσιν και επαναπαυσονται επι την ημεραν του εκλεκτου (Eth^{t} + μου)

(6) Eth^{M} και εδεξαντο επιταγην οι εν τω ουρανω ανω απαντες και την δυναμιν και φωνην μιαν και φως εν ως πυρ (Eth$^{b\ 3\ mss}$ και η δυναμις και φωνη μια και φως εις ως πυρ εδοθη αυτοις)

(7) και την αρχην των φωνων (maqdema qāl) ηυλογησαν ... | Eth και τω πνευματι ζωης (leg. ζωου?)

(8) Eth I λαι ο κυριος των πνευματων εκαθισε τον εκλεκτον επι τον θρονον της δοξης (Eth II + αυτου) | Eth^{q} παντα τα εργα των δικαιων (Eth^{M} αγιων)

(9) κατα την κρισιν της δικαιοσυνης (bafenota (ditt.) kuennanē ṣedq) | Eth II τον κυριου του υψιστου Eth I του κυρ. των πνευματων | Eth I και αγνισουσιν yeqēdesu (Eth II yewēdesu αινησουσιν) το ονομα (Eth II εν τω ονοματι) του κυριου των πνευματων cf. vv. 11, 12.

(10) Eth I και καλεσουσιν πασα (kuello leg. kuellu) η δυναμις των ουρανων και παντες (kuellu) οι αγιοι ανω ... Eth II και καλεσει πασαν (kuello) την δυναμιν των ουρ. και παντας (kuello) τους αγιους ανω ... | $Eth^{b^2\ 5\ mss}$ και (Eth^{M} om.) επι των υδατων

(11) Eth II και αγνισουσιν (yewēdesu leg. yeqēdesu) Eth I om. cf. vv. 9, 12. | $Eth^{a^1\ Garrett\ i\ l}$ εν τω πνευματι της μακροθυμιας Eth^{M} εν μακροθυμια | $Eth^{b\ 2mss}$ ευλογητος αυτος (Eth^{M} om. αυτος)

(12) Eth I παντες οι αγιοι (Eth II + αυτου) | Eth I το ονομα ευλογητον (Eth II αγιον) σου | σφοδρα σφοδρα? zafadfāda 'emḫāyl (leg. waḫāyāla cf. 90.7?) |

(13) Eth II και πασας τας δυναστειας αυτου εφ οσον (ba'amṭāna) εποιησεν Eth I και παν μετρον / παντα αριθμον των εργων αυτου ('amṭāna(nu) gebra(u))

CHAPTER 62

(1) Eth^{tana} και τοτε ('emze) συνεταξε Eth^{M} και ουτως (kamaze) συν. | Eth^{q} ο κυριος των πνευματων (Eth^{M} om. των πνευμ.) $Eth^{e\ ull}$ + (post τοις κραταιοις) ινα δοξαζωσιν | Eth^{M} (συνεταξε) ... τοις κατοικουσιν (yaḫāderewwā) την γην Eth^{tana} Lib. Nat. Tom. 41, 59 τοις κρατουσιν (ye'eḫḫezewwā) της γης cf. v. 3 |

(2) και ο κυριος των πνευματων εκαθισε επι του θρονου της δοξης αυτου Lib. Nat. ibid. και εκαθισε ο εκλεκτος 'ο κυριος των πνευματων' επι του θρον. της δοξ. αυτ. | Eth I και το ρημα του στοματος αυτου αποκτεινει παντας τους αμαρτωλους και και παντες οι αδικοι (Eth II παντας τους αδικους) απο (Eth II pr. και) του προσωπου αυτου απολλουνται cf. Flemming, 71

(3) Eth I και δικαιοσυνη κρινεται ενωπιον αυτου Eth II και οι δικαιοι εν δικαιοσυνη κρινονται ενωπ. αυτ. | Eth^{t} και η δικαιοσυνη ουκ εκλειπει ('iyaḫāleq) ενωπιον αυτου cf. Flemming, 72, $Charles^{1906}$, 112, n. 22

(4) και εδυστοκησεν εν τω τοκετω? (waya'aḍḍebā walid bis om. 1° ditt. cf. Gen. 35.16 (LXX))

(5) Eth^{M} εκεινον (Eth^{u} αυτον) τον υιον γυναικος (walda be'sit) $Eth^{q\ t\ u\ (g)}$ τον υιον ανδρος/ανθρωπου (walda be'si)

(6) Eth II δοξασουσιν και ευλογησουσιν (Eth I ευλογ. και δοξασ.) | Eth II οι βασιλεις κραταιοι (Eth I οι βασ. και οι κρατ. cf. 38.5) | Eth^{M} τον παντοκρατορα τον κρυπτον (Eth^{tana} τον εν κρυπτω)

(7) ο υιος του ανθρωπου (walda 'eguāla 'emma ḥeyāw) Eth^{M} ενωπιον των δυναμεων αυτου (Eth^{u} εν τη δυναμει αυτου) | Eth^{tana} τοις αγιοις και τοις εκλεκτοις εν δεσμοις / φυλακη (moqḥ)

(8) Eth I και σπαρησεται (yezzarrā') η συναγωγη των εκλεκτων και των αγιων (Eth II των αγ. και των εκλ. Eth$^{tana\ u}$ om. και των εκλ. Eth^{t} crpt wayezzēkaru και μνησθησονται (sic) της συν. των αγ. και των εκλ.)

(9) Eth II οι βασιλεις κραταιοι (Eth I οι βασ. και οι κρατ. supra v. 6 et 38.5) | Eth^{q} + (post λατρευσουσιν) και δοξασουσιν

(11) Eth I και παραδωσει αυτους ([ye]mēṭewewwomu) τοις αγγελοις εν τη οργη (bamaqšaft leg. zamaq. της οργης c. Eth II) Eth II και οι αγγελοι της οργης δεξονται αυτους (yetmēṭawewwomu) | Eth^{M} ινα εκδικησωσιν απ αυτων οτι ηδικησαν τα τεκνα και τους εκλεκτους αυτου (Eth^{u} [και] εκδικησουσιν [απ αυτων] ⌜εν οργη⌝ οτι ηδικησαν τα τεκνα [και] τους εκλεκτους αυτου)

(12) Eth$^{tana\ q\ m}$ και (Eth^{M} om.) αγαλλιασονται επ αυτοις | Eth I η ρομφαια αυτου (Eth II η ρομ. του κυριου των πνευματων)

(14) Eth I εδονται (Eth II pr. κατοικησουσιν και) μετα εκεινου / του υιου του ανθρωπου (zeku walda 'eguāla 'emma ḥeyāw)

(15) Eth I και|ενδυσονται (labsu perf. proph.) ενδυματα της δοξης (Eth II της ζωης)

(16) Eth I και τουτο εσται το ενδυμα υμων (Eth II om. το ενδ. υμ.) ενδυμα ζωης παρα ('emḫaba Eth^{M} ḫaba) του κυριου των πνευματων

CHAPTER 63

(1) Eth II οι βασιλεις κραταιοι (Eth I οι κρατ. και οι βασ. cf. 62.6,9.) | Eth^{M} 'emmalā'ekta maqšaftu Eth^{u} 'emmalā'ekta maʿʿatu | Eth I ινα (Eth II pr. και) πεσωσιν

(2) Eth^{M} ο κυριος των βασιλεων (Eth$^{e\ h\ v\ ull}$ κυριων) | Eth^{u} ο κυριος των υψιστων (leʿulān) Eth^{M} ο κυρ. των πλουσιων (bāʿel Eth$^{g\ m}$ beʿulān)

(3) Eth II φανησεται (yebarreh) παν κρυπτον και η δυναμις σου ... (Eth I εν παντι κρυπτω η δυναμις σου) | Eth^{ull} + ενωπιον σου

(5) Eth I τις δωσει ημιν αναπαυσιν ινα δοξαζωμεν και ευχαριστωμεν (Eth II + και ευλογησωμεν) και εξομολογησωμεθα (ne'man) ενωπιον της δοξης σου (Eth II αυτου)

(6) Eth$^{g\ (t^{1}\ tana\ b^{1})}$ (κατα)διωκομεν (nesded Eth^{M} nessaddad)

(7) Eth I ουκ εδοξασαμεν το ονομα του κυριου των πνευματων Eth II ουκ εδοξ. εν τω ονοματι του κυριου των βασιλεων | Eth I και ουκ εδοξασαμεν τον κυριον ημων (Eth II om. ημων) | Eth II εν πασι τοις εργοις αυτου Eth I om. | Eth I και η ελπις ημων ην επι τω σκηπτρω (Eth$^{t\ u\ f}$ θρονω Eth$^{q,\ 7\ mss}$ σκηπ. θρονου) της βασιλειας ημων και επι τη δοξα ημων (Eth II και της δοξης ημων)

(10) Eth^{tana} ου μη κωλυσει (Eth$^{q\ (u\ i)}$ δυνησεται 'itekel leg. 'itekel teklā

δυνησεται κωλυειν) ημας καταβαινοντας απο του βαρους Αδου ('emkebud (leg. 'emkebada) si'ol Eth II 'emlāhbā lakebada si'ol Eth^{q} 'emlebba si'ol leg. westa kebada si'ol εις το βαρυ Αδου?)

(11) εμπροσθεν εκεινου / του υιου του ανθρωπου (zeku walda 'equāla emma ḥeyāw)

(12) Eth^{tana} και τουτο το προσταγμα (Eth II + αυτων Eth I + αυτου) και το κριμα (Eth^{M} + αυτου $Eth^{b\ 7\ mss}$ + αυτων)

CHAPTER 64

(2) Eth II ουτοι εισιν οι αγγελοι οι κατεβησαν εκ του ουρανου (Eth I om. εκ του ουρ.) εις την γην

CHAPTER 65

(1) Eth^{t} εγω Νωε ειδον Eth^{M} ειδεν Νωε

(2) Eth^{t} και εξαρας τους ποδας μου (Eth^{M} αυτου) ... και επορευθην (Eth^{M} επορευθη) ... και εκραξα (Eth^{M} εκραξεν) ... και ελεγον (Eth^{M} ελεγεν Νωε)

(3) Eth I και ελεγον (Eth II ελεγεν) αυτω

(4) Eth^{u} εν τη πτωσει (badaḥḍ) Eth^{M} badeḫra Eth^{tana} dāḫf?

(6) Eth I απο προσωπου του κυριου (Eth^{q} + των πνευματων) | Eth^{v} και την δυναμιν των χωνευοντων χωνευματα πασης κτισεως (feṭrat Eth^{M} γης medr sic)

(8) και δοκιμαζει (yebadder) ο αγγελος

(9) $Eth^{tana\ g\ q}$ χειρι μου (Eth^{M} αυτου)

(10) η κρισις αυτων . . . ουκ εξαριθμησεται ('iyetḫuēlaqu) | δια τους μηνας vide supra, 186, 240

(11) Eth II και ου (Eth I om) σοι

(12) $Eth^{q\ (t)}$ εις βασιλειαν (lamangešt) Eth^{M} εις βασιλεις (lanagašt)

CHAPTER 66

(1) Eth^{M} υπο της γης ($Eth^{m\ t^{1}\ (q)}$ του ουρανου) | Eth^{q} yaḫadderu $Eth^{u,\ 2\ mss}$ yenabberu Eth^{M} yen. wayaḫ. (lectio conflata?) leg. επι παντας τους κατοικουντας

CHAPTER 67

(1) Eth II ιδου (Eth I om.) | Eth^{tana} μοιρα αγαπης δικαιοσυνης (leg. feqra ret')?

(2) Eth$^{g\ q\ u}$ οτε ετελεσαν (wad'a leg. c. Flemming, Charles wad'u?) την εργασιαν (lawe'etu mal'ekt Flemming, Charles^{1906} malā'ekt) Eth^{M} οτε εξηνεγκαν (waḍ'u leg. 'awḍe'u?) την εργασιαν (Eth^{ryl} mal'ekt)

(3) Eth^{M} ου στειρωσει / στειρωθησεται ('iyemakker leg. 'iyemakken c. Charles^{1906}) επι τω προσωπω της γης (Eth$^{b\ 5\ mss}$ om.) | Eth II επι (diba Eth I προ qedma) προσωπω(ου) της γης | Eth^{t} εν τω ονοματι του κυριου ημων

(4) Eth I αποκλεισει Eth II αποκλεισουσιν

(7) Eth II ποταμοι πυρος εξερχονται (Eth I ερχονται)

(8) Eth I τοις βασιλευσιν και τοις κραταιοις και τοις υψιστοις (Eth II om. τοις υψ.) | Eth I εις την ιασιν του σωματος Eth II της ψυχης και του σωματος

(9) Eth I οτι ουκ εστιν ενωπιον του κυριου των πνευματων ος λαλησει λογον κενον Eth II ος λαλησει εν λογω κενω

(10) και αρνουνται το πνευμα του κυριου leg. και αρν. τον κυριον των πνευματων?

(11) Eth I εν τουτοις τοις υδασιν (māyāt Eth II crpt. ditt. mawā'el)

(12) Eth I Μιχαηλ (Eth II pr τον αγιον)

(13) εις ιασιν της σαρκος των αγγελων (lamalā'ekt) | Eth I και εις ηδονην της σαρκος αυτων (latawnēta šegāhomu Eth II lamota šegāhomu εις θανατον της σαρκος αυτων)

CHAPTER 68

(1) Eth$^{tana\ (g)}$ και μετα ταυτα εδωκεν (sc. ο Μιχαηλ) μοι διδαχην (temherta Eth^{M} te'merta cf. Dillmann, *Lex*. 133, Knibb 2,158) παντων των κρυπτων εν βιβλω / γραφη του προπατορος (Eth^{M} ο προπατωρ) μου Ενωχ και τας παραβολας αι εδοθησαν αυτω· και συνεταξατο αυτας εμοι εν τοις λογοις της βιβλου / γραφης των παραβολων

(2) Eth I Μιχαηλ (Eth II pr. ο αγιος) | Eth I τις δυναται υπομειναι την λυπην της κρισεως της ποιηθεισης (Eth II + και μενουσης) | Eth$^{b^2}$ και ου τακησεται (wa'iyetmassawu leg. wa'iyetmassaw Eth^{M} wayetmassaw και τακησονται)

(3) Eth I Μιχαηλ (Eth II pr. ο αγιος) | Eth II τις εστιν ου ουκ εκλυεται (za'iyārāḫāreh̲ Eth I za'iyerasseḥ) η καρδια αυτου | Eth^{tana} ος (Eth^{M} και) εξηλθεν | Eth^{tana} απ αυτων οι απεκριθησαν αυτοις (za'awaše'ewwomu = ענו? Eth^{M} za'awḍe'ewwomu) ουτως

(4) Eth^{tana} και εσται (wakona leg. wayekawwen?) οτε εστηκασιν (qomu (perf. proph.) Eth^{M} qoma) ενωπιον του κυριου των πνευματων | Eth I Μιχαηλ (Eth II pr. ο αγιος) | Eth$^{e\ v\ h^2}$ ουκ εσται αυτοις ελεος (Eth^{M} om.) κατεναντι των οφθαλμων του κυριου | Eth I οτι εικονας ('amsāla Eth II ba'amsāla) του κυριου εποιουν

(5) Eth^{tana} οτι ουδε εικων ('amsāla leg 'i'amsāl Eth^{M} 'imal'ak) ουδε ανηρ λημψεται (yetmēṭawu) την μεριδα αυτου (makfalto Eth^{M} makfaltomu) Eth^{q} το θειον (mal'ekto leg. malakāto?) και την μερ. αυτ.

CHAPTER 69

(1) Eth^{c} και μετα ταυτα το κριμα πτοησει και ταραξει / εκστησει αυτους (yādanaggeḍomu wayāme''e'omu cf. Eth$^{q,\ 2\ mss}$) cf. Dillmann, 38 n. 2 Eth^{M} και μετα τουτο το κριμα πτοησουσιν και ταραξουσιν / εκστησουσιν αυτους (yādanaggeḍewwomu wayāme''ewwomu)

(2) Σεμιαζα = G 6.7 | Eth^{M} Αρεστικιφα Eth^{g} Αρτακιφα | Αρμην | Κωχαβιηλ | Eth^{M} Τυριηλ Eth^{g} Ταρυηλ | Ρυμιαλ | Eth Δανιαλ | Eth^{g} Νηκαηλ Eth^{M} Νυκαηλ | Eth$^{q,\ 5\ mss}$ Βαρακιαλ Eth^{M} Βαρακιηλ | Αζαζηλ | Eth I Αρμαρος Eth II Αρμερες Eth^{u} Αρμανος | Βαταριαλ | Βασασαηλ Eth$^{g\ m\ u\ (q)}$ Βασασιαηλ leg. Βασασιαλ | Ανανηλ Eth^{m} Ανανηιαλ leg. Ανανιαλ | Eth I Τυριηλ Eth II Τυριαλ | Σιμψιηλ | Eth^{M} Ιετριηλ Eth^{m} Ισραηλ | Eth^{M} Τυμαηλ Eth^{g} Ταμυηλ | Eth$^{g\ m\ q}$ Τυριηλ Eth^{M} Ταριηλ | Ρυμιαλ | Eth^{g} Αζαζηλ Eth^{M} Αζηζηλ

(4) Eth I και ουτος εστιν ος επλανησεν τα τεκνα των (Eth II + αγιων) αγγελων

(5) Eth$^{tana\ q}$ εβουλευσατο αυτοις βουλην πονηραν ('amkaromu mekra 'ekuya (cf. LXX Isa. 7.5) Eth^{M} 'ammaromu εγνωρισεν αυτοις? vel crpt)

(7) Eth^{M} εκ της χειρος αυτου (Eth$^{tana\ q}$ αυτων) εξηλθεν (Eth^{a} εξηλθον) επι τους κατοικουντας επι τη γη

(8) Eth^{n} ουτος εδιδαξεν τους υιους των ανθρωπων το πικρον γλυκυ και το γλυκυ πικρον (Eth^{M} το πικρον και το γλυκυ)

(9) Eth$^{ull\ 3\ mss\ (tana\ q)}$ εν μελανι (bamāya Eth^{M} wabamāya ḥemmat) | Eth I πολλοι οι πλανηθεντες Eth II πολλοι επλανηθησαν

(10) Eth I ινα ουτως (lazakamaze leg. bazakamaze c. Charles^{1906}, 124 n. 7) vel εις τουτο [ινα] ουτως cf. Eth II (laze [kama] kamaze)

(11) Eth^{q} δια πολλης (Eth^{M} ταυτης) γνωσεως αυτων Eth^{tana} και δια ταυτης της δυναμεως κατεσθιει ημας (Eth^{M} εμε)

(12) Eth^{M} τα δηγματα της οφεως (neskata 'arwē Eth^{tana} + medr)

(13) Eth$^{q\ (tana\ g)}$ και ουτος εστιν ος ηριθμησεν τω Κασβιηλ το κεφαλαιον των ημερων (zaḫualqua (leg. zaḫuallaqua) lakāsābā'ēl re'sa mawā'el Eth^{M} ḫuelqu lakesbe'ēl) [και] τον ορκον τον κεφαλαιον (re'sa maḥalā) ος εδειξεν τοις αγιοις | Eth I και το ονομα αυτου BIQA (Eth II BEQA) Cf. Charles^{1906}, 124, n. 40, Flemming 84

(14) Eth I και ουτος ειπεν τω Μιχαηλ (Eth II pr αγιω) δειξαι αυτω (Eth II αυτοις) το ονομα το απορρητον | Eth II + ditt. δειξ. αυτ. το ονομ. το απορ. Eth^{M} ινα μνησθησωνται (yezkerewwo) αυτου (Eth$^{m\ (u)\ (tana)}$ του ονοματος του πονηρου (sic) και απορρητου ('ekuy crpt ex 'akā' אכע) (Eth^{tana} om. του πον. και) εν τω ορκω

(15) Eth^{tana} και εθηκεν τουτον τον πονηρον ορκον Eth I εις την χειρα του Μιχαηλ (Eth II pr αγιου)

(16) $Eth^{tana, d}$ και το στερεωμα ⸢αυτου⸣ (ṣenʿu leg. ṣenʿ Eth^{M} ṣanʿu) δια του ορκου αυτου και ο ουρανος εκρεμασθη[σαν] προ του τον κοσμον κτισθηναι

(17) Eth I εκ των κρυπτων (ẖebuʾāta Eth II ẖebuʾāt) των ορεων | Eth^{M} υδατα ηδεα (lāḫeyān māyāt leg. māya ḥeywat υδωρ ζωης / υδατα ζωντα? ($Eth^{u, 3 mss}$ laḥeyāwān māyāt))

(18) Eth^{M} και δια του ορκου εκτισθη (Eth^{q} εθεμελιωθη) η θαλασσα και τα θεμελια αυτης· τω καιρω του θυμου (Eth^{q} οριου) αυτης | Eth^{q} και εταξεν την αμμον αυτη οριον (ʿaqmā Eth^{M} om.) και ουχ υπερβαινει τα ορια αυτης (ʾemwassanā Eth^{M} om.) απο [της κτισεως] του αιωνος (laʿālam leg. ʾem [feṭrata] ʿālam) εως του αιωνος (Eth^{M} απο της κτισεως του αιων. εως του αιων.)

(22) Eth^{tana} και ουτως τοις υδασιν τοις ανεμοις αυτων (Eth^{M} τοις ανεμ. των υδατων τοις ανεμοις) και πασιν πνευμασιν και οδοις αυτων εκ παντων των κλιτων (ẖebrata) πνευματων

(23) εκει (baheyya leg. δι αυτου botu?) | Eth II αι αποθηκαι των φωνων (Eth I αι φωναι) της βροντης | Eth^{q} και τουτω (bazentu Eth^{M} και εκει wabaheyya) | Eth II και αι αποθηκαι της παχνης (Eth II om. αι αποθ.)

(24) $Eth^{q\ n}$ το ονομα Eth^{M} εν τω ονοματι

(25) Eth^{M} συντηρησονται (yetʿāqabu Eth^{q} yetʿawwaq(u) cf. Sir. 13.13 ʿuq συντηρησον) | Eth^{q} συντηρησουσιν (yaʿāqebu) τας οδους αυτων cf. Sir. 2.15 Eth^{M} συντηρησονται (yetʿāqabu) αι οδοι αυτων

(26) το ονομα τουτου / αυτου / του υιου του ανθρωπου (semu laweʾetu walda ʾeguāla ʾemma ḥeyāw)

(27) και η αρχη της κρισεως δοθησεται τουτω / αυτω / τω υιω του ανθρωπου (lotu laweʾetu walda ʾeguāla ʾemma ḥeyāw)

(29) εφανη ουτος / αυτος / ο υιος του ανθρωπου (weʿetu walda beʾsi) | $Eth^{b\ 3\ mss\ (t\ 3\ mss)}$ και ο λογος του υιου του ανθρωπου (wanagaru laweʾetu walda beʾsi (Eth^{n} beʾsit της γυναικος)) ισχυσει (yeṣanneʿ) ενωπιον του κυριου των πνευματων (Eth^{M} crpt και λαλησουσιν (wayenaggeru) τω υιω του ανθρωπου και ισχυσει ενωπ. του κυρ. των πνευμ.)

CHAPTER 70

(1) Eth^{v} ανυψωθη το ονομα του υιου του ανθρωπου (semu lawalda ʾeguāla [ʾemma] ḥeyāw (Eth^{M} crpt semu ḥeyaw)) ⸢προς (Eth^{u} om. Eth^{tana} εμπροσθεν προς) τουτον / αυτον / τον υιον του ανθρωπου⸣ (walda ʾeguāla ʾemma ḥeyāw) προς τον κυριον των πνευματων cf. Flemming, 86

(2) Eth^{M} το ονομα ($Eth^{tana\ m,\ ull}$ + αυτου) εξηλθεν

(3) Eth I ουκ ηριθμηθην (itaḥa(hā)ssabku Eth II crpt ʾitaseḥebku Eth^{u} ʾitasabāḥku)

CHAPTER 71

(1) Eth^{ull} + (post μετα ταυτα) εν εκεινω τω τοπω

(3) Eth I και εξηγαγεν με προς παντα τα κρυπτα και εδειξεν μοι παντα τα κρυπτα του ελεους και ⌜εδειξεν μοι⌝ παντα τα κρυπτα της δικαιοσυνης Eth II και εξηγαγ. με προς παν. τα κρυπ. του ελεους και (Eth^{tana} + εδειξεν μοι παντα) τα κρυπτα της δικαιοσυνης

(5) Eth$^{tana\ t\ (g\ q\ u)}$ και μετεθηκεν το πνευμα μου (Eth^{m} + εφη Ενωχ) και εγω εν τω ουρανω των ουρανων (Eth^{m} + η) Eth II και μετ. το πνευμα του Ενωχ εις τον ουρανον των ουρ. | και εωρακα εκει ... ως οικιαν (botu leg. bēta) οικοδομημενην εν λιθοις χαλαζης cf. 14.10 | ... γλωσσας του πυρος των ζωων (’essāt ḥeyāw leg. ’essāta ḥeyāwān? cf. v. 6)

(6) Eth II κυκλουν κυκλοι (ʿawda (za)yaʿawwed cf. Ec. 1.6 סובב סבב) την οικιαν πυρ (Eth I πυρος) Eth^{M} εκ των τεσσαρων ακρων ταυτης της οικιας (Eth^{tana} της οικ.) | Eth^{q} ποταμοι πληρεις πυρος των ζωων (melu’a ’essāt (leg. ’essāta) ḥeyāwān) Eth^{M} ’essāt ḥeyāw

(10) re’sa mawāʿel Eth$^{ryl\ 2\ mss}$ crpt re’sa maḥalā

(11) Eth^{q} και ηυλογησα αυτον (Eth^{M} om.) | Eth$^{q,\ ryl^2\ f}$ εδοξασα αυτον (Eth^{M} om.)

(13) Eth I χιλιοι και μυριοι αγγελοι (’a’lāf wate’lefita malā’ekt Eth II te’lefita ’a’lāfāt cf. 40.1, 14.22 μυριαι μυριαδες G)

(14) συ ει ο υιος του ανθρωπου (walda beʿsi) | Eth^{M} και εκεινος ο αγγελος Eth$^{g\ m\ t}$ om. ο αγγ. | Eth$^{m\ q,\ 8\ mss}$ εν δικαιοσυνη

(16) Eth II και πας εσται πορευομενος εν ταις οδοις σου (leg. yekawwen yaḥawwer Eth I yaḥawwer)

(17) μετα τουτου / αυτου / του υιου του ανθρωπου (we’etu walda ’eguāla ’emma ’ḥeyāw) και ειρηνη εσται τους δικαιοις και οδος ευθυτητος ⌜τοις δικαιοις⌝ (leg. και εν οδω ευθ. πορευσονται cj. Flemming, 89 yaḥawweru pro laṣādeqān (ditt.)

FOR CHAPTERS 72-82.4-20 SEE APPENDIX A

CHAPTER 80.2-8

(2) Eth^{M} παν εργον επι της γης αποστραφησεται / αλλοιωθησεται (cf. 5.2,3) (yetmayyaṭ) Eth$^{g\ m\ q\ t\ tana}$ αποστρεψει / αλλοιωσει (yemayyeṭ) Eth^{t} + τας οδους αυτου (cf. v. 7) | Eth II και ανεξει/συσχεθησεται (tāqawwem leg. yetqāwam? Eth$^{tana\ g\ q\ t\ u}$ te(ye)qawwem) ο ουρανος

(5a) Eth I και ωφθησεται εν τω ουρανω (Eth II ο ουρανος) 5b και ηξει αβροχια (ʿabār) εν τοις ακροις της Αμαξης της μεγαλης επι δυσμων

(6) και πολλοι αρχοντες των αστερων πλανηθησονται της εντολης / των εντολων

(7) φραχθησεται (yetʿāṣawe(u) = סתם Ithp.)

CHAPTER 81

(2) Eth$^{m\ t\ (g\ tana^{a})}$ και ανεγνων την βιβλον παντων των εργων ανθρωπων

(3) Eth I τον μεγαν (Eth II om.) κυριον τον βασιλεα της δοξης εις τον αιωνα Eth II τον εις τον αιωνα | Eth^{tana} και εκλαυσα (Eth^{M} ηυλογησα) δια τους υιους Αδαμ (ʾAdām Eth II ʿālam) cf. Sir. 40.1

(4) Eth$^{t^{2}}$ και ου μη ευρεθησεται αμαρτια / πλανη κατ αυτον εν τη ημερα της κρισεως (Eth$^{tana^{a}\ g\ u\ (m)}$ om. αμαρτια / πλανη κατ αυτον et leg. ημερα της κρισ. Eth^{M} om. εν τη ημ. της κρισ.)

(5) Eth I οι επτα (Eth II τρεις) αγιοι

(6) Eth II παρα τοις υιοις σου (Eth I τω υιω σου) | Eth^{M} εως παλιν επιταξης (teʾēzez) Eth^{q} εως παλ. παρακαλησης αυτον (tenāzezo)

(9) Eth^{tana} ουκ (Eth^{M} om.) αποθανουνται | Eth$^{tana^{a}}$ ουδε αφανισθησονται (leg. waiyetẖabbeʾu)

CHAPTER 82

(1) Eth^{M} εγραψα σοι (Eth$^{g\ m\ t\ tana^{a}}$ om. σοι) | Eth^{M} και παντα απεκαλυψα σοι | Eth I om. Μαθουσαλα 2° | Eth$^{tana\ ull}$ ινα (Eth^{M} pr. και) δως

(2) Eth II ινα δωσιν τοις τεκνοις αυτων ταις (Eth^{tana} pr. και) γενεαις των γενεων εως του αιωνος (Eth I om. των γεν. εως του αιων.) | Eth^{tana} τω σοφιζομενω (laza yeṭṭabab leg. yeṭabbeb Eth^{M} lazā ṭebab) και δοξασουσιν παντας τους σοφους· και καθευδησει η σοφια (Eth^{M} om. και δοξ. . . . σοφια)

(3) Eth I εν (Eth^{q} pr. και) τη διανοια αυτων (Eth^{tana} υμων) ⌜και⌝ (Eth^{u} om.) ου μη καθευδησουσιν οι συνιεντες (Eth II + αυτην)

CHAPTER 83

(1) Eth I τας ορασεις μου (Eth II om. μου)

(2) Eth^{e} την μητερα σου Εδνα

(3) Eth^{M} ειδον (Eth^{tana} pr. και) εν ορασει τον ουρανον σειομενον και σαλευομενον και πιπτοντα επι την γην

(4) ορη επ ορη(εσιν) κρεμαμενα (yessaqqalu pro מתקלין προσκοπτοντα?)

(6) Eth^{tana} τι εγενετο σοι οτι ουτως κραυγαζεις (Eth^{M} om. εγεν. σοι)

(7) Eth^{M} τα κρυπτα παντων των αμαρτηματων της γης Eth$^{tana\ q\ (g\ m)}$ αμαρτηματα πασης της γης leg. τα κρυπ. των αμαρ. πασης της γης?

(10) Eth$^{tana\ g\ m\ t}$ + (post εδεηθην) και ηρωτησα

(11) και την εκλειψιν των αστερων? (waweḥudāta leg. waweẖdata kawākebt)

CHAPTER 84

(1) τον αγιον ⌜και⌝ τον μεγαν | Eth$^{u, 9 mss}$ τοις υιοις των ανθρωπων Eth^{M} τοις υι. ⌜της σαρκος⌝ των ανθ. Eth^{q} τοις υι. των ανθ. ⌜της σαρκος

(2) Eth^{tana} om. και μεγαλωσυνη σου / Eth$^{tana\ d}$ εις τον αιωνα του αιωνος Eth^{M} εις τον αιωνα και εις τον αι. του αι.

(3) Eth^{M} και ουκ αδυνατει σοι εργον (Eth^{g} παν leg. παν εργον) ουδε εν (wa'i'ahadu ⌜wa'i'ahatti⌝ cf. Charles^{1906}, 162, n. 10) | Eth^{g} απο της καθεδρας του θρονου σου ('emmanbartā leg. 'emnebratā manbārika) Eth^{q} απο του θρονου σου ('emmanbareka)

(4) Eth^{M} και επι την σαρκα των ανθρωπων εσται η οργη σου μεχρι της ημερας της μεγαλης κρισεως Eth^{g} μεχρι της μεγαλης ημερας (leg. ʿelat c. Eth$^{tana\ q}$) της κρισεως

CHAPTER 85

(1) Eth^{g} παντα τον υπνον Eth^{M} παντα

(2) Eth$^{tana\ t,\ n\ v\ ull\ (b)}$ απεκριθη Ενωχ λεγων

(3) Eth^{g} επι (Eth^{M} εκ) της γης | Eth$^{g,\ n\ ull}$ δυο (kel'ētu Eth^{M} kāle') μοσχους

(5) Eth^{M} και ηλθεν (maṣ'at) μετ αυτου αυτη η δαμαλις Eth$^{g\ m\ q\ (tana)}$ και ηλθεν (maṣ'a) μετα (leg. meslehā vel. mesla) ταυτης της δαμαλεως | Eth$^{tana\ g\ m\ t}$ και ηκολουθουν οπισω αυτων

(6) Eth$^{tana\ q}$ επ αυτον (dibēhu Eth^{g} ḫabēhu Eth^{M} sobēhā)

(8) Eth^{M} ετεκεν αλλον βουν λευκον Eth$^{tana\ g\ q\ u}$ ετεκ. δυο βοας λευκους

(10) Eth$^{tana\ g\ m\ q\ t}$ + (post ετερος τον ετερον) πολλοι(ους)

CHAPTER 86

(1) En^{f} 1.1-2 [והא] כוֹכֿבׄ חׄ]ד נפל ...] ביניהון... | edd. 'it arose' yetlēʿāl leg. yetʿallaw vel yetwēlaṭ c. G^{2069} [εγενε]το μετα̣[στραφεις]?

(2) En^{f} 1.2 cf. G^{2069} הא באדין חזי]ת ... [.. ושנין מרעיהו]ן ודיריהון [ושריו] ל]מגח חד לח]ד ...] | Eth^{u} om. και τους μοσχους αυτων | Eth I και ηρξαντο βοαν ετερος εις ετερον (Eth II ετ. συν ετερω) leg.? και ηρξ. κερατιζειν (למגח pro למגעי?) ετ. εις ετ.

(3) En^{f} 1.4-5 [. . . . . . . ת]וריא . . . . [... וׄהא כוכבין שׄ]גיאין | Eth I και εν μεσω τουτων των μοσχων βοες (Eth II pr. και) εγενοντο | Eth II μετ (Eth I pr. και) αυτων (meslēhomu leg. 'amsālu ομοιως αυτω?) και (leg. και c. Eth I) εποιμαινοντο εν μεσω αυτων

(6) Eth$^{tana\ g\ m}$ και φυγειν απ αυτων

CHAPTER 87

(3) ʾemtewledda medr crpt ex ʾemweluda medr = G^{2069} [απο των] υιων της [γης] | Eth^{M} εις ⸢αγιον και⸣ υψηλον τοπον

(4) Eth$^{q, b\ 10 mss}$ και ειπον (Eth^{M} ειπεν) μοι | Eth II παντα τα επερχομενα . . . επι παντας τους βοας (Eth I επι τους β. και παντας αυτων)

CHAPTER 88

(1) Eth$^{tana\ (g\ m\ q)}$ και αυτη η φαραγξ [εστιν] στενη και βαθεια και κεκλεισμενη (ʿeḍew Eth^{M} crpt? ʿeḍub σκληρα) και σκοτεινη

(2) Eth I απεσπασεν την μαχαιραν (Eth II + αυτου) | Eth^{M} και πασα η γη εσεισθη επ αυτοις (dibēhomu Eth^{m} kuellā medr diba medr Eth^{tana} wakuellā medr dibēhomu?)

(3) En^{e}4i2-3 כול כו[כביא שגיאיא... [.... אסר ל]כׄלהון ידין ורגלין ורמא | αστερας μεγαλους | Eth^{M} ελιθοβολησε (wagara Eth^{n} + sayfa? Eth$^{g\ m\ q\ t}$ wawagara Eth^{tana} wawagaru)

CHAPTER 89

(1) En^{e}4i13-16 [וחד מן אר]בׄעתא [אז]ל על חד מן תוריא [חוריא ואלף לה וע]בׄד לה ערב חדה ויתב בגוה [ותלתת תוריא עלל]ו/ יתב]ו עמה לערבא וערבא חפית וכסית [מן עליהון] | Eth^{m} προς εκεινον ταυρον Eth^{M} προς εκεινους ταυρους | Eth^{M} και εδιδαξεν μυστηριον αυτον τρεμοντα (Eth$^{g\ q}$ pr. μη)

(2) En^{e}4i16 [והוית] חזה והא מרזבין שבעה שפכין

(3) En^{e}4i17-18 והא חדרין פתיחו בגוא ארעא ושריו [להבעה ולמסק עליה ו]אׄנה הוית חזה עד ארעא חפית מין Eth^{M} πηγαι Eth^{tana} σχισμαι (ne[q]ʿatāt) pro חדרין | Eth$^{g\ t}$ και εθεωρουν leg. waʾerēʾeyo c. Flemming, 122 (Eth^{u} και ειδον wareʾikewwo)

(4) En^{e}4i19 [וחשוך וערפלא הוו] קאמין עליה | Eth^{M} και εθεωρουν το υψος του υδατος και υψωθη το υδωρ υπερ του βουκολιου Eth^{tana} + εως υψωθη το υδ. υπ. του βουκ. post υπερ leg. και εθεωρουν εως υψωθη το υδ. υπ. του βουκ.?

(5) En^{e}4i19 ותוריא שקעין וטבעין [ואבדין במיא]

(6) En^{e}4i20-21 וערבה פרחה עלא מן מיא וכול תוריא [וערדיא וגמליא] ופיליא ירומו[ן לארעא]

(7) En^{e}4ii1-2 [ועוד חזית חל]מי עד מׄ[רזביא . . . . . מקועי]ן חדריא שכירו וׄ[עמקין אחרנין פתיחו כחדא . . .] και αι σχισμαι της γης [εφραχθησαν] Eth^{tana} και (waʿarayu Eth^{M} om. wa) αβυσσοι (Eth^{M} pr. και) αλλαι (leg. kāleʾāt) αμα ανεωχθησαν (ʿarayu . . . tafatḥa leg. tafatḥu c. Eth^{M})

(8) En^{e}4ii2-4 [ומיא שפכין] ונחׄתין בגוהון עד פתׄ[יחא ארעא . . . וערבא] תקנת [ע]ל ארׄעא Eth^{tana} και ερρει (ʾawḥaza leg. yeweḥḥez) το υδωρ [και] κατεβαινε (leg. yewarred) εις αυτας Eth^{M} crpt? και ηρξατο (ʾaḫaza) το υδωρ καταβαινειν εις αυτας

(9) Eth I om. και εις μελας

(10) Eth I αετοι (’anserta) pro Eth II γυποι (’awseta)

(11) En^{d} 2 i 24-25 = En^{e} 4 ii 11 למדבר אלן לאלן . . . וערדין . . .

(12) En^{d} 2 i 25-26 cf. En^{e} 4 ii 12-14 ועגלא [. . . חזי]ר אכום ודכר די ען . . . אמ[רין תרי] עשר | Eth^{M} εγεννησεν υας πολλους Eth^{q} υας / υειους υγιειας (dāẖen sic. cf. Eth^{tana} ’arāwit deẖna θηρια υγιη!)

(13) En^{d} 2 i 28-29 + En^{e} 4 ii 15 [. . . . מ]נהון לערדיא וערדיא יהבו[. . .]

(14) En^{d} 2 i 29 + En^{e} 4 ii 16-17 ודבר דכרא ל[ח]ד עשר אמריא כולהון . . . לות דביא [והשתגאו] και ο κυριος ηγαγε

(15) En^{e} 4 ii 17-18 [ודביא] שריו למלחץ לענ]א . . . . בנהר דמ[שקע מין Eth$^{t\ u,\ 11\ mss}$ εκφοβειν (Eth^{M} φοβεισθαι) αυτους cf. 14.9

(16) En^{e} 4 ii 20 [וחד] אמר נת[יק מן דביא . . . תקיפא]ית עד נחת מ[רא די ענא] Eth$^{i\ ryl}$ + ’emṣerḥ post ο κυριος των προβατων

(18) Eth I απηντησεν αυτω (Eth II τω προβατω)

(22) Eth II το προσωπον αυτου ενδοξον και δεινος και εντιμος η ορασις αυτου cf. 89.30 (Eth I το προσ. α. ενδ. και εντιμον και δεινον ιδειν)

(27) En^{e} 4 iii 14-15 [כל ד]ביא רדפין לענא . . . ומיא חפו עליהון

(28) En^{e} 4 iii 16-19 [. . . ואזל]ו צדיותא אתר זי[. . . .] ועיניהון התפתחו[. . . . רעא] [להון ויהב לה]ון מ[יא ועשבא ודכרא אזל ודבר להון] Eth^{m} ουτε δενδρον ουτε υδωρ ουτε χορτος Eth^{tana} και ουκ ηνοιξαν τους οφθαλμους αυτων | Eth^{tana} εδωκεν αυτοις χορτον και υδωρ | Eth$^{g\ t}$ επορευετο pr. απο τοτε (’emza leg. ’emze pro Eth^{M} ’enza?)

(29) En^{e} 4 iii 19 + En^{d} 2 ii 27 [וס]לק לר[אש כיף] חד ר(א)ם

(30) En^{d} 2 ii 29 [. . . לקובל] ענא וחזיה תקיף ורב וד[חיל . . .] Eth I και η ορασις αυτου μεγαλη και (Eth II om. μεγ. και) δεινος / φοβερα και δυνατη

(31) En^{c} 4 1 [וזעקן לאמרא די בתריה דכרא דן] די הוה ביניהון | En^{c} 4 2 [וכולהון הווא [דחלין] לא יכלין אנחנא למק[ם] לקובל [מרנא] Eth^{tana} και εκραζον οπισθεν του προβατου (Eth$^{g\ m\ q}$ + ⸢μετ αυτων⸣) [ο] ηγειτο αυτων και [οπισθεν] του δευτερου προβατου ο ην εν μεσω αυτων και ελεγον | Eth^{tana} ου δυναμεθα στηναι (Eth^{M} om.) εμπροσθεν του κυριου ημων ουδε εμβλεπειν αυτον

(32) En^{c} 4 3-4 [ואמרא דן די דבר להון ת]נא וסלק לראש כפא דן וענא שריוא לאת[חשכה עיניהון ולמטעא מן אורחא דאחזית ל]הון ואמרא לא ידע בהון

(33) En^{c} 4 4-5 ומרא ענא רגז על[יהון רגוז רב . . . . . כפא] דן ואתא על ענא ואשכח [כול שגאהון מת]חשכין עיניהון וטעין] Eth I και επλανηθησαν (Eth II + ⸢απο των οδων αυτου⸣)

(34) En^{c} 4 6 [. . . מן ק]ודמוהי ול[הוון] צבין למ[ת]ב לדיריהון

(35) En^{c} 4 7-9 [. . . ואתא ע]ל[יהון ומן ענא] קטל טעיתא ושריו למ[רעד . . .] 7
8 אתיב אמרא דן לכול ענא טעיתא לדיריה[ון ותב אמרא דן למרא . . .]
9 [לאתנחא] ולמבגל ולמגעה ולזעקה ומרא ענא (ומא ע ms)
Eth^{tana} και ανεστρεψεν αυτα προς τας μανδρας αυτων (Eth^{M} και ανεστρεψαν προς τας μανδ. αυτ.)

(36) Eth I ειδον εν τη ορασει (Eth II ειδ. εκει την ορασιν leg. ειδ. εκεινην την ορ.?) | En^{c} 4 10 [... עד די א]מ̊ר̊[א דן] אתהפך והוא אנוש ועבד משכן [למרא]

(38) Eth$^{\text{n ull}}$ + τριακοντα ημερας post εζητησαν αυτο

(39) παντα (kuellomu) προβατα (sic) leg. αλλα (kāle'ān) προβατα c. Flemming, 127 vel. δυο kel'ē προβατα cf. Charles^{1906}, 173, n. 23.

(42) Eth$^{\text{ull (g n)}}$ c. G^{vat} και ηρξαντο οι κυνες και οι αλωπεκες και οι υες κατεσθιειν τα προβατα μεχρι ου ηγειρεν ('anše'a Eth^{M} tanše'a ex v. 41) ⌜αλλο προβατον⌝ ο κυριος των προβατων ενα εξ αυτων κριον ηγεισθαι αυτων

(43) και ο κριος ουτος ηρξατο κερατιζειν εντευθεν κακειθεν τους κυνας και τους αλωπεκας και τους υας μεχρι ου παντας απωλεσεν En^{d} 2 iii 27 [בקר]נ̇ו̇ה̇[י] = G^{vat} En^{d} 2 iii 28 חז̊י̊ר̊ין שגיא[ין]

(44) και το προβατον ου (leg. zatafatḥa c. Charles^{1906}, 174 n. 17 = G^{vat}) ηνοιγησαν οι οφθαλμοι αυτου «και» εθεασατο | En^{d} 2 iii 28-29 [... חזה] לדכרא די ענ[א] | οτι αφηκεν την δοξαν αυτου | και ηρξατο κερατιζειν ταυτα τα προβατα και κατεπατησεν αυτα και επορευθη απρεπως / ανοδια En^{d} 2 iii 30 (teste Milik), באורח]א לא תקינא] = G^{vat} ανοδια

(45) ος αφηκεν την δοξαν αυτου

(46) om. σιγη | και εποιησεν αυτο αρχοντα post κριον | Eth$^{\text{m, 5 mss}}$ cf. G^{vat} εθλιβον (yāṣeḥḥebewwomu Eth^{M} yāṣehheqewwomu = εθορυβαζον 14.8 G)

(47) και ανεστη ο κριος ο δευτερος και post επεδιωκεν | εως ου κατεβαλον οι κυνες τον κριον τον πρωτον

(48a) Eth I ο δευτερευων εκεινω τω κριω (daḫārāwi lazeku ḥargē) Eth^{M} ο εσχατος κριος (zeku ḥargē daḫārāwi)

(49) + και οι υες post αλωπεκες | εφοβουντο και εφυγον απ αυτου

(48b) Eth$^{\text{q m (g)}}$ ηγειρεν (naš'omu)

(50) και ο οικος ⌜ην⌝ μεγας και πλατος ⌜και⌝ ωκοδομηθη τοις προβατοις Eth^{M} και πυργος υψηλος (Eth$^{\text{m t, 8 mss}}$ + και μεγας ωκοδομηθη) επι τω οικω τω κυριω των προβατων Eth^{tana} ⌜οικοδομη (māḫnaṣa leg. māḫnaṣ crpt ex māḫfad?) υψηλη επι τω οικω⌝ και πυργος υψηλος και μεγας ωκοδομηθη επι τω οικω (οικοδομη ... οικω ditt.)

(51) και παλιν ἐιδον | Eth^{M} παλιν 2° (Eth$^{\text{q u, 4 mss}}$ om.) | Eth^{M} τον οικον αυτων (Eth^{tana} αυτου)

(52) Eth^{M} και εκραξεν επι τοις προβατοις Eth^{tana} προς τα προβατα | Eth^{u} ανεβιβασεν αυτον και κατωκισεν αυτον παρ εμοι (ḫabēya Eth^{M} leg. ḫabēya post ανεβ. αυτον)

(54) Eth^{u} οτι (kama) pro οτε (soba) leg. (ειδον) οτι οτε? | Eth^{M} εν ταις μανδραις (Eth^{q} φυλακαις (bamawā'elihomu leg. bamu'ālehomu) αυτων) | Eth^{tana} εως εφωνουν αυτους (yeṣēwe'ewwomu) ... τουτους τους φωνευτας?

(57) Eth I και εβοων τω κυριω των λεοντων | Eth^{q} απο παντων των λεοντων και των θηριων

(61) και εφωνησε τον αλλον (lakāle' Eth$^{m\ t}$ pr. lakuellomu cf. Charles[1906] 178 n. 25).

(63) Eth$^{b\ 8\,mss}$ ποσα απολλυουσι + καθ εαυτους (ex. v. 62?) | Eth^{m} ποσα ⸢πεμπουσιν και⸣ αποδιδοασιν | Eth$^{tana\ g\ q}$ 'ewaṭṭenomu leg. 'emaṭṭenomu = (ινα) λογιζωμαι αυτα Eth^{M} 'emaṭṭewomu (ινα) αποδω αυτα cf. 61.2, Flemming, 131, Charles[1906], 179 n. 13.

(68) Eth^{M} και εν καθ εν ο αλλος (leg. kāle'u pro lakāle'u) γραψει (yeṣeḥḥef Eth$^{u\ (g)}$ yeṣṣaḥḥaf γραφησεται) εν γραφη | lakāle'u bamaṣḥaf 2° om. (ditt.) Eth^{q} + yeṣeḥḥef

(69) Eth I om. σφοδρα (bezuḫa ṭeqqa) (ex. v. 67?)

(71) Eth$^{tana\ g\ m\ t}$ εκ της χειρος αυτου Eth^{M} εν τη χειρι αυτου

(73) Eth^{q} + ως προτερον post ωνομαζετο (ditt.?)

(75) Eth$^{q\ (u)}$ εως διεσπαρησαν εις παντα τα αγρια προβατα (kuello 'abāgeʿa gadām acc. loci) Eth^{M} εως διεσπ. παντα τα προβατα εις τον αγρον

(76) Eth$^{tana\ (t)}$ ανεγνω τα μεγαλα αυτων (ʿabiyāta) τω κυριω (Eth^{t}) των προβατων Eth^{M} ḫaba 'abyāta 'egzi'a 'abāgeʿ ('read (it) out) in the dwelling of the Lord of the sheep' (Knibb)

CHAPTER 90

(1) τριακοντα και επτα ποιμενες | Eth^{tana} εκαστος τον καιρον αυτου (babagizēhu Eth^{M} babagizēhomu) | leg. c. Flemming εκαστος ποιμην (kuellu nolāwi (Eth I) Eth II kuellu nolāweyān) εν τω καιρω αυτου

(2) Eth$^{g\ q\ t}$ εν τω οραματι μου (Eth^{M} om. μου) Eth^{tana} τοις οφθαλμοις μου

(3) Eth^{M} εγω εκραξα (ṣārḫku Eth$^{g\ q\ (m)}$ crpt. naṣṣarku 'I looked' (Charles)

(4) απο των κυνων ('em'aklāb crpt ex 'elku 'abāgeʿ?) om.

(6) Eth I και ιδου (Eth II om.) | Eth II μικροι (= זעירין?) (Eth I om.) αρσενες

(7) Eth^{m} παρηνωχλησαν αυτοις ('asreḫewwomu leg. yāsarreḫewwomu cf. Charles[1906], 183, n. 2 Eth^{M} crpt 'i(ye)ṣarreḥewwomu)

(8) ενα των αρσενων

(9) Eth^{tana} ελεπτυνον (yādaqqeqewwomu) Eth^{M} κατεβαλλον (yāwaddeqewwomu) τα κερατα αυτων

(11) Eth^{g} γυπες και αετοι και κορακες ⸢οτε επετασαν επι τους αρσενας και ηρπαζον ενα των αρσενων και συνετριβον τα προβατα και κατησθιον⸣ add. ex v. 8 Eth^{g} + και εως διηρπαζον αυτα? | και εσιγησε τα προβατα crpt ex וחש ענא(?)

(12) Eth^{q} το κερας αυτων

(14) Eth^{g} και εβοηθησεν αυτῳ + και εσωσεν αυτο | εποιμαινετο αυτο παντη (leg. wa'arʿayo pro wa'ar'ayo et cf. Zech. 11.7 et contra supra, v. 10.)

(15) Eth^{m} οτι ('esma pro 'eska) ηλθεν ο κυριος ˹των κυριοτητων και˺ των προβατων (sic.) | Eth^{tana} εν σκοτει (ṣelmat) Eth^{M} εν σκια (ṣelālot)

(16) Eth$^{tana\ g\ q}$ και ηλθεν μετ αυτων παντα τα αγρια προβατα | Eth^{q} και εξηλθον (pro ηλθον) παντες αμα | Eth^{M} και αλληλοις εβοηθουν (tarāde'u) Eth^{m} και συνηχθησαν (tagābe'u) leg. και συνηχθ. και αλληλ. εβοηθ.? cf. Est. 9.16

(17) Eth^{g} οτι ('esma = Eth^{tana} pro 'eska) ηνοιχθη [το βιβλιον] εν τω λογω κυριου ˹εν˺ τη απωλεια

(20) kuello leg. kāle'u ο αλλος crpt ex חוירא cf. Charles^{1906}, 185, n. 25, Flemming, 137

(21) Eth I τους επτα λευκους ανθρωπους τους πρωτους Eth II om. ανθ. per hmt (sab' sab'ātu | Eth^{M} 'emkokab (Eth^{g} lakokab) qadāmāwi zayeqaddem | leg. παντας (kuello pro 'em'elku) τους αστερας cf. Charles, 213

(23) Eth^{g} και ιδου παντες δεδεμενοι ενωπιον αυτου

(24) Eth^{tana} πυρ φλογος ('essāt walāhb)? Eth^{ull} πυρ φλεγον ('essāt zayeleḥḥeb)

(28) Eth^{tana} μετεθηκαν (mēṭewwo Eth^{M} ṭam'o immersit eum cf. Eth^{m} tasaṭma submersus fuit (sic); Dillmann, 285, em. ṭomo, Charles^{1906}, 187 n. 1 em. ṭomewwo 'they folded up'; cf. Eth$^{n\ p\ (y)}$ ṭawamo et infra taṭawma et Charles, 214.

(29) leg. 'anš'a, cf. Sir. 69.13? | Eth$^{b\ (m)\ ryl\ 5\ mss}$ [και] ο κυριος των προβατων ην εν μεσω αυτου Eth^{M} και παντα τα προβατα ην εν μεσ. αυτ.

(31) Eth^{tana} εις το μεσον των προβατων των ανεγκλητων (za'enbala (Eth^{M} 'enbala) ˹tekun˺ kuennanē (Eth^{m} zekuellu kuennanē))

(34) Eth^{g} εκληθησαν (taṣawwe'u) Eth^{M} εκλεισθησαν (ta'aṣwu)

(38) ρημα (nagar) = ראמא

(42) Eth I εμνησθην του υπνου (Eth II + μου) του πρωτου

CHAPTER 91

(1) Eth^{g} παντα τα τεκνα της μητρος σου και συναγαγε μοι τους αδελφους σου

(2) Eth^{M} παντας (Eth^{tana} om.) τους αδελφους αυτου

(3) Eth$^{tana\ g\ q,\ 2\ mss}$ διελαλειτο πασιν τοις υιοις της δικαιοσυνης (Eth^{M} την δικαιοσυνην Eth^{t} + αυτου Eth^{m} om. τοις υιοις et leg. την δικ.) | Eth I τεκνα του Ενωχ Eth II τεκνα μου | Eth I αγαπητοι Eth II + μου | Eth^{m} αγαπατε ˹τον θεον ημων εν αληθεια˺ pro Eth^{M} αγαπ. την αληθειαν

(4) Eth$^{g\ u}$ om. per hmt? και μη κοινωνητε μετα των καρδιας δισσης Eth^{tana} μετα των πορευομενων εν καρδια δισση

(5) οτι δει / μελλει αδικια ενισχυειν (yeṣanne' hellāwē leg. yeṣanne' hallawo c. Charles, 226)

(6) Eth$^{tana\ q\ t}$ και κρατησει / κατακυριευσει ολης (sc. της γης) εργα ανομιας και ⌜εργα⌝ αδικιας και πονηριας (te'eḫḫez kuello Eth^{M} tet'aḫḫaz kuellu(ā) cf. Dillmann, *Lex.*, 768 obtinebunt (intrs.) opera justitiae, Charles, Knibb 'will prevail'.) | δευτερον vel δισσως (bakā'ebata)

(7) Eth^{M} και οτε Eth$^{m\ t}$ και απο τοτε | Eth I εν πασιν εργοις Eth II και παντα εργα | Eth^{g} om. απο του ουρανου | Eth^{g} εξελευσεται ο αγιος ⌜κυριος επι την γην⌝ | Eth^{g} ινα ποιηθηται η κρισις επι την γην

(8) Eth$^{b\ 5\ mss}$ om. και 2°

(9) Eth^{g} και (Eth^{M} om.) οι πυργοι

(11-17) vide infra, 327f.

(18) Eth^{g} και δειξω υμιν πασας τας οδους της δικαιοσυνης En^{g} 1 ii 18-19 ארחת קשטא . . . . די תדעון מהׄ [יתעבד]

(19) Eth$^{tana\ q\ t}$ ακουσον μου (Eth^{M} om. μου) En^{g} 1 ii 19-21 [. . . בארחת] קושטא למהך בהון וׄ[. . . ע]ל דׄי אבד לסוף אבדנׄא

CHAPTER 92, 93.1-2

(1) En^{g} 1 ii 21-22 [ספר חנוך די כת]ב ויהב למתׄ[ושלח ברה] Eth^{g} η γεγραμμενη υπο του Ενωχ επιστολη | En^{g} 1 ii 22 [חנוך ספר מהיר וח]כים אנשא Eth^{M} (Ενωχ) γραμματευς οξυγραφος (ṣaḥāfi zakuellu temherta Eth$^{tana\ g\ m}$ te'merta) και σοφωτατος (leg. ṭabiba) ⌜παντων⌝ ανθρωπων | En^{g} 1 ii 23 [ובחׄי]ר בׄנׄיׄ אנשא ודין כל ארעא] Eth^{M} ενδοξος και (Eth^{g} om.) κριτης πασης της γης | En^{g} 1 ii 24 [ולדריא אחריא לכול י]תבי ארעא]

(2) En^{g} 1 ii 25 [לא תה]וׄן בבאשתא את[ון] Eth^{M} ο μεγας (Eth$^{tana\ g\ q,\ 4\ mss}$ + και) αγιος

(3) Eth^{m} η δικαιοσυνη pro ο δικαιος | Eth$^{tana\ m}$ εκ του υπνου αυτων | Eth^{g} και η ɔοφια εξαναστησει και διαβησεται η δικαιοσυνη cf. 91.10

(4) Eth$^{tana\ t}$ και κρινει (yekuēnen Eth^{M} yekawwen) | Eth^{q} + εν αληθεια | Eth$^{g\ m\ q\ t\ (tana)}$ εν φωτι (baberhān Eth^{M} berhān) αεναω

(5) En^{g} 1 iii 16-17 [בח]שׄוכא [לעלם]

(93.1) Eth$^{g\ (q)}$ και μετα (Eth^{q} 'emdeḫra) το δεδωκεναι (wahabani Eth^{M} kona leg. kona wahabani) [την επιστολην αυτου] ανελαβε Ενωχ την παραβολην αυτου λεγων ('aḥaza yetnāgar cf. 1.2) ⌜εκ των γραφων αυτου⌝ En^{g} 1 iii 18 [ובתר די יהב הוא ספריה נסב ח]נׄוך מתלה אמר

(2) και περι των εκλεκτων του αιωνος και περι του φυτευματος της δικαιοσυνης και αληθειας (Eth$^{g\ m\ t}$ om. της δικ. και Eth^{q} om. και της αληθ.⁾ En^{g} 1 iii 19-20 [ועל בחירי עלם די על]ו מן נצבת יצבתא | En^{g} 1 iii 20-21 [. . . .]ת [ב]נׄי אנה הוא חנוך אחזׄי εγω αυτος Ενωχ εν τω φαινεσθαι μοι κατ οναρ ουρανιον | En^{g} 1 iii 21-22 [ומן] ממר עׄיׄרׄיׄן וׄקדשין אנה כלא ידעת [ובלוחות שמיא כול]א קר[ית ואתבונ]ת και απο των λογων των αγιων αγγελων εγνων και απο των ουρανιων γραφων διενοουμην

CHAPTER 93.11-14

(11) En^{g} 1 v 15 [מנו הוא כול בני אנשא די יכ]ל̊ ינדע מנ̊[ו טעם שמיא . . . . ד]י יכל ישמע ק̊ל̊י קדשא οτι τις εστι παντων των υιων των ανθρωπων ος δυναται ακουειν την φωνην του αγιου και ου ταρασσεται; η τις δυναται διανοεισθαι τα διανοηματα αυτου; η τις δυναται ιδειν παντα τα εργα του ουρανου; Eth$^{m\ (t)\ ryl^{mg}}$ παντα τα εργα καλα (šannāy crpt ex samāy vel. leg. παν. τα εργα καλα του ουρανου?)

(12) En^{g} 1 v 18 + [או אודע] זיוא די אנ[ון י]תנ[ון] | Eth II και τις εστιν ος δυναται ⸢γνωναι τα εργα του ουρανου και⸣ ιδειν την πνοην αυτου η το πνευμα αυτου (Eth I om. αυτου 1° 2°) και δυναται εκδιηγεισθαι En^{g} 1 v 19 [למתב למת̊[ניה . . .

(13) En^{g} 1 v 20 [או מנו הוא [מן כול בני א]נוש די יכל [וינדע מה הוא] אורכה ופתיה די ארעא כולה או [מנו הוא . . . די תחזה לה כולה משחתה] וצרתה ποσον εστι το πλατος και το μηκος της γης και τινι εδειχθη παντα τα μετρα αυτης;

(14) En^{g} 1 v 22 ומנו הוא כול אנוש די יכ̊[ל ינדע . . . ומה] הוא רומהון והיכה אנון סמכי[ן] ποσον εστι το υψος αυτων και εφ ω εστερεωθησαν

CHAPTER 93.3-10, 91.11-17

(3) En^{g} 1 iii 22-24 ו̊ב̊[אדין] נסב חנוך מתלה ושמר . . . ש̊ב̊יע א̊[תילדת בשבוע] קדמי ועד עלי קשטא כת̊[ר]. . . Eth^{M} και Ενωχ (Eth$^{tana\ m\ q\ t}$ om.) νυν ανελαβεν την παραβολην αυτου (ʾaḫaza ʾenka yetnāgar cf. 1.2) ⸢εκ των γραφων λεγων⸣ | Eth^{M} μεχρις ου η κρισις και η δικαιοσυνη επεσχον (taʿaggaša Eth^{g} + ετι ʾenka)

(4) En^{g} 1 iii 24-25 ומן בתרה יקום שבוע תנין די בה שקרא וחמסא יצמח[ון] Eth^{M} και εξαναστησει (Eth^{g} και ελευσεται και εξαν.) μετα με εν τω δευτερω σαββατω μεγαλη κακια και δολος βλαστησει | Eth^{M} διαθηκην / νομον ποιησει Eth$^{q\ u,\ n\ ull}$ διαθηκη / νομος ποιηθησεται

(5) Eth I και μετ αυτον εξελευσεται (Eth II ελευσεται) εις (Eth II om.) φυτον | waʾemdeḫrēhu crpt ex ואחרתיה(?)

(6) Eth^{g} ορασις (Eth^{M} ορασεις) των αγιων και της δικαιοσυνης (Eth^{M} των δικαιων) | Eth^{M} η αυλη (Eth^{q} την αυλην) [της σκηνης] ποιηθησεται (Eth^{q} ποιησει)

(8) Eth^{e} το αγιον (bēta maqdas) cf. frg. Copt. templum (Milik 81, Dexinger, 113) | Eth$^{m\ t\ (q)}$ πας ο λαος και οι ηγεμονες της δυναμεως (kuellu zamada šerwa [leg. wasarāwita] ḫayl cf. Jer. 41 (LXX 48). 13, 16) Eth^{M} šerw ḫeruy 'the whole race of the chosen root' (Knibb)

(9) En^{g} 1 iv 11 [ור]בי[ו עבדוהי]?

(10) En^{g} 1 iv 12-13 [. . . י]תבחרון ב[חירי]ן̊ ל̊שהד̊י̊ קשט מן נ[צבת] קשט על̊[מ]א̊ די שבעה פ̊[עמי]ן̊ חכמה̊ ומדע תתיה̊[ב להון] Eth^{M} οι εκλεκτοι δικαιοι Eth$^{tana\ g\ (t),\ 2\ mss}$ οι εκλεκτοι της δικαιοσυνης | om. εις μαρτυρας της δικαιοσυνης | οις δοθησεται επταπλασιον επιστημη / διδαχη ⸢πασης της κτισεως αυτου⸣

(11) En^{g} 1 iv 14 ולהון עקרין אשי חמסא ועבד שקרא בה למעבד [כלה?] και απο τοτε αποκοψεται τα θεμελια της ανομιας και οι αμαρτωλοι απολουνται μαχαιρα· απο των

βλασφημων αποκοψονται εν παντι τοπω και οι λογιζομενοι την αδικιαν και οι ποιουντες βλασφημιας απολουνται ρομφαια

(12) En^{g} 1 iv 15-17 ומן בתרה יקום שבוע תמיני קשוט דבה תתיה[ב חרב] לכול קשיטין למעבד דין קשוט מן כול רשיעין ויתיהבון בידהון Eth^{M} και μετα ταυτα εσται αλλο σαββατον ογδοον της δικαιοσυνης και δοθησεται αυτω μαχαιρα του ποιηθηναι κρισιν και δικαιοσυνην απο των αδικων και αποδοθησονται οι αμαρτωλοι εις τας χειρας των δικαιων

(13) En^{g} 1 iv 17-18 ועם סופה יקנון נכסין בקשוט ויתבנא היכל מלכות רבא ברבותא לכול דרי עלמין . . . κτησονται οικους εκ της δικαιοσυνης αυτων | ο οικος του μεγαλου βασιλεως cf. P.O. VI. 3,443 (175) 5, Milik, 268 οικος της μεγαλης βασιλειας | Eth$^{g\ m\ (tana)}$ εν δοξα (basebḥat Eth^{M} lasebḥat)

(14) En^{g} 1 iv 19-22 ומן בתרה שבוע תשעי יק[ום די בה] ד[ין קשוט] יתגלא לכול בני ארעא כולה וכול עב[די רשעה יפקו]ן מן כול ארעא כולה וירמון לבור [עלם ויחזון אנושא] כלהון לארח קשט עלמא και μετα ταυτα εν τω σαββατω τω εννατω εν αυτω κρισις δικαια αποκαλυφθησεται παντι τω κοσμω και παντα τα εργα των ασεβων εξελευσεται εκ πασης της γης και γραφησεται (yeṣṣaḥḥaf) εις την αεναον απωλειαν και παντες οι ανθρωποι επιβλεψουσιν εις τας οδους της αληθειας

(15) En^{g} 1 iv 22-23 ומן [בתרה שבוע עשרי יקום דבשבי]עה דין עלמא וקץ דינא רבא [יתעבד מן כולהון עירין די שמיא] Eth^{M} και μετα ταυτα εν τω σαββατω τω δεκατω εν τω εβδομω μερει εν αυτω κρισις αεναος ⌜και⌝ ποιηθησεται εκ των εγρηγορων ⌜και⌝ (Eth$^{tana\ t,\ i\ n}$ om.) του ουρανου αεναου· (κρισις) μεγαλη η εκδικηθησεται (leg. c. Eth$^{m\ u}$ zayetbēqal) εκ του μεσου των αγγελων Eth^{g} om. και ποιηθ. . . . αεναου et leg. κρισις αεναος μεγαλη η εκδικηθησεται (leg. zayetbēqal) εκ του μεσου των αγγελων Eth$^{tana\ t}$ εκ παντων των αγγελων

(16) En^{g} 1 iv 23-25 ושמין קדמין בה יעברון ושמ[ין חדתין יתחזון וכול שלטני] שמיא ז[הר]ין [ודנחין לכול עלמי]ן [שבעא פעמין] και ο πρωτος ουρανος απελευσεται (yewaḍḍe' Eth$^{b\ vat\ 71}$ ελευσεται yemaṣṣe' = Rev. 21.1) και ουρανος καινος φανησεται και πασαι αι δυναμεις των ουρανων φωτισθησονται επτακις εως του αιωνος

(17a) En^{g} 1 iv 25-26 [ומן בתרה ש]בעין שגי [די לא] איתי סוף לכול מ[נינהון . . . טבא וקוש]טא יעבדון και μετα ταυτα σαββατα πολλα εσται οις ουκ εστιν αριθμος / αναριθμητα εις τον αιωνα εν αγαθοτητι και εν δικαιοσυνη· και η αμαρτια απο τοτε ου μη φωνησεται ('itetbahhal) leg. φανησεται (cf. 92.5) εως του αιωνος

(17b?) En^{g} 1 ii 13-17 [וקשיטי]א [יתעירון מן שנתהון ויקומו]ן ויהלכ[ון באורחות קשטא ועו]לה תשב[ת כולה] ותנוח אר[עא מן חמסא עד] כול דרי עלמין cf. Eth. 91.10, 92.3

CHAPTER 94

(1) En^{g} 1 v 24-25 וכען לכון אנה אמר בני [. . . .]ארחת קשט[א . . . .]

(2) Eth^{q} οδοι των τρυφων (feg' pro gef' crpt? cf. 13.2, 102.10) Ca. 7.7 (LXX)

(3) Eth I εν τοις οδοις της κακιας Eth II + και της αδικιας | Eth^{g} ινα μη απολησθε ως οι ζητουντες την κακιαν (ως οι ζητ. την κακ. pro αλλα ζητειτε v. 4)

(4) Eth^{g} om. αλλα ζητειτε cf. v. 3 fin. | Eth^{tana} ζητειτε υμιν και εκλεγεσθε ζωην αγαθωσυνης ḫiruta Eth^{M} ḫerita 'a life that is pleasing' (Knibb)

(5) επιλαμβανεσθε των εντολων μου)? (Eth^{M} te'eḫhezu Eth^{g} bate'zāzu leg. te'eḫḫezu bate'zāzeya) εν τη μελετη της καρδιας υμων (Eth^{M} baḫellinnā lebbekemu)

(11) Eth$^{tana,\ g\ m\ q\ t}$ οι δικαιοι σου

CHAPTER 95

(1) Eth$^{g\ (tana\ q)}$ τις δωσει μοι υδωρ (Eth^{M} om.) τοις οφθαλμοις μου ινα γενωνται νεφελη (πηγη?) υδατος

(2) Eth^{g} τις δωσει υμιν (Eth^{M} + ινα ποιητε) την εχθραν και την κακιαν και ευρησει υμας τους αμαρτωλους η κρισις

(3) Eth^{g} ινα ποιη (sc. ο κυριος) κατ αυτους κρισιν

(4) Eth I ουαι υμιν οι αναθεματα(ι) αναθεματιζετε ινα (Eth II + μη) λυθη | Eth^{tana} add. (post δια την αμαρτιαν υμων) το ρημα της φαρμακειας (nagaru ḫaba leg. ḫabr) απερριφη (tagadfa hallo)?

(6) Eth^{m} οι κολλωμενοι (yādlewewwā) τη αδικια (cf. Rom. 12.9)

(7) Eth^{M}παραδοθησεσθε και διωχθησεσθε Eth^{g} παραδωσετε και διωξετε Eth$^{g\ (m\ tana)}$ εν τη αδικια Eth^{M} οι της αδικιας ('ella 'āmaḍā)

CHAPTER 96

(1) Eth$^{t\ u}$ tafaššeḥu 'rejoice' crpt ex tasaffawu ελπιζετε

(2) Eth^{M} εγερθησονται (yetnašše'u Eth^{tana} crpt yetkaššatu) (Eth^{g} yetnassatu) | Eth^{M} εισελευσεσθε (tebawwe'u Eth$^{g\ (tana)}$ yetbawwe'u εισελευσονται)

(4) Eth$^{tana\ q}$ εαν ('emma) pro Eth^{M} οτι ('esma) (Eth^{g} 'esma 'emma)

(5) πινετε ισχυν ριζης της πηγης (sic) crpt ex חוליא ראשי יין(?)

(6) οι πινετε υδατα παντοτε [εκλιποντα](?)

(8) Eth$^{tana\ q\ t}$ εως ('eska) pro Eth^{M} οτι ('esma)

CHAPTER 97

(2) Eth I αγγελοι του ουρανου (Eth II om. του ουρανου)

(4) Eth$^{tana\ g\ q}$ και υμεις εσεσθε (Eth^{M} pr ουκ) ως αυτοι

(5) Eth$^{tana\ m\ q\ t,\ 12\ mss}$ την προσευχην των δικαιων Eth^{M} των αγιων

(6) Eth I του μεγαλου (Eth II + και) αγιου | και καταισχυνθησεται το προσωπον υμων G^b crpt κατα προσωπον υμων | Eth^{g^1} c. G^b αναφελει (yaḫādeg) τα εργα της ανομιας $Eth^{tana\ g\ q\ t}$ yegaddef Eth^M yetgaddaf 'every deed will be rejected' (Knibb) | παντα τα εργα κατισχυοντα (zaṣanʿa) εν τη ανομια

(7) Eth^M ων (Eth^{g^1} c. G^b om.) το μνημοσυνον

(8) $Eth^{g^1, h}$ c. G^b χρυσιον και αργυριον Eth^M αργ. και χρυσ. | α ουκ ην εν δικαιοσυνη | και κεκτημεθα παν ο εαν θελησωμεν cf. v. 9 G^b

(9) Eth^M και νυν ποιησωμεν (Eth^{g^1} + παν) ο εαν ενθυμηθωμεν (ḫallayna) | οτι αργυριον τεθησαυρικαμεν και πεπληρωκαμεν τους θησαυρους ημων | και ως υδωρ πολλα τα αγαθα (ḫarasta 'husbandmen' crpt ex ḫerāta αγαθα) των οικιων ημων cf. Nickelsburg, 93.

(10) πεπλανησθε (ḫasatkemu crpt ex seḥetkemu c. G^b cf. Nickelsburg, ibid. | Eth^{tana} cf. G^b ου μη παραμεινη υμιν ο πλουτος υμων (Eth^M om. υμων) | Eth cf. G^b αναβησεται (yaʿāreg) | Eth^M παραδοθησεσθε Eth^{tana} εσεσθε

CHAPTER 98

(1) Eth^M και τοις ($Eth^{g\ g^1\ q\ t}$ pr. ουχι c. G^b) αφροσι | πολλα (bezuḫa) leg. πολλας ... [ανομιας] c. G^b (bezuḫa [ʿammaḍā])

(2) Eth I οτι καλλος περιθησεσθε εφ υμας υμεις οι ανδρες υπερ γυναικας και χρωματα υπερ παρθενους (Eth II + bamangel om. crpt ex 'emdengel? cf. Knibb, 2, 230) | Eth I και αργυριον (berur Eth II crpt baberur) και χρυσιον και πορφυρα και τιμια και βρωματα [και τα αγαθα εν ταις οικιαις αυτων cf. G^b?] ως υδωρ εκχυθησεται G^b εσονται ... εις βρωματα crpt ex להון למכלא(?)

(3) δια τουτο επιστημην και φρονησιν ουκ εχουσιν | και εν αυτω pro ουτω cf. G^b C-B 34 | om. παντων | και εν ατιμια και εν σφαγη και εν ερημωσει μεγαλη

(4) ουτως αμαρτια (leg. δουλεια?) ουκ απεσταλη επι την γην

(5) και ατεκνια γυναικι ουκ εδοθη αλλα δια τα εργα των χειρων ατεκνος αποθανειται

(6) Eth I c. G^b κατα του αγιου (Eth II + και) μεγαλου | παντα τα πονηρα εργα υμων | και ουκ εσται υμιν εργον αδικον κεκαλυμμενον ουδε αποκεκρυμμενον

(7) Eth^M και μη ειπητε τη καρδια υμων οτι ου γινωσκετε ουδε βλεπετε ('itā'ammeru wa'iterē'eyu Eth^{ull} c. G^b ου γινωσκουσιν ουδε βλεπουσιν) παντα τα αμαρτηματα ($Eth^{tana,\ ull}$ + ημων) εν (Eth^{ull} pr. οτι) τω ουρανω | om. θεωρειται ουδε | απογραφεται ημεραν εξ ημερας ενωπιον του υψιστου

(8) Eth^M τα αδικηματα α αδικειτε ($Eth^{g^1, n}$ c. G^b om. α αδικ.) | Eth^M μεχρι της ημερας (Eth^m + του θανατου υμων και) της κρισεως υμων

(9) Eth^m των φρονιμων ου μη γινωσκητε ('ita'ammerewwomu) $Eth^{tana\ g^1\ g}$ επελαθεσθε (terasseʿewwomu) Eth^q ('iterʿeyewwomu ου μη ιδητε) Eth^M ου μη ακουσητε (tesammeʿewwomu) | om τα δε κακα [περιεξει] υμας

(10) οτι ετοιμοι εστε υμεις | αλλα απελθοντες αποθανεισθε | $Eth^{tana\ q}$ οτι ου

γινωσκετε και εαν (baza tadallawkemu) ετοιμασθητε (baza cf. Gen. 40.15) Eth^{M} bēzā οτι ου γινωσκετε λυτρον / εξιλασμα? | και εις ημεραν αναγκης και στενοχωριας

(11) Eth I ποθεν υμεις εσθιετε εν αγαθοις και πινετε και εμπιπλασθε; η μην ('esma) απο παντων των αγαθων α περισσευει ο κυριος (Eth II + υμων) ο υψιστος επι (Eth^{tana} + πασης) της γης; και ουκ εσται υμιν ειρηνη

(12) ουα υμιν οι αγαπωντες τα εργα της αδικιας | δια τι υμιν ελπιζετε τα καλα | γινωσκετε | om. νυν | παραδοθησεσθε και τραχηλοκοπησουσιν υμας και αποκτενουσιν υμας και ου μη φεισονται υμων (leg. wa'iyemehhekukemu pro wa'iyemeḥḥerukemu cf. 95.5 Nickelsburg, 94.)

(13) Eth^{M} οτι ταφος υμιν ου μη ορυγη ('iyetkarray $Eth^{tana\ g\ g^1}$ crpt wa'iyetra''ay)

(14) οι ακυρωσιν | οτι ου μη

(15-16) Eth^{M} λογους ασεβων (rasi'ān) sed leg. res'ān ασεβειας / πλανησεως Eth^{m} αδικων ('āmāḍeyān) | Eth^{tana} οτι αυτοι γραφουσι τα ψευδη αυτων ινα ⌜αποπλανηθωσιν και⌝ αλλους αποπλανησωσιν (⌜yerse'u wa⌝yārasse'ewwo labā'ed) Eth^{M} ινα ακουσωσιν και μη επιλαθωνται την αφροσυνην ('iyerse'ewwā la'ebad) sic. Eth om. πλανασθε υμεις αυτοι

CHAPTER 99

(1) και τα ρηματα (lanagara leg. lagebra εργα?) ψευδη δοξαζουσιν και τιμωσιν | σωτηρια αγαθα?

(2) καταπατηθησονται

(3) προεχεσθαι και διδοτε αυτας διαμαρτυριαν | $Eth^{tana\ t}$ εισαγαγωσιν (yāba(wwe)'ewwon Eth^{M} crpt yānabberewwo)

(4) τοτε συνταραχθησονται λαοι και ανασταθησονται πατριαι των εθνων | Eth^{d} c. G^{b} | $Eth^{m\ t}$ c. G^{b} εν ημερα απωλειας της αδικιας (Eth^{M} om. της αδικ. $Eth^{tana\ q}$ leg. των αμαρτωλων pro της αδικ.)

(5) αι συλλαβουσαι (cf. G^{b} et LXX Hos. 2.7) εκβαλουσιν (Eth^{tana} pr. και leg. yāwaḍḍe'u c. Nickelsburg, 95) | Eth I cf. G^{b} και εξαρπασουσιν (yemaššeṭu) τα νηπια αυτων και εγκαταλειψουσιν αυτα (Eth II ⌜τα νηπια αυτων⌝) και εξ εαυτων (Eth^{g^1} c. G^{b} om. εξ εαυτ.) εκτρωσουσιν τα τεκνα αυτων και θηλαζουσαι (leg. yāṭabbewu) εγκαταλειψουσιν / ριψουσιν (yegaddefewwomu) τα νηπια αυτων (Eth II αυτα) | και ου μη επιστρεψουσιν προς αυτα ουδε μη φεισονται (cf. 98.12) των αγαπητων αυτων

(6) G^{b} om. v. 6. $Eth^{t^2,\ e\ 3\ mss}$ εις ημεραν της εκχυσεως του αιματος

(7) και οι (leg. c. Eth^{q} wa'ella Eth^{M} om.) λατρευοντες λιθοις και οι γλυφοντες εικονας χρυσους / χρυσας και αργυριου / αργυρας και ξυλου / ξυλινας και οστρακου / οστρακινας | και οι λατρευοντες πνευμασιν ακαθαρτοις και δαιμονιοις και πασαις ταις πλαναις / πασιν τοις ειδωλοις | $Eth^{q\ t\ (g\ tana)}$ ου κατ επιστημην ('ibatemhert Eth^{M} crpt 'ibameḥrāmāt?) | και παν βοηθημα ου μη ευρηθησεται απ αυτων

(8) ενεκα της αφροσυνης της καρδιας αυτων et + (G^b om. per hmt) και τυφλωθησονται οι οφθαλμοι αυτων εν φοβω καρδιας αυτων | και εν τοις οραμασιν των ενυπνιων αυτων (9) εν αυτοις πλανηθησονται και φοβηθησονται

(9) οτι παντα τα εργα επι ματαιω εποιησαν και ελατρευσαν τοις λιθοις και επι μιας απολουνται

(10) μακαριοι παντες οι υπολαμβανοντες (yetmēṭawu) τους λογους της φρονησεως / σοφιας | Eth crpt τας οδους του υψιστου | Eth I c. G^b εν οδοις δικαιοσυνης αυτου (Eth II om. αυτου) | και 5° οτι

(11) G^b om. ουα υμιν οι κλινοντες κακα κατα των πλησιων υμων οτι εν τω Αδη αναιρεθησονται

(12) G^b om. $Eth^{g\ g^1\ tana\ (q),\ i}$ ουα υμιν οι ποιουντες ζυγα (lamasfarta(t) Eth^M lamašarrata θεμελια) της αδικιας και του δολου και οι ευοδουμενοι? (Eth^{g^1} yāmakkeru Eth^M yāmarreru $Eth^{g\ q\ tana}$ yā'ā(a)mmeru) επι της γης οτι εν αυτη εξαναλωθησονται (yetwēde'u)

(13) ουαι υμιν | εν κοποις αλλων | $Eth^{tana\ t,\ x}$ και πασαν οικοδομην υμων εκ πλινθων και λιθων αδικιας [ποιειτε·]ουα υμιν (Eth^{g^1} 'elē lakemu Eth^M 'ebelakemu cf. Nickelsburg, 95f.)

(14) Eth^{tana} c. G^b θεμελιωσιν (mašarrata Eth^M masfarta) | Eth^M και διωξεται πνευματα αυτων οπισω πλανησεως $Eth^{g\ q\ t}$ και διωξονται οπισω πνευματος πλανησεως

(16) om. τοτε | εκστρεψει | + (post την δοξαν υμων) και δωσει πονηριαν εις την καρδιαν υμων | και επεγερει το πνευμα του θυμου αυτου et om. καθ υμων | $Eth^{g^1\ (tana\ q\ t)}$ + επι την καρδιαν υμων | απολλυειν | Eth I παντες οι αγιοι και δικαιοι Eth II παντες οι δικ. και αγ.

CHAPTER 100

(1) και τοτε εν ενι τοπω πατερες συγκερατισθησονται τοις υιοις αυτων και αδελφοι συν αδελφοις πεσουνται εν απωλεια εως ρεη ως ποταμος εκ του αιματος αυτων

(2) Eth I οτι ανθρωπος ουκ αφεξει την χειρα αυτου απο των υιων αυτου ουτ απο των υιων υιων (Eth II + εν ελεει) αποκτειναι αυτους | om. απο του αγαπητου αυτου | και ο αμαρτωλος (Eth II ẖāṭe' Eth I crpt laẖāṭe') ουκ αφεξει την χειρα αυτου απο του αδελφου του εντιμου | φονευθησονται (⸢wa⸣yetqātalu $Eth^{3\ mss}$ om. wa) | om. επι το αυτο

(3) εν τω αιματι | Eth^M και το αρμα μεχρι του υψους αυτου (malʿeltā) $Eth^{tana\ g\ g^1\ q}$ 'eska ʿelata leg. lāʿla malʿelt(ā) μεχρι επανω του υψους αυτου? καταποθησεται

(4) και τοτε καταβησονται om. καταδυνοντες et εν ημερα εκεινη | Eth II και συστρεψουσιν εις ενα τοπον παντας οιτινες εβοηθουν (Eth I crpt yāwarredewwā) τη αδικια | Eth II ο υψιστος εγερθησεται εν ημερα εκεινη (Eth I + της κρισεως

Eth$^{tana\ q}$ + της κρισ. της μεγαλης) |· Eth$^{t\ M}$ c. G^{b} εκ παντων ('emkuellomu Eth$^{g\ g^1\ m\ q}$ 'emmā'kalomu εκ μεσου) των αμαρτωλων cf. Nickelsburg, 97.

(5) και δωσει / ταξει φυλακην | εκ των αγιων αγγελων | Eth$^{tana,\ 7\ mss}$ και (Eth^{M} om.) τηρησουσιν αυτους | ⌜παντα⌝ τα κακα και ⌜πασα⌝ αμαρτια | απ εκεινου (sc. καιρου) 'emmani leg. 'amēhu vel. 'em'amēhu(ā) cf. Nickelsburg, 97f., 98 n. 43 et infra v. 6 | υπνον μακρον | Eth^{M} και ουκ εσονται οι φοβουμενοι (zayefarrehu) Eth^{tana} οι εκφοβουντες (zayāstafarrehu) cf. G^{b}

(6) και τοτε | απ εκεινου wa'emmani leg. wa'amēhu (ut supra, v. 5) | παντας τους λογους | της αδικιας αυτων

(7) φλεξητε | και κομιεισθε

(9) G^{b} ουαι υμιν . . . επι τοις εργοις του στοματος υμων crpt et ditt. | Eth II ουαι υμιν τοις αμαρτωλοις οτι (Eth I om. c. G^{b}) επι τοις λογοις του στοματος υμων | Eth$^{tana(t)}$ emgebra [qedus?]

(10) Eth I και νυν γινωσκετε οτι απο των αγγελων αναζητησει (Eth II αναζητησουσιν οι αγγελοι) τα εργα υμων | En$^{g\ (tana)}$ απο του ουρανου (Eth^{M} εν τω ουρανω) και (Eth^{M} om.) απο του ηλιου και απο της σεληνης και απο των αστερων περι των αμαρτιων ημων οτι επι τη γη εποιειτε κρισιν προς τους δικαιους

(11) Eth^{M} και διαμαρτυρησεται (Eth$^{tana\ t}$ διαμαρτυρησονται) εφ υμιν (cf. Dt. 4.26) πασαν νεφελην και ομιχλην και δροσον και ομβρον οτι παντες εσονται κωλυομενοι αφ υμων ινα μη καταβωσιν εφ υμας· και εγρηγορηθησονται (Eth$^{t,\ b}$ wa'iyehēlewu leg. wayeḫēlewu Eth$^{g\ q,\ y}$ wayeḫēleyu Eth^{M} 'iyeḫēleyu) επι ταις αμαρτιαις υμων

(12) και νυν διδοτε δωρα ομβρω ινα μη κωλυθη καταβηναι υμιν και δροσος εαν δεξηται αφ υμων χρυσιον και αργυριον [διαγραψατε] ινα καταβη | om. και νεφελη . . . καταβωσιν

(13) Eth^{tana} οταν επιρριψη εφ υμας χιων και παχνη και ψυχος αυτων και ⌜παντες⌝ οι ανεμοι και (Eth^{M} του παγετου) ο παγετος και πασαι αι μαστιγες αυτων τοτε ου δυνησεσθε υποστηναι εμπροσθεν αυτων

CHAPTER 101

(1) κατανοησατε τον ουρανον (crpt ex τοινυν G^{b} cf. Nickelsburg, 120) υμεις παντες υιοι των ανθρωπων (Eth crpt weluda samāy leg. weluda sab') «και» παντα τα εργα του υψιστου· και φοβεισθε αυτον και μη ποιειτε το πονηρον (Eth^{m} + επι της γης) εναντιον αυτου

(2) τον ομβρον και την δροσον | καταβηναι επι την γην εινεκα υμων

(3) και εαν | Eth$^{g^1\ m}$ εν (πασιν) τοις εργοις υμων Eth^{M} και επι παντα τα εργα υμων | ουχι υμεις οιτινες εσεσθε δεομενοι αυτου | διοτι G^{b} δια τι ελαλειτε (om. τω στοματι υμων) επι / κατα την δικαιοσυνην αυτου μεγαλα και σκληρα· και ουκ εσται υμιν ειρηνη

(4) και ουχ ορατε τους ναυκληρους ως σειομενα υπο του κλυδωνος και σεσαλευμενα υπο των ανεμων τα πλοια αυτων· και χειμαζομενοι

(5) και δια τουτο / G^{b} ουτως φοβουνται διοτι παντα τα υπαρχοντα αυτων [και] τα

αγαθα εκβαλλουσιν (yewaḍḍe' leg. yāwaḍḍe'u c. G^b) επι τη θαλασσα μεθ εαυτων και μεριμνωσιν εν τη καρδια αυτων (Eth^gr ⸢šannāy⸣ yeḥēleyu balebbomu Eth^M crpt wašannāya 'iyeḥēleyu balebbomu) οτι η θαλασσα καταπιεται αυτους και απολουνται εν αυτη

(6) + (post υδατα αυτης) και πασα η κινησις αυτου (ḥusatā) | Eth^e c. G^b συνεστησατο ('aqama leg. ʿaqama cf. Charles[1906], 211, n. 24 Eth^M ḥatama εσφραγισατο? παντα τα εργα αυτης (Eth I + και υδατα) και συνεδησεν πασαν αυτην αμμω

(7) Eth I και απο της εμβριμησεως αυτου φοβειται και ξηραινεται (Eth II ξηρ. και φοβ.) | και παντες οι ιχθυες αυτης αποθνησκουσιν και παντα τα γινομενα εν αυτη· και υμεις οι αμαρτωλοι οι επι της γης ου φοβουνται αυτον

(8) ουχι αυτος εκτισεν τον ουρανον και την γην και παντα τα εν αυτοις· και τις εδωκεν επιστημην ⸢και σοφιαν⸣ πασιν τοις κινουμενοις ⸢επι τη γη και⸣ τοις επι τη θαλασσα· ουχι αυτοι οι ναυκληροι την θαλασσαν φοβουνται· και οι αμαρτωλοι τον υψιστον ου φοβουνται

CHAPTER 102

(1) και τοτε εαν | πυρ σκληρον leg. [τον κλυδωνα] πυρ[ος της καυσεως] σκληρον? | και που σωθησεσθε | και οταν δω / εκβαλη φωνην αυτου εφ υμας (2) ουχι εσεσθε συνσειομενοι και φοβουμενοι; | om. ηχω ... συνταρασσομενην | και παντες οι φωστηρες σειομενοι εν φοβω μεγαλω και πασα η γη συνσειομενη και τρεμουσα και συνταρασσομενη·

(3) Eth I και παντες οι αγγελοι συντελουντες το συνταχθεν αυτοις και ζητουντες αποκρυβηναι απο προσωπου της δοξης της μεγαλης (Eth II του μεγαλου της δοξης) και τρεμοντα τα τεκνα της γης και σειομενα· και υμεις αμαρτωλοι επικαταρατοι εις τον αιωνα· ουκ εστιν υμιν χαιρειν

(4) μη φοβεισθε υμεις αι ψυχαι των δικαιων και θαρσειτε υμεις οι αποθανοντες ευσεβεις

(5) Eth I c. G^b και μη λυπεισθε ... λυπης Eth II + (post ψυχαι υμων) εν μεγαλη αναγκη και εν στεναγμω | ουκ απαντηθη τω σωματι / τη σαρκι υμων | om. επει αι ημεραι ... επι της γης | αλλα υπομενετε την ημεραν ('enka baʿelat leg. 'anuẖu laʿelat?) εν η [εσται] η κρισις των αμαρτωλων (konkemu ẖāṭe'āna leg. [yekawwen] kuennanēhomu laẖāṭe'ān?) και την ημεραν (leg. (laʿelat) της καταρας και της οργης Cf. Charles 254, Charles[1906] 212 n. 36, 39.

(6) και οταν αποθανητε ερουσιν περι υμων οι αμαρτωλοι οτι ως απεθανομεν (kama motna leg. kamāna ομοιως ημιν cf. v. 7) απεθανον οι ευσεβεις

(7) ιδου ομοιως ημιν (kamāna) απεθανοσαν εν λυπη και εν σκοτια· και τι περισσον αυτοις υπερ ημων;

(8) απο του νυν ισωθημεν (taʿarrayna sic. leg. σωθησονται / σωθητωσαν) και τι αναστησονται / αναστητωσαν και τι οψονται [το φως] εις τον αιωνα οτι ιδου αυτοι απεθανον και απο του νυν εις τον αιωνα ου μη οψονται το φως

(9) ⸢λεγω υμιν αμαρτωλοι⸣ καλως υμας (leg. ημας c. G^{b}) φαγειν και πειν om. τοιγαρουν | Eth I c. G^{b} αρπασαι . . . αγαθας Eth II tr. και λωποδυτειν post πειν

(10) Eth I ιδετε (re'ikemewwomunu) τους δικαιους ως εγενετο η τελευτη αυτων (Eth II + ειρηνη) οτι πασα αδικια ουχ ευρεθη επ αυτοις εως ημερας θανατου αυτων και απωλοντο

(11) εν αναγκη pro μετ οδυνης

CHAPTER 103

(1) Eth^{tana} και νυν εγω ομνυω υμιν τοις δικαιοις εν τη δοξα του μεγαλου και εν τη τιμη (leg. bakebru) της βασιλειας αυτου και εν τη μεγαλωσυνη αυτου ⸢ομνυω υμιν⸣

(2) οτι εγω επισταμαι | Eth^{M} c. G^{b} το μυστηριον τουτο Eth$^{g\ g^{1}\ m\ q}$ om. τουτο | και ανεγνων | Eth$^{g\ q}$ c. G^{b} τας πλακας Eth^{M} εν ταις πλαξιν Eth$^{g^{1}}$ εκ των πλακων | Eth II την γραφην των αγιων cf. G^{b} Eth I την γραφ. αγιαν | και ευρηκα pro G^{b} και ανεγνων | Eth^{tana} c. G^{b} περι υμων Eth^{M} περι αυτων

(3) οτι παντα αγαθα | Eth II ητοιμασται υμιν (lomu leg. lakemu) Eth I c. G^{b} om. υμιν | + (post ευσεβων) και αγαθα πολλα δοθησεται υμιν αντι του κοπου υμων και το μερος υμων κρειττον του μερους των ζωντων

(4) Eth I και ζησονται τα πνευματα υμων των αποθανοντων ευσεβων και αγαλλιασονται (G^{b} om. και ζησ. . . . αγαλλ.) και χαιρησονται και ου μη απολωνται τα πνευματα αυτων (Eth II om. και ου μη απολ.) | μη ουν | και νυν μη φοβεισθε τους ονειδισμους αυτων

(5) Eth I cf. G^{b} ουαι υμιν οι νεκροι των αμαρτωλων (Eth II o. υμιν τοις αμαρτωλοις) οταν αποθανητε εν πλουτω / μαμμωνα της αδικιας υμων (Eth II εν ταις αμαρτιαις υμων) | ερουσιν εφ υμιν οι ομοιοι υμων | Eth$^{g^{1}}$ πασας τας ημερας αυτων ιδον αγαθα (Eth^{M} om.) | om. εν τη ζωη αυτων

(6) και νυν απεθανον εν αγαθοις (C-B ευθηνια) και εν πλουτω· αναγκην και φονον ουκ ιδον εν τη ζωη αυτων και ενδοξως απεθανοσαν G^{b} om. και νυν . . . ιδον per hmt cf. C-B 65, Nickelsburg, 126 | και η κρισις ουκ εγενηθη αυτοις εν τη ζωη αυτων

(7b 8) και κακαι (crpt ex εκει) εσονται [εν] αναγκη μεγαλη (Eth^{g} mendābē (Eth^{tana} pr. wa) 'abiya Eth^{M} mendābēhomu) | + (post αι ψυχαι υμων) και η κρισις η μεγαλη εσται εις πασας γενεας μεχρι του αιωνος G^{b} om. per hmt

(9) μη ειπητε τοις δικαιοις και τοις οσιοις οι οντες εν τη ζωη αυτων | + (post εκοπιασαμεν) και πασαν θλιψιν ειδομεν και κακα πολλα ευρηκαμεν και ανηλωμεθα και ολιγοι εγενηθημεν και ολιγοψυχουμεν | om. και αντιλημπτορα ουχ ευρηκαμεν sed cf. v. 10.

(10) Eth II και απολωλαμεν και ουκ ην ο βοηθων ημιν εν λογω η εν εργω (Eth I om. η εν εργω)· ησθενησαμεν / ηδυνατησαμεν (se'enna) και ουδενα [αντιλημπτορα] ουχ ευρηκαμεν· και συντετριμμενοι και απολωλαμεν και απηλπισμεθα ειδεναι σωτηριαν ημεραν εξ ημερας G^{b} om. και απολωλαμεν . . . ουδενα

(11) και ηλπισαμεν | και του κοπου ημων ου κεκυριευκαμεν | και οι ανομοι εβαρυναν εφ ημας τον ζυγον αυτων

(12) Eth^{tana} και κυριευουσιν ημων οι εχθροι ημων και τοις (wa'ella leg. wala'ella) εγκεντριζουσιν ημας και περικυκλουσιν ημας (waya'awweduna) (Eth^{M} om. και περικ. ημας) ⌜και τοις εχθροις ημων⌝ εκυψαμεν τους τραχηλους ημων και ουκ ηλεησαν ημας

(13) και εζητησαμεν πορευεσθαι απ αυτων οπως αναφυγωμεν και αναψυχωμεν και ουχ ευρηκαμεν που φυγαδευσομεν και σωθησομεθα απ αυτων

(14) Eth^{M} και ενετυγχανομεν προς τους αρχοντας (Eth$^{g^1}$ + και εβοησαμεν (waga'aru leg. waga'arna) εν τη αναγκη ημων) | Eth^{M} και εκραξαμεν (Eth^{g} + και εβοησαμεν) κατα τους κατεσθιοντας ημας | om. και βιαζομενους ημας | και τας κραυγας ημων ουκ επειδεν | om. και τας εντευξεις ημων ουκ απεδεξαντο

(15) om. και ουκ αντελαμβανοντο ... εφ ημας | + (G^{b} om. per hmt) και αντελαμβανοντο των βιαζομενων ημας και κατεσθιοντων ημας και ολιγωντων ημας και εκρυπτον την αδικιαν αυτων ουδε αφειλον αφ ημων τον ζυγον αυτων των κατεσθιοντων ημας· και διεσπειρον ημας | και απεκτεινον ημας | om. και εις ολιγους ... πεφονευμενων ημων | + και εκρυπτον τον φονον ημων | και ουκ αναμιμνησκουσιν οτι τας χειρας αυτων επηραν εφ ημας om. περι των αμαρτωλων ... αυτων post αναμιμ. G^{b} om. οτι τας χειρας ... εφ ημας

CHAPTER 104

(1) Eth II + (post υμιν) οι δικαιοι Eth I c. G^{b} om. | Eth$^{g^1}$ ενωπιον των αγγελων του μεγαλου | Eth^{M} + (post μεγαλου) και απογραφεται τα ονοματα υμων ενωπιον της δοξης του μεγαλου Eth$^{4\ mss}$ om. c. G^{b} (om. per hmt?)

(2) Eth crpt οτι παλαι (baqadāmi) επαλαιωθητε cf. Nickelsburg, 129 | και νυν αναλαμψετε ως φωστηρες του ουρανου και φανειτε και αι θυριδες του ουρανου ανοιγησονται υμιν

(3) και η κραυγη υμων [ακουσθησεται και] η κρισις [υμων ην] κραζετε και φανειται υμιν (waṣerāḫā zi'akemu kuennanē ṣereḫu crpt 'and cry your cry for judgement' Knibb; cf. Nickelsburg, 130 | οτι εξ αρχοντων εκζητηθησεται πασα η θλιψις υμων (Eth$^{g\ m\ t,\ î}$ kuellu Eth^{M} kuello mendābēkemu) και εκ παντων οιτινες συνελαβοντο τοις βιαζομενοις υμας

(4) G^{b} om.

(5) Eth I τι (menta Eth II 'enta Eth^{tana} 'enza) μελλετε ποιεν τα κακα; μη ('akko sed leg. 'ekuyā 'akko cf. G^{b}) μελλετε αποκρυβηναι εν τη ημερα της κρισεως της μεγαλης; και μη ευρεθησετε ως οι αμαρτωλοι και κρισις αιωνιος εσται εξ υμων εις πασας τας γενεας του αιωνος; om. σκυλ[λ]ησεσθε

(6) και νυν μη φοβεισθε | Eth I c. G^{b} ευοδουμενους (yeddēlawu bafenotomu Eth II crpt yeddēlawu bafetwatomu 'prospering in their desires' Knibb) | om. παντων | + (post αυτων 2°) | οτι μετοιχοι γενησεσθε των αγγελων (Eth I laḫērānā των αγαθων crpt ex αγγελων Eth II laḥarrā samāy cf. Flemming, 164, Charles^{1906}, 217 n. 39) G^{b} om. οτι μετ ... αγγελ. per hmt

(7) Eth^{tana} και εαν (’emmasa Eth^{M} crpt ’esma) ειπητε υμεις αμαρτωλοι· ου μη εκζητηθωσιν μηδε απογραφωσιν πασαι αι αμαρτιαι ημων· απογραφουσιν πασας τας αμαρτιας υμων εξ ημερων / καθ εκαστην ημεραν

(8) εγω αποδεικνυω

(9) Eth^{M} του αγιου και ($Eth^{g\ q\ t}$ om.) μεγαλου | $Eth^{g\ g^1\ (q)}$ μη λογιζησθε (’itaḥāsebewwā Eth^{M} c. G^{b} μη δοτε επαινον ’itesabbeḥewwo) | οτι ου γινεται παντα τα ψευδη υμων και πασα η πλανη υμων εις δικαιωμα αλλα εις αμαρτιαν μεγαλην

(10) Eth I και νυν γινωσκω τουτο το μυστηριον οτι τους λογους της αληθειας αναστρεφουσιν και αλλασσουσιν πολυ (Eth II πολλοι) αμαρτωλοι και λαλησουσι πονηρα | γραφας Eth^{tana} + μου | κατα τους λογους αυτων (pro G^{b} επι τοις ονομασιν αυτων)

(11) εν ταις φωναις / γλωσσαις αυτων G^{b} εν τοις ονομασιν αυτων ditt. | μητε αλλοιωσωσιν μητε αφελωσιν (’iyāḥaṣṣeṣu) απο των λογων μου | παντα α προτερον διαμαρτυρουμαι / διεμαρτυρησαμην περι αυτων

(12) και μυστηριον δευτερον γινωσκω | δικαιοις και φρονιμοις | αι βιβλοι (om. μου) | εις χαραν και εις αληθειαν και εις σοφιαν πολλην

(13) Eth II ʳκαι αυτοις δοθησονται βιβλοιˀ (ditt.) και (Eth I c. G^{b} + αυτοι) πιστευσουσιν αυταις | En^{c}5i20 ישמחון | αγαλλιασονται (yetḫāšayu) παντες οι δικαιοι $Eth^{g\ g^1\ q\ t,\ b\ 10\,mss}$ ανταποδοθησεται (yet῾aššayu) πασιν τοις δικαιοις | οι μαθησονται απ αυτων

CHAPTER 105

(1) Eth I και τοτε ʳλεγει κυριοςˀ φωνησουσιν (Eth II pr. οτι) και διαμαρτυρησονται τοις υιοις της γης εν τη σοφια αυτων· αποδειξατε αυτοις οτι υμεις εστε διδασκαλοι / ηγουμενοι αυτων και εκδικηται / αρχηγοι (῾esēyāta leg. ῾asāyeyān) επι πασαν την γην | En^{c}5i21-22 בבני ארע]א . . . . אנ]תון תהוון [. . . . פורע]נין
[על כולה ארעא]

(2) οτι εγω και ο υιος μου κοινωνησει αυτοις (neddēmar meslēhomu cf. Sir. 13.2) εις τον αιωνα εν ταις οδοις της αληθειας εν τη ζωη αυτων· και ειρηνη εσται υμιν· χαιρετε υιοι της αληθειας αμην

CHAPTER 106

(1) και μετα χρονον ελαβεν ο υιος μου Μαθουσαλα τω υιω αυτου Λαμεχ γυναικα και συνελαβεν εξ αυτου και ετεκεν παιδιον | En^{c}5i26-27 . . . וקרית שמה למ[ך
נסב מתוסלח לה אנ]תה והיא . . . om. και εκαλεσεν ... (2) παιδιον

(2) και ην το σωμα αυτου λευκον ως χιων και πυρρον ως ροδον και το τριχωμα της κεφαλης αυτου ως εria λευκα ʳκαι η κομη αυτουˀ· και ωραιοι οι οφθαλμοι αυτου και οτε ανεωξεν τους οφθαλμους αυτου ελαμψεν πασαν την οικιαν ωσει ηλιος ʳκαι σφοδρα ελαμψεν πασα η οικιαˀ En^{c}5i28 ...ו]שמוק

(3) και οτε ανεληφθη / ανεστη G^b | και ελαλησεν τω κυριω της δικαιοσυνης cf. v. 11

(4) Λαμεχ ο πατηρ αυτου | $Eth^{q\,(t)}$ om. vs. 4-11

(5) εγω εγεννησα τεκνον αλλοιον | $Eth^{m^2,\ b\ 4\ mss}$ c. G^b αλλα Eth^M και | ομοιον τοις τεκνοις των αγγελων | και τα ομματα | $Eth^{tana\ g^1}$ c. G^b και (Eth^M om.) ενδοξον το προσωπον αυτου

(6) εξ αγγελων | om. αυτον | μηποτε θαυμασιον ποιηθησεται

(7) και νυν παραιτουμαι σε πατερ μου και δεομαι σου | και ακουσον απ αυτου την αληθειαν οτι μετα των αγγελων εστιν η κατοικησις αυτου

(8) και οτε ηκουσεν Μαθουσαλα τους λογους του υιου αυτου | οτι ηκουσεν εκει ειναι με | και εκραξεν pro G^b και ειπεν μοι et om. πατερ μου . . . ηκε [προς] εμε | διοτι ('esma) pro G^b δια τι | om. τεκνον 2°

(9) και απεκριθη μοι λεγων | Eth I c. G^b δι αναγκην (ṣāheq leg. ṣāʿeq) μεγαλην ηλθον προς σε (om. πατερ) Eth II crpt δια λογον (nagar) μεγαν | $Eth^{tana\ m}$ και δια ορασιν θαυμαστον ωδε ηγγισα (baze qarabku Eth^M bazaqarabku)

(10) Eth II και νυν πατερ μου επακουσον μου οτι εγεννηθη Λαμεχ τω υιω μου τεκνον ου (Eth I και) ουκ εστιν η εικων αυτου και ο τυπος αυτου ομοιος τυπω των ανθρωπων | + (post ακτισιν) και ανεωξεν τους οφθαλμους αυτου και ελαμψεν πασαν την οικιαν cf. v. 2

(11) $Eth^{g,\ b\ 4\ mss}$ c. G^b απο ('emwesta Eth^M westa) των χειρων | τον κυριον του ουρανου

(12) Eth^M ο πατηρ αυτου Λαμεχ Eth^{g^1} cf. G om. ο πατ. αυτ. | ου πιστευει οτι εξ αυτου εστιν αλλα υπολαμβανει εξ αγγελων του ουρανου· και ιδου ηλθον προς σε ινα υποδειξης μοι την αληθειαν

(13) και απεκριθην εγω Ενωχ λεγων αυτω· ανακαινισει ο κυριος καινα (Eth^u om.) επι της γης En^c 5 ii 17 בל יכ֯(ה֯)[רת מריא מלה על ארעא] | και τουτο ηδη (Eth^{g^1} om.) τεθεαμαι εν ορασει και εσημανα σοι οτι εν τη γενεα Ιαρεδ του πατρος μου παρεβησαν τον λογον κυριου εκ υψους του ουρανου ('emmalʿelta $Eth^{e\ h\ k}$ malʿelta Eth^n malā'ekta samāy Eth^{tana} (pro εκ υψ. του ουρ.) απο της διαθηκης του ουρανου 'emšerʿāta samāy) En^c 5 ii 17 ביומי ירך אבי] En^c 5 ii 18 עברו

(14 + 17a) En^c 5 ii 18 ועב[e]רין . . . ו]שריו למע[ל] | + (post εξ αυτων) και εγεννησαν τεκνα εξ αυτων G^b om. per hmt | 17a και τικτουσιν (yewalledu leg. yewalledā) επι της γης γιγαντας ('ella yārbeḫ) ου του πνευματος αλλα της σαρκος

(15) και εσται απωλεια μεγαλη επι πασης ($Eth^{g^1,\ 5\ mss}$ om. c. G^b) της γης και κατακλυσμος εσται και απωλεια μεγαλη επι ενιαυτον ενα

(16) Eth^{tana} και (Eth II om.) εσται τοδε το παιδιον (zewald Eth^M zewe'etu wald leg. οτι το παιδ. zawe'etu wald) το γεννηθεν υμιν αυτο καταλειφθησεται επι της γης και τρια αυτου τεκνα σωθησεται μετ αυτου αποθανοντων παντων των ανθρωπων των επι της γης· ⸢σωθησεται αυτο και τα τεκνα αυτου⸣ En^c 5 ii 20-21 [ודן עלימא] די֯ יליד [לכו]ן֯ . . . [ותלתת בנוה]י יפ[לטון . . .] ארעא

(17) (17a vide supra, v. 14, 17b cf. G^b supra, v. 15) και εσται οργη μεγαλη επι της γης και καθαρισθησεται η γη απο πασης φθορας om. και πραυνει την γην (ותנוח ארעא?) απο της ουσης εν αυτη φθορας G^b om. και καθαρ. η γη leg. και

καταπαυει η γη και καθαρ. απο της ουσ. εν αυτ. φθορ.? En^c 5 ii 22 [ותתד]כא ארעא מן חבלא [ר]בא]

(18) Eth^M και νυν λεγε τω υιω σου Λαμεχ οτι το γεννηθεν τεκνον αυτου (Eth^g1 σου c. G^g) εστιν δικαιως Eth^g g1 tana ṣādeq δικαιος En^c 5 ii 22 בקשוט | και καλεσον το ονομα αυτου Νωεον En^c 5 ii 23 [קר]ׄי שמה | υμιν | om. εφ ου αν καταπαυσεται et + (post καταλειμμα) και αυτος και τα τεκνα αυτου σωθησονται απο της φθορας της ερχομενης επι την γην απο παντων των αμαρτηματων και απο πασης της αδικιας η εσται συντελουμενη επι της γης εν ταις ημεραις αυτου En^c 5 ii 24-25 ... [ו]יפלט הוא [די] להוא ביומ[והי] ......

(19) και μετα ταυτά εσται η αδικια πλειων της συντελουμενης προτερον επι της γης οτι γινωσκω τα μυστηρια των αγιων οτι αυτος ο κυριος υπεδειξεν μοι και εμηνυσεν μοι και εν ταις πλαξιν του ουρανου ανεγνων [ταυτα] En^c 5 ii 25-27 [רשע] תׄקיף ...... ד]ׄי ידע אנה ברׄזׄ[הון שמיא די]קדישין אחיוני ואחזיוני ... שמיא קרית

CHAPTER 107

(1) En^c 5 ii 27-29 וחזית כתיב בהון די [ד]רׄ מן דר יבאש בהׄון ובאש להׄוא [עד די יקומון] דרי קושׄטׄא ובאישתא ורשעה יסוף וחמסא יכלא מן ארעא ועׄ[ד די יאתן טבן] עליהון και τεθεαμαι c. En^c G^b τοτε τεθ. om. και ειδον τοδε | οτι γενεα γενεας αδικησει (tet'ēbes) | Eth^M c. G^b αλλαξει (tetlāḥas) Eth^q t2 b 8 mss αρθησεται (tetgaḥḥaš cf. En^c) | Eth^M παντα τα αγαθα ηξει (Eth^b 3 mss pr. ουχ) επ αυτην Eth^g1 επι την γην cf. G^b

(2) En^c 5 ii 29-30 וכען אזל נא עדׄ למׄךׄ ברך . . . . די עלימא דן ברה הואה בקשוט ולא בכדבין απερχου τεκνον μου | om. και 2°

(3) Eth II^(m) οτι το μυστηριον (zaḥebu' Eth I baḥebu', cf. G^b) εδηλωσεν αυτω ολον (gebra crpt ex gab'a Eth^m nagaro ditt.) | Eth^(g m tana) επεστρεψεν και εδηλωσεν αυτω (Eth^M om. εδηλ. αυτω) | και εκαλεσαν το ονομα τουτου του παιδιου Νωε οτι ευφρανει την γην απο πασης απωλειας Eth^g1 οτι σωσει και ευφρανει

CHAPTER 108

(1) Eth^M επιστολη / γραφη αλλη (Eth^tana (m) + του Ενωχ) ην εγραψεν (Ενωχ Eth^tana (m) om.) τω υιω αυτου Μεθουσαλα και τοις μετ αυτον ερχομενοις και τοις τηρουσιν τον νομον επ εσχατων ημερων

(2) Eth^g g1 οι εποιησατε αγαθον (Eth^M om.) και υπεμεινατε εν ταυταις ταις ημεραις εως συντελεσθησονται οι κακοποιουντες ⌜και συντελεσθησεται η δυναμις των κακοποιουντων⌝

(3) Eth^M υμεις δη υπομεινετε εως αφανιζεται η αμαρτια οτι εξαλειφθησεται τα ονοματα αυτων εκ των βιβλων των αγιων Eth^g (g1) εκ της βιβλου της ζωης και εκ της βιβλου του αγιου (Eth^g1 εκ των αγιων βιβλων) και το σπερμα αυτων απολειται εις τον αιωνα και τα πνευματα αυτων αποκτενουνται και κραξουσιν και αλαλαξουσιν εν τοπω ερημω αορατω (Eth^g1 εν τοπ. κριματος / βασανισμου (dayn)) και εν πυρι καυθησονται οτι ουκ εστιν εκει γη

(4) και ιδον εκει ως νεφελην αθεωρητον διοτι εκ του (Eth$^{g^1}$ + πολλου) βαθους αυτου (Eth^{tana} σκοτους αυτ.) ουκ (Eth^{tana} om.) εδυναμην επισκοπειν (lāʿla naṣṣero) και φλογα πυρος ιδον φλεγομενην ενδοξως (sebuḥ cf. Exod. 15.1 Dillmann, *Lex.*, 357) και κυκλουμενα ως ορη ενδοξα και σαλευομενα ωδε και ωδε

(5) Eth II και ηρωτησα ενα των αγιων (Eth$^{ryl\ f}$ sebuḥān leg. qeddusān vel. om. c. Eth^{u}) αγγελων των μετ εμου λεγων αυτω· τι εστιν τουτο το ενδοξον οτι ουκ εστιν ουρανος αλλα φλοξ πυρος μονον φλεγομενη και φωνη κραυγης και κλαυθμου και αλαλαγμου και οδυνης ισχυρας

(6) Eth^{M} και ειπεν μοι· ουτος ο τοπος ον ορας ωδε βληθησονται (Eth$^{tana\ g\ q}$ εξωσθησονται (yetwassadu)) τα πνευματα των αμαρτωλων και των βλασφημων και των κακοποιουντων και των αλλοιωντων παντα α ελαλησεν εν στοματι των προφητων ο κυριος | Eth I παντα (Eth II om.) α μελλει γενεσθαι

(7) Eth^{M} οτι εσται εξ αυτων (Eth$^{g^1}$ υμων) γεγραμμενα και εγκεκολαμμενα (cf.103.2) ανω εν τω ουρανω ινα αναγνωσιν αυτα οι αγγελοι και επιγινωσκωσιν α μελλει ερχεσθαι τοις αμαρτωλοις και τοις πνευμασιν των πραων / οσιων (cf. 25.4) και τοις ταπεινουσιν την σαρκα αυτων· και κομιουνται παρα (Eth$^{g\ g^1}$ baḫaba Eth^{M} diba) του θεου· και τοις επηριαζομενοις υπο των πονηρων ανθρωπων

(8) και τοις αγαπωσιν τον θεον ουτε χρυσιον ουτε αργυριον ουτε ⸢ηγαπησαν⸣ παντα τα αγαθα τα εν τω κοσμω αλλα εδιδουν τα σωματα αυτων τη οδυνη

(9) και τοις απο του γενεσθαι αυτους ουκ επιθυμουσιν την βρωσιν την επιγειαν αλλα ελογιζοντο εαυτους ως πνοην αφανιζομενην· και ταυτα ετηρησαν και πολυ επειρασεν αυτους ο κυριος | Eth I και ηυρεθη τα πνευματα αυτων καθαρα (Eth II baneṣḥ σωφρονως / εν σωφροσυνη Dillmann *Lex.*, 100f.) ινα ευλογησωσιν το ονομα αυτου

(10) Eth^{M} και πασας τας ευλογιας αυτων διηγησαμην εν ταις βιβλοις και ανταπεδωκεν αυτοις (Eth^{m} (post διηγησ.) τον μισθον αυτων κατα κεφαλην αυτων | οτι ουτοι ηυρεθησαν αγαπωντες τον ουρανον η την πνοην αυτων την εν τω κοσμω (Eth$^{g^1}$ zabaʿālam) και καταπατηθεντες υπο πονηρων ανθρωπων και ακουσαντες παρ αυτων ονειδισμους και βλασφημιας και ατιμαζομενοι (Eth^{tana} nabaru εμενον) ευλογουντες με (Eth$^{g^1\ q}$ om. με Eth^{tana} ημας)

(11) και νυν καλεσω τα πνευματα των αγαθων απο της γενεας του φωτος και αλλοιωσω τους γεννηθεντας εν τω σκοτει οι ουκ εν τη σαρκι αυτων την τιμην εκομισαντο ως ην αξιον της πιστεως αυτων

(12) και εξαξω / οδηγησω εν φωτι λαμπρω τους αγαπωντας το αγιον ονομα μου και καθισω εκαστον επι τον θρονον της τιμης αυτου

(13) και φωτισθησονται εις χρονους αναριθμητους οτι δικαιοσυνη (leg. frt ṣādeq δικαιον pro ṣedq) το κριμα του θεου οτι τοις πιστοις την πιστιν δωσει (yehub) εν τοις κατοικητηριοις των οδων της αληθειας

(14) και οψονται τους εν τω σκοτει γεννηθεντας εκβαλλομενους εις σκοτος των δικαιων φωτισθεντων

(15) και κραξουσιν οι αμαρτωλοι και οψονται αυτους εκλαμποντας και πορευσονται αυτοι που εγραφησαν αυτοις ημεραι και χρονοι

APPENDIX A

THE 'ASTRONOMICAL' CHAPTERS OF THE ETHIOPIC BOOK OF ENOCH (72 TO 82)

Translation and Commentary by Otto Neugebauer

With Additional Notes on the Aramaic Fragments
by Matthew Black

SUMMARY

Ethiopic literature has preserved the '*Book of Enoch*' which is, as we know, closely related to the Jewish sect that is represented in the '*Dead Sea Scrolls*'. Ten chapters of this work are concerned with *astronomical* concepts of a rather primitive character (variation in the length of daylight, illumination and rising amplitude of the moon, wind-directions, etc.), dominated by simple arithmetical patterns.

The present paper gives a new translation of these chapters, followed by notes where the meaning of the text is not self-explanatory. An appendix, with additional notes, deals with related source material in the Qumran astronomical scrolls.

INTRODUCTION

It has long been recognized that the astronomical chapters of the Book of Enoch constitute a composition of their own without much direct contact with the other parts of the treatise. This does not mean, however, that the astronomical book is unrelated to the rest of the Book of Enoch. On the contrary, its contents reflect faithfully, but in greater detail, the simple concepts that prevailed in the communities which produced the Enochian literature.

I do not think, however, that one should consider the astronomical chapters as a literary unit composed by one author who followed some stylistic reasoning. It seems obvious to me that the text, as we have it, consists of two major versions, both covering essentially the same material, to which are added several still more fragmentary pieces. What we have is not the work of one author (or 'redactor') but a conglomerate of closely related versions made by generations of scribes who assembled, to the best of their knowledge, the teaching current in their community about the structure and the laws of the cosmos. It is also important to note that purely calendaric rules on fasts and feasts are conspicuously absent, in marked contrast to the later Ethiopic 'computus' of Judaic and Christian origin.

It is, of course, possible that there existed originally one treatise written to codify the astronomical doctrines of a religious sect. Such a treatise would then have reached us only in several more or less modified versions, two of which are reflected in the present chapters 72 to 76 and 77 to 79,1 respectively. Fragments from additional versions are preserved in 79,2 to 80,1, while the description in 82 of the angelical hierarchy of the stars evidently belongs to a quite different source. Furthermore it should be remembered that innumerable fragments of Enochian 'astronomy' (concerning the variation of the length of daylight, the 'gates,' the winds, etc.) are scattered all through the Ethiopian 'computus'-treatises.

The several chapters of our treatise are grouped around only a small number of topics: solar year and lunar months, winds, the hierarchy of the stars, always hemmed in by a rigid schematism unrelated to reality. First the reader is told about the division of the 'solar' year into four seasons of 30 + 30 + 31 days each. Then comes the variation of the length of daylight, based on a linear progression with extrema in the ratio 2:1. Then the lunar phases are also described by a linear pattern, assuming day 14 or day 15, respectively, as full moon dates. The variable illumination of the moon is expressed either in terms of the moon's illuminated area (from 1 to 14 parts) or in relation to the sun's brightness, thus increasing from 1/98 (i.e. 1/7·14) to 1/7 (at full moon). Finally lunar months, alternatingly full and hollow, are related to the (schematic) solar year (but without any trace of a cyclic adjustment) and to the rising and setting in the 'gates' at the eastern, respectively western, horizon. Unrelated to these gates are the twelve gates for the winds, four of which are beneficial, while eight bring discomfort and destruction. In contrast the stars are astronomically totally insignificant, being nothing but a replica of the division of the solar year. Neither constellations nor the zodiac nor planets are ever mentioned. This remains the rule also for the Ethiopic computus until the Arab conquest.

The search for time and place of origin of this primitive picture of the cosmic order can hardly be expected to lead to definitive results. The use of 30-day schematic months could have been inspired, e.g., by Babylonian arithmetical schemes (of the type of 'Mul-Apin'), or by the Egyptian calendar. But the number and location of the epagomenal days was obviously chosen under the influence of the Jewish seven-day week and has no parallel elsewhere. The linear pattern for the variation of the length of daylight as well as the ratio 2:1 of its extrema suggests an early Babylonian background. But there is no visible trace of the sophisticated Babylonian astronomy of the Persian or Seleucid-Parthian period.

Dillmann's statement[1] that the astronomical part of the Book of Enoch is based on concepts extant in the Old Testament is simply incorrect: the Enoch year is not an old semitic calendaric unit; the schematic alternation between hollow and full months is not a real lunar calendar, and there exists no linear scheme in the Old Testament for the length of daylight, or patterns for 'gates', for winds, or for 'thousands' of stars, related to the schematic year. The whole Enochian astronomy is clearly an *ad hoc* construction and not the result of a common semitic tradition.

SUMMARY OF THE CONTENTS OF CH. 72 TO 82

First version:	72 to 76 (with 74 probably being an intrusion)
72,2-5:	Gates and Windows; winds drive the chariot of the sun (cf. 73,2 and also 18,4)
6-36:	length of daylight, $M:m = 12:6$; year of $4 \cdot 91^d = 364^d$
37:	brightness and size of sun and moon (cf. 73,3; 78,3,4)
73,1-3:	winds drive the chariot of the moon (cf. 72,5); brightness of sun and moon (cf. 72,37; 78,4)
4-8:	increase of the area of illumination and of brightness of the moon from day 1 to day 14 (incomplete); first visibility on the preceding day 30 (i.e., after a hollow month)
74,1-4:	illumination of the moon during 15 days (i.e., for a full month)
5-9:	Gates and moon rise (incomplete)
10-16:	garbled description of an octaeteris
75,1-7:	stars ('thousands') and seasons (cf. 82,4-20)
8,9:	circumpolar stars
76,1-13:	the 12 gates of the winds and their qualities (cf. the short version 33 to 36)
14:	concluding words to Methuselah (cf. 79,1)
Second version:	77 to 79,1
77:	Mythical geography
78,1:	two-division of the year (cf. 78,15,16; 79,4,5)
2-5:	lunar phases; size and brightness of sun and moon (cf. 72,33-37); Gates
6-14:	lunar visibility, waning moon; hollow and full months (cf. 73,4-8 and 74,1-4)
15,16:	two-division of the lunar year (cf. 78,1; 79,4,5)
17:	visibility of the moon during night and daytime
79,1:	concluding words to Methuselah (cf. 76,14)
Additional Fragments:	79,2 to 80,1; 80,4-20 (80,2 to 82,3 intrusion: apocalyptic)
79,2,3:	Gates and lunar phases
4,5:	two-division of the Enoch-year (cf. 78,1,15,16) and Enoch epact
6,80,1:	concluding speech of Uriel
82,4-20:	hierarchy of stars ('thousands'), their leaders during the Enoch year (cf. 75,1).

It seemed tempting to utilize in this commentary to the astronomical chapters of the Book of Enoch the numerous parallels and variants found in

[1] Dillmann, 220.

the Ethiopic 'computus' treatises.[2] Since, however, practically all of these texts are unpublished and since only a detailed study could bring order and relative completeness to this huge mass of material,[3] I have usually abstained from referring to such 'secondary' sources, though they may well contain information more reliable than the Book of Enoch in its present condition. I made good use, however, of the possibility of discussing my interpretations of the text with Professor Ephraim Isaac at the Institute for Advanced Study in Princeton.

In many ways every student of Enoch is indebted to Dillmann's pioneering work. When deviating from it in some technical details, however, I did not find it necessary always to quote Dillmann's translation and notes. In particular I did not refer to all the cases where the mix-up between the 'gates' in the horizon and the modern concepts of orbital motion in the ecliptic produced misleading explanations. The insight into its archaic primitiveness is the key to understanding Enochian 'astronomy.'

CHAPTER 72

(1) Book on the Motion of the Luminaries of the Heaven, how each one of them stands in relation to their number, to their powers and their times, of their names and their origins and their months, as the holy angel Uriel, who is their leader, showed to me when he was with me. And he showed to me their whole description as they are, and for the years of the World to eternity, until the creation will be made anew to last forever.

(2) This is the first law of the luminaries: the light (called) Sun has its exit among the gates of heaven in the east and it sets among the gates of heaven in the west.

(3) And I saw six gates from which the sun rises and six gates where the sun sets; and (also) the moon rises and sets in these gates, as well as the leaders of the stars together with those which they lead. Six (gates) are in the east and six in the west and all of them are arranged in sequence. And there are many windows to the right and to the left of these gates.

(4) And first comes out the great light called Sun and its roundness is as the roundness of heaven and it is all filled with fire that illuminates and heats.

(5) And the chariot in which it rises the winds drive. And the sun goes down from the heaven and it turns toward north in order to travel toward the

---

[2] Cf. for these texts my *Ethiopic Astronomy and Computus*, *Österr. Akad. d. Wiss. Phil.-Hist. Kl. SB* 347 (1979) (hereafter EAC).

[3] The majority of printed catalogues deals only in a very unreliable fashion with treatises of this type.

east; and it is guided in such a way that it enters in the (proper) gate and shines (again) in heaven.

(6) In this way (the sun) emerges in the first month from the great gate, the fourth of these six gates in the east.

(7) And in this fourth gate from which the sun emerges in the first month there are twelve window-openings from which flames come forth when (these windows) are opened in their (proper) times.

a. (8) When the sun rises in the sky it rises from this fourth gate (during) 30 days; and the sun sets exactly in this (fourth) gate in the west. (9) And these days the day increases over the (preceding) day and the night decreases from the (preceding) night during 30 days. (10) And on this (30th) day the day is two ninths, (i.e. two) 'parts', longer than the night, the day being exactly 10 parts and the night exactly 8 parts. (11) And the sun rises from the fourth gate and sets in the fourth (gate).

b. (Then) the sun moves to the fifth gate in the east, for 30 days, and it rises from it and it sets in the fifth gate. (12) And then the day increases two parts and the day amounts to eleven parts and the night decreases and amounts to seven parts.

c. (13) And (the sun) returns to the east and enters the sixth gate and it rises and sets in the sixth gate (during) 31 days according to its (the gate's) characteristics (for the season). (14) And during these days the day increases over the night (until) the day is twice (as long as) the night, such that the day amounts to twelve parts and the night decreases and amounts to six parts. (15) Then the sun sets out to shorten the day and to lengthen the night.

d. And when the sun returns to the east it enters the sixth gate and it rises from it and it sets (in it during) 30 days. (16) And when the 30 days are completed the day has decreased exactly one part and the day is eleven parts and the night is seven parts. (17) And then the sun leaves this sixth gate in the west and

e. travels toward east to rise in the fifth gate (during) 30 days and it sets also in the west in the fifth gate. (18) And on this day the day has decreased two parts and the day is ten parts and the night is eight parts. (19) And the sun rises from the fifth gate and it sets in the fifth gate in the west.

f. (And then) it rises in the fourth (during) 31 days (according to) its (the gate's) characteristics (for the season), and it sets in the west. (20) On this day the day equals the night and they are the same and the night is nine parts and the day is nine parts. (21) And the sun rises from this (fourth) gate and it sets in the west.

g. And (then) it returns to the east and it rises from the third gate (during) 30 days and it sets in the west in the third gate. (22) And on these days the night increases over the day and the night increases over the (preceding) night and the day decreases from the (preceding) day until 30 days and the night is exactly ten parts and the day eight parts. (23) And the sun rises from this third gate and it sets in the west in the third gate.

h. And (then) it returns toward the east and the sun rises (during) 30 days in the second gate in the east and it sets also in the second gate in the western sky. (24) And on this day the night is eleven parts and the day seven parts. (25) And in these days the sun rises from this second gate and sets in the west (also) in the second gate.

i. And (then) it returns to the east to the first gate (during) 31 days and it (also) sets in the first gate in the western sky. (26) And on this day the night has increased to become twice (the length of) the day and the night is exactly 12 parts and the day is 6 parts.

k. (27) [And the sun has (thus) completed its appearances (in all gates) and then returns to these (same) appearances and it rises (again) in all its gates (during) 30 days and it sets opposite to them in the West.]

l. (28) And (during) these days the night has decreased by a ninth part (of its mean length), that is by one part; and the night consists of eleven parts and the day of seven parts.

m. (29) And the sun returns and enters the second gate in the east [and it returns to these appearances] during 30 days, rising and setting (in the second gate). (30) And in these days the night decreases in its length and the night is ten parts and the day eight parts. (31) And in these days the sun rises from the second gate and sets in the west.

n. And (then) it returns to the east and it rises in the third gate (during) 31 days and it sets in the western sky. (32) And in these days the night decreases and it is 9 parts and the day is 9 parts and the night equals the day. [And the year is exactly 364 days (long)].

(33) And the length of the day and the night and the shortness of day and night vary with the circuit of the sun,

(34) because its course becomes longer day after day or shorter night after night.

(35) And this is the rule for the circuit of the sun, when it returns (to the east) and rises (again). This great luminary is called 'sun' for all eternity.

(36) And what rises is the great luminary and it is named according to its appearance as the Lord has commanded.

(37 And it rises and similarly it sets and it does not diminish (in brightness) and it does not rest, but travels day and night. And its light is seven times as bright as the (light of full) moon but with respect to their size the two are equal.

Table I*

[a] ⊙ rises and sets in G_4 for 30^d; d increases, n decreases. Finally $d = n + 2$, thus $d = 10$, $n = 8$. Sunrise and sunset in G_4.

[b] ⊙ returns to E, in G_5 for 30^d; d increases by 2 to $d = 11$, n decreases to $n = 7$.

[c] ⊙ returns to E, enters G_6; rises and sets in G_6 for 31^d, as appropriate: d increases to $d = 2n$, thus $d = 12$, $n = 6$.
⊙ begins to shorten d and lengthen n:

[d] ⊙ returns to E, enters G_6; rises and sets in it for 30^d. After completion of these 30^d, d has decreased by 1, thus $d = 11$, $n = 7$.
⊙ leaves G_6 in the W and

[e] travels to E, to rise and to set in G_5 for 30^d. Finally d has decreased by 2, thus $d = 10$, $n = 8$. Sunrise and sunset in G_5.

[f] ⊙ rises in G_4 for 31^d, as appropriate, sets in W. (Finally) $d = n$, thus $n = 9$, $d = 9$. Sunrise and sunset in G_4.

[g] ⊙ returns to E, rises and sets in G_3 for 30^d. Now n increases over d; n increases, d decreases, until (after) 30^d $n = 10$, $d = 8$.
Sunrise and sunset in the W in G_3.

[h] ⊙ returns to E, rises in G_2 for 30^d, sets in the W in G_2. Finally $n = 11$, $d = 7$. Sunrise and sunset in the W in G_2.

[i] ⊙ returns to E, [rises and] sets in the W in G_1 for 31^d. Finally n has increased to $n = 2d$, thus $n = 12$, $d = 6$.

[k] {⊙ completed its appearances (in all gates), now returns to these appearances, rising in all gates for 30^d and sets in the W.}

[l] [⊙ rises in G_1 for 30^d Finally] n has decreased by 1, thus $n = 11$, $d = 7$.

[m] ⊙ returns to E, [rising] in G_2 returns to appearances, rising and setting for 30^d; n decreases, thus (finally) $n = 10$, $d = 8$. Sunrise and sunset in the W in G_2.

[n] ⊙ returns to E, rising in G_3 for 31^d, setting in the W; n decreases to 9, thus $d = 9$ and $n = d$. {The year is 364^d long.}

* ⊙ ... sun, G ... gate (its number as subscript), E/W ... East/West, d ... day, n ... night { } ... intrusion

Notes to Chapter 72. Sun and Moon

The composition of this chapter is very simple: its core is formed by twelve strictly parallel verses that describe the variation of the length of daylight and night during the year. To this tabulation is added a general introduction (2 to 7) about the six 'gates' on the eastern and western horizon where the sun rises and sets. Similarly the tabulation is followed by some general remarks (33 to 37) about the sun and its role in the universe.

This structure of the chapter has been obscured by dividing the text into twenty-five verses unrelated to the original tabulation. I have therefore compiled a table (opposite) which shows the original pattern. Needless to say, there are many small variations from sentence to sentence. A serious disturbance occurred in verse 28 where a gloss (27) intruded into the text (with a repercussion still visible in 29). But Table I makes it easy to restore the basic scheme for each month:

> 'The sun returns (from the preceding gate) to the east and enters the next gate in which it rises—and sets in the west—for 30 (or 31) days. During that time the days increase/decrease and the nights decrease/increase such that the day becomes... (parts), the night... (parts). Sunrise—and sunset in the west—takes place in this gate'.

Twelve such sentences are the exact equivalent of our Table II or of the graph Fig. 1. Similar verbal presentations of tabular material are not only found frequently in Ethiopic computus texts but also in Aramaic fragments (Milik, Enoch, pp. 278-281). I have no doubt that the same genesis underlies also the next chapter.

Table II

month	in gate	during days,	ending in ... of daylight,	... of night
month 1	in gate 4	during 30 days,	ending in 10^p of daylight,	8^p of night
2	5	30	11	7
3	6	31	12	6
4	6	30	11	7
5	5	30	10	8
6	4	31	9	9
7	3	30	8	10
8	2	30	7	11
9	1	31	6	12
10	1	30	7	11
11	2	30	8	10
12	3	31	9	9

(1) Preamble, giving a summary of topics concerning the celestial luminaries. The expression ḥezabihomu, literally 'their tribes, populations', obviously refers to the hierarchical grouping of the stars. Similarly šelṭānomu means 'their powers', exercised by the stars over the division of the year, the seasons and the epagomenal days. Cf. for all these influences 75,1-7 and 82,4-20.

(2,3) In the course of the year sun and moon rise and set in six 'gates' on the eastern, respectively western, horizon. To the right and to the left of these gates are 'windows', presumably for the stars (whereas in 36,2,3 'small gates' are assigned to them). It should be noted that 'right' and 'left' are not the same as 'south' and 'north' since these associations are reversed with the change of direction of the observer.

Verse 3. 'Arranged in sequence' probably refers to the numbering of the gates from one to six or from south to north.

(4) The 'roundness' (kebabu) of the sun corresponds to the roundness of the heavenly cupola. Neither 'Umkreis' (Dillmann) nor 'disc' (Knibb) are suitable descriptions of the sky.

(5) Here we are told that the chariot of the sun (and of the moon, cf. 72,2) is blown by winds, and that the sun, after setting in the west, returns via the north to the east. Independent of this motion of the luminaries is the (daily) rotation of the heaven, i.e. of the stars, which is also caused by winds (18,4). A slightly different picture is found in 41,5 where the sun and moon are said to come from, and return to, 'chambers' (mazāgebt). Similar differences are recognizable for the stars: 'windows' in 72,2,3, 'gates' in 36,2,3.

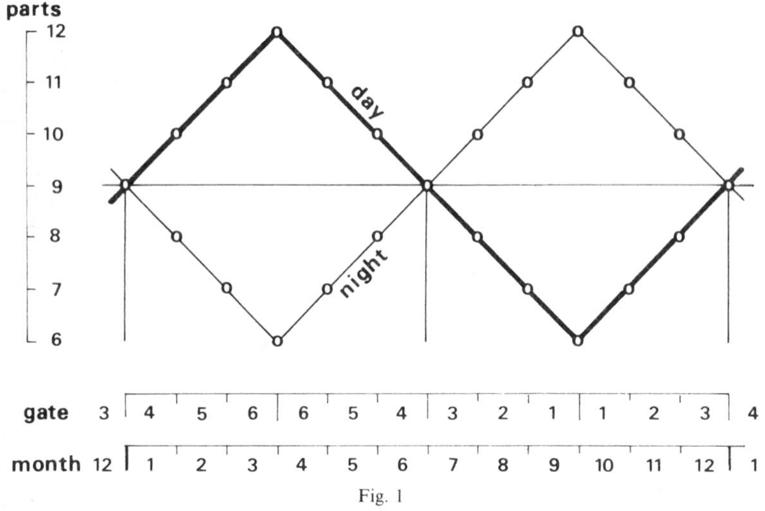

Fig. 1

(6 to 32) The essential content of these verses is summarized in Table II and in the graph of Fig. 1. Note that the linear pattern for the length of daylight and night ignores the epagomenal days (as is admitted in 75,1).

The variation of the length of daylight and night during the solar year is here described by a numerical sequence that alternates, with constant difference, between a maximum M and a minimum m. If one chooses the simplest increment, i.e. if one makes the monthly difference '1 part' (p), then one has $M = m + 6$. If one furthermore assumes that $M = 2m$, then one finds finally $m = 6^{p}$, $M = 12^{p}$ (and always daylight + night $= 18^{p}$). This is exactly what we have in our text.

The use of such an alternating sequence (known as 'linear zigzag function') suggests a Babylonian origin, since functions of this type play a fundamental role in Babylonian astronomy. For the length of daylight we find in cuneiform texts two ratios: one $M{:}m = 3{:}2$, the other (in earlier texts, e.g. in the series 'Mul-Apin') $M{:}m = 2{:}1$. In the first case the units of time are 'large hours' (i.e. 4 of our hours), in the second case we deal with 'manas', i.e. weights of water, outflowing from a cylindrical water clock.[4]

[4] Cf. Neugebauer, The Water Clock in Babylonian Astronomy, Isis 37 (1947), pp. 37-43.

In our texts the 'parts' are never connected with any meteorological unit, neither hours, nor weights or volumes. Hence borrowing from Mesopotamia remains only a possibility, though supported by another (slightly conjectural) feature, an 18-division of the day (counted in manas) in Mul-Apin.[5]

Of course a borrowing from comparatively early Babylonian material cannot be used as a chronological criterion for the time of composition of the astronomical section of the Book of Enoch. Methods of this kind have a life-span of many centuries and easily survive the development of more advanced methods.

(6 to 11) In month 1 the sun emerges from gate 4 and the length of daylight increases during this month from 9^p to 10^p. Note that this implies that the year begins at the vernal equinox (as in Babylonia) whereas the Ethiopic calendar follows the Alexandrian year that begins with the month Thoth, roughly September.

Why is gate 4 called 'large'? Dillmann[6] thinks of a comparison with the 12 windows which eject flames (72,7). Are other gates not provided with such windows?

Verse 11. After 30 days the sun returns (yegabbe') from gate 4 in the west (via the north) to gate 5 in the east.

(13 and 19) The translation of te'merta zi'ahā as 'its sign' is misleading since it could be taken as a reference to zodiacal signs[7] (which do not exist in Enoch's astronomy). The purpose of this remark, however, is to explain that 31 days of the sun's risings in the same gate is indicative for the position of the equinoxes and solstices. The Greek equivalent of te'mert is σημεῖον used in a technical sense,[8] in particular in relation to meteorological and calendric dates (ἐπισημαίνειν).[9] Hence we may say that the rising of the sun is a specific gate is 'indicative' or 'characteristic' for the seasons. Cf. also 75,6 and 82,16 and 19.

Verse 19. As Fig. 1 (p. 394) shows, the autumnal equinox occurs when the sun rises at the beginning of gate 3 in month 7.

(27) Dillmann rendered 'ar'estihu by 'Bahnabschnitte' (hence Knibb: 'division of journey'—instead of 'orbital segments'). It seems to me, however, that no reference to the sun's 'orbit'[10]—at any rate a much too modern concept—is intended. In my opinion what is meant is simply the appearances (literally the 'heads', the 'beginnings') in the consecutive gates. This interpretation is supported by a variant in Ṭānā 9: 'ar'ayāhu, indicating something like 'appearances'. Cf. also next section.

(33 to 35) The variability of the length of daylight and night is caused by the variability of the sun's positions in the gates (cf. verse 27), i.e. by the changing rising amplitude, as is indeed the case.

A number '60' (of risings and settings?) in verse 35 is omitted in several manuscripts (among them Ṭānā 9). This seems to be the better version.

(37) The sun's brightness never changes (in contrast to the moon) and it is sevenfold

[5] This connection between Mul-Apin and the Book of Enoch was suggested many years ago by A. J. Sachs (cf. Neugebauer, *l.c* note 4, p. 40).

[6] Dillmann, 222.

[7] So expressly by Charles.

[8] Cf. e.g., also Matth. 16,3 σημεῖα τῶν καιρῶν 'the signs of the time'.

[9] Cf. e.g., Ptolemy's work Φάσεις ἀπλανῶν ἀστέρων καὶ συναγωγὴ ἐπισημασιῶν (Opera minora, pp. 2-67). Cf. also RE Suppl. 7 col. 176-198 [Rehm].

[10] Not to ask: the daily orbit? the yearly orbit? what are 'Bahnabschnitte'?

the moon's greatest brightness[11] (cf. 73,3 and 78,4). Their apparent sizes, however, are equal (cf. 78,3).[12]

CHAPTER 73

(1) And after the rule (concerning the sun) I saw another rule about the smaller luminary, called Moon.

(2) And its roundness is as the roundness of heaven and the chariot on which it travels is driven by winds; and light is given to it in measure.

(3) And each month its place of rising and of setting varies (through all gates) but its days are as the days of the sun. And when its light is evenly spread (over its disc) then it amounts to one seventh of the light of the sun.

(4) And thus (the lunar month), begins, when (the moon) itself moves away (from the sun) toward east on the 30th day, and (when) on this day it becomes visible, it is for you the beginning of the (lunar) month on the thirtieth day, (when the moon is setting) together with the sun in the gate from which the sun rises, (5) (but) at a distance (from the sun) of half of a seventh part.

And its whole disc is empty (i.e.) without light, excepting its seventh part of a fourteenth part (i.e. 1/98) of the light (of the sun).

(6) And on (this) day (the moon) takes on a seventh part of one half (i.e. 1/14) of its light, and (thus) its light is the seventh of a seventh part and one half of it (i.e. 1/98 of the light of the sun).

(7) (The moon) sets with the sun and when the sun rises, the moon rises with it and it takes on one half part (of 1/7) of its light. And in this night, at the beginning of the (lunar) day, which is the first day of the month, the moon sets with the sun, and it is dark in this night.—A seventh of a seventh part and one half.

(8) And the moon rises and comes out on this day with exactly the seventh part (of its total light) and recedes from the rising of the sun and it (the moon) is illuminated during the remaining (part) of its day a sixth(?) and a seventh part (of the light of the sun).

---

[11] Dillmann, 226 suggests a derivation of this ratio from Isaiah 30.26. This passage (and similarly Enoch 91,16), however, does not compare the sun with the moon but deals with some future events in the universe.

[12] This is very nearly correct, as is common knowledge in ancient astronomy (based on evidence from solar eclipses).

Notes to Chapter 73. The Moon's Variable Illumination

The original arrangement in this chapter was probably similar to the arrangement in the preceding chapter: a central tabulation preceded (and perhaps also followed) by general remarks. In the extant version, however, only the introduction is preserved (verses 1 to 3) while the tabulation breaks off after day 2. No doubt originally all days until full moon (day 14) had been listed. Instead we find now a disorganized chapter (74) which obviously does not belong to the original composition.

(1 to 3) The moon's chariot is driven by winds (cf. 72,5) and the roundness of the moon is as the roundness of the heavens.[13] Light is given to the moon (from the sun) and produces at full moon one-seventh of the sun's brightness (cf. 72,37 and 78,4). The rising and setting points of the moon change rapidly, but the number of 'days' in a lunar calendar is the same as the corresponding number of solar days (73,3); for example, day 14 has the same distance from day 1 in a lunar calendar as a solar day 14 from solar day 1, in spite of the variability of the moments of moon-rise and moon-set in relation to sun-set.—For 73,2 cf. also 78,4.

(4 to 8) In these verses we have a fragmentary description of a linear scheme for the increasing illumination of the moon during the first half of the lunar month. This increase is expressed in two scales: first, in absolute terms from 1^p to 14^p (hence proportional to the illuminated area),[14] and, secondly, in terms of solar brightness, hence increasing from $1/14 \cdot 1/7 = 1/98$ on the first day to 1/7 at full moon (cf. 73,3). Our text represents only a fragment of this scheme, which concerns the first two days. But the whole scheme is preserved in several computus texts (cf. EAC, p. 196) the only difference being that a 'full month', i.e. a 15-day increase, is contemplated.

Some trouble has been caused by an unfortunate terminology used in this section: the term ṣebāḥ 'morning' here stands for 'day' (as we sometimes count 'summers' as 'years', or 'winters' (keramt) in Ethiopic). To retain in astronomical context the literal meaning of an idiom of this type leads to senseless translations; e.g., 'on that night at the beginning of its morning, at the beginning of the moon's day',[15] instead of 'on that night, at the beginning of the (lunar) day, which is the first day of the month'.

(4) To say that a lunar month begins on 'day 30' characterizes its predecessor as a hollow month.[16] At the beginning of the new month the moon has obtained enough (easterly) elongation from the sun to be visible at sunset. At conjunction, however, the moon is still nearer to the sun and thus rises and sets (invisibly) in the same gate as the sun. (Of course, all this is only thus simple in the schematic lunar calendar which ignores, of necessity, all complexities of the actual lunar motion).

The conclusion of this verse is incorrectly assigned in part to the next verse. It contains the statement that the elongation of the moon from the sun at the evening of first visibility is 1/14 of the total elongation (reached at full moon). The use of reḫeqa in the technical sense of 'elongation' is well attested in computus treatises. We read, e.g., in BM Add 24995 (28^a II,1): 'On the second day (the moon) recedes (yereḫḫeq) from the sun and becomes visible at 8 kekros and illuminates 2 parts of 15 (of its

---

[13] Knibb's MS has 'sun' instead of 'heaven'. The parallel with 72,4 shows that 'heaven' is the better version.

[14] The 'parts' ('eda) here have nothing to do with the 'parts' (kefla) in 72,6 to 32.

[15] Knibb, 2, 172 (73,7). Cf. also *Gen.* 50,3 and *Num.* 13,25 in the Ethiopic Bible (E. Isaac). 'Mornings' for 'days' is also well attested in computus texts.

[16] This is standard terminology in Babylonian astronomy, cf. F. X. Kugler, *Die Babylonische Mondrechnung*, Freiburg 1900, p. 36.

greatest) light (at full moon) and 1 part of 98 (parts) of the light of the sun'. Incidentally, this close parallelism supports our conclusion that verses 73,4 to 8 are only a fragment of a complete table for the moon's illumination, both absolute and in relation to the sun.

(5) The numerical data for day 1 are: darkness of the moon's disc excepting 1/2·1/7 of its area that shines with the brightness of 1/7·1/14 (= 1/98) of the sun's light.

(6 to 8) The numbers in these verses are obviously corrupt as the many variants show, in part probably caused by the usual confusion of sixes and sevens. Both translation and notes are therefore only tentative and show not much more than that we are dealing with the description of the moon's increasing illumination. The text ends abruptly after verse 8.

(6) The daily increment of the moon's illuminated area is 1/14. Its brightness on the first day of the lunar month amounts to 1/7·1/7·1/2 (= 1/98) of the sun's light.

(7 and 8) On day 1 the moon is still near conjunction and therefore (nearly) rises and sets at the same time as the sun (cf. 73,4). The number 1/7·1/7·1/2 at the end of verse 7 is perhaps a meaningless duplication from verse 6.

Turning to day 2 (in verse 8) the moon's illuminated area is 2/14 = 1/7. It follows again a remark about the increasing elongation, but one should expect a motion 'away from the sun toward east' (as in verse 4) instead of a 'receding from the rising sun'. Perhaps this is simply a scribal error. Why the 'remaining part of the day' is mentioned in the present context I do not know. For the brightness of the crescent on day 2 one should expect 1/7·1/7 (of the sun's brightness) and, indeed, some variants contain these numbers.

CHAPTER 74

(1) And I saw another circuit and (another) rule for it (the moon), whereby according to that rule it produces the cycle of the months.

(2) All this showed to me Uriel, the holy angel, who is the leader of all of them. And I wrote down their positions as he showed them to me and I wrote down their respective months and the phases of their illumination until full moon on the fifteenth day.

(3) And in steps (of fractions) of sevenths (lit. single seventh parts) the full moon is completed in the east and in steps (of fractions) of sevenths complete darkness is reached in the west.

(4) In certain months (the moon) changes (the location of) its settings (with the sun, but) in certain months it goes its own individual way.

(5) In two months (the moon) sets with the sun in these two middle gates, that is in the third and fourth gate.

(6) (The moon) comes out (from the same gate) during seven days and it turns and moves back to the gate from which the sun rises, and it completes its light. And (the moon) recedes from the sun and enters for eight days the sixth gate from which the sun rises.

(7) And when the sun rises from the fourth gate (the moon) comes out

(from the sixth gate) during seven days until it rises from the fifth (gate) and it returns again during seven days to the fourth gate and it completes its light and it recedes (from the sun) and it enters the first gate (during) eight days.

(8) And again it returns (after) seven days to the fourth gate from which the sun rises.

(9) Thus I saw their positions when the months begin at sunset.—It seems pointless to attempt to give an accurate translation of the confused nonsense which some scribes produced from some trivial arithmetical relations (for which cf. the notes on p. 401). Readers who wish to see some rendering of these scrambled verses may look up Charles, 149-161 or Knibb, 173-4.

(10) We are dealing with five ('solar') years of 364 days each.

(11) Five lunar years fall short of five solar or sidereal years by (50 days, similarly three lunar years by) 30 days.

(12) In this way the length of the lunar years is not too long and not too short by a single day in all eternity in relation to the years of 364 days each.

(13) Three years are 1092 days long, five years 1820 days, thus eight years 1912 days.

(14 to 16) Three lunar years are 1062 days long, thus 30 shorter than three solar years. Similarly for five and eight years.

(17) And the year is correctly completed in relation to its position within (the era of) the World and to the positions of the sun that rises and sets in its gates for 30 days (each).

Notes to Chapter 74. The Lunar Year

This chapter contains a fragmentary description of the shift from gate to gate of sun and moon, based on a simple arithmetical scheme that is well known from computus treatises.[17] The present text, however, covers only the discussion for the first month. The remaining tabulation is replaced by a badly bungled attempt to describe an octaeteris that would relate a lunar year to the Enoch-year. It is quite evident that these verse (10 to 17) are a later addition.

(1 and 2) A reference to the angel Uriel supports our impression that this chapter was originally not connected with the preceding or following chapter. Also full moon is here associated with day 15, not with day 14, as in chapter 73 (but cf. 78,6,7).

(2 to 4) The text as it stands is not very clear. What was intended to be expressed may be formulated as follows: Enoch writes down the pattern for the gates traversed by sun and moon during the lunar year. In each month the moon is waxing and waning: first its light increases until 1/7 of the sun's brightness is reached at full moon, visible in the east when the sun sets in the west; then the moon returns to darkness at conjunction which normally takes place in the same gate with the sun, though occasionally the moon may appear in an adjacent gate (as can actually be the case).

[17] Cf. Table III (18), taken from EAC, p. 160. The positions of the numbers are not rigidly the same in all manuscripts; 7 and 8 as well as 1 and 2 can interchange places, as long as the proper totals 29 or 30 are preserved.

(5) The gates 3 and 4 correspond to the equinoxes (cf. Fig. 1, p. 394).

(6) We have here a general description of the relationship between the days of a lunar month and the gates: the moon comes out through one of the outermost gates during seven (or eight) days; there it turns and moves back to the gate from which the sun rises during this month, and its light becomes full (at sunset); then the moon recedes again from this gate.

At the end of this verse the words 'enters for eight days the sixth gate' do not belong here and should be deleted.

Table III

Months	1	2	3	4	5	6	7	8	9	10	11	12	1	Months
Gates														Gates
4	2													4
5	2	2												5
6	8	8	4	4										6
5	2	2	2	2	2									5
4	1	1	2	2	1	2								4
3	1	1	1	1	1	1	2							3
2	2	2	2	2	2	2	2	2						2
1	8	7	8	7	8	7	8	7	4	4				1
2	2	2	2	2	2	2	2	2	2	2	2			2
3	1	1	1	1	1	1	1	1	2	2	1	2		3
4	1	1	2	2	1	1	1	1	1	1	1	1	2	4
5		2	2	2	2	2	2	2	2	2	2	2	2	5
6			4	4	8	8	8	8	8	7	8	8	8	6
5					2	2	2	2	2	2	2	2	2	5
4						1	1	1	1	1	1	1	1	4
3							1	2	2	2	1	1	1	3
2								1	2	2	2	2	2	2
1									4	4	8	7	8	1
2											2	2	2	2
3												1	1	3
4													1	4
Days	30	29	30	29	30	29	30	29	30	29	30	29	30	Days

(7 to 9) The tabulation begins with month 1 at the vernal equinox. Conjunction takes place in gate 4, then the moon's rising and setting moves on to gate 6 for a period of seven days. Moving back[18] to gate 4 we have full moon, followed by another delay (of 8 days) in gate 1. Thus the complete scheme for this month would look about as follows (cf. also Table III):

gates:	4	5	6	5	4	3	2	1	2	3	4	
	sun				sun						sun	
during days:	[2]	[2]	7	[2]	[1]	[1]	[2]	8	[2]	[1]	[1]	total: 29

The continuation of this tabulation is omitted just as in the preceding chapters.

(10 to 17) These verses constitute an abortive attempt to describe an octaeteris. The scribe had obviously only a very vague idea of the working of such a cycle, remembering only a separation of 8 years into two groups, one of 5 years (with 2

[18] The assignment of seven days to the return to gate 4 (instead of 2 days) is a scribal error, perhaps caused by a similar passage in verse 8.

intercalary full months—hence his 30 days in 74,11), and one of 3 years (with 1 intercalary month).[19] However instead of operating with Alexandrian years he assumes Enoch years; and because this does not lead to any reasonable relationship, he ends up with some correct but irrelevant numerical identities, based on the comparison of 5 + 3 Enoch years with 5 + 3 lunar years:

$5 \cdot 364 = 1820$ days	$5 \cdot 354^{d} = 1770 = 1820 - 50$ days
$3 \cdot 364 = 1092$ days	$3 \cdot 354^{d} = 1092 = 1820 - 30$ days
$8 \cdot 364 = 2912$ days	$8 \cdot 354^{d} = 2832 = 2912 - 80$ days

I surmise that this whole group of verses is a late addition, written under the influence of some computus treatise, where a mix-up of Alexandrian and Enoch years is quite common.

CHAPTER 75

(1) And their leaders, at the head of (each) thousand (stars), who are appointed (to rule) over the whole creation and over all stars (have to do also) with the four additional (days), without deviating from their positions, corresponding to the computus of the year. And they render service (also) on these four days which are not counted in the computus of the year.

(2) And with respect to these (four days) people err since these luminaries do true service (also) in the (following) positions of the cosmos: once in the first gate and once in the third gate and once in the fourth gate and once in the sixth gate, so that the accuracy (of return) of the world is achieved after 364 (days) (with respect to the) positions of the cosmos.

(3) Thus the signs, the times, the years, and the days were shown to me by the angel Uriel whom the eternal Lord of glory has appointed (to rule) over all the heavenly luminaries in heaven and in the world, such that they rule on the face of the sky and are seen from the earth and are made the guides of day and night, (namely) the sun and the moon and the stars and all the servants who return on all chariots of heaven.

(4) Likewise Uriel showed me twelve openings, openings in the disc of the chariots of the sun in the sky, from which come forth over the earth the rays of the sun and its heat when they are opened at the proper time.

(5) And (there are openings) for the winds and for the wind (that brings) dew, when the openings of heaven are opened at the boundaries (of the earth).

(6) I have seen twelve gates in the heaven at the boundaries of the earth from which come out sun and moon and stars, and all the works of heaven from the east and from the west.

(7) And (I saw) many window openings to the right and to the left and

---

[19] Cf. EAC, p. 83ff.

(each) one window emits heat at its time according to the gates from which the stars rise, as they are ordered, and in which they set according to their (the gates') numbers.

(8) And I saw chariots in the heaven travelling in the world above the gates, where the stars revolve which never set.

(9) And one of these (circuits) is larger than all of them and it is one which circles the whole world.

Notes to Chapter 75. The Stars

(1,2) The stars convey cosmic order to the calendar by their organization, which agrees exactly with the divisions of the Enoch year (ḥasāba ʿāmat), including the epagomenal days at the end of each season. Some people commit an error by ignoring these epagomenal days; this could refer to the lunar calendar of the Jews (which has no intercalary days), or to the Egyptian calendar (with five epagomenal days at the end of the year), or even to the schematic year of 'Mul-Apin', which contains only twelve 30-day months.

In fact, however, the epagomenal days are 'not counted in the computus of the year' since it would disturb the linearity of the scheme for the variation of the length of daylight (cf. note to 72,6 to 32, p. 394). This admission of a contradiction between theory and practice is obviously due to a gloss that intruded into the text.

The divisions between the seasons are marked by the rising of the sun in one of the following gates: winter solstice in gate 1, the equinoxes in gates 3 and 4 (autumnal and vernal equinox respectively, as is seen from the trend in the variation of the length of daylight—cf. 72,6 to 32, p. 394, Fig. 1), the summer solstice in gate 6. Cf. also 82,6 and 74,5. At these points the cosmos returns accurately to its previous position, 364 days earlier. Hence the (assumed) symmetry of the seasons of the solar year is taken as the ultimate basis for the calendar, and the stars reflect the same order.

(3) The angel Uriel shows to Enoch all the things 'about the signs and about the times' (late'mert wala'azmān); this could refer to the role of the stars as indicators of the climatic changes from season to season (cf. above, 72,13 p. 12).

Sun, moon, and stars move 'on the face of heaven' and thus are visible from the earth. The 'servants[20] who return (yaʿawedu—not, in this case, 'revolve') on all chariots of heaven' are perhaps responsible for the return of the celestial bodies from their settings in the west to the eastern gates via the north (cf. 72,5; 78,5).

(4) Once more Uriel explains the purpose of the gates and of the chariots: 'twelve openings in the disc of the chariot of the sun ... from which the rays of the sun ... and heat ... come out'. This picture has no parallel in the rest of the text and I suspect some confusion with the gates traversed by the sun (cf. also verse 7) or with the twelve windows from which flames are ejected (72,7).

(5) Probably an intrusion,[21] in part duplicating verse 6.

(6, 7) Again 12 gates, east and west, but now not only for sun and moon but also for the stars. This makes little sense since the stars rise in all points of the eastern horizon. Then there are 'windows' to the right and to the left (cf. 72,3,7) from which

[20] Dillmann's 'dienstbare Geschöpfe' (in his time an idiom reminiscent of household help) became 'serving creatures' in Knibb's translation.

heat comes out — a moment before (in 75,4) the openings in the solar chariot performed this function — and also stars.

In verse 6 the 'works of heaven' (gebrāta samāy) probably means the meteorological phenomena connected with the seasons. The 'numbers' in verse 7 probably refer to the numbering of the gates, thus guaranteeing the proper positions of risings and settings.

(8, 9) There are chariots (obviously for stars) 'above[22] the gates' for those stars which never set, i.e. circumpolar stars. One of their circuits is the greatest, encircling the whole (always visible) world.[23]

CHAPTER 76

(1) And I saw at the boundaries of the earth twelve gates, open to all winds, from where the winds come out and blow over the earth.

(2) Three of them are open at the front of heaven (i.e. in the east) and three in the west, and three at the right side of heaven, and three on the left side.

(3) And the three first ones are in the direction of east, and (then) three are in the direction of north, and then those on the left, in the direction of south, and three in the west.

(4) From four of them winds of blessing and prosperity come out, (but) through eight of them come winds (causing) calamities; when they are sent out they bring devastation over the whole earth and the water on it, and to all that inhabit it, to all that is in the water or on dry land.

(5) And the first wind that comes out from these gates is called easterly. From the first gate in the direction of east, inclined toward south, devastation, drought, and heat and destruction come out.

(6) And in the second gate, the middle one, (the wind) comes out straight; and from it rain and fruitfulness and prosperity and dew come out. And from the third gate, in the direction toward north, cold and drought come out.

(7) And then the winds in the direction from south come out from three gates. First, from the first gate, that is inclined toward east, a hot wind comes out.

(8) And from the middle gate, next to it, beautiful fragrance and dew and rain and prosperity and health come out.

(9) And from the third gate, in the direction toward west, dew and rain and locusts and devastation come out.

---

[21] Cf. Dillmann, 233/4.

[22] Some manuscripts add here 'and below them', which makes no sense. Unfortunately Knibb accepted this version (following Dillmann but not Flemming).

[23] Dillmann's 'durchkreutzt die ganze Welt' (hence Knibb's 'goes round through the whole world') is senseless. Obviously Dillmann was not familiar with the concept 'greatest always visible circle'. In his notes (234) he even considers the 'Morgenstern' or the Great Bear. In Greek astronomy this circle is known as the 'arctic circle'.

(10) And then the winds in the direction from north, (also) called baḥr (Sea), ... From the seventh gate, (inclined) toward east, dew and rain, locusts and devastation come out.[24]

(11) And from the middle gate, in a straight direction, health and rain and dew and prosperity come out. And from the third gate, (inclined) toward west, mist and hoar-frost and snow and rain and dew and locusts come out.

(12) And then the fourth (group of) winds, in the direction toward west: from the first gate, in the direction of north, dew and rain and hoar-frost and cold and snow and frost come out.

(13) And from the middle gate dew and rain, prosperity and blessing come out. And from the next gate, in the direction toward south, drought and devastation, burning and destruction come out from it.

(14) And thus (the description of) the twelve gates in the four (quarters) of heaven is completed; and I have shown to you, my son Methuselah, all their laws, (and all their) calamities and benefactions.

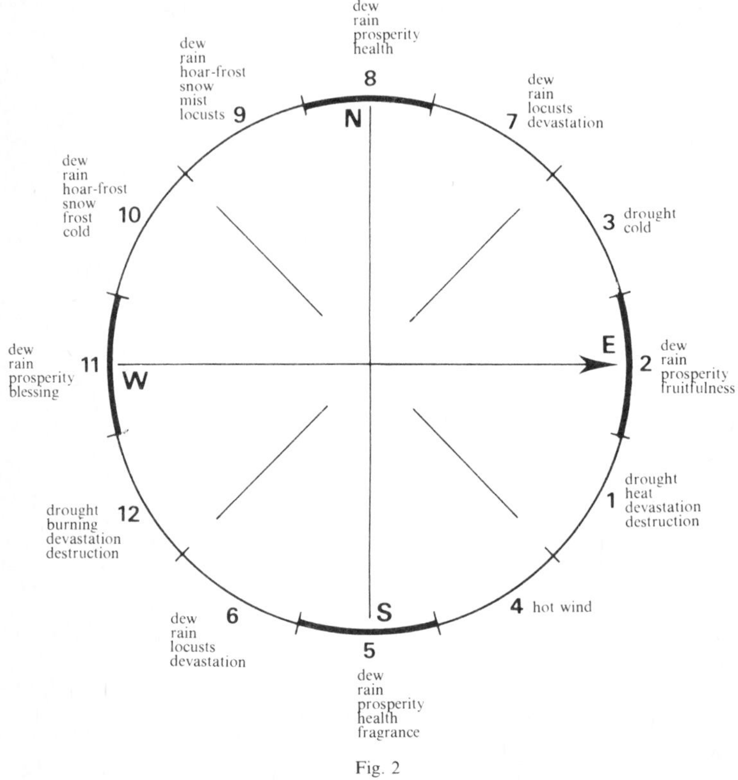

Fig. 2

[24] The text of this verse is corrupt but there is no doubt about the essential points, the order of the gates and the quality of the winds. A remark 'inclined toward south' (and similarly

Notes to Chapter 76. The Winds

The Ethiopian rose of winds consists of a sequence of twelve openings, again called 'gates', which encircle the whole horizon. The winds from the cardinal directions are supposed to be beneficial, in contrast to the winds from the remaining eight gates that bring discomfort and devastation (cf. Fig. 2).[25] Lists of this type are also found in many Ethiopic 'computus' treatises. An abridged version is preserved in 34,2 to 36,1 of the Book of Enoch.

One might think that long experience with climatic conditions would be condensed in such lists. In fact, however, we have her again only a schematic pattern, as far removed from empirical data as the arithmetical schemes for the length of daylight or the shadow tables. In all these cases scribal tradition has wiped out any connection with reality, if one ever existed.

The purely schematic character of qualities enumerated in the present section is easily recognizable in spite of some omissions or additions.[26] This is quite obvious in the case of winds from the cardinal directions. All of them bring 'dew, rain, prosperity' to which is added one more gift:

E: fruitfulness S and N: health W: blessing.

Only the southern wind has a fifth quality ('fragrance').

This list for the principal wind-directions strongly suggests that each wind should be associated with exactly four qualities. This is indeed confirmed for the destructive winds, listed here in the order of the text:

E 1 drought, heat, devastation, destruction
  3 drought, cold
S 4 hot wind
  6 dew, rain, locusts, devastation
N 7 dew, rain, locusts, devastation
  9 dew, rain, hoar-frost, snow (mist, locusts)
W 10 dew, rain, hoar-frost, snow (frost, cold)
  12 drought, burning, devastation, destruction.

Excepting E 3 and S 4 and the rather senseless additions (shown in parentheses) in N 9 and W 10, we always have exactly four qualities mentioned. But this list reveals one more structural pattern. The third wind in one group has the same qualities as the first wind in the next: S 6 = N 7, N 9 = W 10, W 12 = E 1, thus closing the cycle. Only E 3 and S 4 are exceptions, which is not surprising since both entries are obviously defective.

Summarizing these regularities we can now say that the twelve-wind arrangement contains four beneficial gates and only four qualities for the remaining eight gates. This suggests an historical evolution from an 8-point rose of winds to a 12-point arrangement, the latter probably recommending itself by the formal similarity to the 6 + 6 'gates' for the risings and settings of sun and moon. Both eight-division and

'... toward north' in the next verse) makes no sense. The cause of all this trouble is probably the replacement of the eight-point rose of winds by twelve points, i.e. the duplication of the intermediate directions. Cf. the commentary (p. 405).

[25] The accuracy of the 12-division of the horizon as shown in Fig. 2 should not be taken seriously. No numerical data are ever associated in our texts with these 'gates'.

[26] The order of the qualities listed in Fig. 2 for each individual wind can differ from the order (or rather disorder, which tends to obscure parallelisms) in the text.

twelve-division are well known in hellenistic and Roman schemes, e.g., in geographical or architectural context.[27]

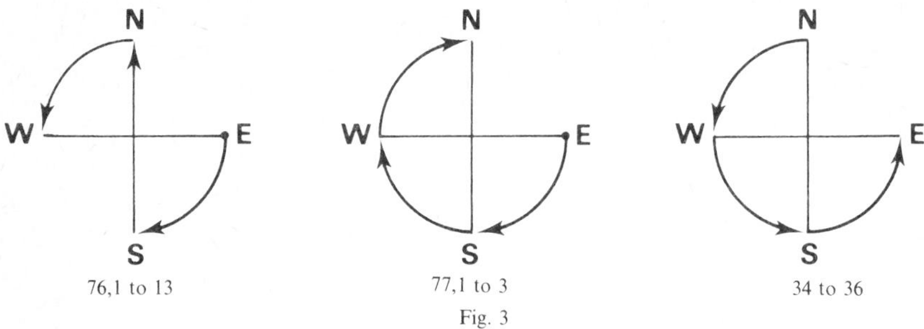

76,1 to 13 77,1 to 3 34 to 36

Fig. 3

Finally, a remark should be made about the order in the text of enumerating the gates for the winds. Since little consistency is found in computus texts in matters of orientation, it is not surprising to meet the same situation in the present treatise. Fig. 3 illustrates two instances from the 'astronomical' book (Ch. 76 and 77) and compares them to the short version in Ch. 34 to 36. In 76 and 77 the speaker faces east, the 'front of heaven' (gaṣṣa samāy), thus north is 'left' and south is 'right'. In 34 to 36, however, the enumeration starts with north and proceeds counter-clockwise. In the astronomical sections the enumeration is either clockwise or E-S-N-W, as shown in Fig. 3, revealing one of the differences between the two versions to which 76 and 77, respectively, belong.

On the other hand the existence of an overall common background is visible in the fact that the only qualities associated with a wind-direction in 34 to 36 are 'dew, rain, hoar-frost, snow (hail)', ascribed in 34,2, to the north (in general) and in our chapter to N 9 = W 10 (cf. above).

Concluding words to Methuselah (76,14) mark the end of the first version of the 'astronomical' book, similar to 79,1 for the second version.

Verse 7 is corrupt, as is evident from the restriction to only one quality (heat = moq, in some MSS misread to mot = death). Several computus treatises have here nafās meweq zasemu netug (or natig[28]), 'hot wind called netug/natig'. A similar gloss disturbed verse 10 by giving the north wind a special name, bāḥr = sea or ocean. With bāḥr is commonly associated as its counterpart the wind libā, i.e. λίψ a southerly wind. Perhaps netug is a substitution for λίψ and is derived from νότος, the south wind.

In 77,2 netug is assigned to the west and derived from the meaning 'diminish.' I suspect, however, that the transformation of this wind from the SE to the W is caused by this etymology, rather than explaining it. The shift from the eight-point rose of winds (still reflected in the restriction to only eight names—cf. EAC, p. 199, Fig. 4) to the twelve-point scheme (above Fig. 2) could only augment the confusion of directional terminology.

In fact the name bāḥr, 'north', being unrelated to any great sea, may also be the result of a learned scribal interpretation of βόρεας. All these explanations are glosses, introduced by zasemu 'which is called ...', based on the same principle of assimilation of foreign words to semitic roots, exemplified by the change of τῶν θεῶν to ṭentyon and explained by the scribes as meaning 'at the beginning'.

[27] Cf. (with caution) RE 8A, 2 cols. 2351f. and 2378 for the 12-division and col. 2364 for 8-division [Böker].

[28] Cf. Littmann, Zeitschr. f. Assyriol. 16 (1902), p. 384.

CHAPTER 77

(1) The first quarter is called east because it is the chief (quarter); and the second called south because there the Most High descends {and there in particular descends the one blessed in eternity}[29].

(2) And the quarter in the west is called netug (diminished) because all celestial luminaries decrease there and go down.

(3) And the fourth quarter which is called north is divided into three parts. The first of them is the habitat for man, the second (contains) the oceans and gorges and forests and rivers, darkness and mist. The third part (contains) the Garden of Justice.

(4) I saw seven high mountains, higher than all mountains on the earth, and hoar-frost comes from them. {And days and seasons and years traverse them.}[29]

(5) I saw seven rivers on the earth, greater than all other rivers; one of them, coming from the west, sheds its waters into the Great Sea.

(6) And two (of them) come from the north to the sea and shed their waters into the Erythrean Sea in the east.

(7) And the remaining four (rivers) come from the northern side toward the sea, two to the Erythrean Sea, and two empty into the Great Sea—and (some) say: into the desert.

(8) I (also) saw seven large islands in the sea and on land; two on land and five in the Great Sea.

Notes to Chapter 77. Mystical Geography

(1 to 3) The cardinal directions are here enumerated in the order E-S-W-N, in contrast to 76 and 33 to 36 (cf. Fig. 3, p. 406). As in 76,2 we are facing east. For the wind netug, cf. the note to 76,7 (p. 406).

The description in verse 1 of the south as 'where the Most High descends' is explained in an Amharic (unpublished) commentary as a reference to Mt. Sinai (communication by Prof. Ephraim Isaac). In verse 3 one might expect the human habitat to be located between ocean and paradise, not to the west of the ocean.

(4 to 8) In verse 4 the traditional translation of yaḫalef[30] as 'schwinden dahin' (Dillmann, Flemming), 'pass away and vanish' (Charles), makes no sense in the present context. I therefore suggests a translation 'days and seasons and years traverse (them)", meaning that the sun, the cause of days, and years, traverses the space above the mountains.

The tendency of this geography is too mythological to allow accurate identifications. It seems plausible, however, to take the Erythrean Sea for the Persian Gulf into

---

[29] Probably a gloss.

[30] Some MSS (but not Ṭānā 9) have here wayaḥawwer: 'and proceed,' 'go', etc.

which the Euphrates and Tigris empty their waters. The two islands 'on land' (verse 8) could be land between rivers, as Dillmann suggested (238), Mesopotamia and Meroe. Milik assumed[31] a Greek version meaning 'near (ἐπὶ) land.'

CHAPTER 78

(1) The names for the sun are as follows: the first 'Oryārī and the second Tomās.

(2) The moon has four names: the first name is 'Asonyā, the second 'Eblā, the third Benāsē, and the fourth 'Erā'e.

(3) These are the two great luminaries. Their roundness is as the roundness of the heavens and the amount of the roundness of the two is the same.

(4) In the disc of the sun is combined seven times what is of light in the moon, and according to measure (light) is injected (from the sun into the moon) until the seventh part (of the light of the sun) has been transmitted.

(5) And they set and they enter the gates in the west, and they return via the north to the eastern gates and (thus) come out (again) at the front of the sky (i.e. in the east).

(6) And when the moon rises it becomes (first) visible on the sky when it contains the light of one-half of one-seventh part (of its total) and in fourteen (steps) it completes its full light.

(7) And fifteen (parts of) light are put into it until in fifteen (days) its light is completed according to the character of the year and it makes fifteen parts while the moon is at its fourteenth part.

(8) And when (the moon) is waning it decreases on the first day (to) fourteen parts of its light, on the second day it decreases to thirteen parts, on the third it decreases to twelve parts and on the fourth it decreases to eleven parts, and on the fifth it decreases to ten parts and on the sixth it decreases to nine parts, and on the seventh it decreases to eight parts, and on the eighth it decreases to seven parts, and on the ninth it decreases to six parts, and on the tenth it decreases to five parts, and on the eleventh it decreases to four parts, and on the twelfth it decreases to three parts, and on the thirteenth it decreases to two (parts), and on the fourtheenth it decreases to half of one-seventh of its total light, and on the fifteenth is consumed what remains of the total.

(9) And in certain months the moon (is visible) for each one of twenty-nine days and at times for twenty-eight days.

---

[31] Milik, *Chronique d'Égypte* 46 (1971), p. 333.

(10) Then Uriel showed me another rule how light is put into the moon and where it is put into from the sun.

(11) The whole time in which (the illumination of) the moon progresses, light is transmitted to it, facing the sun, until the fourteenth day when its light is complete. And when the (lunar disc) is completely aflame (then) its light in the sky is complete.

(12) On the first day it is called New Moon because on the day light appears on it (for the first time).

(13) And (the light) becomes exactly complete on the day when the sun sets in the west and when in the east (the moon) rises at (the beginning of) night. And the moon is illuminated all night until the sun rises opposite it and the moon is seen opposite the sun.

(14) And where the light for the moon entered it, there again it wanes until all its light is consumed and the days of the (lunar) month are used up and the moon's disc remains empty without light.

(15 And in three months their duration amounts to thirty days and in three months it amounts to twenty-nine days each, in which it makes its recession, in the first time and in the first gate, in 177 days.

(16) And in the time of its waxing it becomes visible in three months for thirty days each, and it becomes visible in three (other) months for twenty-nine days each.

(17) At night it is visible for twenty (nine) (days) each like a man and at day-time (it is) like the sky because without its light there is nothing else on it.

(79,1) And now, my son, I have shown you everything and completed is (the story about) the law of all the stars in the sky.

Notes to Chapter 78. Lunar Phases

Much in this chapter duplicates the preceding versions.

(1) A two-division of the year is reflected in the attribution of two names to the sun. For the etymology, cf. Charles, 106, notes; also Milik Chronique d'Égypte 46 [1971] 338, commentary to line 7.

(2) Corresponding to its four major phases the moon is given four names; cf. Charles, 166/7, notes.

(3 to 5) Equality of the apparent diameters of sun and moon (cf. 72,37) and brightness 7:1 (cf. 72,37 and 73,3). Setting of the luminaries in the western gates and return via the north (cf. 72,5 and 75,3) to the east.

(4) The transfer of light 'according to measure' (here and in 73,2) probably means the steady increase of illumination, proportional to time.

(6 to 9) Verses 6 and 7 deal in a general fashion with the waxing moon, allowing either day 14 or day 15 as full moon date. What the character of the 'year' has to do in this context (verse 7) I do not understand. Error for 'month' (hollow/full)?

Verse 8 concerns the waning moon, describing the day-by-day decrease of the moon's illuminated area in terms of 'parts', from 14 on the first day to invisibility on the 15th day. For a fragment of a Greek version cf. (Milik, *Chronique d'Égypte* 46 (1971), 339, and below, 420.

Verse 9 offers the possibility that a 'month' may contain 29 or 28 days (of visibility), being either full or hollow.

(10 to 14) Once more a general description of the lunar phases, introduced by Uriel. In verse 11 one should not say that the moon is 'opposite' the sun during the whole time of waxing in order to avoid misinterpretation as 'opposition' (which is the proper term in verse 13). Prof. E. Isaac suggests, therefore, translating baqedma ḍaḥay as 'facing the sun'.

Verse 14 states correctly that darkness of the waning moon begins on the same side (the western rim) on which the illumination of the waxing moon begins.

(15, 16) The lunar year is schematically divided into two halves, each containing three full and three hollow months, thus a total of 177 days. Apparently in analogy to the halves of a lunar month these two halves of the lunar year are denoted as 'waning' and 'waxing' (why in that order?). Cf. also 79,3,4.

(17) During 2[9] nights,[32] when the moon is visible, 'it looks like a man'. At daytime, however, the moon is invisible, apparently because it has no corporality, being only a receptacle of the solar light. Cf. also the Aramaic version.[33]

79(1) Final words of Methuselah, similar to the end of the first version (76,14).

CHAPTER 79

(2) And he (Uriel) showed me every one of their rules for every day and for every season with its power, and for every year, and about the places of exit (i.e. the gates), and concerning the rules for every month and for every week;

(3) and the decrement of the moon that accumulated in the sixth gate, because in the sixth gate its (the moon's) light is completed. The beginning of the decrement

(4) that accumulates (is) in the first gate (and it is counted) at its (proper) time, (i.e.) when 177 days are completed — or, after the reckoning with weeks, 25 (weeks) and two days.

(5) And how it (the moon) falls behind in relation to the sun — or, after the reckoning with the stars, exactly five days in one single (period of) time, (i.e. half a lunar year) and when this position which you see has been traversed (by the stars).

(6) Such is the appearance and the picture for each luminary shown to me by the great angel Uriel who is their leader.

---

[32] The reading 20 in the MSS is obviously a scribal error, unfortunately not emended by Dillmann and thus retained ever since. The Aramaic version has no number.

[33] Milik, 295/5, below 418. Cf. also Claire Préaux, La lune dans la pensée grecque, Ch. III. (Académie Royale de Belgique, Mémoires de la Classe des Lettres, 2ᵉ sér. t. 61, fasc. *4* [1973]).

CHAPTER 80

(1) And in these days the angel Uriel spoke to me and said to me: see, I have shown to you, O Enoch, everything and I have revealed to you everything to be seen about the sun, the moon, and everything about those who guide the stars in heaven and all who turn (back) their works, (and about) their times, and their places of exit (i.e. the gates).

Notes to Chapter 79. The Lunar Year

(2) 'He showed me' refers, of course, to the angel Uriel, not to Methuselah, who was addressed in 79,1. Cf. also 79,6-80,1.

The 'power' of each season (lit. 'time') refers to the stars which during stretches of 91 days represent the seasons, as we are told in 82,15 to 20.

(3, 4) The verses 3 to 5 assume a two-division of the lunar year, similar to 78,15,16. Details remain obscure since the text is obviously corrupt.

In the present context 'sixth gate' and 'first gate' do not refer to the numbering of the gates from south to north but probably mean here 'a sixth gate' and 'a first gate', thus describing an interval of six gates traversed by the sun, i.e. the time of half a lunar year (cf. 78,15).

During this time the lunar months develop a 'decrement' (tāḥṣāṣita) with respect to the calendar months. How the changing illumination of the moon got involved with this problem I do not know. Actually this whole chapter is only an expanded (and therefore more obscure) version of 78,15.

(5) One half lunar year is now compared to one half of the Enoch-year. The latter is considered to be a 'sidereal' year. Indeed, 'the law of stars' is identical with the order of the Enoch-year (cf. 75,1,2 and 82). The difference in question is, of course, 5 days; cf. also 74,10 to 17.

The 'position traversed' (by the stars) marks the completion of the two seasons; cf. 82,9 and 10.

(80,1) Concluding speech of the angel Uriel, referring to his teaching about sun, moon, and stars (cf. also 82,7,8), i.e. the topics which constitute the core of the 'astronomical' Book of Enoch. The remaining topics (including the winds?) appear to be later accretions.

The 'turning back' probably refers to the return of the celestial bodies to the east after their setting; cf. for parallel 75,3 (p. 19).

(80,2 to 82,3) is an intrusion of non-astronomical material: apocalyptic and again concluding words to Methuselah.

CHAPTER 82

(4)Blessed are all the righteous ones, blessed are those who walk in the path of righteousness and do not err, like the sinners, in counting all their days in which the sun travels in the sky, entering in and coming out from the doors for thirty days, together with the leaders of the thousands of the orders of the stars, together with the four (days) that are added in order to separate the intervals (of the year, i.e.) the four intervals, the parts of the year, which lead them and with which they make their entry on four days.

(5) There are people who err concerning them (the epagomenal days) by not counting them in the reckoning of the year, for such people err and do not know them correctly,

(6) although they belong to the computus of the year and are truly recorded forever: one in the first gate and one in the third and one in the fourth and one in the sixth (gate) and the year is completed in 364 days.

(7) For (this) is true and the computation exact as (here) recorded, since (everything) concerning the luminaries, the months and the festivals and the (years) and the days Uriel has shown to me and revealed it as he was ordered by the Lord, of the whole creation of the world, and about the host of heaven.

(8) And he had power in heaven over night and day, so as to make light visible to men, sun and moon and stars and all the powers of heaven which revolve in their circuits.

(9) And this is the law of stars which set in their (proper) places, and at their times and their festivals and at their months.

(10) And these are the names of their leaders, who watch them that they enter at their times, who guide them in their positions and their order, in their times and their months and their powers and their positions.

(11) Their four leaders who separate the four parts of the year enter first; after them (enter) the twelve leaders of the orders who separate the months; and the 360 heads over thousands (of stars) are the ones who separate the days: and for the four epagomenal days those are the leaders who separate the four parts of the year.

(12) And concerning these heads over thousands: always one (of the four main leaders) is placed at the position between the leaders (of thousands) and their followers; but these (single) leaders separate (the seasons).

(13) And these are the names of the leaders who separate the four fixed parts of the year: Melk'ēl, Hel'ememēlēk, Mel'ēyal, Nārēl.

(14) And the names of those whom they lead are 'Adnār'ēl, 'Iyāsusa'ēl, and 'Iyelumē'ēl. These three follow the leaders of the orders (of thousands); (then again) one (of the four main leaders) follows the three leaders of the orders which (in turn) follows after those (main) leaders (who are placed) at the positions which separate the four seasons of the year.

(15 At the beginning of the year Melk'ēl rises first and rules—to whom is (also) given the name 'Southern Sun'. And the total of days during which he exercises his power is 91 days.

(16) And these are the signs of the days which are to be seen on earth in the days of his period of rulership: sweat, heat, and dryness(?). And all trees

bear fruit, and leaves appear on all trees (and there will be) good harvest, and rose-flowers which blossom in the fields; but the trees of winter are withered.

(17) And these are the names of the leaders who are the subordinates: Berke'ēl, Zēlebsā'ēl, and another one who is added, (as) head of thousands, called Hēloyāsēf; and completed are the days of rulership (over his season) with this one.

(18) And the second leader after him is Hel'ememēlēk whom they (also) call 'Luminous Sun'; and the total of the days of his light are 91 days.

(19) And these are the signs of these days on earth: heat and drought; and the trees bring their fruit to ripeness and maturity and make their fruit dry; and the sheep mate and become pregnant; and men gather all the fruits of the earth, and everything which is in the fields, and the vats of wine. (And this) will take place in the days of his rulership.

(20) And these are the names, and the orders and the subordinates, the leaders of thousands: Gēda'ēyāl, Kē'ēl, and Hē'ēl and the name of one who is added to them as a head of thousands called 'Asfā'ēl, and completed are the days of rulership with this one.

Notes to Chapter 82. The Hierarchy of the Stars

As before, e.g. in 75,1 and 2, the arrangement of the stars follows exactly the pattern of the Enoch-year.

(4,5) There are 'leaders of thousand (stars)', responsible for each month of 30 days and four leaders of higher rank who are associated with the four epagomenal days. Again people are mentioned who do not deal correctly with these days (cf. 75,2).

In verse 4 'anqaṣ 'door' (also 'division,' 'chapter' and 'cycle') probably means here not the gates but the division between the seasons.

(6) The epagomenal days are associated with gates 1, 3, 4, 6, i.e. with the solstices and equinoxes (cf. 75,2).

(7,8) All this is handed down on the authority of the angel Uriel, who is set over the luminaries, sun, moon, and stars. Cf. the similar epilogue in 80,1.

What follows in the remaining verses of Chapter 82 is obviously an addition taken from a different source.[34] It contains one of those lists of freely invented names which enhance the authority of cosmologic revelations. The text as we have it is slightly in disorder, which is not surprising for a meaningless list of (angel-) names. Nevertheless it seems to be clear that the original structure was simple enough: first are listed the leaders of the four seasons, then the twelve subordinate commanders for the single months. The text is incomplete only in the case of the last season, perhaps owing to an early mutilation of the manuscript.

(9,10) Introductory remarks to the following (but not 'headings' as Dillmann, 249, says).

(11) Different ranks are given to the 'leaders' (marāḥeyān) of the stars. There are

---

[34] Also Dillmann, 248, assumes the 'Unächtheit' of 82,9 to 20.

four leaders of the seasons, associated with the four epagomenal days, each one on duty for 91 days; then follow four groups of three leaders of the 'orders' (šerʿātāt) corresponding to the twelve months of 30 days each; finally, the 'leaders of the thousand (stars)', concerned with the 360 single days. These leaders are presumably angels. Dillmann expressed their ranks by hellenistic titles: 4 toparchs, 12 taxiarchs, 360 chiliarchs.[35]

(12,13) The names of the leaders of the seasons, represented by the epagomenal days, are:

Melkiel Helememelek Meleyel Narel

(14,15) The first season, spring, is ruled over by Melkiel. The subordinate leaders are:

Adnarel Iyasusael Iylumiel.

It is the function of the fixed stars to signal by their (heliacal) rising the beginnings of months and seasons. The cyclic order of these phenomena is expressed in the text by indicating that the season-stars can be considered either as preceding or following the three months of each season. The second half verse of 14 seems to express in a clumsy fashion the fact that it is immaterial in a circular sequence of stars to distinguish between leading and following positions.

(1) and 18) The names 'Southern Sun' and 'Luminous Sun', associated with seasonal leaders, do not fit very well, respectively, spring and summer, where they are mentioned.

(16,19) The 'signs' are here clearly referring to climatic and agricultural characteristics (cf. 72,13).

(17,18) The second season has the following subordinate rulers:

Berkeel Zelebsael Heloyasef.

(20) The subordinate rulers of the third season are:

Gedaeyal Keel Heel.

Only the name Asfael is preserved for the last season.[36] Here the text ends abruptly.

Although the text of this verse, as we have it, is in disorder, the preserved words nevertheless suffice to show that the original version was an exact parallel to 82,17: the names of two subordinate leaders are given, followed by a third one who brings the three months of the season to a conclusion.

Additional Notes on the Aramaic Fragments

by Matthew Black

Chapter 72

(1) With šelṭānomu cf. שלטן 82.10, Enastr[b] 28.2, Milik, Enoch, p. 295 (ובש]לטֹנהון[?).

(27) Milik, Enoch, p. 282, equates 'ar'estihu with חרתיה 'its sections' at Enastr[b] iii.2.

(4-8) This linear scheme is found in the Aramaic fragments (Milik, Enoch, p. 278f.; EAC, p. 195f.). That the Ethiopic texts go back ultimately to this Aramaic Enoch is

[35] Dillmann, 52, notes; 248, n. 1.

[36] Charles, Enoch, 178, n. 20, considers the name Asfael as 'merely an inversion' of Heloyaseph (of the second season). There is, of course, no reason visible for such a reduction of the number of leaders (not to mention the inept execution of the 'inversion').

amply demonstrated by the surviving fragment of En 78.6-8, 9-12 etc. (Milik, Enoch, p. 292f.). The terminology of these texts has given rise to a certain amount of confusion in the Ethiopic, e.g. the fractions 1/2 of 1/7th (1/14th) or 6/7ths and 1/2 (13/14ths) (see note on 73.6). For the general pattern of the Aramaic texts, consult Milik, Enoch, p. 274f. and EAC, p. 196, n. 6.[37]

Chapter 73

(6) The correct translation of 1/14 is found at 78.6, manfaqa sābeʿtaʾeda, lit. 'half of a seventh part': so Enastr^{b} 6.8, פלג שביע חד (Milik, Enoch, p. 284).

Chapter 76

(3-10) are preserved (fragmentarily) at Enastr^{c} 1.ii.1 (Milik, Enoch, p. 284f.).

(3) Enastr^{c} 1.ii.1 (Milik, Enoch, p. 285) has 'and the three (gates) which are after them are on the north (lit. the left)': (ותלתה די בתריהון על שמאל). This corresponds to the clause in Ethiopic, 'and then those on the left'. Charles bracketed this clause (Charles, Enoch, p. 163) as 'nonsense'. It is undoubtedly original: the preceding clause 'and (then) three are in the direction of the north' seems to be a doublet. In this verse the Ethiopic text follows the order E N S W which does not correspond to the order E S N W in verses 5-14;[38] for a discussion of this problem and possible explanations, see Charles Enoch, p. 163, Martin Le Livre d'Hénoch, p. 176, Knibb, Enoch, 2, p. 176. (Milik claims (p. 286) that it is the order [S] N [W] which is found in the Aramaic, but there is no evidence in the fragment for this—only the north is mentioned.)

(4-5) Cf. Enastr^{c} 1.ii.2, Milik, Enoch, p. 285. The Aramaic has a fuller form of text. For these destructive winds, see above, p. 23.

(6) relates to the favourable East wind (above, p. 404f.). Enastr^{c} 1.ii.5: '... by the second gate comes forth the east wind, the chief (of winds) ([נפקא רוח קדים קד[מ]יא)'. Milik reconstructs (p. 285) רוח קדים קדימא, 'the east-east wind', comparing line 6 רוח קדים גרבה, 'the east-north wind', which comes from the third eastern gate. But 'east-east' is a meaningless tautology. At 77.1 Eth. the wind is called 'east' because it is qadāmāwi, 'chief, first': here Enastr^{c} 1.ii.15 reads קדמיא and the line is restored by Milik: '[And they call the east (quarter) East] because it is the first (בדי הוא קדמיא) (Eth. has translated 'quarter' as 'wind': see below, note on 76.13, 77.1.) This east wind is said to be 'in the middle', i.e. between the two destructive winds of gates 1 and 3, and according to the Eth. it 'comes forth in a straight line', i.e. blowing due E-W, unlike the two other east winds which are deflected to the south or the north. (See Flemming-Radermacher, p. 99.)

---

[37] The expression rendered in EAC 'keeping (in darkness) a remainder of 2/7 (= 4/14)' is a curious one in the original. It occurs twice in this passage Enastr^{b} 7.iii.4 and 8: thus line 4 נפק . . . ושלט בשאר יממא דן שביעין תרין ופלג . . . rendered by Milik 'it (the moon) emerges ... and it keeps during the rest of this day two sevenths (parts of its light) and a half'. Milik explains ושלט (p. 282), 'lit. 'and it reigns (over such and such a fraction of its light)'. May we not rather have here the lost root שלט behind Heb. שֶׁלֶט 'a shield', and meaning 'to cover' velare? (cf. ThWNT s.v. ἐξουσία Bd. II., p. 570 (Foerster).) We should then translate: 'it emerges and covers during the rest of this day 5/14'.

[38] See EAC p. 198 for Ethiopic directional terminology.

(13) Enastr^{c} 1.ii.14, Enastr^{b} 23.1 (cf. Milik, Enoch, pp. 228, 289, 290) [א]בדן ומות וחרבן [. . .] 'devastation, death, [heat?] and destruction'. Cf. 76.5 where several manuscripts read mot ('death') for moq ('heat'); mot may have fallen out of Eth. by haplography. (The reading מות seems reasonably certain.)

(14) Enastr^{c} 1.ii.14, Enastr^{b} 23.2 'And (the description of) the twelve gates of the four quarters of heaven (רוחי שמיא) is completed; their full number and explanation I have shown to you, my son Methuselah'. Eth.'s 'four gates' (ḫāwāḫew) is a scribal error for nafāsāt = רוחות 'quarters' (Flemming Henoch, p. 103). (Milik reads שלמהון ופרשהון as a hendiadys 'their complete explanation'.)

Chapter 77

(1) Cf. Enastr^{c} 1.ii.13-20, Enastr^{b} 23, Milik, Enoch, p. 287f.

The first quarter is called east That nafās translated 'region, quarter' is clear from Enastr^{b} 23.4 'And the great quarter (רוח רבא) (they call) the west quarter רוח מערבא ...

because it is the chief (quarter) Enastr^{c} 1.ii.14 ('The east they call East) because it is the chief (quarter) (בדי הוא קדמיא). For the word-play on קדם, Dillmann, Henoch, p. 236: קדמיא can mean both 'first, chief' or 'in front'—The East is 'in front'.

The Most High descends Enastr^{b} 23.2 בדיל לתמן דאר רבא 'because there the Great One dwells'. The Aram. assumes an etymology from Heb. דר רם; the translator reads the text as ירד רבא καταβαίνει ὁ ὕψιστος. Cf. Dillmann, Henoch, p. 236, Knibb, Enoch, p. 179.

(2) **because all celestial luminaries decrease there** ... Enastr^{b} 23.5 is defective in this clause: I suggest with Eth. בדי [לתמן חסרו נהור]י שמיא[39]. The reading מאין '(celestial) bodies' is corrected three times at Enastr^{c} 1.ii.17,18: the first correction מאנין gives the true reading (the word occurs again at line 17 מנאין [(written מאאן)]: '(celestial) bodies setting and bodies entering'. (Milik reads interrogative מנ(א)ין 'whence' but translates 'there'.) The whole verse then reads: 'And the West is called the great quarter, because there the heavenly luminaries wane, (celestial) bodies setting and (celestial) bodies entering, and all the stars; and on this account it is called West (lit. 'setting').' Presumably 'great' because it has to accommodate all the heavenly host after they set.

(3a) Enastr^{c} 1.ii.19, Enastr^{b} 23.6-9 has a much longer text, relatively fully preserved; for the text see Milik, Enoch, pp. 288, 289. I read Enastr^{c} 1.ii.18 בדי מאנין זרחין 'because (celestial) bodies arise' as above at Enastr^{b} 23.7. 'And the north (they call) North because in it all celestial bodies (lit. vessels) hide and assemble and revolve and proceed to the East of heaven. And the east (they call) East because from there the celestial bodies (מאני שמיא) arise; and also (they call it) *mizrah* because (there celestial) bodies arise (דנחין, זרחין), moons ... to appear ...' (For Milik's conjectural supplement, see p. 288.)

(3b) For the second part of the verse, cf. Enastr^{c} 1.ii.19, Enastr^{b} 23.9; on the analogy of v. 4 perhaps supplement [וחזית תלת פלגות] 'And I saw three divisions of the earth, one of them for the traffic of men, and one of them for [all seas and rivers], and one of them for the deserts [and the ...] and the Paradise of righteousness'. Milik supplies (after 'for the deserts') ולשבע 'and for the Seven (ultra-terrestrial regions)', a fascinating but unsupported guess (p. 291).

---

[39] Syr. ܣܗܪܐ ܒܪ ܚܣܝܪ (P Sm col 1341) luna decrescens.

(4) A few letters only are preserved in the first part of the verse (Milik, p. 289), but [ונ]חת עליהון תלגׂא is certain and 'snow comes down upon them'.

Chapter 78

(1) Milik's observation (Chronique d'Égypte, 46 (1971), p. 338) that the two names for the sun correspond to the two seasons of the year seems correct, but the order is not the 'dry' season followed by the 'wet' season, but probably the other way round (cf. Charles, Enoch, p. 167), winter and spring (early summer), the 'wet' season', Oryārēs (אורי חרס) followed by summer, the 'dry' or 'hot' season (Tomās חמה?). 'Asonyā ('Asenyā) for the moon may be connected with Accadian sin, sen 'moon' (cf. sivan, month of the moon-god, Sinai, etc.). 'Eblā can only be לבנה 'the white one', 'Ērā' ירח 'moon', and Benāsē is probably corrupt in the first syllable (Dillmann) unless it stands for בן־אנש, 'man' (cf. 78.17).

(6-8) cf. Enastr[c] 1.iii.3-9 (Milik, Enoch, p. 292). In line 1 בשמיא (corresponding to basamāy) is barely identifiable in the photograph and nothing else is now recoverable from the traces of other letters visible. At line 2 [... משל]מין בכול יום ויום ? עד יום ארבעת עשר ומש]למין בה כול נהורה] is all that can be recovered with certainty, but the Aram. text is evidently fuller than Eth.: the meaning seems to be that 'they (the added fractions) fill up (the light) each day and complete in it (the 14th day) all its light'. The last few words can be confidently restored from verse 7 where Enastr[b] 1.iii.5 has preserved a text: עד יום חמשת עשר ומשלמין בה נהורׂה 'Until the 15th day and they (the fractions) complete in it its light.'

(7) cf. Enastr[c] 1.iii.6, read and translated by Milik: ודבר ירחיא בפלגי שבעין 'And it (the moon) accomplishes (lit. guides) (its) phases by halves of sevenths.' The reading דבר is certain (cf. Knibb): the word occurs again at 78.5 as a noun (Enastr[b] 26.3., Milik, p. 294); cf. Tg. דברא Jud. 5.21. If we assume that ירח 'month' can mean 'phase of the moon' (Milik), an alternative construction would be to take דבר as a noun and render 'and the course of the moon's phases is by halves of sevenths'.

(8) Enastr[c] 1.iii.8 reads [וביומא רב]יעי חד מן חד עשׂ[ר־יא] 'and on the fourth day (the moon decreases) by one part from eleven parts, i.e. to ten parts (not as Milik 'eleven parts'). This firm text enables us to restore the earlier sequence which should be correctly rendered: 'And on the first day (the moon decreases) by one part from fourteen parts', i.e. to thirteen parts (not as Milik fourteen parts).

(10) **Then Uriel showed me another rule** Milik equates the fragment Enastr[b] 25.1-4 with this verse, but the identification is doubtful. All that is now visible in the photograph are the words: [. . . ח]שבון אחרן אחזית לה די אזל [. . .]. 'Another calculation I was shown with regard to it (the moon?).' חשבון 'calculation' is certainly right (cf. 79.1 Enastr[b] 26.7) and אחזית is to be construed as an inner passive of the Ophal (cf. Milik, Enoch, p. 202). (Milik supplements ['And Uriel demonstrated to me] a further calculation by having shown it unto me that ...', but this is forcing the syntax to support the identification.) In line 1 שניא 'years' is visible but there is nothing corresponding in Eth. There are several other passages where similar words are found, e.g. 73.1 '... I saw another law', 74.1 'I saw another course, a law ...'. (The verb אזל is probably auxiliary, but the main verb is lost.)

(17) **At night it is visible ... nothing else in it.**This verse is fragmentarily preserved at Enastr[b] 26.4-6 but in a longer form of text. (Enastr[b] 26.3 is reproduced in Eth. at 79.3, but Enastr[b] 26.4-6 go together and belong to the text behind 78.17 Eth.) Enastr[c] 26.4-5 reads: [. . .] בה כדמות חזי דמי כדי נהורה בה הא[יר בלילא מן קצ[ת דמי חזוא דן כדמות אנש

... in it(?) it resembles the likeness of a mirror when the light shines on it. On some nights (מן קצת) this appearance resembles the image of a man'. There is a play on the noun חזי, חזיא 'mirror' (Heb. חזיא מראה Tg. Ex. 38.8) and חזוא 'vision, appearance, (Milik's 'like an image of vision' makes little sense). I take האיר as a Haphel of אור (cf. the use of אורתא for 'moon-light'; אר Hoftijzer, p. 23). Has this line 4 fallen out of the original behind Eth. by hmt or a similar form of scribal error (note the common phraseology of lines 4 and 5). מן קצת = *partim* (Dan. 2.42) must refer in this context to the appearance of 'the man in the moon' for only a part of the full times of the moon's waxing or waning. There is nothing in the Aram. text about 'twenty (nine) days', which could have arisen in the Eth. text from Greek κατεικάζει = דמי read as κατ' εἴκοσι, baba'ešrā. All that remains of the rest of the verse in Aram. is ובימםא מן [. . .]ה בלחודוהי. The second phrase is read as a repeated מן קצת by Milik. The last phrase looks like נהורה בלחודוהי 'light by itself'; perhaps the original read 'and in the day-time, for part (of the time), it resembles the sky for there is no light in it by itself: ובימםא מן [קצת דמי כדמות שמיא די לית בה נהור]ה בלחודוהי. The meaning would be that the moon 'resembles the sky' in the sense that its now invisible disk is blue 'like the sky'. (There is no evidence to support Milik's 'like the sun in the sky'.)

Chapter 79

(1) Enastr[b] 26.6 וכען מחזא אנה לך ברי 'And now I am showing to you, my son ...'

(3-5) Fragments of the original of these verses have been preserved at Enastr[b] 26.2-4: בתרעא שתיתיא בה 'in the sixth gate ...' [. . .וי]ומין תרן ומחסר מן דבר שמשא '[4](twenty five weeks) and two days. [5]And it (the moon) falls behind the course of the sun ...' (For the rest of this fragment see on 78.17.)

Chapter 82

(9) Fragments of verses 9-13 are preserved at Enastr[b] 28.1-5 (Milik, Enoch, p. 295). It is pointless to try to reconstruct an original text on the basis of the few words and phrases preserved: the most we can do is to identify the terminology and its Eth. equivalent. Thus lines 1 and 2 preserve five nouns two of which occur in Eth. verse 9, two in verse 10: למעדיהון לחדשיהון לדגליהון . . . [ובש]לטנהון לכל מסרתהון. The first two words correspond to Eth's 'their festivals and at their months'. (There is no astronomical justification for taking מעדין as 'signs of the Zodiac', Milik, p. 295, 187f.) Eth's 'months' should be understood, in the light of Aram. חדשיהון as 'their new moons', closely associated with 'festivals'. The last term דגליהון probably corresponds to šer'ātātihomu of verses 10, 11 in the sense of τάγματα 'order, classes', especially in a military sense 'battalions'. See Milik, Enoch, p. 147, Knibb, Enoch, 2, p. 188f. The two terms in line 10 corresponds to šelṭānātihomu 'their powers' and either to meqwāmātihomu or makānātihomu, 'their positions' or 'their places'. See Milik, p. 187 for מסרת. At lines 3 and 4 the fragment has [. . .]ן ראשין ד[and [מפ]רשין בי[. . .] corresponding to Eth. marāḥeyān zašer'ātāt 'leaders of the orders' (ראשין ד[דגלין] probably translated first as ταξιάρχαι); the remaining fragment corresponds to 'ella yelēleywwomu la'awwrāḫ, 'who separate the months' ([דימפ]רשין בין ירחיא). Line 5 corresponds to verse 13 [וא]לן שמהת 'and these are the names ...'

(16ff.) A description of spring, summer and winter occurs at Enastr[d] i.1f. (Milik, Enoch, p. 296) with expressions recalling En 2.1-5 and En 3 (En[a] 1.ii.3ff.) and

En 82.16ff. Milik detects in this piece the missing original of the description of autumn and winter which should have followed 82.20. Certainly some original Aramaic description of the seasons, preserved more fully at En[a] 1.ii.3ff., has served to provide the foundation of the poetic account of the seasons at En 2 and 3. En[d] [. . .]ומטר מחתין על ארעא וזרע[ונין . . .] line 2 '... and rain descend upon the earth, and plants(?).' Cf. Eth. En 2.3 '... and clouds and dew and rain rest upon it (the earth)'. (Has מחתין, Aph. ptc. נחת been read as מניחין Aph. ptc. נוח = 'aʿrafa?)

line 3 [. . .]עשב ארעא יעא ונפק ויעל[י. . .]
'... the herbage of the earth sprouts, comes forth and blossoms.' Cf. 82.16.

lines 4-6[40] ושתוא הוה ועלי כל אילנוא [מתבישין ונפלין ברא מן ארבעה]עשר אילנין די לא חזה להון [לאתעריא/לאתערטלה . . .] על[יהו]ן מתקימין
'But winter comes and the leaves of all the trees [wither and fall except for four]teen trees from whom it is unseemly [to be stripped bare ...] their leaves abide ... Cf. Eth. En 3 '... the trees appear withered ... with the exception of fourteen trees which are not stripped bare (but) which abide with the old (foliage) till the new appears after two or three years.' (cf. En[a] 1.ii.5-6) See above, 111.

APPENDIX B

APOCALYPSIS HENOCH GRAECE ADDENDA ET CORRIGENDA

I 2 ἁγιολόγων leg. ἀπὸ | λόγων 5 πιστεύσουσιν leg. (ἐπι)σεισθήσονται | ασωσιν leg. ζητήσουσιν?

V 6 leg. κατάλυσις [καὶ] κατάρα | 7 + (post ἐκλεκτοῖς) ἔσται | 8 leg. κατὰ λήθην 9 leg. εἰρήνῃ 2°

VI 2 n. del *after them* | 4 n. 2° leg. 1° | 7n. leg. *dekadarchs* | 8 leg. δεκα(δ)άρχαι

VII 3 ὡς δὲ leg. ὥστε

IX 5, 6 Sync² leg. Sync. | ἔγνωσαν ἄνθρωποι leg. ἐγνώρισεν ἀνθρώποις | 9 Sync. leg. κιβδήλους (ms κιβδηλα) [καὶ αἷμα πολὺ] ἐπὶ τῆς γῆς ⸢τῶν ἀνθρώπων⸣ ἐκκέχυται

X 7 ὅλῳ ᾧ ἐπάταξαν leg. ὅλως ὃ ὑπέδειξαν?

XIV 9 leg. ⁹καὶ εἰσῆλθον et del ⁹post ἐξεπέτασάν με 11 leg. ὁ οὐρανός 21 αὐτοῦ 2° leg. αὐτόν

XVI 1 Sync. + (post τελεσθήσεται) ἐφ' ἅπαξ ὁμοῦ τελεσθήσεται

XVII 4 leg. ὑδάτων ζώντων

XVIII 2 leg. ²καὶ τὸν λίθον et del. ²post γῆς 3 leg. ³καὶ αὐτοί et del. ³post βαστάζοντας 12 leg. ἦν pro ᾖ 15 κυλιόμενοι ms: leg. κυκλούμενοι?

---

[40] Restored from En 2.2, En 3.

XX 2 G^2 leg. ὁ εἷς τῶν ἁγίων ἀγγέλων 4 $G^{1,2}$ ἐκδίκων leg. ἐκδιώκων 5 $G^{1,2}$ ἐπὶ τῷ χάῳ leg. ἐπὶ τῷ λαῷ n. del. G^2 5 λαῷ

XXI 3 G^2 del. ὁμοῦ 7 G^2 leg. μεγάλου

XXII 3 ἐκρίθησαν leg. ἐκτίσθησαν (ms εκριθησαν) 7 leg. ἀφανισθῇ 9 οὕτως leg. οὗτος 10 οὕτως leg. οὗτος 11 ἣν leg. ἵν' ᾖ 12 οὕτως leg. οὗτος 13 οὕτως leg. οὗτος | + (post ἁμαρτωλοί) ὅσοι ἀσεβεῖς

XXIV 2 leg. ἕν τῷ ἑνί bis

XXV 5 leg. βοράν n. del βορραν G βοραν

XXVII 3 leg. εὐσεβεῖς n. εὐσεβεῖς cj. Charles ms ἀσεβεῖς

XXX 1 leg. ὕδατος ἀενάου (ms ενω) 2 δένδρα leg. δένδρον | σχίνῳ leg. σχοίνῳ (ms σχυνω)

XXXI 1 leg. σαρραυ (ms σαρραν) 2 leg. πλήρη ἐκ στακτῆς (ms πληρη εξαυτης) 3 ἀμυγδάλων leg. ἀμυγδάλου (ms αμυγδαλω)

XXXII 1 σχίνου leg. σχοίνου (ms σχυνω) 2 ἐπ' ἄκρων leg. ἐκ μακρῶν (ms επακρων) 4 κερατίᾳ leg. κερατέᾳ (ms κερατι)

P. Oxy. XVII 2069 (frg. 3 recto)[1]

LXXVII 7-LXXVIII 1

1 [. . . εἰς τὴν] ἐρυθρὰν θ[άλασσαν]]
2 [. . . δύο] εἰς τὴν μ[εγάλην θάλασσαν]
3 [καὶ ἐπεχύθη[2] ἐκεῖ ὕδα]τα πολλὰ (ms. πολυ) ε[ἰς]
4 [τὸ λεγόμενον Μα]να[β]δηρ[α] . . .]
5 [καὶ ἑπτὰ εἶδον νήσους μεγάλας ἐν τῇ] θαλάσσῃ [καὶ ἐν τῇ γῇ]
6 [πέντε ἐν τῇ μεγάλῃ θαλάσσῃ καὶ δύο ἐν] τῇ ἐρυθρᾷ θ̣[αλάσσῃ.]
7 [$^{78.1}$ καὶ τὰ ὀνόματα τοῦ ἡλίου οὕτως· τὰ πρῶ]τα καλε[ῖται]

P. Oxy XVII 2069 (frg. 3 recto)

LXXVIII 8

1 [καὶ ἐν τῇ πέμπτῃ] ἡμέρᾳ το̣[ῦ μηνὸς ἐλαττοῖ ἓν δέκατον]
2 [ὅλου τοῦ φωτ]ό̣ς vac. καὶ ἐν τ[ῇ ἕκτῃ ἡμέρᾳ τοῦ μηνὸς ἐλαττοῖ]
3 ἓν ἕνατο]ν ὅλου τοῦ̣ [φωτός vac.]
4 [καὶ ἐν τῇ ἑβδόμῃ] ἡμέρᾳ τ[οῦ μηνὸς ἐλαττοῖ ἓν ὄγδοον]
5 [ὅλου τοῦ φω]τ̣ος vac. καὶ [ἐν τῇ ὀγδόῃ ἡμέρᾳ τοῦ μηνὸς]
6 [ἐλαττοῖ ἓν ἕ]βδομον ὅλ̣[ου τοῦ φωτός]

---

[1] Vide supra, 258, 410.

[2] leg. c. Eth. yessawwaṭu

P. Oxy. XVII 2069 (fgr 1 recto)

LXXXV 10-LXXXVI.2

10 ....[ἕ]τερος τὸν ἕτερον.

LXXXVI I [καὶ πάλιν] ὤν ἀναβλέψας τ[οῖς ὀφθαλμοῖς μου ἐν] ὕπνῳ εἶδον τὸν οὐρανὸν ἐπάνω] καὶ ἐθεώρουν [καὶ ἰδοῦ ἀστὴρ εἷς ἔπεσεν] ἐκ τοῦ οὐρανοῦ [εἰς τὸ μέσον τούτ]ων τῶν μεγάλω[ν βοῶν καὶ ἐγένε]το μετα[στραφεὶς εἰς ταῦρον καὶ ἤσθιεν καὶ ἐποιμ]άν[ετο ἐν μέσῳ αὐτῶν]. (frg 2 recto) [2] [καὶ τότε εἶδον τοὺς βοὰς ἐκείνους μεγάλους καὶ μελάνας καὶ ἰδοῦ πάντες ἠλ]λοίασ[αν τὴν μάνδραν αὐτῶν καὶ] τὴν νομὴν [αὐτῶν καὶ τοὺς μόσχους αὐτῶν] καὶ ἤρξα[ντο κερατίζειν ὁ ἕτερος τὸν ἕ]τερ[ον]

P. Oxy XVII 2069 (frg I verso)

LXXXVII 1 [καὶ ὁ] ἕτερος [καταπιεῖν τὸν ἕτερον κα]ὶ ἤρξατο πᾶσα [ἡ γῆ βοᾶν]. 2 [καὶ πάλιν ὤ]ν ἀναβλέψας [τοῖς ὀφθαλμοῖς μου] ε[ἶ]ς τὸν οὐρανὸν [καὶ ἐθεώρουν ἐν τῷ ὁ]ράματι vac. καὶ ἰ[δοῦ εἶδον ἐξερχόμενο]ν ἐκ τοῦ οὐρανοῦ [ὡς ὅμοια τοῖς ἀ]νθ[ρώπ]οις [λευκοῖς καὶ οἱ τέσσαρες ἐξ]ώδευ[σαν] (frg 2 verso) [ἐκεῖθεν καὶ οἱ τρεῖς μετ' αὐτῶν. 3 καὶ οἱ τρεῖς οἱ] ἤ[σ]α[ν] ἐ[ξ]ερχό[μενοι ὕστερον ἐκράτησαν] τῆς χειρός μ[ου καὶ ἐπῆράν με ἀπὸ τῶν] υἱῶν [τῆς γῆς ...] α τ

LXXXIX 43 [ἐλυμήνα]το leg. [ἤρξα]το | n. leg. Kirkpatrick

XCII C-B,11 [...]α Ἑνὼχ γρα[μματεὺς] [ὀξὺς/ταχὺς τῶν] ἔργων καὶ σ[οφώτατος τῶν ἀνθρώπων]

XCVII 10 [ἀναπτήσεται] leg. [ἀναβήσεται] vel. [ἀποβήσεται]

XCVIII 4 leg. [... καὶ ὀμνύω ὑμῖν ὅτι ἡ δουλεία] ἐπὶ τὴν [γῆν οὐκ ἀπεστάλη ...] | 5 leg. καὶ στείρα γυναικὶ οὐκ ἐδόθη [ἀτεκνία] | leg. δούλον εἶναι ἢ δούλην | leg. ὁμοίως 2° | 8 leg. [8]ἀπὸ τοῦ [νῦν] 12 leg. κα[λῶν ἔχετε ὑμῖ]ν

XCIX 4 + (post ταραχθήσονται) καὶ ἀνασταθήσονται 5-6 ἐκσπάσουσιν leg. ἐξαρπάσουσιν 7 εὕρητε leg. εὑρή[σε]ται 8 ms καὶ καρδίας 9 ἐλαεργ[ήσατε] leg. ἐλατρ[εύσατε] 13 χά[ρις] leg. χ[αίρειν] 16 ἐκτρίψει leg. ἐκ[σ]τρέψει

C 7 φυλάξητε leg. φλέξητε 8 [περιέχει] leg. [λήμψεται] 10 leg. [ἐν πυρὶ] φλεγομένῳ [καυθήσεσθε]

CI 5 χειμαζ[όμενοι πάν]τες leg. χειμαζ[όμενοι οὕ]τως

CII 8 + (post αἰῶνα) καλῶς

CIII 2 ἀναγκαίαν· ἔγνων leg. ἁγίων καὶ ἀνέγνων 9 εἴπητε leg. εἴπετε 11 τῶ[ν ὀψωνίων] leg. [τοῦ κόπου] ἡμῶν 12 περικ[λεί]ουσιν leg. περικ[ύκλ]ουσιν 13 leg. [13]ἐζητήσαμεν πο[ρεύεσθαι]

CIV 5 [μὴ φοβεῖσθε] leg. [τί μέλλετε ποιεῖν]? om. [ἀλλ' ὑμεῖς οἱ ἁμαρτωλοί] 10 ἀντι[γράφουσιν] leg. ἀνα[στρέφουσιν]

CV deest Graece

CVI 1 ἔλαβεν leg. ἔλαβον del. n. CVI.1 ἔλαβεν G^{b} ἔλαβον 5 μου leg. μοι 7 [ἐρώτησον] leg. [ἄκουσον] 8 [εἶδ]εν leg. [ἤκουσ]εν 18 καὶ ὁσίως [καὶ] κάλεσον αὐτοῦ τὸ ὄνομα [Νωε] leg. καὶ κάλεσον αὐτοῦ τὸ ὄνομα [Νω]εον | καταπαύσητε leg. καταπαύσεται | leg. ἀπὸ [πάντων τῶν ἁμαρτημάτων συντελουμένων ἐπὶ τῆς γῆς] 19 leg. ὑπέδειξάν μοι καὶ ἐμήνυ[σάν] μοι

CVII 1 κακ[ίων ἔσται] leg. κακ[ώσει αὐτούς] καὶ εἶδον τόδε leg. καὶ εἶδον τότε 2 leg. ἀπότρεχέ τε

APPENDIX C

ANALYSIS OF PREFERRED READINGS

Since Greek and Aram. evidence exists for Chh. 1-36 and from 97-108, these sections are dealt with first.[1] There follows an analysis of the preferred variant readings in the 'Parables' where no Greek or Aram. frgs. are extant, but where the existence of a Greek *Vorlage* is not in doubt.

I CHAPTERS 1-36

List 1: Eth^{tana} alone or with other ms support

1.1 Eth$^{\text{tana u}}$, 3, 6, 9 (tris + Eth$^{\text{tana, e}}$); *2*.1; Ch. *3* Eth$^{\text{tana t}}$; Ch. *4*; 5.3 Eth$^{\text{tana t}^{1}\text{ u}}$; *6*.4; *8*.2 Eth$^{\text{tana (g)}}$ c. G; *9*.8 Eth$^{\text{tana q}}$, Eth$^{\text{tana m}}$ c. G, 9 Eth$^{\text{tana g}}$ c. G; *10*.2 c. G, 4 c. G, 16, 20 cf. G; *11*.2 c. G; *12*.1 c. G, 3 c. G, 4 c. G + Eth$^{\text{tana (g)}}$; *13*.3 Eth$^{\text{tana g u}}$ c. G; *14*.2 c. G, 4; *15*.2 (bis), Eth^{tana}, Eth$^{\text{tana q t u, 8 mss}}$ 5 Eth$^{\text{tana g t u}}$ (bis), 6 Eth$^{\text{tana (m)}}$, 11; *16*.1 Eth$^{\text{tana (q)}}$ cf. G; *17*.3; *18*.9 cf. G; *19*.1 + Eth$^{\text{tana (m t), b}}$, 2 Eth$^{\text{tana q (g a}^{1}\text{)}}$; *21*.1 Eth$^{\text{tana u}}$, 2, 3, 6 (bis), 7 Eth$^{\text{tana m u (t}^{1}\text{)}}$, 9; *22*.2 Eth$^{\text{tana m g t}^{1}\text{ u}}$, 7 Eth$^{\text{tana q, (2 mss)}}$, 8 Eth$^{\text{tana g m q}}$ c. G, 9 Eth$^{\text{tana g q}}$ c. G, 13; *24*.5, 6 c. G; *25*.6 c. G; *27*.2 c. G; *28*.2 c. G; *29*.1 c. G; *30*.1 *32*.2 c. G.

List 2: Eth I

1.2, 4, 7; *2*.2 c. G; *5*.5 c. G, 6; *6*.3, 8; *7*.1; *8*.2; *9*.3, 4, 6; *10*.1 c. G, 3 c. G, 12, 17 c. G, 19; *12*.2, 3, 5; *13*.2, 4, 10; *14*.19, 21 c. G, 22 cf. G, 23; *15*.5, 7 c. G(bis); *18*.5, 10, 14 c. G; *20*.7; *21*.5 c. G, 7 c. G; *22*.3, 7, 11, 12; *25*.4; *26*.2 c. G; *27*.5 c. G; *28*.1, 3; *29*.2; *33*.4; 36.4(bis).

[1] For convenience I include in this section chh. 80.2-8, 81, 82.1-3 (the non-astronomical portions of chh. 72-82 the 'Astronomical' Chapters') and 83-96, since a Greek *Vorlage* can be assumed for these chapters.

List 3: Eth I mss

4.1 Eth^{u}; *5*.1 Eth$^{\text{g u}}$, 2 Eth^{m}; *7*.5 Eth^{q}; *10*.2 Eth^{m} c. G, 9 Eth$^{\text{g, 2 mss}}$ c. G, 10 Eth^{q}, 12 Eth^{u} c. G, 16 Eth^{q}, Eth$^{\text{g t u (q)}}$, 20 Eth$^{\text{g q u (m)}}$ c. G, 22 Eth$^{\text{u, n}}$ c. G; *12*.5 Eth$^{\text{g (tana q u)}}$; *13*.6 Eth$^{\text{t u, n}}$ c. G; *15*.3, 5 Eth$^{\text{g q u}}$, 11 Eth$^{\text{g q}}$ c. G, Eth$^{\text{m, 3 mss}}$, 12 Eth$^{\text{m t}^1\text{ u}}$ c. G; *16*.1 Eth^{g}, Eth$^{\text{g (q t u tana)}}$; *17*.1 Eth$^{\text{g m q (t}^2\text{)}}$, 2 Eth^{g}, 6 Eth$^{\text{m q t}}$ c. G; *18*.4 Eth$^{\text{m, 2 mss}}$, 6 Eth^{g}, 7 Eth^{g} cf. G; *19*.1 Eth^{m} c. G; *21*.9 Eth$^{\text{q, 3 mss}}$ c. G; *22*.4 Eth$^{\text{m, ull}}$, 9 Eth^{q} c. G; *23*.2 Eth^{g}; *24*.4 Eth$^{\text{q u, ull}}$, Eth$^{\text{g m u}}$ c. G; *25*.6 Eth$^{\text{g q, 3 mss}}$; *26*.3 Eth$^{\text{q (tana)}}$.

List 4: Eth II/Eth^{M} or Eth II/Eth^{M} mss

1.2; *2*.1; *6*.6 (bis); *9*.1 c. G, 4 Eth$^{\text{b + 2 mss}}$, 6 Eth$^{\text{n a}^1}$ cf. G, 7 Eth$^{\text{t}^2}$, 11 Eth^{n} c. G; *10*.3 Eth^{n}, 22; *13*.2; *14*.15 Eth^{M}, 19; *16*.1 Eth$^{\text{b 2 mss}}$ c. G; *18*.4 Eth$^{\text{b + 6 mss}}$, 11, 15 Eth$^{\text{b}^1\text{ + 2 mss}}$ c. G; *20*.2 Eth^{M}, 4 Eth^{M}; *22*.6 Eth^{ull} cf. G, 13 Eth$^{\text{d y}}$ c. G; *24*.3 Eth$^{\text{h o b}^1\text{ (g)}}$; *25*.2(bis) Eth^{M}, 7; *27*.2 Eth^{M}; *34*.1 Eth^{M}; *36*.2 Eth^{v}.

The picture of the textual character of these early chh. which emerges from a scrutiny of the above data, in particular in List 1, shows Eth^{tana}—the newcomer to Eth I—in a position of unchallenged predominance as *facile princeps* among Eth mss for this section of Enoch; and of the 60 readings listed 22 agree with G. The text of Eth^{g}, which was accorded by Flemming and Charles a position of undisputed priority, must now give place to that of Eth^{tana}. Eth^{g} does not even come a close second, although it does occupy the second place, with 9 out of the 60 readings going with Eth^{tana} (3 with Eth^{tana} only) to which we must add the 50 readings in List 2, since all of these have Eth$^{\text{g/g}^1}$ as a component. Moreover, in List 3, Eth I mss, Eth^{g} alone has 5 preferred readings with 11 others mostly in combinations with Eth$^{\text{m, q, t, u}}$.

It must be emphasised, however, that Eth^{tana} and Eth^{g} together have by no means a monopoly of preferred Eth I readings. Thus in List 3, Eth^{q} occurs 4 times alone with a preferred reading plus 12 times in combinations mostly with Eth$^{\text{m, u, t, g}}$; Eth^{m} occurs 3 times alone plus 8 times, mostly with Eth$^{\text{g, q, t, u}}$. *Any of the mss of Eth I, singly or in groups, may preserve the superior reading.*

As Flemming noted (above, 3) few of these Eth I readings are found in Eth II mss (List 1, 1.9, 19.1,2, 22.7; List 3, 10.9, 22, 13.6, 15.11, 18.4, 21.9, 22.4, 24.4, 25.6).[1] Nevertheless, Eth I readings do not have a complete monopoly of preferred readings, as appears from List 4. But the relative paucity of

[1] I have followed Charles in separating Eth I mss from those of Eth II by a comma (e.g., List 3, 25.6); where only 'mss' are reported in the editions I give this contraction, usually with the number of mss (e.g. above List 3 at 21.9). The mss can be identified in the Textual Notes or in the Charles^{1906} edition.

examples (29) in this list is conclusive evidence for the superiority of the Eth I text; it clearly belongs to the first recension of the Eth. text, based on a ms which is at times close to G, and providing a fuller and better text. At the same time 'in some cases there is in β most probably a survival of the original text where it has been lost in the present representatives of α'.[1] But these cases are comparatively few, and for the most part consist of slight variations such as the insertion or omission of καὶ (18.4, 15). In fact, by far the greater number of Eth II variations are of the inner-Ethiopic variety, even in some cases where they are construable and make some sense, frequent omissions, expansions, misunderstandings, due to the carelessness and at times the presumption of medieval Eth. scribes.[2] Only occasionally is an important variant preserved (e.g. 10.22, 14.15, 24.3, 34.1, 36.2).

II Chapters 80.2-8, 81, 82.1-3, 83-108

The discovery and publication of the Chester Beatty papyrus of the last chapters of Enoch (97-107.3, but without Ch. 105) led to an earlier renewed interest in Enoch studies,[3] and, in particular, to the reopening of the textual problem by G. W. E. Nickelsburg in his article 'Enoch 97-104: A Study of the Greek and Ethiopic Texts'.[4]

Nickelsburg's main conclusion (153) is that the Eth. is a translation of the same Greek version as has survived in the Chester Beatty papyrus text, though the latter is, unfortunately, the product of a careless scribe and is marred 'by more than three dozen haplographies, often sizable ones'. Otherwise, 'it appears to be quite reliable in the material it reproduces' (153). This agrees in the main with my own estimate of the character and value of the papyrus text, though I would add and underline that, while both G^b and Eth. suffer from haplographies and dittographs, with their consequent omissions and duplications, not to mention other forms of corruption, Eth. (and the Greek text it is rendering) has, more often than not, preserved the omissions in G^b (whether occurring by haplography, lacunae in the ms or for other reasons).[5] It is an indispensable supplement to the Greek papyrus text.

On the Eth. side, the situation is no different from that elsewhere in the Eth. ms tradition: there is wide-spread corruption, although here 'haplographies', mostly omissions by hmt (or hma) are comparatively few compared with other parts of the book.

---

[1] Cf. Charles[1906], xxii, cf. Flemming, X. List 3 above includes instances noted by Charles, who also gives 6.6(?) and 22.9(?) as examples.

[2] Cf. Flemming X (foot).

[3] E.g., in addition to the work of C-B, J. Jeremias, 'Beobachtungen zu neutestamentlichen Stellen an Hand des neugefundenen griechischen Henoch-Textes' in ZNW 38 (1939), 115-24.

[4] *Armenian and Biblical Studies*, Sion, Supp. I, Jerusalem, 1976.

[5] E.g. at 97.6; 98.12; 99.4, 8, 10-12, 16; 100.1-2, 10-11; 101.7-8; 102.5, 8-9; 103.3-4; 103.6, 12, 15; 104.7; 106.7.

The main conclusion of the study in this connection is stated as follows: 'By far, the most reliable single Eth. manuscript is t (Eth^{t}) ... Forming a close second to t *are the manuscripts of* β (Eth II)' (italics mine) (153). So far as these chapters are concerned the judgements of Charles, that the evidence of Eth II is 'late and secondary' ('on the whole disastrous')—judgements 'based largely on chapter 1-36—are wrong ... in those parts of Enoch for which we have only the Eth., we would seem to be following the most reliable course if we accepted the joint readings of t and β (Eth^{t} and Eth II)...' (155).

The occasional survival of an original correct reading in Eth II mss was not overlooked by Charles or Flemming, and has again been emphasised by Knibb.[1] But the reassessment proposed by Nickelsburg goes far beyond this caution of other editors: it amounts to the recognition of the majority or canonical text of Enoch as virtually the foundation of a modern critical edition.

The crucial evidence on which this conclusion rests is set out in tabular form (104). In a detailed study of 33 variants Nickelsburg notes that the whole Eth. tradition is corrupt in 10 of these: out of the 23 cases remaining, Eth^{t} (an Eth I ms and a close ally of Eth^{u}) has 17 readings correct, Eth II 16, whereas Eth$^{g\ g^{1}}$ and Eth^{m} (a related group and the foundation of the Charles and Flemming critical editions) have each 6 readings correct, Eth$^{q\ u}$ each 7. This evaluation of Eth^{t} is reinforced later by noting its relative freedom from corruption and haplographies, a feature of the entire Eth. ms tradition. On this basis the conclusion is reached that Eth^{t} and Eth II contain the largest number of correct readings for En. 97-104.

A scrutiny of the 17/16 variants attributed to Eth^{t}/ Eth II reveals that 7 of these variants are also attested by Eth I mss other than Eth^{t} (No. 5 by Eth$^{g\ m\ q\ t\ u}$, No. 7 by Eth$^{m\ q\ t\ u}$, No. 14 by Eth$^{g\ q\ t\ u}$, No. 18 by Eth$^{g\ m\ t\ u}$, No. 25 by Eth$^{g\ g^{1}\ t^{2}}$, No. 27 by Eth^{u}, No. 28 by Eth^{m}). These cannot then be classed as Eth II readings; all that we can conclude is that, in 6 of these readings of the older form of Eth I text, Eth II has got the text right. In 7 further cases Eth II goes with Eth^{t} only: Nos. 4, 15, 16, 17, 22, 25, 27. We again have to do with an Eth I correct reading (even if it has the support of one ms only of Group I) which is correctly reproduced in the Eth II mss. Nickelsburg himself rightly rejects the suggestion that Eth^{t} could belong to Eth II. It is no doubt possible, however, to interpret this evidence as supporting, in these cases, the superiority of the Eth II text—but only if we are prepared to give greater weight to the numbers of Eth mss than to the quality of the ms support. If we take the latter view, the above analysis leaves 2 correct readings only which are supported exclusively by Eth II mss, No. 23, Eth^{e}, No. 26 Eth$^{o\ b^{1}}$, scarcely a

---

[1] 2, 35, and above, Introduction, 3.

sufficient basis for so far-reaching a reassessment of the critical conclusions of Charles and Flemming, even though these were reached without the benefit of the Greek papyrus text of chapters 94-104.

The following are the preferred readings in this section.

List 1: Eth^{tana} alone or with ms support

81.3, 9; 82.2; *83.3*, 6; 85.6 Eth$^{\text{tana q}}$; *88.1* Eth$^{\text{tana (g m q)}}$; *89.4*, 7, 8, 31 (bis), 35, 71 Eth$^{\text{tana g m t}}$, 76 Eth$^{\text{tana (t)}}$; *90.1*, 9, 15, 28, 31; *91.3* Eth$^{\text{tana g q, 2 mss}}$, 6 19, Eth$^{\text{tana q t}}$; *92.4* Eth$^{\text{tana t}}$; *94.4*; *97.4* Eth$^{\text{tana g q}}$, 5 Eth$^{\text{tana m q t, 12mss}}$, 10; *98.10* Eth$^{\text{tana q}}$, 15; *99.3* Eth$^{\text{tana t}}$, 14 c. G^{b}; *100.5*, 9 Eth$^{\text{tana t}}$ c. G^{b}, 13; *103.1*, 2 c. G^{b}, 12; *104.7*; *106.5* Eth$^{\text{tana g}^{1}}$ c. G^{b}, 9 Eth$^{\text{tana m}}$, 16.

List 2: Eth$^{g\ g^{1}}$ alone or with other ms support

84.3, 4; *85.1*, 3 Eth^{g}, Eth$^{\text{g, n ull}}$, 5 Eth$^{\text{g m q (tana)}}$; *90.2* Eth$^{\text{g q t}}$, 14, 34; *91.9*; *92.4* Eth$^{\text{g m q t (tana)}}$; *93.1* Eth$^{\text{g (q)}}$, 6; *91.13* Eth$^{\text{g m (tana)}}$; *95.1*, 2, 7 Eth$^{\text{g (m tana)}}$; *97.6* Eth$^{g^{1}}$ c. G^{b}, 7 Eth$^{g^{1}}$ c. G^{b}, 8 Eth$^{g^{1}\text{, h}}$ c. G^{b}; *98.8* Eth$^{g^{1}\text{ n}}$ c. G^{b}; *99.12* Eth$^{\text{g }g^{1}\text{ tana (q) i}}$, 13; *100.10*; *101.3* Eth$^{g^{1}\text{ m}}$, 5 Eth$^{g^{1}}$; *103.2* Eth$^{\text{g q}}$ c. G^{b}, 5 Eth$^{g^{1}}$, 8 Eth$^{\text{g (tana)}}$; *104.3* Eth$^{\text{g m t, i}}$, 9 Eth$^{\text{g q t}}$, Eth$^{\text{g }g^{1}\text{ (q)}}$, 13 Eth$^{\text{g }g^{1}\text{ q t, 11 mss}}$; *106.11* Eth$^{\text{g, b 4mss}}$, 18 Eth$^{g^{1}}$; *108.2* Eth$^{\text{g }g^{1}}$, 3 Eth$^{\text{g }(g^{1})}$, 7 Eth$^{\text{g }g^{1}}$.

List 3: Eth I

*80.5*a; *81.3*, 5; *83.1*; *86.2*, 3; *88.2*; *89.18*, 33, 36, 48a; *90.1*, 6, 21, 42; *91.3*(bis), 7; *93.5*; *94.3*; *95.4*; *97.2*; 6; *98.2*, 6, 11; *99.5*, 10 c. G^{b}; *100.2*, 10; *101.7*; *102.5*, 9 c. G^{b}, 10; *103.3* c. G^{b}, 4, 5; *104.1*, 5, 6, 13 c. G^{b}; *105.1*; *106.9*; *108.6*, 9.

List 4: Eth I mss

81.4 Eth$^{t^{2}}$; *84.1* Eth$^{\text{u, 9 mss}}$; *87.4* Eth$^{\text{q, b 10 mss}}$; *89.15* Eth$^{\text{t u, 11 mss}}$, 46 Eth$^{\text{m, 5 mss}}$, 52 Eth^{u}, 75 Eth$^{\text{q (u)}}$; *90.7*, 16 Eth^{m}; *93.8* Eth$^{\text{m t (q)}}$; *98.9* Eth^{m}; *99.4* Eth$^{\text{m t}}$ c. G^{b}, 7 Eth^{q}, Eth$^{\text{q t (g tana)}}$; *100.4* Eth$^{\text{t M}}$ c. G^{b}, 11 Eth$^{\text{t,b}}$; 106.5 Eth$^{m^{2}\text{, b 4mss}}$ c. G^{b}; 107.1 Eth$^{\text{q }t^{2}\text{ b, 8 mss}}$.

List 5: Eth II/Eth^{M} or Eth II/Eth^{M} mss

81.6; *82.1*(bis), 2; *87.4*; *88.3* Eth^{M}; *89.1* Eth^{M}, 22, 42 Eth$^{\text{ull (g n)}}$, 68 Eth^{M}; *90.3* Eth^{M}, 6, 29 Eth$^{\text{b (m) 6mss}}$; *91.3* Eth^{M}; *92.1* Eth^{M}, 2 Eth^{M}; *93.4*, 6 Eth^{M}; *95.7* Eth^{M}; *96.2* Eth^{M} (bis); *98.1* Eth^{M}, 13 Eth^{M}; *99.14* Eth^{M}; *100.4*, 11 Eth^{M}; *101.6* Eth^{e} c. G^{b}; *103.2* Eth^{M} c. G^{b}, Eth II, 10,14 Eth^{M}; 104.1 Eth^{M}; 106.12 Eth^{M}; *107.1* Eth^{M}, 3; *108.1* Eth^{M}, 5, 6 Eth^{M}, 10 Eth^{M}.

The main conclusion to be drawn from this set of data is not fundamentally different from that reached in the analysis of Chh. 1-36. In this section, however, Eth$^{\text{g, g}^1}$ either individually or together, or each or both with other mss (38), have only slightly fewer preferred readings than Eth^{tana}, again, either by itself or with other supporting mss (42). The essential situation, however, is the same as in Chh. 1-36: Eth I, represented chiefly, in these chapters by Eth^{tana} and Eth$^{\text{g}^1\text{ g}}$, is the oldest and best form of text.

It will be noted that in List 4, preferred readings of Eth I mss, there is one case (100.4) where the reading adopted is supported by Eth^{t} only and the majority text (Eth^{M}), i.e. by all other Eth mss. The situation is similar to the Eth^{t} readings which are found supported by Eth II and claimed by Nickelsburg as examples of the better reading being preserved in Eth II (see above, 3, 425).

Occasionally is List 5, Eth II/Eth^{M} or a single Eth II/Eth^{M} ms, or several such mss, preserve an important reading (e.g., at 90.29, 98.13, 101.6 (= G^{b})), but, for the most part, we have again to do in this List 5 with minor variations, as in List 4 in Chh. 1-36.

III Chapters 37-71 The Book of the Parables

There is general agreement that the Eth. translators of the 'Parables' made use of a Greek version, whether or not a semitic original may also have been available to them (above, 4, 185). No Greek fragment of the 'Parables', however, has ever been found; and this makes the task of retroversion of the Ethiopic into Greek appear less certain and less reliable than where a Greek version is extant and available for comparison and reconstruction. Nevertheless, the task is less difficult and its results less precarious than might at first appear. Milik was of the opinion that 'the Greek original of the book was certainly composed in metrical poetry' (92): this is less certain than the comparative ease with which the Eth. goes back into 'Biblical' (or 'Septuagint') Greek, almost certainly for the reason that the original *Grundschrift* of the book was a semitic one, in my opinion, a Hebrew original.

A reconstruction and collation of the preferred readings gives the following picture:

List 1: Eth^{tana} alone or with other ms support

38.2, 3; *39*.13(?); *41*.5, 8; *44*.1; *45*.1 Eth$^{\text{tana (n)}}$; *46*.6; *47*.2 Eth$^{\text{tana g m, 4 mss}}$, 4 Eth$^{\text{tana m}}$; *52*.6; *53*.2 Eth$^{\text{tana t}^1}$, 5, 6, 7 Eth$^{\text{tana, b 3 mss}}$; *54*.8 (bis); *55*.3 Eth$^{\text{tana g m}}$, 4 Eth$^{\text{tana g m t}^1}$; *56*.6; *60*.5 Eth$^{\text{tana t}}$, 11 Eth$^{\text{tana g m}}$, 14, 18, 21 Eth$^{\text{tana g (q), 4 mss}}$, 24 (bis), 25; *61*.3; *62*.1 (bis); *63*.10, 12; *65*.9 Eth$^{\text{tana g q}}$; *68*.1

Eth$^{\text{tana (g)}}$, 3 (bis), 4, 5; *69*.5 Eth$^{\text{tana q}}$, 11, 16 Eth$^{\text{tana, d}}$, 22; *70*.2 Eth$^{\text{tana m, ull}}$; *71*.5 Eth$^{\text{tana t (g q u)}}$.

List 2: Eth^{g} alone or with other ms support

41.9 Eth$^{\text{g a}^1}$; *43*.2 Eth$^{\text{g q t}}$(?); *46*.8 Eth$^{\text{g t}^1}$; *49*.3 Eth$^{\text{g q u}}$; *50*.2; *52*.1 Eth$^{\text{g m t}^1\text{ u}}$; *54*.4, 7 Eth$^{\text{g m}}$; *56*.1 Eth$^{\text{g (m u)}}$, 2; *63*.6 Eth$^{\text{g (t}^1\text{ tana b}^1\text{)}}$; *67*.2 Eth$^{\text{g q u}}$.

List 3: Eth^{q} alone or with other ms support

48.6; *54*.1; *56*.7, 8; *58*.6; *60*.21 Eth$^{\text{q, 6 mss}}$; *61*.1; *62*.1; *65*.6, 12 Eth$^{\text{q (t)}}$; *69*.13 Eth$^{\text{q (tana g)}}$, 18 (bis), 24 Eth$^{\text{q, n}}$, 25; *71*.6.

List 4: Eth I

39.6 (bis), 11; *40*.2, 9 (bis), 10; *41*,4, 8; *42*.2; *43*.4; *45*.3 (tris), 6; *46*.2, 3; *48*.4, 5, 9, 10; *52*.3, 5, 6, 7; *54*.2; *56*.3; *57*.2; *58*.4; *60*.1, 2, 3, 4, 5, 12 (bis), 19; *61*.2, 3, 4, 8, 9, 10, 12 (bis); *62*.2, 3, 8, 11, 12, 14, 15, 16; *63*.1, 5, 7 (tris); *65.3; 67*.4, 8 (bis) , 9, 11, 12, 13; *68*.2 (bis), 3, 4 (bis); *69*.4, 13, 14, 15, 17; *70*.3; *71*.3, 13.

List 5: Eth I mss

39.1 Eth^{m}; *41*.7 Eth$^{\text{u (g q, 2 mss)}}$; *48*.1 Eth$^{\text{m, 3 mss}}$; *53*.7 Eth^{m}; *55*.3 Eth$^{\text{m (q)}}$; 59.2 Eth^{m}; *60*.6 Eth^{u}, 19 Eth$^{\text{t}^1\text{ u}}$; *63*.2 Eth^{u}; *65*.1 Eth^{t}, 2 Eth^{t}, 4 Eth^{u}(?).

List 6: Eth II/Eth^{M} or Eth II/Eth^{M} mss

37.2 Eth^{M}; *38*.2, 5; *39*.3, 4, 5, 7 Eth^{M} (bis), 12 Eth^{M}, 13 Eth^{v}(?); *40*.1, 2 Eth$^{\text{c v}}$, 3 Eth$^{\text{d y}}$(?); *41*. 2 Eth^{M}, 3 Eth^{M}, 8 Eth^{M}(?); *43*.1 Eth^{M}, 2 Eth^{M}; *45*.4 Eth^{M}; *47*.1 Eth^{M}; *48*.7 Eth^{M}; *50*.3; *51*.1 Eth^{M}, 3; *52*.2 Eth^{M}, 6; *53*.1; *54*.5 Eth$^{\text{a}^1\text{ Garrett}}$, 7 Eth^{M}, 10 Eth^{M}; *55*.2; *56*.1 Eth^{M}, 5 Eth^{M} (bis), 7 Eth^{n}, Eth$^{\text{t}^2}$; *58*.5; *60*.8 Eth^{M}, 22; *61*.5 Eth^{M}, 6 Eth^{M}, 9, 10 Eth$^{\text{b}^2\text{ 5 mss}}$, 11 Eth II, Eth$^{\text{a}^1\text{ Garrett i 1}}$, 13; *62*.5 Eth^{M}, 6 (bis), 9; 11; *63*.1, 2 Eth^{M}, 3; *64*.2; *65*.6 Eth^{v}, 11; *66*.1 Eth^{M}; *67*.1; *68*.2 Eth$^{\text{b}^2}$, 3, 4 Eth$^{\text{e v h}^2}$; *69*.1 Eth^{c}, 8 Eth^{n}, 23 (bis), 29 Eth$^{\text{b 3 mss (t, 3 mss)}}$; *70*.1 Eth^{v}, *71*.6, 14 Eth^{M}.

Chh. 37-71 are about 6 pages longer than Chh. 1-36 (1-36 c. 19 pages, 37-71 c. 25 pages in Dillmann's edition of the text), thus accounting for the somewhat longer lists of preferred readings. The textual situation, however, is strikingly similar, with one important difference—there are more preferred readings coming from the Eth II/ Eth^{M} grouping than in the other two sections. Nevertheless, the total number of Eth I preferred readings in Lists 1, 2, 3, 4, 5 (164) again shows the superiority of that textual recension to Eth II/ Eth^{M} preferred readings (70). Perhaps the medieval scribal 'revision', which produced the 'disastrous' vulgate text, was less thorough in the Parables, so that many more correct readings escaped the scribal Verschlimmbesserung of the text. Again it is noteworthy that List 1 (Eth^{tana} and its allies) has most preferred readings (45), while List 2 and 3 (Eth^{g} and its allies and Eth^{q} and its allies) have each 12 and 16 respectively.

INDEX OF AUTHORS

BIBLICAL INDEX

OLD TESTAMENT

Genesis

Exodus

Leviticus

Numbers

Deuteronomy

Psalms

Proverbs

Jeremiah

Lamentations

Ezekiel

INDEX TO APOCRYPHA

INDEX TO PSEUDEPIGRAPHA

INDEX TO QUMRAN TEXTS

TALMUD AND LATER JEWISH LITERATURE

JOSEPHUS

PHILO AND PSEUDO-PHILO

PATRISTIC REFERENCES

CLASSICAL GREEK AND LATIN LITERATURE

SUBJECT INDEX